Polytopes and Discrete Geometry

CONTEMPORARY MATHEMATICS

764

Polytopes and Discrete Geometry

AMS Special Session
Polytopes and Discrete Geometry
April 21–22, 2018
Northeastern University, Boston, MA

Gabriel Cunningham
Mark Mixer
Egon Schulte
Editors

AMS AMERICAN MATHEMATICAL SOCIETY

2020 *Mathematics Subject Classification.* Primary 05C10, 05C85, 05E45, 20D06, 51M20, 52A35, 52B10, 52B11, 52B15, 57M60.

For additional information and updates on this book, visit
www.ams.org/bookpages/conm-764

Library of Congress Cataloging-in-Publication Data

Names: Cunningham, Gabriel, 1982- editor. | Mixer, Mark, 1981- editor. | Schulte, Egon, 1955- editor.

Title: Polytopes and discrete geometry : AMS Special Session on Polytopes and Discrete Geometry, April 21-22, 2018, Northeastern University, Boston, Massachusetts / Gabriel Cunningham, Mark Mixer, Egon Schulte, editors.

Description: Providence, Rhode Island : American Mathematical Society, [2021] | Series: Contemporary mathematics, 0271-4132 ; volume 764 | Includes bibliographical references.

Identifiers: LCCN 2020042991 | ISBN 9781470448974 (paperback) | 9781470464202 (ebook)

Subjects: LCSH: Combinatorial geometry. | Convex polytopes. | AMS: Combinatorics {For finite fields, see 11Txx} – Graph theory {For applications of graphs, see 68R10, 81Q30, 81T15, 82B20, 82C20, 90C35, 92E10, 94C15} – Planar graphs; geometric and topological aspect | Combinatorics {For finite fields, see 11Txx} – Graph theory {For applications of graphs, see 68R10, 81Q30, 81T15, 82B20, 82C20, 90C35, 92E10, 94C15} – Graph algorithms [See also 68R10, 68W05]. | Combinatorics {For finite fields, see 11Txx} – Algebraic combinatorics – Combinatorial aspects of simplicial complexes. | Group theory and generalizations – Abstract finite groups – Simple groups: alternating groups and groups of Lie type [See also 20Gxx]. | Geometry {For algebraic geometry, see 14-XX} – Real and complex geometry – Polyhedra and polytopes; regular figures, division of spaces [See also 51F15]. | Convex and discrete geometry – General convexity – Helly-type theorems and geometric transversal theory. | Convex and discrete geometry – Polytopes and polyhedra – Three-dimensional polytopes. | Convex and discrete geometry – Polytopes and polyhedra – n-dimensional polytopes. | Convex and discrete geometry – Polytopes and polyhedra – Symmetry properties of polytopes. | Manifolds and cell complexes {For complex manifolds, see 32Qxx} – Low-dimensional topology – Group actions in low dimensions.

Classification: LCC QA640.3 .A47 2021 | DDC 516/.13–dc23

LC record available at https://lccn.loc.gov/2020042991

∞ The paper used in this book is acid-free and falls within the guidelines
established to ensure permanence and durability.
Visit the AMS home page at `https://www.ams.org/`

10 9 8 7 6 5 4 3 2 1   26 25 24 23 22 21

Contents

Preface

This volume focuses on developments in the fields of discrete and convex geometry. It contains the proceedings of the Special Session on Polytopes and Discrete Geometry at the American Mathematical Society meeting held from April 21-22, 2018, at Northeastern University, Boston, Massachusetts. While this volume is aimed at researchers in discrete and convex geometry and researchers who work with abstract polytopes or C-groups, the editors believe that junior mathematicians, including graduate students and post-doctoral fellows could benefit greatly from a glimpse into these research areas. This volume offers access to various current topics and research problems in these fields. Specifically,

- The paper *The cd-index: a survey* by Margaret M. Bayer is a survey of the cd-index of Eulerian partially ordered sets; it discusses inequalities on the cd-index, connections with other combinatorial parameters, computation, and algebraic approaches.
- The paper *d-dimensional self-dual polytopes and Meissner polytopes* by Tibor Bisztriczky and Déborah Oliveros presents a construction of a class of convex self-dual d-polytopes for $d > 2$, and examines conditions under which they are involutory self-dual, and have metric embeddings. Such a polytope generates a Reuleaux polytope and the paper explores the relation between this Reuleaux polytope and Meissner polytopes (modified ball-polytopes of constant width).
- The paper *On the ranks of string C-group representations for symplectic and orthogonal groups* by Peter A. Brooksbank determines the ranks of abstract regular polytopes whose automorphism group is $PSp(4, F_q) \cong \Omega(5, F_q)$, and comments on the regular ranks of higher-dimensional symplectic and orthogonal groups.
- The paper *Perfect colorings of regular graphs* by Joseph Ray Clarence Damasco and Dirk Frettlöh characterizes the color adjacency matrices of perfect colorings of graphs, and in particular, connected graphs. Then, it determines the lists of all color adjacency matrices corresponding to perfect colorings of 3-regular, 4-regular and 5-regular graphs with two, three and four colors, and all perfect colorings of the edge graphs of the Platonic solids with two, three and four colors, respectively.
- The paper *Tverberg theorems over discrete sets of points* by J. A. De Loera, T. A. Hogan, F. Meunier, and N. H. Mustafa discusses Tverberg-type theorems with coordinate constraints and determines the $m-$Tverberg number, when $m \geq 3$, of any discrete subset $\mathbb{R}^2$. It also presents improvements on the upper bounds for the Tverberg numbers of $\mathbb{Z}^3$ and $\mathbb{Z}^j \times \mathbb{R}^k$

and an integer version of the well-known positive-fraction selection lemma of J. Pach.

- The paper *The vertices of primitive zonotopes* by Antoine Deza, Lionel Pournin, and Rado Rakotonarivo provides geometric and combinatorial properties for primitive zonotopes, and shows that the logarithm of the complexity of convex matroid optimization is quadratic. It also gives a sharp asymptotic estimate for the number of vertices of a primitive zonotope that can be seen as an intermediate between the permutahedra of types A and B.
- The paper *Barycenters of points in polytope skeleta* by Michael Gene Dobbins and Florian Frick classifies n-tuples of dimensions $(k_1, \ldots, k_n)$ that sum to nk such that, for a given point p in an nk-polytope, there are n points from faces of these prescribed dimensions whose barycenter is p. It also investigates the weighted analogue of this question.
- The paper *Two families of locally toroidal regular 4-hypertopes arising from toroids* by Maria Elisa Fernandes, Dimitri Leemans, Claudio Alexandre Piedade, and Asia Ivić Weiss, presents two infinite families of locally toroidal hypertopes of rank 4 that are constructed from regular toroids of types $\{4,3,4\}_{(s,s,0)}$ and $\{3,3,4,3\}_{(s,0,0,0)}$. The Coxeter diagram of the first of the two families is star-shaped and the diagram of the other is a square. In both cases the toroidal residues are regular toroidal maps of type $\{3,6\}$.
- The paper *Self-polar polytopes* by Alathea Jensen investigates the existence, construction, facial structure, and practical applications of self-polar polytopes, as well as the place of these polytopes within the broader class of self-dual polytopes.
- The paper *Isomorphisms of maps on the sphere* by Ken-ichi Kawarabayashi, Pavel Klavík, Bojan Mohar, Roman Nedela, and Peter Zeman describes a modified linear-time algorithm solving the isomorphism problem for spherical maps. The algorithm described can also be used to determine (in linear time) the group of orientation-preserving symmetries of a spherical map.
- The paper *Some enumeration relating to intervals in posets* by Jim Lawrence considers an iterative construction using the intervals of a poset, and shows that the functions giving the number of elements of a given rank in the kth iteration are polynomials in 2^k.
- The paper *String C-group representations of almost simple groups: A survey* by Dimitri Leemans aims at giving the state of the art in the study of string C-group representations of almost simple groups. It also suggests a series of problems and conjectures to the interested reader.
- The paper *Orientation-reversing symmetry of closed surfaces immersed in euclidean 3-space* by Undine Leopold and Thomas W. Tucker considers finite groups of isometries G of $\mathbb{E}^3$ and closed surfaces S such that a G-general position immersion of S realizes a restricted Riemann-Hurwitz equation for the orientation-preserving subgroup G^+ of G, extending results of the authors for the orientation-preserving case $G = G^+$.

- The paper *Realizations of the 120-cell* by Peter McMullen provides the realization spaces of the 120-cell $\{5, 3, 3\}$. This completes such a classification for all the classical regular polytopes.
- The paper *Prescribing symmetries and automorphisms for polytopes* by Egon Schulte, Pablo Soberón, and Gordon Ian Williams studies the groups for which it is possible to find a convex polytope with that group as automorphism group with additional geometric conditions on the action of the group or its subgroups. In particular, it proves that for every abelian group G of even order and an involution s of G, there is a centrally symmetric convex polytope whose automorphism group is G and such that s corresponds to the central symmetry.
- The paper *The versatile rhombic triacontahedron and crystallography* by Marjorie Senechal and Jean E. Taylor shows that subsets of the rhombic triacontahedron tile $\mathbb{R}^3$ and correspond to the combinatorial types of lattice Voronoi cells. By relaxing the hypothesis of convexity in the classification of parallelohedra, it provides a uniform description of periodic approximants to a large class of quasicrystals.
- The paper *Tilings with congruent edge coronae* by Mark D. Tomenes and Ma. Louise Antonette N. De Las Peñas discusses properties of a normal tiling of the Euclidean plane with congruent edge coronae, and proves that the congruence of the first edge coronae is enough to say that the tiling is isotoxal.

During our session, we also had the pleasure of having five high-school students give a talk. Their paper *The geometry of H_4 polytopes* by Tomme Denney, Da'Shay Hooker, De'Janeke Johnson, Tianna Robinson, Majid Butler and Sandernish Claiborne has been accepted for publication in another journal and will appear soon.

Please enjoy this volume.

Gabe Cunningham
Mark Mixer
Egon Schulte

Contemporary Mathematics
Volume **764**, 2021
https://doi.org/10.1090/conm/764/15355

The *cd*-index: A survey

Margaret M. Bayer

Abstract. This is a survey of the *cd*-index of Eulerian partially ordered sets. The *cd*-index is an encoding of the numbers of chains, specified by ranks, in the poset. It is the most efficient such encoding, incorporating all the affine relations on the flag numbers of Eulerian posets. Eulerian posets include the face posets of regular CW spheres (in particular, of convex polytopes), intervals in the Bruhat order on Coxeter groups, and the lattices of regions of oriented matroids. The paper discusses inequalities on the *cd*-index, connections with other combinatorial parameters, computation, and algebraic approaches.

1. Early history

The history of the *cd*-index starts with the combinatorial study of convex polytopes. Over one hundred years ago Steinitz proved the characterization of the face vectors of 3-dimensional polytopes [**92**]. Interest in the number of faces of convex polytopes in higher dimensions grew with the development of linear programming from the 1950s on. While the affine span of the face vectors of d-dimensional polytopes is known to be given just by the Euler equation, a full characterization of the face vectors of polytopes of dimensions 4 and higher still eludes us. The major breakthrough on this question came through the definition of the Stanley–Reisner ring and the discovery of the connection between polytopes and toric varieties. As a result the face vectors of simplicial polytopes were characterized by Billera and Lee [**15**] and Stanley [**88**], following a conjecture of McMullen [**72**].

In the 1970s Stanley, in studying balanced complexes and ranked partially ordered sets [**87**], broadened the focus from counting the elements of each rank to counting chains in the poset with elements from a specified rank set. Bayer and Billera applied this perspective to convex polytopes, and initiated the study of "flag vectors" of polytopes [**3**]. They found the "generalized Dehn–Sommerville equations," a complete set of equations defining the affine span of the flag vectors of d-polytopes; the dimension turns out to be a Fibonacci number. As with face vectors, a complete characterization of flag vectors of polytopes is unknown.

Shortly after the proof of the generalized Dehn–Sommerville equations, Fine (see [**6**]) found a way to encode the flag vectors in the most efficient way, with the *cd*-index. The generalized Dehn–Sommerville equations apply not just to convex polytopes, but to all Eulerian posets. Likewise, the *cd*-index is defined for Eulerian

2010 *Mathematics Subject Classification.* Primary 05-02; Secondary 05E45, 06A08, 52B05.
Key words and phrases. *cd*-index, polytope, Eulerian poset, Bruhat order, peak algebra.

posets, which include the face posets of regular CW spheres, intervals in the Bruhat order on finite Coxeter groups, and the lattices of regions of oriented matroids.

There are two main issues for research on *cd*-indices. One is the question of the nonnegativity of the coefficients, or, more generally, inequalities on the *cd*-index, for Eulerian posets or for particular subclasses. The other (related) issue is the combinatorial interpretation of the coefficients, either directly in terms of the poset or in terms of other combinatorial objects. In the last thirty-five years, much research has been carried out on these issues.

2. Basic definitions

DEFINITION 2.1. An *Eulerian poset* is a graded partially ordered set such that each (nonsingleton) interval $[x, y]$ in the poset has an equal number of elements of even and odd rank.

DEFINITION 2.2 (from [**91**]). A finite *regular CW complex* is a finite collection of disjoint open cells σ in Euclidean space such that each σ is homeomorphic to an open ball of some dimension n and its boundary is homeomorphic to a sphere of dimension $n-1$, which is the union of lower dimensional cells. If the complex is homeomorphic to a sphere, it is called a *regular CW sphere*.

(For a precise definition that does not assume Euclidean space, see [**55**].)

Björner [**17**] showed that the face posets of CW complexes are exactly the posets with a unique minimum element $\hat{0}$ and at least one other element, and for which the order complex of every open interval $(\hat{0}, x)$ is homeomorphic to a sphere. (The *order complex* of a poset is the simplicial complex whose faces are the chains of the poset.)

The set of closed cells of a regular CW sphere, ordered by inclusion, along with the empty set and an adjoined maximum element, forms an Eulerian poset. A convex polytope is a regular CW sphere. The face lattice of an n-dimensional polytope is a rank $n+1$ Eulerian poset. (In what follows we generally do not distinguish between a polytope and its face lattice.)

DEFINITION 2.3. For P a graded poset of rank $n{+}1$ and $S \subseteq [n] = \{1, 2, \ldots, n\}$, the *$S$-flag number* of P, denoted $f_S(P)$, is the number of chains $x_1 \prec x_2 \prec \cdots \prec x_s$ of P for which $\{\text{rank}(x_i) : 1 \le i \le s\} = S$. (By convention, $f_\emptyset(P) = 1$.) The *flag vector* of P is the length 2^n vector $(f_S(P))_{S \subseteq [n]} \in \mathbf{N}^{2^n}$.

When the poset is the face lattice of a polytope, the indexing set is typically shifted to represent dimensions of the faces, rather than ranks in the poset. The restriction of the flag vector to terms indexed by singleton sets is the *f-vector* (or *face vector*) of P.

Sommerville [**86**] proved the Dehn–Sommerville equations for f-vectors of simplicial polytopes by applying Euler's formula to each interval in the face lattice. The same method gives equations on flag vectors, known as the *generalized Dehn–Sommerville equations*.

THEOREM 2.4 ([**3**]). *The affine dimension of the flag vectors of rank $n+1$ Eulerian posets is $F_n - 1$, where (F_n) is the Fibonacci sequence (with $F_0 = F_1 = 1$). The affine hull of the flag vectors is given by the equations*

$$\sum_{j=i+1}^{k-1} (-1)^{j-i-1} f_{S\cup\{j\}}(P) = (1 - (-1)^{k-i-1}) f_S(P),$$

where $i \leq k-2$, $i, k \in S \cup \{0, n+1\}$, *and* $S \cap \{i+1, \ldots, k-1\} = \emptyset$.

This affine space is spanned by the flag vectors of convex polytopes. Bases of polytopes were given by Bayer and Billera [**3**] and by Kalai [**63**].

In the mid-1980s, Jonathan Fine discovered a compact way to represent these equations. To see this, we first need the transformation to the flag h-vector.

DEFINITION 2.5. Let P be a rank $n+1$ Eulerian poset with flag vector $(f_S(P))_{S \subseteq [n]}$. The *flag h-vector* of P is the vector $(h_S(P))_{S \subseteq [n]} \in \mathbf{N}^{2^n}$, where

$$h_S(P) = \sum_{T \subseteq S} (-1)^{|S \setminus T|} f_T(P).$$

This transformation is invertible: $f_S(P) = \sum_{T \subseteq S} h_T(P)$. The flag h-vector has algebraic meaning through the Stanley–Reisner ring of the order complex. For a convex polytope (and more generally, a balanced Cohen–Macaulay complex), the entries in the flag h-vector are nonnegative [**87**].

The flag h-vector can be represented by a polynomial in noncommuting variables a and b. Associate with $S \subseteq [n]$ the monomial $u_S = u_1 u_2 \cdots u_n$, where $u_i = a$ if $i \notin S$ and $u_i = b$ if $i \in S$. Then the *ab*-polynomial is $\Psi_P(a,b) = \sum_{S \subseteq [n]} h_S u_S$. Here is an equivalent formulation for the *ab*-polynomial: associate to each chain $x_1 \prec x_2 \prec \cdots \prec x_s$ of P with rank set S the monomial $w_1 w_2 \cdots w_n$, where $w_i = a - b$ if $i \notin S$ and $w_i = b$ if $i \in S$. Then $\Psi_P(a,b)$ is the sum of these monomials over all chains of P.

Fine's inspiration was to see that when P is a convex polytope, the *ab*-polynomial can be written as a polynomial with integer coefficients in the noncommuting variables c and d, where $c = a + b$ and $d = ab + ba$.

DEFINITION 2.6. Let P be a rank $n+1$ poset. The *cd-index* of P is the polynomial $\Phi_P(c,d)$ such that $\Phi_P(a+b, ab+ba) = \Psi_P(a,b)$, if such a polynomial exists.

The *cd*-index of a rank $n+1$ poset is considered a homogeneous polynomial (in noncommuting variables) of degree n by assigning degree 1 to c and degree 2 to d. A straightforward recursion shows that the number of *cd*-words of total degree n is the Fibonacci number F_n. It is easy to see from the definition that the *cd*-index of the dual of a poset (reverse the order relation) is obtained from the *cd*-index of the poset by reversing all the *cd*-words.

THEOREM 2.7 ([**6**]). *Let P be a graded poset. Then P has a cd-index if and only if the flag f-vector of P satisfies the generalized Dehn–Sommerville equations. In this case the coefficients of the cd-index are integers.*

We will sometimes refer to the affine space of coefficients of the *cd*-words of fixed degree as the *generalized Dehn–Sommerville space*.

The definition of the *cd*-index gives a way of computing it from the flag h-vector, and hence from the flag f-vector. Here are formulas for several low ranks. (Meisinger's dissertation [**73**] has many useful tables, including flag number formulas for the *cd*-index up through rank 9.)

rank 3: $cc + (f_1 - 2)d$
rank 4: $ccc + (f_3 - 2)cd + (f_1 - 2)dc$
rank 5: $cccc + (f_4 - 2)ccd + (f_2 - f_1)cdc + (f_1 - 2)dcc + (f_{13} - 2f_3 - 2f_1 + 4)dd$

TABLE 1. Some *cd*-indices

n-gon	$cc + (n-2)d$
tetrahedron	$ccc + 2cd + 2dc$
cube	$ccc + 4cd + 6dc$
octahedron	$ccc + 6cd + 4dc$
4-simplex	$cccc + 3ccd + 5cdc + 3dcc + 4dd$
4-cube	$cccc + 6ccd + 16cdc + 14dcc + 20dd$

Table 1 gives the *cd*-indices of some familiar polytopes.

Fine believed that the coefficients of the *cd*-index of a convex polytope are always nonnegative. This appears as a more general conjecture in [**6**].

CONJECTURE 2.8. *The coefficients in the cd-index of every regular CW sphere are nonnegative.*

The conjecture turns out to be true. In the next section we will look at the great body of work addressing this conjecture.

Note that while Fine did not publish anything about the *cd*-index, his calculations involving the *cd*-index inspired his work on an alternative approach to flag vectors [**49, 50**].

Stanley [**90**] noted that it is sometimes useful to write the *cd*-index as a polynomial in c and e^2, where $e = a - b$ and thus $e^2 = c^2 - 2d$. Purtill [**80**] showed that if P is a convex polytope, then $\Phi_P(c,d)$ can be written as a polynomial in the noncommuting variables c, d, and $-e^2 = 2d - c^2$ with nonnegative coefficients.

Next, let us consider another important parameter for convex polytopes. The *h-vector* of a simplicial polytope is the result of a certain linear transformation on the f-vector. This transformation was noted by Sommerville [**86**], but its significance was not understood for decades. For simplicial polytopes, the h-vector has interpretations in shellings of polytopes, the Stanley–Reisner ring and the toric variety associated with the polytope. Unfortunately, the particular transformation from f-vector to h-vector does not give a meaningful vector if the polytope is not simplicial. This problem was resolved by consideration of the toric variety associated with a nonsimplicial, rational polytope, and, in particular, its intersection homology. Stanley [**89**] gave the definition as follows. In the case of simplicial polytopes, this h-vector specializes to the aforementioned h-vector. In the general case it is sometimes referred to as the "toric h-vector."

The recursive definition uses the following notational conventions. For an Eulerian poset P, write $\hat{0}$ for the unique minimal element and $\hat{1}$ for the unique maximal element. Denote by P^- the poset $P \setminus \{\hat{1}\}$. We use interval notation in a poset; in particular $[\hat{0}, t) = \{s \in P : \hat{0} \preceq s \prec t\}$. The rank of an element t of P is denoted $\rho(t)$. Finally, $k_{-1} = 0$.

DEFINITION 2.9. Families of polynomials f and g in a single variable x are defined by the following rules:

- $f(\emptyset, x) = g(\emptyset, x) = 1$
- If P is an Eulerian poset of rank $n+1 \geq 1$, and if $f(P^-, x) = \sum_{i=0}^{n} k_i x^i$, then $g(P^-, x) = \sum_{i=0}^{\lfloor n/2 \rfloor} (k_i - k_{i-1}) x^i$.
- If P is an Eulerian poset of rank $n+1 \geq 1$, then
  $f(P^-, x) = \sum_{t \in P^-} g([\hat{0}, t), x)(x-1)^{n-\rho(t)}$

When $f(P^-, x) = \sum_{i=0}^{n} k_i x^i$, let $h_i = k_{n-i}$, and call $(h_0, h_1, \ldots, h_n)$ the (toric) h-vector of P.

In the case where P is the face lattice of a rational convex n-polytope, this h-vector has many of the same properties as the h-vector of a simplicial polytope; in particular, it is nonnegative ($h_i \geq 0$) and symmetric ($h_i = h_{n-i}$).

THEOREM 2.10. *For Eulerian posets, the toric h-vector is the image of a linear transformation of the flag vector. Equivalently, it is the image of a linear transformation of the cd-index.*

A recursive proof of this proposition was given in [**6**]. Explicit formulas for the *cd*-index–h-vector map were given in [**4**]. Hetyei [**59**] was able to simplify the *cd*-index–h-vector relation by introducing the "short toric polynomial." The paper [**4**] also gave a combinatorial approach involving lattice paths, due to Fine. A method for computing the toric h-vector (as well as the *cd*-index) of a convex polytope by sweeping a hyperplane through the polytope was given by Lee in [**70**].

3. Inequalities

3.1. Nonnegativity. As mentioned before, Fine believed that the *cd*-index has nonnegative coefficients for all polytopes, and Bayer and Klapper conjectured nonnegativity for all regular CW spheres. Here is a sequence of results and conjectures on nonnegativity. We say the *cd*-index of a poset is *nonnegative* when all its coefficients are nonnegative, and we denote this by $\Phi_P(c, d) \geq 0$.

THEOREM 3.1 ([**80**] Purtill 1993). *The cd-indices of the following polytopes are nonnegative.*

- *polytopes of dimension at most 5*
- *simple polytopes*
- *simplicial polytopes*
- *quasisimplicial polytopes (all facets are simplicial)*
- *quasisimple polytopes (all vertex figures are simple)*

THEOREM 3.2 ([**90**] Stanley 1994). *The cd-indices of S-shellable CW spheres are nonnegative. In particular the cd-index of every convex polytope is nonnegative.*

(The class of S-shellable CW spheres includes all convex polytopes.)

An important class of posets is the class of Cohen–Macaulay posets. These are the posets whose order complexes have Cohen–Macaulay Stanley–Reisner rings. A poset is *Gorenstein** if and only if it is Cohen–Macaulay and Eulerian.

CONJECTURE 3.3 ([**90, 91**]).

- *The cd-index of every Gorenstein* poset is nonnnegative.*
- *The cd-index of every Gorenstein* lattice is coefficientwise greater than or equal to the cd-index of the Boolean lattice (simplex).*

Stanley [**90**] showed that if a Gorenstein* poset is also simplicial (all of its intervals up to an element $x \neq \hat{1}$ are Boolean lattices), then its *cd*-index is nonnegative. Moreover, he showed that the inequalities of this theorem would imply all linear inequalities satisfied by the flag f-vectors of all Gorenstein* posets, and all those satisfied by the smaller class of S-shellable CW spheres.

In the most general case, Bayer [**2**] determined which *cd*-coefficients are bounded for all Eulerian posets.

THEOREM 3.4.

(1) *For the following cd-words w, the coefficient of w as a function of Eulerian posets has greatest lower bound* 0 *and has no upper bound:*
   (a) $c^i dc^j$, *with* $\min\{i,j\} \leq 1$,
   (b) $c^i dcd \cdots cdc^j$ *(at least two d's alternating with c's, i and j unrestricted).*

(2) *The coefficient of c^n in the cd-index of every Eulerian poset is* 1.

(3) *For all other cd-words w, the coefficient of w as a function of Eulerian posets has neither lower nor upper bound.*

In particular, there are Eulerian posets with some negative *cd*-coefficients. Easy examples in odd rank are obtained as follows. Take two copies of the "smallest" rank n Eulerian poset, namely, the poset with two elements at each rank, all pairs of elements of different ranks comparable. Identify the top and bottom elements of the two copies. For rank $2k+1$, the *cd*-index is $2c^{2k} - (c^2 - 2d)^k$ [**43**].

Before Stanley's conjecture was proved by Karu for all Gorenstein* posets, there was progress on special cases. Reading [**83**] used the proof of the Charney–Davis Conjecture for dimension 3 (Davis and Okun [**23**]) and a convolution formula to prove the nonnegativity of the coefficients of certain *cd*-words for Gorenstein* posets. Novik [**77**] proved the nonnegativity of certain *cd*-coefficients for odd-dimensional simplicial complexes that are Eulerian and Buchsbaum (a weakening of Cohen–Macaulay), in particular for odd-dimensional simplicial manifolds. Hetyei [**58**] constructed a set of polyspherical CW complexes having nonnegative *cd*-indices. (The face posets of these complexes are Gorenstein* posets.) Hsiao [**60**] constructed an analogue of distributive lattices having nonnegative *cd*-indices. (These are Gorenstein* posets.)

Karu pursued a proof of nonnegativity of the *cd*-index for complete fans using methods from algebraic geometry, and was able to extend his proof to the general Gorenstein* case.

THEOREM 3.5 ([**64**] Karu 2006). *The cd-index of every Gorenstein* poset is nonnegative.*

A complete characterization of the *cd*-indices of Gorenstein* posets is presumably beyond reach, but Murai and Nevo [**75**] obtained the characterization for rank 5.

3.2. Monotonicity. In this section we consider comparisons among the *cd*-indices of different posets.

Billera, Ehrenborg and Readdy studied the *cd*-indices of zonotopes (polytopes arising as the Minkowski sum of segments) and, more generally, of the lattice of regions of oriented matroids. In analogy to our notation for nonnegativity of the *cd*-index, we write $\Phi_Q(c,d) \geq \Phi_P(c,d)$ to mean that the coefficient of each *cd*-monomial in the *cd*-index of Q is greater than or equal to the corresponding coefficient in the *cd*-index of P.

THEOREM 3.6 ([**12**]). *Let Q_n be the n-dimensional cube, and Q_n^* its dual poset.*

- *If the rank $n+1$ poset P is the lattice of regions of an oriented matroid, then $\Phi_P(c,d) \geq \Phi_{Q_n^*}(c,d)$.*
- *If Z is an n-dimensional zonotope, then $\Phi_Z(c,d) \geq \Phi_{Q_n}(c,d)$.*

In this context the *cd*-index can be modified to the *c*-2*d*-index, because the coefficient of every word containing k ds is a multiple of 2^k.

Nyman and Swartz fixed the dimension and number of zones and found the zonotopes with minimum and maximum *cd*-indices.

THEOREM 3.7 ([**78**]). *For fixed r and n, let $\mathcal{H}_L$ be an essential hyperplane arrangement with underlying geometric lattice the rank r near pencil with n atoms, and let $\mathcal{H}_U$ be an essential hyperplane arrangement with underlying geometric lattice the rank r truncated Boolean lattice with n atoms. Let Z_L and Z_U be the zonotopes dual to $\mathcal{H}_L$ and $\mathcal{H}_U$. Then for any r-dimensional zonotope Z with n zones,*

$$\Phi_{Z_L}(c,d) \le \Phi_Z(c,d) \le \Phi_{Z_U}(c,d).$$

Ehrenborg [**29**] gave additional inequalities for the *cd*-index of zonotopes.

The following result of Billera and Ehrenborg is analogous to an inequality on the toric g-vector of rational polytopes, conjectured by Kalai [**63**] and proved by Braden and MacPherson [**21**].

THEOREM 3.8 ([**11**]). *Let P be a polytope and F a face of P. Let P/F be the polytope whose face lattice is the interval $[F,P]$ of the face lattice of P. Denote the pyramid over a polytope Q by $\mathrm{Pyr}(Q)$. Then*

- $\Phi_P(c,d) \ge \Phi_F(c,d) \cdot \Phi_{\mathrm{Pyr}(P/F)}(c,d)$
- $\Phi_P(c,d) \ge \Phi_{\mathrm{Pyr}(F)}(c,d) \cdot \Phi_{P/F}(c,d)$

In particular, if F is a facet of P, then $\Phi(P) \ge \Phi(\mathrm{Pyr}(F))$, so among all polytopes having F as a facet, the one with minimum *cd*-index is the pyramid over F. Repeated application of this shows that the simplex minimizes the *cd*-index among polytopes.

Billera and Ehrenborg were also able to show the upper bound theorem for *cd*-indices of polytopes. (For the upper bound theorem for f-vectors, see [**72**].)

THEOREM 3.9 ([**11**]). *Let P be an r-dimensional polytope with n vertices, and let $C(n,r)$ be the cyclic r-polytope with n vertices. Then*

$$\Phi_P(c,d) \le \Phi_{C(n,r)}(c,d).$$

Ehrenborg and Karu proved a decomposition theorem for the *cd*-index of a Gorenstein* poset, resulting in the following inequalities.

THEOREM 3.10 ([**37**]). *Let B_n be the Boolean lattice of rank n.*

- *If P is a Gorenstein* lattice of rank n, then $\Phi_P(c,d) \ge \Phi_{B_n}(c,d)$.*
- *If P is a Gorenstein* poset, and Q is a subdivision of P, then $\Phi_Q(c,d) \ge \Phi_P(c,d)$.*

In his master's thesis, Dornian proved the following.

THEOREM 3.11 ([**25**]). *Let Π be the face poset of a simplicial $(d-1)$-sphere, let $S_d(k)$ be a stacked d-polytope, and suppose each has a triangulation as a shellable d-ball with k simplices. Then $\Phi_\Pi \le \Phi_{S_d(k)}$.*

3.3. Other inequalities. Stanley [**90**] showed that for each *cd*-word $w \ne c^n$ there is a sequence of Eulerian posets whose *cd*-indices (normalized) tend to w. This can be seen as a strengthening of the fact that coefficients of *cd*-words have no upper bound (Theorem 3.4). Another proof of this by Bayer and Hetyei is in [**5**], where some extreme rays of the closed cone of flag f-vectors of Eulerian posets are given.

The nonnegativity of the *cd*-index can be translated into inequalities on the flag h-vector and flag f-vector. Some simpler inequalities can also be extracted. Stanley [**90**] considered the comparison of two entries of the flag h-vector. The result depends on a function of sets that looks mysterious, but makes more sense when visualizing how a *cd*-word (with d of degree 2) "covers" an interval of integers. For $S \subseteq [n]$, let $\omega(S) = \{i \in [n-1] : \text{exactly one of } i \text{ and } i+1 \text{ is in } S\}$. Stanley [**90**] showed that the following theorem follows from the nonnegativity of the *cd*-index for all Gorenstein* posets.

THEOREM 3.12. *Let S and T be subsets of $[n]$. The following are equivalent.*

- $\omega(T) \subseteq \omega(S)$
- *For every Gorenstein* poset P of rank $n+1$, $h_T(P) \leq h_S(P)$.*

In particular, the largest entries in the flag h-vector for Gorenstein* posets are h_S, for $S = \{0, 2, 4, \ldots\}$ and $S = \{1, 3, 5, \ldots\}$. Readdy [**81**] showed that in the case of the crosspolytope, the maximum h_S occurs only for these sets.

For the specific case of the simplex (Boolean lattice), Mahajan [**71**] looked at inequalities among the coefficients of the *cd*-index. He found, for example, that for the simplex the coefficient of any *cd*-word of the form udv is greater than or equal to the coefficient of $uccv$. Furthermore the maximum coefficient is, for n even, the coefficient of cd^jc with $j = (n-2)/2$ and, for n odd, the coefficient of $cdcd^jc$ (and that of cd^jcdc, which is the same) with $j = (n-5)/2$.

Ehrenborg [**30**] gave a method for lifting any *cd*-inequality to give inequalities in higher ranks. For flag vectors of rational polytopes, a main source of inequalities was the nonnegativity of the g-vector (as in Definition 2.9) and a form of lifting of these by convolution (Kalai [**63**]). Stenson [**94**] showed that the inequalities described in this section for *cd*-indices give flag vector inequalities that are not implied by the g-vector convolution inequalities.

Murai and Yanagawa [**76**] defined *squarefree P-modules*, a generalization of the Stanley–Reisner ring, and used it to generalize the *cd*-index to a class of posets they call *quasi CW posets.* They were then able to prove that the coefficient of w for a Gorenstein* poset is less than or equal to the product of coefficients of associated *cd*-words having a single d. As a consequence, they get upper bounds on the *cd*-index of Gorenstein* posets when the number f_i of rank i elements is fixed for all i.

4. Computing the *cd*-index

4.1. Specific polytopes and posets. Certain polytopes and posets have particularly nice *cd*-indices, often connected to other combinatorial objects. We will generally not define the associated combinatorial objects; the interested reader can find details in the references.

Purtill's early results on nonnegativity of the *cd*-index (Theorem 3.1) resulted from studying CL-shellings of polytopes [**80**]. In this study he showed that the *cd*-index of the simplex is the (noncommutative) André polynomial of Foata and Schützenberger [**51**]; the André polynomial is a generating function for permutations satisfying certain descent properties. Purtill also extended the notion of André permutations to signed permutations, defined signed André polynomials, and showed that the signed André polynomial is the *cd*-index of the crosspolytope (the dual of the cube). Note that reversing the monomials in the signed André polynomial gives the *cd*-index of the cube. Thus, in the case of the simplex, crosspolytope and cube, each coefficient in the *cd*-index can be computed by counting (signed) permutations with certain descent patterns.

Subsequently, Simion and Sundaram [**95**] defined the *simsun* permutations, also counted by the André polynomials. Hetyei [**56**] gave an alternative set of permutations, which he called *augmented André permutations*, that give the *cd* coefficients for the cube (and thus for the crosspolytope). Billera, Ehrenborg and Readdy [**12**] gave formulas for the *cd*-indices of the simplex, cube, and crosspolytope, with summations over all permutations. Ehrenborg and Readdy [**45**] applied the connection in the opposite direction, and used the *ab*-index of the simplex and crosspolytope to study the major index of permutations and signed permutations.

Inequalities for the *cd*-index of zonotopes were given in Section 3.2. Billera, Ehrenborg and Readdy [**13**] showed that n-dimensional zonotopes span the generalized Dehn–Sommerville space and that they generate as an Abelian group all integral polynomials of degree n in c and $2d$. Bayer and Sturmfels [**7**] showed that the flag vector of an oriented matroid is determined by the underlying matroid. Billera, Ehrenborg and Readdy [**12**] gave an explicit formula for the *cd*-index of an oriented matroid in terms of the *ab*-index of its lattice of flats. In particular, this gives formulas for the *cd*-indices of zonotopes and of essential hyperplane arrangements. Ehrenborg, Readdy and Slone [**46**] extended this to affine and toric hyperplanes. In another direction it was extended to "oriented interval greedoids" by Saliola and Thomas [**84**].

Ehrenborg and Readdy gave recursive formulas for the *cd*-index of the simplex and the cube [**39**]. They also gave recursive formulas for the *cd*-index of the lattice of regions of the braid arrangements A_n and B_n [**40**]. Jojić [**61**] then gave the *cd*-index of the lattice of regions of the arrangements D_n in terms of those of A_n and B_n.

Hsiao [**60**] gave a general construction of a class of Gorenstein* posets, based on a signed version of the construction of a distributive lattice from the order ideals of a general poset. For the resulting "signed Birkhoff posets" he gave a combinatorial description of the *cd*-index in terms of peak sets of linear extensions of the underlying poset.

Two other combinatorial computations of the *cd*-index of a simplex (Boolean lattice) were given by Fan and He [**47**] (based on methods from the Bruhat order (Section 5)), and by Karu [**66**] (counting certain integer-valued functions).

4.2. Operations on posets. Among the tools used in the study of *cd*-indices are results about the effect on the *cd*-index of various operations on posets. The methods used to develop many of these involve the coproduct, introduced by Ehrenborg and Readdy [**39**]; we postpone discussion of that until Section 6.

The most straightforward effect on the *cd*-index occurs for the join of two posets.

DEFINITION 4.1. Given graded posets P and Q, the *join* $P * Q$ of P and Q is the poset on the set $(P \setminus \{\hat{1}\}) \cup (Q \setminus \{\hat{0}\})$ with $x \preceq y$ in $P * Q$ in the following cases:

- $x \preceq y$ in $P \setminus \{\hat{1}\}$
- $x \preceq y$ in $Q \setminus \{\hat{0}\}$

- $x \in P \setminus \{\hat{1}\}$ and $y \in Q \setminus \{\hat{0}\}$.

THEOREM 4.2 ([**90**]). *If P and Q are Eulerian posets, then so is $P * Q$, and*

$$\Phi_{P*Q}(c,d) = \Phi_P(c,d)\Phi_Q(c,d).$$

The *pyramid* of a poset P is the Cartesian product $\mathrm{Pyr}(P) = P \times B_1$, where B_1 is the two-element chain. The *prism* of a poset P is the "diamond product," $\mathrm{Prism}(P) = P \diamond B_2 = (P \setminus \{\hat{0}\}) \times (B_2 \setminus \{\hat{0}\}) \cup \{\hat{0}\}$, where B_2 is the Boolean lattice on two elements. The dual operation to the prism operation takes P to $\mathrm{Bipyr}(P)$. (The terms come from the polytope context.) These operations produce Eulerian posets from Eulerian posets. Ehrenborg and Readdy computed the effect of these operations on the *cd*-index. They expressed this in terms of a couple of derivations on *cd*-words. We show one set of formulas; for others see [**39**]. Define a derivation D on *cd*-words by $D(c) = 2d$ and $D(d) = cd + dc$.

THEOREM 4.3 ([**39**]). *Let P be an Eulerian poset. Then*

- $\Phi(\mathrm{Pyr}(P)) = \frac{1}{2}\left[\Phi(P)\,c + c\,\Phi(P) + D(\Phi(P))\right]$
- $\Phi(\mathrm{Prism}(P)) = \Phi(P)\,c + D(\Phi(P))$
- $\Phi(\mathrm{Bipyr}(P)) = c\,\Phi(P) + D(\Phi(P))$

Ehrenborg and Readdy [**39**] also described the effect on the *cd*-index of other operations on polytopes: truncation at a vertex, gluing polytopes together along a common facet (in particular, performing a stellar subdivision of a facet), and taking a Minkowski sum with a segment. They also gave a formula for the *ab*-index (flag h-vector) of the Cartesian product of arbitrary polytopes. Ehrenborg and Fox [**32**] gave recurrences for the *cd*-index of the Cartesian product and free join of polytopes. Slone [**85**] gave a lattice path interpretation of the diamond product of two *cd*-words. Fox [**52**] extended this interpretation to all *cd*-words. Ehrenborg, Johnson, Rajagopalan and Readdy [**36**] gave formulas for the *cd*-index of the polytope resulting from cutting off a face of a polytope and for the *cd*-index of the regular CW complex resulting from contracting a face of the polytope. S. Kim [**68**] showed how the *cd*-index of a polytope can be expressed when a polytope is split by a hyperplane. Wells [**96**] generalized the idea of bistellar flips to (polytopal) PL-spheres and computed the effect on the *cd*-index.

For a fixed graded poset P, one can form the poset $I(P)$ of all closed intervals ordered by inclusion. Jojić [**62**] studied this poset, showed that if P is Eulerian then $I(P)$ is Eulerian, and computed the *cd*-index of $I(P)$ in terms of that of P. Jojić also computed the effect on the *cd*-index of the "E_t-construction" of Paffenholz and Ziegler [**79**].

Hetyei [**57**] introduced the Tchebyshev transform on posets. He used it to construct a sequence of Eulerian posets (one in each rank) with a very simple formula for the *cd*-coefficients. The *ce*-index (a variation of the *cd*-index) of this poset is equivalent to the Tchebyshev (Chebyshev) polynomial. Ehrenborg and Readdy [**42**] continued the study of the Tchebyshev transform on general graded posets. They showed that the Tchebyshev transform (of the first kind) preserves these poset properties: Eulerian, EL-shellable and Gorenstein*. In the Eulerian case, they computed the *cd*-index of $T(P)$ in terms of that of P and showed that nonnegativity of the *cd*-index of P implies nonnegativity of the *cd*-index of $T(P)$. They showed that a second kind of Tchebyshev transform is a Hopf algebra endomorphism on the Hopf algebra of quasisymmetric functions (see Section 6).

4.3. Shelling components. Stanley [**90**] decomposed the *cd*-index of an n-dimensional simplex (Boolean lattice) into parts based on a shelling of the simplex, and used the parts for a formula for the *cd*-index of a simplicial Eulerian poset. A simplicial Eulerian poset is an Eulerian poset such that for every $x \neq \hat{1}$, the interval $[\hat{0}, x]$ is a Boolean lattice. Note that the h-vector of a simplicial Eulerian poset is defined by the transformation from the f-vector (mentioned in Section 2) for simplicial polytopes.

Let $\sigma_0, \sigma_1, \ldots, \sigma_n$ be any ordering of the facets of the n-simplex Σ^n; it is a shelling order. Let $\hat{\Phi}_i^n(c,d)$ be the contribution to the *cd* index of Σ^n from the faces added when σ_i is shelled on. (For details see [**90**].) We refer to these as the *shelling components* of the *cd*-index.

THEOREM 4.4.

- *For all* i, $0 \le i \le n-1$, $\hat{\Phi}_i^n(c,d) \ge 0$
- *If* P *is a simplicial Eulerian poset of rank* $n+1$ *with* h*-vector* $(h_0, h_1, \ldots, h_n)$, *then* $\Phi_P(c,d) = \sum_{i=0}^{n-1} h_i \hat{\Phi}_i^n(c,d)$.

As a consequence, Stanley proved the nonnegativity of the *cd*-index for Gorenstein* simplicial posets before Karu's proof for general Gorenstein* posets. Stanley conjectured, and Hetyei [**56**] proved formulas for the shelling components $\hat{\Phi}_i^n(c,d)$ in terms of André permutations and in terms of simsun permutations. Ehrenborg and Readdy [**39**] gave a compact recursion for these shelling components, and Ehrenborg [**31**] gave more recursions for them.

Ehrenborg and Hetyei [**34**] developed the analogous results for cubical Eulerian posets, that is, Eulerian posets whose lower intervals are isomorphic to the face lattice of a cube. Billera and Ehrenborg [**11**] gave a formula for the contribution of each facet in a shelling of a polytope. Lee [**70**] described a dual approach: the calculation of the *cd*-index by "sweeping" a hyperplane through the polytope, keeping track of the contribution at each vertex.

5. Bruhat order

The original motivation for the *cd*-index came from the combinatorial study of convex polytopes, but Reading began the study of the *cd*-index for another important class of Eulerian posets: intervals in the Bruhat order of Coxeter groups. In short, for v and w elements of a Coxeter group, $v \prec w$ if and only if some reduced word representation of v is a subword of a reduced word for w. An interval in the Bruhat order is Eulerian and shellable, and hence Gorenstein*. For example, the *cd*-index of the Bruhat order of $\mathcal{S}_4$ is $c^5+c^3d+2c^2dc+2cdc^2+dc^3+2cd^2+dcd+2d^2c$. For more information on Bruhat order in our context, see [**18, 82**].

Reading [**82**] gave a recursive formula for the *cd*-index of a Bruhat interval. He showed that Bruhat intervals span the generalized Dehn–Sommerville space, and gave an explicit basis.

THEOREM 5.1. *The set of cd-indices of Bruhat intervals of rank* r *spans the affine span of cd-indices of Eulerian posets of rank* r.

The Bruhat order of a universal Coxeter group contains intervals isomorphic to the face lattices of certain polytopes, the duals of stacked polytopes. Reading conjectured that these have maximum *cd*-indices.

CONJECTURE 5.2 ([**82**]). *Let (W, S) be a Coxeter system, and let $[u, v]$ be an interval in the Bruhat order of W with u of length k and v of length $n + k + 1$. Then the cd-index of $[u, v]$ is coefficientwise less than or equal to the cd-index of a dual stacked n-polytope with $n + k + 1$ facets.*

In particular the cd-index of an interval $[1, v]$ is less than or equal to the cd-index of a Boolean lattice.

The *cd*-index can be found in the peak algebra of quasisymmetric functions [**14**] (see Section 6). Billera and Brenti [**10**] used this to define for Bruhat intervals the *complete cd-index*, a nonhomogeneous polynomial in c and d, whose homogeneous part of top degree is the *cd*-index. They used this to give an explicit computation of the Kazhdan–Lusztig polynomials of the Bruhat intervals for any Coxeter group. They conjectured that all coefficients of the complete *cd*-index are nonnegative for all Bruhat intervals.

Besides the top degree terms, whose nonnegativity follows from Karu's theorem, the nonnegativity of certain coefficients in the complete *cd*-index of Bruhat intervals have been verified [**20, 48, 65**]. Blanco [**19**] used CL-labeling due to Björner and Wachs [**18**] to describe the computation of the complete *cd*-indices of dihedral Bruhat intervals (those isomorphic to intervals in a dihedral reflection subgroup) and Bruhat intervals in universal Coxeter groups.

Blanco [**20**] defined the shortest path poset in a Bruhat interval, and showed that if the poset has a unique maximal rising chain then it is a Gorenstein* poset. Y. Kim [**69**] studied the uncrossing partial order of matchings on $[2n]$, which is isomorphic to a subposet of the dual Bruhat order of affine permutations. He gave a recursion for the *cd*-indices of intervals in this poset.

6. Algebras

The h-vector of a simplicial polytope and the flag h-vector of Eulerian posets have interpretations in the Stanley–Reisner ring of the polytope or of the order complex of the poset. The *cd*-index is not found naturally in this ring. It turns out that other algebras are better habitats for the *cd*-index. There is an extensive literature on these algebras, and this survey will only touch the surface. For a deeper look, the reader is directed to (in chronological order) [**27, 39, 16, 8, 1, 41, 14, 10, 54, 66, 22**]. See Billera [**9**] for a survey of some of these connections.

Perhaps the beginning of the story is [**27**], where Ehrenborg gave a Hopf algebra homomorphism from the Hopf algebra of posets (the "reduced incidence Hopf algebra") to the Hopf algebra of quasisymmetric functions. The homomorphism gives some (not all) known results on flag vectors of Eulerian posets.

An important concept underlying some of the algebraic structures is the convolution of flag numbers, introduced by Kalai [**63**]. The entries of the flag vector, f_S^n are considered as functions from rank n graded posets to nonnegative integers. A convolution product is defined: for P a poset of rank $n + m$,

$$f_S^n * f_T^m(P) = \sum_{\substack{x \in P \\ \rho(x)=n}} f_S^n([\hat{0}, x]) f_T^m([x, \hat{1}]) = f_{S \cup \{n\} \cup (T+n)}^{n+m}(P).$$

Ehrenborg and Readdy [**39**] described a coproduct on the vector space spanned by graded posets by

$$\Delta(P) = \sum_{\substack{x \in P \\ \hat{0} < x < \hat{1}}} [\hat{0}, x] \otimes [x, \hat{1}],$$

and a coproduct on the noncommutative polynomial ring $k\langle a, b\rangle$ by

$$\Delta(u_1 u_2 \cdots u_n) = \sum_{i=1}^{n} u_1 \cdots u_{i-1} \otimes u_{i+1} \cdots u_n.$$

(Recall that the *ab*-polynomial of a graded poset has as its coefficients the flag h-numbers, and these coefficients can be written in terms of the flag f-vector.) They showed that the *ab*-index is a Newtonian coalgebra map between the resulting coalgebras, and showed that this map takes the subalgebra spanned by Eulerian posets to the subalgebra $k\langle c, d\rangle$. They used this to derive the formulas for the effect of various operations on the *cd*-index (see Section 4.2).

Billera and Liu [**16**] introduced a graded algebra of flag operators on posets. (In [**14**] it is referred to as the "algebra of forms on Eulerian posets.") The flag number functions f_S^n span a graded vector space over $\mathbf{Q}$, $A = \oplus_{n \geq 0} A_n$, with $A_n = \{\sum_{S \subseteq [n-1]} \alpha_S f_S^n : \alpha_S \in \mathbf{Q}\}$. With the convolution product, A becomes a graded algebra, and can be generated by the trivial flag operators $f_\emptyset^j$. Billera and Liu determined the two-sided ideal of A of elements that vanish for all Eulerian posets. Write $A_\mathcal{E}$ for the quotient of A by this ideal.

THEOREM 6.1 ([**16**]).

- *As graded algebras, $A \cong \mathbf{Q}\langle y_1, y_2, \ldots\rangle$, the free graded associative algebra on generators y_i of degree i. The isomorphism is determined by $f_\emptyset^j \mapsto y_j$.*
- *As graded algebras, $A_\mathcal{E} \cong \mathbf{Q}\langle y_1, y_3, y_5, \ldots\rangle$.*

The *ab*-polynomial of an Eulerian poset is in the (noncommutative) polynomial ring $A_\mathcal{E}\langle a, b\rangle$. Billera and Liu gave another proof of the existence of the *cd*-index for Eulerian posets by proving

THEOREM 6.2 ([**16**]). *As a polynomial with coefficients in $A_\mathcal{E}$, the ab-polynomial of every Eulerian poset is in $A_\mathcal{E}\langle c, d\rangle$.*

The authors also described one-sided ideals of the graded algebra A representing flag vector relations on simplicial polytopes and on cubical polytopes, and gave dimension arguments from the resulting A-modules.

Next we look at the *peak algebra*, introduced by Stembridge [**93**] in the study of enriched P-partitions, and described in terms of the *cd*-index in [**14**], following [**8**]. For $T = \{t_1, t_2, \ldots, t_k\} \subseteq [n]$, let

$$M_T = \sum_{1 \leq i_1 < i_2 < \cdots < i_k} x_{i_1}^{t_1} x_{i_2}^{t_2 - t_1} x_{i_3}^{t_3 - t_2} \cdots x_{i_k}^{t_k - t_{k-1}}.$$

This is a quasisymmetric function in x_i ($i \geq 1$); the sum is over all increasing k-tuples. For a *cd*-word $w = c^{n_1} d c^{n_2} d \cdots c^{n_k} d c^m$ of degree n, define $S_w \subseteq [n]$ by $S_w = \{n_1 + 2, n_1 + n_2 + 4, \ldots, n_1 + n_2 + \cdots + n_k + 2k\}$

DEFINITION 6.3 (from [**14**]). Let $\mathcal{Q}$ be the algebra of quasisymmetric functions over $\mathbf{Q}$ in the variables $x_1, x_2, \ldots$. The *peak algebra* is the subalgebra Π of $\mathcal{Q}$ generated by the elements $\Theta_w = \sum_{\substack{T \\ S_w \subseteq T \cup (T+1)}} 2^{|T|+1} M_T$ for each *cd*-word w.

The two algebras have natural coproducts that make them Hopf algebras.

THEOREM 6.4 ([**8**]). *The algebra of forms on Eulerian posets and the peak algebra are dual Hopf algebras.*

Following [**27**], Billera, Hsiao and van Willigenburg [**14**] considered the quasisymmetric representation of the flag f-vector, $F(P) = \sum_{S\subseteq[n]} f_S(P)M_S$, and showed that when written in terms of Stembridge's basis $\{\Theta_w\}$ the coefficients are the *cd*-index of P (modified by factors of 2). Then an Eulerian poset has a nonnegative *cd*-index when its peak quasisymmetric function has a nonnegative representation in terms of this basis.

Aguiar [**1**] took an algebraic approach to constructing the *ab*-index of a poset using a morphism from the algebra of all posets to the noncommutative polynomial algebra $k\langle a, b\rangle$, considered as infinitesimal Hopf algebras. This perspective enabled him to consider generalizations of the *ab*-index to weighted posets and the relative *ab*-index. Using a map involving the zeta and Möbius functions of a poset, he found the infinitesimal Hopf algebra corresponding to Eulerian posets, that is, the *cd*-index.

Ehrenborg and Readdy [**41**] gave yet another proof of the existence of the *cd*-index for Eulerian posets through the homology of their Newtonian coalgebra generated by a and b, mentioned above.

Karu [**66**] gave an algebraic formulation of a conjecture of Murai and Nevo [**74**] and proved it for a special case. The statement depends on a representation of *cd*-words by 0/1-vectors: $\mathrm{mdeg}(w)$ is the 0/1-vector obtained from w by replacing each c by 0 and each d by 10. For w a *cd*-word and $v = \mathrm{mdeg}(w)$, write $\Phi_{P,v}$ for the coefficient of w in $\Phi_P(c, d)$. For vectors $v \in \mathbf{Z}^n$ that are not equal to $\mathrm{mdeg}(w)$ for all *cd*-words w, let $\Phi_{P,v} = 0$.

THEOREM 6.5 ([**66**]). *Let P be the poset of a Gorenstein* simplicial complex of dimension n. Then there exists a standard $\mathbf{Z}^n$-graded k-algebra $A = \oplus_v A_v$ such that $\Phi_{P,v} = \dim A_v$.*

Murai's and Nevo's conjecture was that this theorem holds for all Gorenstein* posets.

Fine [**50**] suggests a successor to the *cd*-index. Like the *cd*-index, it provides a special basis for the flag vector ring, which he calls $\mathcal{R}$. The successor is a basis that, conjecturally, has this property: the structure coefficients for product and pyramid (which he calls cone) are all nonnegative integers. See also Theorem 4.3 and [**32**] for product and pyramid in the *cd*-index. If $\mathcal{R}$ were the representation ring of some algebraic object $\mathcal{G}$ that satisfies Schur's lemma, then the existence of such a basis would be immediate.

7. Related parameters

Here we mention some generalizations of the *cd*-index to broader settings.

Ehrenborg and Readdy [**38**] considered a generalization of the lattice of the cube. For various r, take the posets M_r with r minimal elements and a unique maximum element, then take the Cartesian product and add a unique minimum element. The result they called an $\mathbf{r}$-cubical lattice, where $\mathbf{r}$ is the vector of the various values of r in the factors. They defined a generalized *cd*-index for these lattices, and showed that the coefficients count a class of permutations generalizing the André permutations.

Ehrenborg [**28**] considered a relaxation of the Eulerian condition. A poset P is k-*Eulerian* if every interval of rank k is Eulerian. He showed that the *ab*-index of a k-Eulerian poset can be written in terms of c, d and $e^{2k+1} = (a-b)^{2k+1}$. He also related the k-Eulerian posets to an ideal in the Newtonian coalgebra.

Ehrenborg, Hetyei and Readdy [**35**] considered *Eulerian level posets*, infinite posets with a certain uniformity at each level and whose finite intervals are Eulerian. For these posets they extended the *cd*-polynomial to a *cd*-series.

As mentioned in Section 3.3, Murai and Yanagawa [**76**] considered a class of "quasi CW posets" and defined an extended *cd*-index for this class. Grujić and Stojadinović [**54**] developed an analogue of the *cd*-index for "building sets" (see [**24**]) by introducing a Hopf algebra of building sets and mimicking the known Hopf algebra construction of the *cd*-index.

Ehrenborg, Goresky and Readdy [**33**] extended the definition of the *cd*-index to "quasi-graded posets" with a generalization of the notion of Eulerian. These posets arise from Whitney stratified manifolds. They studied in particular the *cd*-indices of semisuspensions. Ehrenborg and Readdy [**44**] applied this to study manifold arrangements.

Dornian, Katz and Tsang [**26**] introduced the "mixed *cd*-index" of a strong formal subdivision of an Eulerian poset. This is defined in terms of the local *cd*-index of Karu [**64**], and maps to the "mixed h-polynomial" of [**67**], a multivariate polynomial invariant of subdivisions of polytopes.

Murai and Nevo [**74**] related the *cd* index of a class of Eulerian posets to the f-vector of a simplicial complex. The relation is via the γ-vector, first introduced by Gal [**53**]. Here the set of S^*-shellable spheres is a subset of Stanley's S-shellable spheres, and includes convex polytopes.

THEOREM 7.1 ([**74**]). *Let P be an $(n-1)$-dimensional S^*-shellable regular CW sphere, with cd-index $\Phi_P(c,d)$. Define δ_i by $\Phi_P(1,d) = \delta_0 + \delta_1 d + \cdots + \delta_{\lfloor n/2 \rfloor} d^{\lfloor n/2 \rfloor}$. Then there exists an $\lfloor n/2 \rfloor$-colored simplicial complex Δ such that $\delta_i = f_{i-1}(\Delta)$ for $0 \le i \le \lfloor n/2 \rfloor$.*

This is a survey of the *cd*-index, but as mentioned in Section 2, of great interest in the study of Eulerian posets and, in particular, of convex polytopes, is the toric h-vector. The toric h-vector contains much less information than the *cd*-index. However, Lee was led by his study of sweeping a hyperplane through a polytope to an extension of the toric h-vector that is equivalent to the *cd*-index [**70**].

8. Conclusion

The introduction of the *cd*-index opened up many directions of research on Eulerian posets. There are many specific open questions, but the overriding issue is to find combinatorial interpretations for the coefficients, beyond those in special cases mentioned here. I hope that this survey will become outdated soon, because of significant research advances.

Acknowledgments

The referees were generous with corrections and suggestions, which improved the paper. I also wish to thank Louis Billera, Jonathan Fine, Eran Nevo and Richard Stanley, who gave helpful feedback on a draft of this paper.

References

[1] M. Aguiar, *Infinitesimal Hopf algebras and the* **cd***-index of polytopes*, Discrete Comput. Geom. **27** (2002), no. 1, 3–28, DOI 10.1007/s00454-001-0049-4. Geometric combinatorics (San Francisco, CA/Davis, CA, 2000). MR1871686

[2] Margaret M. Bayer, *Signs in the cd-index of Eulerian partially ordered sets*, Proc. Amer. Math. Soc. **129** (2001), no. 8, 2219–2225, DOI 10.1090/S0002-9939-00-05831-7. MR1823903

[3] Margaret M. Bayer and Louis J. Billera, *Generalized Dehn-Sommerville relations for polytopes, spheres and Eulerian partially ordered sets*, Invent. Math. **79** (1985), no. 1, 143–157, DOI 10.1007/BF01388660. MR774533

[4] Margaret M. Bayer and Richard Ehrenborg, *The toric h-vectors of partially ordered sets*, Trans. Amer. Math. Soc. **352** (2000), no. 10, 4515–4531, DOI 10.1090/S0002-9947-00-02657-X. MR1779486

[5] Margaret M. Bayer and Gábor Hetyei, *Flag vectors of Eulerian partially ordered sets*, European J. Combin. **22** (2001), no. 1, 5–26, DOI 10.1006/eujc.2000.0414. MR1808080

[6] Margaret M. Bayer and Andrew Klapper, *A new index for polytopes*, Discrete Comput. Geom. **6** (1991), no. 1, 33–47, DOI 10.1007/BF02574672. MR1073071

[7] Margaret Bayer and Bernd Sturmfels, *Lawrence polytopes*, Canad. J. Math. **42** (1990), no. 1, 62–79, DOI 10.4153/CJM-1990-004-4. MR1043511

[8] Nantel Bergeron, Stefan Mykytiuk, Frank Sottile, and Stephanie van Willigenburg, *Noncommutative Pieri operators on posets*, J. Combin. Theory Ser. A **91** (2000), no. 1-2, 84–110, DOI 10.1006/jcta.2000.3090. In memory of Gian-Carlo Rota. MR1779776

[9] Louis J. Billera, *Flag enumeration in polytopes, Eulerian partially ordered sets and Coxeter groups*, Proceedings of the International Congress of Mathematicians. Volume IV, Hindustan Book Agency, New Delhi, 2010, pp. 2389–2415. MR2827977

[10] Louis J. Billera and Francesco Brenti, *Quasisymmetric functions and Kazhdan-Lusztig polynomials*, Israel J. Math. **184** (2011), 317–348, DOI 10.1007/s11856-011-0070-0. MR2823980

[11] Louis J. Billera and Richard Ehrenborg, *Monotonicity of the cd-index for polytopes*, Math. Z. **233** (2000), no. 3, 421–441, DOI 10.1007/s002090050480. MR1750930

[12] Louis J. Billera, Richard Ehrenborg, and Margaret Readdy, *The c-2d-index of oriented matroids*, J. Combin. Theory Ser. A **80** (1997), no. 1, 79–105, DOI 10.1006/jcta.1997.2797. MR1472106

[13] Louis J. Billera, Richard Ehrenborg, and Margaret Readdy, *The cd-index of zonotopes and arrangements*, Mathematical essays in honor of Gian-Carlo Rota (Cambridge, MA, 1996), Progr. Math., vol. 161, Birkhäuser Boston, Boston, MA, 1998, pp. 23–40. MR1627367

[14] Louis J. Billera, Samuel K. Hsiao, and Stephanie van Willigenburg, *Peak quasisymmetric functions and Eulerian enumeration*, Adv. Math. **176** (2003), no. 2, 248–276, DOI 10.1016/S0001-8708(02)00067-1. MR1982883

[15] Louis J. Billera and Carl W. Lee, *A proof of the sufficiency of McMullen's conditions for f-vectors of simplicial convex polytopes*, J. Combin. Theory Ser. A **31** (1981), no. 3, 237–255, DOI 10.1016/0097-3165(81)90058-3. MR635368

[16] Louis J. Billera and Niandong Liu, *Noncommutative enumeration in graded posets*, J. Algebraic Combin. **12** (2000), no. 1, 7–24, DOI 10.1023/A:1008703300280. MR1791443

[17] A. Björner, *Posets, regular CW complexes and Bruhat order*, European J. Combin. **5** (1984), no. 1, 7–16, DOI 10.1016/S0195-6698(84)80012-8. MR746039

[18] Anders Björner and Michelle Wachs, *Bruhat order of Coxeter groups and shellability*, Adv. in Math. **43** (1982), no. 1, 87–100, DOI 10.1016/0001-8708(82)90029-9. MR644668

[19] Saúl A. Blanco, *The complete* **cd***-index of dihedral and universal Coxeter groups*, Electron. J. Combin. **18** (2011), no. 1, Paper 174, 16. MR2836809

[20] Saúl A. Blanco, *Shortest path poset of Bruhat intervals*, J. Algebraic Combin. **38** (2013), no. 3, 585–596, DOI 10.1007/s10801-012-0416-7. MR3104730

[21] Tom Braden and Robert MacPherson, *Intersection homology of toric varieties and a conjecture of Kalai*, Comment. Math. Helv. **74** (1999), no. 3, 442–455, DOI 10.1007/s000140050098. MR1710686

[22] Francesco Brenti and Fabrizio Caselli, *Peak algebras, paths in the Bruhat graph and Kazhdan-Lusztig polynomials*, Adv. Math. **304** (2017), 539–582, DOI 10.1016/j.aim.2016.09.001. MR3558217

[23] Michael W. Davis and Boris Okun, *Vanishing theorems and conjectures for the ℓ^2-homology of right-angled Coxeter groups*, Geom. Topol. **5** (2001), 7–74, DOI 10.2140/gt.2001.5.7. MR1812434

[24] C. De Concini and C. Procesi, *Wonderful models of subspace arrangements*, Selecta Math. (N.S.) **1** (1995), no. 3, 459–494, DOI 10.1007/BF01589496. MR1366622

[25] Patrick Dornian, *Subdividing the cd-index of Eulerian posets*, Master's thesis; arXiv:1908.00039 [math.CO], 2019.

[26] Patrick Dornian, Eric Katz, and Ling Hei Tsang, *Poset subdivisions and the mixed cd-index*, arXiv:2001.01875 [math.CO], 2020.

[27] Richard Ehrenborg, *On posets and Hopf algebras*, Adv. Math. **119** (1996), no. 1, 1–25, DOI 10.1006/aima.1996.0026. MR1383883

[28] Richard Ehrenborg, *k-Eulerian posets*, Order **18** (2001), no. 3, 227–236, DOI 10.1023/A:1012296719116. MR1867234

[29] Richard Ehrenborg, *Inequalities for zonotopes*, Combinatorial and computational geometry, Math. Sci. Res. Inst. Publ., vol. 52, Cambridge Univ. Press, Cambridge, 2005, pp. 277–286. MR2178324

[30] Richard Ehrenborg, *Lifting inequalities for polytopes*, Adv. Math. **193** (2005), no. 1, 205–222, DOI 10.1016/j.aim.2004.05.005. MR2132764

[31] Richard Ehrenborg, *Prisms and pyramids of shelling components*, Ann. Comb. **22** (2018), no. 4, 769–779, DOI 10.1007/s00026-018-0412-2. MR3884755

[32] Richard Ehrenborg and Harold Fox, *Inequalities for* **cd**-*indices of joins and products of polytopes*, Combinatorica **23** (2003), no. 3, 427–452, DOI 10.1007/s00493-003-0026-z. MR2021159

[33] Richard Ehrenborg, Mark Goresky, and Margaret Readdy, *Euler flag enumeration of Whitney stratified spaces*, Adv. Math. **268** (2015), 85–128, DOI 10.1016/j.aim.2014.09.008. MR3276590

[34] Richard Ehrenborg and Gábor Hetyei, *Flags and shellings of Eulerian cubical posets*, Ann. Comb. **4** (2000), no. 2, 199–226, DOI 10.1007/s000260050007. MR1770690

[35] Richard Ehrenborg, Gábor Hetyei, and Margaret Readdy, *Level Eulerian posets*, Graphs Combin. **29** (2013), no. 4, 857–882, DOI 10.1007/s00373-012-1173-z. MR3070061

[36] R. Ehrenborg, D. Johnston, R. Rajagopalan, and M. Readdy, *Cutting polytopes and flag f-vectors*, Discrete Comput. Geom. **23** (2000), no. 2, 261–271, DOI 10.1007/PL00009499. MR1739609

[37] Richard Ehrenborg and Kalle Karu, *Decomposition theorem for the* **cd**-*index of Gorenstein posets*, J. Algebraic Combin. **26** (2007), no. 2, 225–251, DOI 10.1007/s10801-006-0055-y. MR2335713

[38] Richard Ehrenborg and Margaret Readdy, *The* **r**-*cubical lattice and a generalization of the* **cd**-*index*, European J. Combin. **17** (1996), no. 8, 709–725, DOI 10.1006/eujc.1996.0062. MR1415312

[39] Richard Ehrenborg and Margaret Readdy, *Coproducts and the cd-index*, J. Algebraic Combin. **8** (1998), no. 3, 273–299, DOI 10.1023/A:1008614816374. MR1651249

[40] R. Ehrenborg and M. A. Readdy, *On flag vectors, the Dowling lattice, and braid arrangements*, Discrete Comput. Geom. **21** (1999), no. 3, 389–403, DOI 10.1007/PL00009428. MR1672984

[41] Richard Ehrenborg and Margaret Readdy, *Homology of Newtonian coalgebras*, European J. Combin. **23** (2002), no. 8, 919–927, DOI 10.1006/eujc.2002.0620. MR1938348

[42] Richard Ehrenborg and Margaret Readdy, *The Tchebyshev transforms of the first and second kind*, Ann. Comb. **14** (2010), no. 2, 211–244, DOI 10.1007/s00026-010-0057-2. MR2653890

[43] Richard Ehrenborg and Margaret Readdy, *On the non-existence of an R-labeling*, Order **28** (2011), no. 3, 437–442, DOI 10.1007/s11083-010-9182-0. MR2851357

[44] Richard Ehrenborg and Margaret Readdy, *Manifold arrangements*, J. Combin. Theory Ser. A **125** (2014), 214–239, DOI 10.1016/j.jcta.2014.03.003. MR3207473

[45] Richard Ehrenborg and Margaret Readdy, *A poset view of the major index*, Adv. in Appl. Math. **62** (2015), 1–14, DOI 10.1016/j.aam.2014.09.005. MR3277463

[46] Richard Ehrenborg, Margaret Readdy, and Michael Slone, *Affine and toric hyperplane arrangements*, Discrete Comput. Geom. **41** (2009), no. 4, 481–512, DOI 10.1007/s00454-009-9134-x. MR2496314

[47] Neil J. Y. Fan and Liao He, *The complete* **cd**-*index of Boolean lattices*, Electron. J. Combin. **22** (2015), no. 2, Paper 2.45, 18. MR3359948

[48] Neil J. Y. Fan and Liao He, *On the non-negativity of the complete* **cd**-*index*, Discrete Math. **338** (2015), no. 11, 2037–2041, DOI 10.1016/j.disc.2015.05.005. MR3357790

[49] J. Fine, *The Mayer-Vietoris and IC equations for convex polytopes*, Discrete Comput. Geom. **13** (1995), no. 2, 177–188, DOI 10.1007/BF02574036. MR1314961

[50] J. Fine, *Linear homology in a nutshell*, `arXiv:1908.00039` [math.CO], 2019.

[51] Dominique Foata and Marcel-Paul Schützenberger, *Nombres d'Euler et permutations alternantes (unabridged version)*, Tech. report, University of Florida; available at Séminaire Lotharingien de Combinatoire, 1971, https://www.mat.univie.ac.at/~slc/books/foaschuetz1.pdf.

[52] N. Bradley Fox, *A lattice path interpretation of the diamond product*, Ann. Comb. **20** (2016), no. 3, 569–586, DOI 10.1007/s00026-016-0323-z. MR3537919

[53] Światosław R. Gal, *Real root conjecture fails for five- and higher-dimensional spheres*, Discrete Comput. Geom. **34** (2005), no. 2, 269–284, DOI 10.1007/s00454-005-1171-5. MR2155722

[54] Vladimir Grujić and Tanja Stojadinović, *Hopf algebra of building sets*, Electron. J. Combin. **19** (2012), no. 4, Paper 42, 25, DOI 10.37236/2413. MR3007177

[55] Patricia Hersh, *Regular cell complexes in total positivity*, Invent. Math. **197** (2014), no. 1, 57–114, DOI 10.1007/s00222-013-0480-1. MR3219515

[56] G. Hetyei, *On the cd-variation polynomials of André and Simsun permutations*, Discrete Comput. Geom. **16** (1996), no. 3, 259–275, DOI 10.1007/BF02711512. MR1410161

[57] Gábor Hetyei, *Tchebyshev posets*, Discrete Comput. Geom. **32** (2004), no. 4, 493–520, DOI 10.1007/s00454-004-1115-5. MR2096745

[58] Gábor Hetyei, *Polyspherical complexes*, Ann. Comb. **9** (2005), no. 4, 379–409, DOI 10.1007/s00026-005-0265-3. MR2205031

[59] Gábor Hetyei, *The short toric polynomial*, Trans. Amer. Math. Soc. **365** (2013), no. 3, 1441–1468, DOI 10.1090/S0002-9947-2012-05659-5. MR3003270

[60] Samuel K. Hsiao, *A signed analog of the Birkhoff transform*, J. Combin. Theory Ser. A **113** (2006), no. 2, 251–272, DOI 10.1016/j.jcta.2005.03.003. MR2199274

[61] Duško Jojić, *Weighted derivations and the cd-index*, Discrete Comput. Geom. **39** (2008), no. 4, 678–689, DOI 10.1007/s00454-008-9067-9. MR2413153

[62] Duško Jojić, *The cd-index of the poset of intervals and E_t-construction*, Rocky Mountain J. Math. **40** (2010), no. 2, 527–541, DOI 10.1216/RMJ-2010-40-2-527. MR2646456

[63] Gil Kalai, *A new basis of polytopes*, J. Combin. Theory Ser. A **49** (1988), no. 2, 191–209, DOI 10.1016/0097-3165(88)90051-9. MR964383

[64] Kalle Karu, *The cd-index of fans and posets*, Compos. Math. **142** (2006), no. 3, 701–718, DOI 10.1112/S0010437X06001928. MR2231198

[65] Kalle Karu, *On the complete* **cd**-*index of a Bruhat interval*, J. Algebraic Combin. **38** (2013), no. 3, 527–541, DOI 10.1007/s10801-012-0413-x. MR3104727

[66] Kalle Karu, *M-vector analogue for the* **cd**-*index*, J. Combin. Theory Ser. A **137** (2016), 257–272, DOI 10.1016/j.jcta.2015.09.002. MR3403523

[67] Eric Katz and Alan Stapledon, *Local h-polynomials, invariants of subdivisions, and mixed Ehrhart theory*, Adv. Math. **286** (2016), 181–239, DOI 10.1016/j.aim.2015.09.010. MR3415684

[68] Sangwook Kim, *Flag enumerations of matroid base polytopes*, J. Combin. Theory Ser. A **117** (2010), no. 7, 928–942, DOI 10.1016/j.jcta.2010.02.002. MR2652103

[69] Younghwan Kim, *The cd-indices of intervals in the uncrossing partial order on matchings*, Electron. J. Combin. **26** (2019), no. 1, Paper No. 1.40, 22. MR3934371

[70] Carl W. Lee, *Sweeping the* **cd**-*index and the toric h-vector*, Electron. J. Combin. **18** (2011), no. 1, Paper 66, 20. MR2788683

[71] Swapneel Mahajan, *The cd-index of the Boolean lattice*, http://www.math.iitb.ac.in/ swapneel/Bool.pdf, 2002.

[72] P. McMullen and G. C. Shephard, *Convex polytopes and the upper bound conjecture*, Cambridge University Press, London-New York, 1971. Prepared in collaboration with J. E. Reeve and A. A. Ball; London Mathematical Society Lecture Note Series, 3. MR0301635

[73] Günter Meisinger, *Flag numbers and quotients of convex polytopes*, Ph.D. thesis, Universität Passau, 1993.

[74] Satoshi Murai and Eran Nevo, *On the cd-index and γ-vector of S*-shellable CW-spheres*, Math. Z. **271** (2012), no. 3-4, 1309–1319, DOI 10.1007/s00209-011-0917-4. MR2945609

[75] Satoshi Murai and Eran Nevo, *The flag f-vectors of Gorenstein* order complexes of dimension 3*, Proc. Amer. Math. Soc. **142** (2014), no. 5, 1527–1538, DOI 10.1090/S0002-9939-2014-11897-1. MR3168460

[76] Satoshi Murai and Kohji Yanagawa, *Squarefree P-modules and the* **cd**-*index*, Adv. Math. **265** (2014), 241–279, DOI 10.1016/j.aim.2014.07.037. MR3255460

[77] Isabella Novik, *Lower bounds for the cd-index of odd-dimensional simplicial manifolds*, European J. Combin. **21** (2000), no. 4, 533–541, DOI 10.1006/eujc.1999.0352. MR1756160

[78] Kathryn Nyman and Ed Swartz, *Inequalities for the h-vectors and flag h-vectors of geometric lattices*, Discrete Comput. Geom. **32** (2004), no. 4, 533–548, DOI 10.1007/s00454-004-1137-z. MR2096747

[79] Andreas Paffenholz and Günter M. Ziegler, *The E_t-construction for lattices, spheres and polytopes*, Discrete Comput. Geom. **32** (2004), no. 4, 601–621, DOI 10.1007/s00454-004-1140-4. MR2096750

[80] Mark Purtill, *André permutations, lexicographic shellability and the cd-index of a convex polytope*, Trans. Amer. Math. Soc. **338** (1993), no. 1, 77–104, DOI 10.2307/2154445. MR1094560

[81] Margaret A. Readdy, *Extremal problems for the Möbius function in the face lattice of the n-octahedron*, Discrete Math. **139** (1995), no. 1-3, 361–380, DOI 10.1016/0012-365X(94)00141-5. Formal power series and algebraic combinatorics (Montreal, PQ, 1992). MR1336848

[82] Nathan Reading, *The cd-index of Bruhat intervals*, Electron. J. Combin. **11** (2004), no. 1, Research Paper 74, 25. MR2097340

[83] Nathan Reading, *Non-negative cd-coefficients of Gorenstein* posets*, Discrete Math. **274** (2004), no. 1-3, 323–329, DOI 10.1016/j.disc.2003.07.001. MR2025096

[84] Franco Saliola and Hugh Thomas, *Oriented interval greedoids*, Discrete Comput. Geom. **47** (2012), no. 1, 64–105, DOI 10.1007/s00454-011-9383-3. MR2886091

[85] Michael Slone, *Homological combinatorics and extensions of the cd-index*, ProQuest LLC, Ann Arbor, MI, 2008. Thesis (Ph.D.)–University of Kentucky. MR2711666

[86] D. M. Y Sommerville, *The relations connecting the angle sums and volume of a polytope in n dimensions*, Proc. Roy. Soc. London, Ser. A **115** (1927), 103–119.

[87] Richard P. Stanley, *Balanced Cohen-Macaulay complexes*, Trans. Amer. Math. Soc. **249** (1979), no. 1, 139–157, DOI 10.2307/1998915. MR526314

[88] Richard P. Stanley, *The number of faces of a simplicial convex polytope*, Adv. in Math. **35** (1980), no. 3, 236–238, DOI 10.1016/0001-8708(80)90050-X. MR563925

[89] Richard Stanley, *Generalized H-vectors, intersection cohomology of toric varieties, and related results*, Commutative algebra and combinatorics (Kyoto, 1985), Adv. Stud. Pure Math., vol. 11, North-Holland, Amsterdam, 1987, pp. 187–213, DOI 10.2969/aspm/01110187. MR951205

[90] Richard P. Stanley, *Flag f-vectors and the cd-index*, Math. Z. **216** (1994), no. 3, 483–499, DOI 10.1007/BF02572336. MR1283084

[91] Richard P. Stanley, *A survey of Eulerian posets*, Polytopes: abstract, convex and computational (Scarborough, ON, 1993), NATO Adv. Sci. Inst. Ser. C Math. Phys. Sci., vol. 440, Kluwer Acad. Publ., Dordrecht, 1994, pp. 301–333. MR1322068

[92] Ernst Steinitz, *Über die Eulersche Polyderrelationen*, Archiv. Math. Phys. **11** (1906), no. 3, 86–88.

[93] John R. Stembridge, *Enriched P-partitions*, Trans. Amer. Math. Soc. **349** (1997), no. 2, 763–788, DOI 10.1090/S0002-9947-97-01804-7. MR1389788

[94] Catherine Stenson, *Families of tight inequalities for polytopes*, Discrete Comput. Geom. **34** (2005), no. 3, 507–521, DOI 10.1007/s00454-005-1193-z. MR2160052

[95] Sheila Sundaram, *The homology of partitions with an even number of blocks*, J. Algebraic Combin. **4** (1995), no. 1, 69–92, DOI 10.1023/A:1022437708487. MR1314560

[96] Daniel J. Wells, *General flips and the cd-index*, ProQuest LLC, Ann Arbor, MI, 2010. Thesis (Ph.D.)–University of Kentucky. MR2992617

Department of Mathematics, University of Kansas, 1460 Jayhawk Blvd, Lawrence, Kansas 66045-7594

Email address: `bayer@ku.edu`

URL: `http://bayer.ku.edu`

Contemporary Mathematics
Volume **764**, 2021
https://doi.org/10.1090/conm/764/15357

d-dimensional self-dual polytopes and Meissner polytopes

Tibor Bisztriczky and Déborah Oliveros

ABSTRACT. We present a construction of a class of convex self-dual d-polytopes $\mathcal{P}$ for $d > 2$, and examine conditions under which $\mathcal{P}$ is involutory self-dual and has a metric embedding. Such a $\mathcal{P}$ generates a Reuleaux polytope $\mathcal{B}$, and we explore the relation between $\mathcal{B}$ and Meissner polytopes (modified ball-polytopes of constant width).

1. Introduction and basic terminology

Bodies of constant width have been objects of interest for centuries. Besides the circlular disk, the most widely known example is the Reuleaux triangle, constructed as the intersection of three disks with center at the vertices of the regular triangle with edges of length say, 1, and whose boundary consists of three congruent circular arcs of radius 1. The Reuleaux triangle, can be generalized to Reuleaux polygons, which are plane convex figure of constant width whose boundary consists of a finite (necessarily odd) number of circular arcs of radius 1, say. As in the case of the Reuleaux triangle, the Reuleaux tetrahedron is the intersection of four balls centered at the vertices of a regular tetrahedron. It is well known, however, that the obtained convex body does not have constant width. In 1912, Meissner and Schilling [**10**] showed how to modify the Reuleaux tetrahedron, in order to obtain two convex bodies of constant width, they replace a neighbourhood of three edges by curved patches formed by a piece of a surface of revolution of a circular arc (see [**14**], [**9**] for further details). Besides the Meissner bodies and the classical bodies of revolution of Reuleaux polygons, no other examples of constructions of bodies of constant width were known until recently, when Montejano and Roldan-Pensado [**11**] gave a generalization of the Meissner and Schilling construction to generate examples of bodies of constant width called Meissner polyhedra.

Their construction contains the following main ingredients:

1. They begin with a metric embedding of an involutory self-dual, 3-connected, planar graph G; that is, a 3-dimensional embedding of G, such that the distance between every vertex and its dual facet is exactly one.
2. Based on the set V_G of vertices of the already embedded graph G, they define a ball polyhedron $P_G = \bigcap_{x \in V_G} B(x)$, called a Reuleaux polyhedron, and show that such ball polyhedron is a standard ball polytope in the sense of [**2**] and the set of 0-singular points of P_G is V_G.

2010 *Mathematics Subject Classification.* Primary 5201, 52A15, 52B11.

Supported by Proyecto PAPIIT 107218, 106318 and CONACYT Ciencia Básica 282280.

3. Finally, they obtain bodies of constant width by a "surgery" that consists of replacing a neighbourhood of one of the edge arcs of each pair of dual edges of G by curved patches formed by a piece of the surface of revolution of a circular arc.

The main motivation for this paper is to generalize these bodies of constant width to higher dimension. This paper does not yet settle this issue, but we take a first step in this direction by finding interesting problems and results along the way. We were surprised to find out that not much is known about self-dual polytopes in dimensions 4 and higher and not much has been done towards describing boundary complexes of ball polytopes in higher dimensions (see [**6**]).

Given a convex d-polytope $\mathcal{P}$, we will denote by $\mathcal{F}(\mathcal{P})$ and $V(\mathcal{P})$, the set of facets and vertices of $\mathcal{P}$, and by $\mathcal{L}(\mathcal{P})$ its face-lattice. Two polytopes $\mathcal{P}_1$ and $\mathcal{P}_2$ are said to be combinatorially equivalent ($\mathcal{P}_1 \simeq \mathcal{P}_2$) if their face-lattices are isomorphic, or equivalently, if there exists a one-to-one inclusion-preserving mapping Ψ from $\mathcal{L}(\mathcal{P}_1)$ to $\mathcal{L}(\mathcal{P}_2)$. For each d-polytope $\mathcal{P}$, it is easy to establish by a polarity argument the existence of a d-polytope $\mathcal{P}^*$, called the dual of $\mathcal{P}$, with the property that their face-lattices are anti-isomorphic; that is, there exists a one-to-one inclusion-reversing map Φ from $\mathcal{L}(\mathcal{P})$ to $\mathcal{L}(\mathcal{P}^*)$. A d-polytope is said to be self-dual if it is combinatorially equivalent to its polar dual; that is, if $\mathcal{P} \simeq \mathcal{P}^*$ via Φ. Futhermore we will say that $\mathcal{P}$ is involutory self-dual if $\Phi^2 = id$. It is known, that not all self-dual polytopes are involutory (see [**5**], [**7**]). Self-dual 3-polytopes have been studied extensively (see [**12**] and the references therein) and the 3-dimensional spherical and projective polytopes have been fully characterized (see [**1**]). In higher dimension, however, not much appears to be known. In [**13**], some constructions are studied together with the complexity aspects of self-duality. In Section 2, we show (Theorem 2.1) that given a self-dual d-polytope $\mathcal{P}$, there exists a self-dual $(d+1)$-polytope $\mathcal{P}'$, and if $\mathcal{P}$ is involutory self-dual then $\mathcal{P}'$ is involutory self-dual. In Section 3, we provide several infinite families of examples, and in Section 4, we show that if the base polytope $\mathcal{P}$ is involutory self-dual, has a metric embedding, and its vertices lie on a sphere with center at the origin and radius $r \leq 1$ then $\mathcal{P}'$ is involutory self-dual and has a metric embedding. Finally we show in Section 5 that the construction defined in [**11**] does not extend to $\mathbb{R}^4$ for the Reuleaux 4-simplex.

A d-polytope $\mathcal{P} \subset \mathbb{R}^d$ is the convex hull of a finite number of points in $\mathbb{R}^d$, or equivalently, the intersection of a finite number of half spaces. Throughout this paper, we will assume the reader is familiar with the basic notions of convex polytopes (see [**4**] ,[**15**], [**3**] for excellent references and basic terminology).

Let Y be a set of points in $\mathbb{R}^d$. We will denote the convex hull and the affine hull of Y by $[Y]$ and $\langle Y \rangle$, respectively. If Y is finite with elements $y_1, y_2, \dots, y_n$, we let $[Y] = [y_1, y_2, \dots, y_n]$ and $\langle Y \rangle = \langle y_1, y_2, \dots, y_n \rangle$. We use the coordinate system $(x_1, \dots, x_d)$ for $\mathbb{R}^d$, and a simpler (x, y, z, w) for $\mathbb{R}^4$.

2. Prismoidal constructions of self-dual polytopes

Let $m \geq d \geq 3$ and $\mathcal{Q}_{1m} = [y_{11}, y_{12}, \dots, y_{1m}]$ be a convex $(d-1)$-polytope in $\mathbb{R}^d$ with the vertex set $V(\mathcal{Q}_{1m}) = \{y_{11}, y_{12}, \cdots y_{1m}\}$. Assume that $\langle \mathcal{Q}_{1m} \rangle$ is the hyperplane $x_d = 0$, and that the origin $(0, 0, \dots, 0)$ is an interior point of $\mathcal{Q}_{1m}$.

For $k \geq 2$ and $i = 2, \dots k$, let $\mathcal{Q}_{im} = [y_{i1}, y_{i2}, \dots, y_{im}]$ be a positive homothetic copy of $\mathcal{Q}_{(i-1)m}$ via φ_i so that $\varphi_i(y_{(i-1)r}) = y_{ir}$ for $r = 1, \dots, m$ and the following holds:

1) $\langle \mathcal{Q}_{1m} \rangle$ and $\langle \mathcal{Q}_{im} \rangle$ are parallel distinct hyperplanes in $\mathbb{R}^d$.
2) $\langle \mathcal{Q}_{bm} \rangle$ separates $\mathcal{Q}_{am}$ and $\mathcal{Q}_{cm}$ for $1 \leq a < b < c \leq k$, and
3) $\mathcal{R}_{km} := [\mathcal{Q}_{1m}, \mathcal{Q}_{2m}, \ldots \mathcal{Q}_{km}]$, the convex hull of the union of the $\mathcal{Q}_{im}$'s, has km vertices.

We will call $\mathcal{R}_{km}$, a *k-layered d-prismoid.* Note that $\mathcal{R}_{km}$ has the property that

4) $[y_{(i-1)r}, y_{ir}]$, $r = 1, \ldots, m$ are the edges of $\mathcal{R}_{km}$ that intersect $\mathcal{Q}_{(i-1)m}$ and $\mathcal{Q}_{im}$, and
5) $[y_{(i-1)a}, y_{(i-1)b}, y_{ia}, y_{ib}]$ are 2-faces (4-gons) of $\mathcal{R}_{km}$ if $[y_{1a}, y_{1b}]$ is an edge of $\mathcal{Q}_{1m}$

Assume next, that $\mathcal{Q}_{1m}$ is a self-dual polytope via an anti-isomorphism ϕ_1 on the face-lattice of $\mathcal{Q}_{1m}$ that maps y_{1r} to $\bar{y}_{1r}$. Then, we extend this anti-isomorphism for every $i = 2, \ldots k$ to ϕ_i (on the face-lattice of $\mathcal{Q}_{im}$) as the map y_{ir} to $\bar{y}_{ir}$ defined by

$$y_{ia} \in \bar{y}_{ib} \text{ if and only if } y_{(i-1)a} \in \bar{y}_{(i-1)b}.$$

Then we observe that $\mathcal{Q}_{im}$ is self-dual with

6) $\mathcal{F}(\mathcal{Q}_{im}) = \{\bar{y}_{i1}, \bar{y}_{i2}, \ldots, \bar{y}_{im}\}$ for $i = 1, 2, \ldots, k$,
7) $\phi_i(\phi_i(y_{ir})) = \phi_i(\bar{y}_{ir}) = \bar{\bar{y}}_{ir} \in V(\mathcal{Q}_{im})$,
8) $y_{ir} \in \bar{y}_{is}$ if and only if $\bar{\bar{y}}_{is} \in \bar{y}_{ir}$, and
9) $\mathcal{F}(\mathcal{R}_{km}) = \{\mathcal{Q}_{1m}, \mathcal{Q}_{km}\} \cup \{[\bar{y}_{(i-1)r}, \bar{y}_{ir}] | i = 2, \ldots k, r = 1, \ldots, m\}$.

Finally let $\mathcal{P}_{km} = [y_{00}, \mathcal{Q}_{1m}, \mathcal{Q}_{2m}, \ldots \mathcal{Q}_{km}]$ be the d–polytope with the property that $y_{00} \in \mathbb{R}^d$ is a point beyond $\mathcal{Q}_{1m}$ and beneath all other facets of $\mathcal{R}_{km} = [\mathcal{Q}_{1m}, \mathcal{Q}_{2m}, \ldots \mathcal{Q}_{km}]$; cf. [**4**] for the definitions of beyond and beneath, and Figure 1 for an example of $\mathcal{P}_{23}$ and $\mathcal{P}_{33}$ in $\mathbb{R}^3$.

Clearly, $\mathcal{F}(\mathcal{P}_{km}) = \{\mathcal{Q}_{km}\} \cup \{[y_{00}, \bar{y}_{1r}], [\bar{y}_{(i-1)r}, \bar{y}_{ir}] | i = 2, \ldots k, r = 1, \ldots, m\}$. For $j = 1, 2, \ldots, k-1$ and $r = 1, 2, \ldots, m$, we let $Y_{00} = \mathcal{Q}_{km}$, $Y_{kr} = [y_{00}, \bar{y}_{1r}]$ and $Y_{jr} = [\bar{y}_{(k-j)r}, \bar{y}_{(k-j+1)r}]$.

Now we are ready to show our Main Result.

THEOREM 2.1. *$\mathcal{P}_{km}$ is a self-dual polytope via the anti-isomorphism Φ induced by the anti-isomorphism ϕ_1, and $\Phi(y_{00}) = y^*_{00} = Y_{00}$, $\Phi(y_{jr}) = y^*_{jr} = Y_{jr}$ for $j = 1, \ldots, k$ and $r = 1, \ldots, m$. Furthermore, $y_{ja} \in y^*_{ja}$ for some j and a if and only if $y_{1a} \in \bar{y}_{ia}$.*

PROOF. We begin by observing that 6) to 9) implies that for $i = 1, 2, \ldots, k$,

if $y_{ir} \in V(\mathcal{P}_{km})$ then $\bar{y}_{ir} \in \mathcal{F}(\mathcal{P}_{km})$ and $\bar{\bar{y}}_{ir} \in V(\mathcal{P}_{km})$; furthermore, $y_{ir} \in \bar{y}_{is}$ if and only if $\bar{\bar{y}}_{is} \in \bar{y}_{ir}$ if and only if $y_{1r} \in \bar{y}_{1s}$ if and only if $\bar{\bar{y}}_{1s} \in \bar{y}_{1r}$ (with $\bar{y}_{00} = Y_{00}$ and $\bar{\bar{y}}_{00} = y_{00}$).

Then observe that for $1 \leq r, s \leq m$, $\bar{y}_{is} \subset Y_{js}$ if and only if $i + j \in \{k, k+1\}$ if and only if $\bar{y}_{jk} \subset Y_{ir}$.

Hence follows that $y_{ir} \in y^*_{js} = Y_{js}$ if and only if $y_{ir} \in \bar{y}_{is} \subset Y_{js}$ if and only if $\bar{\bar{y}}_{is} \subset \bar{y}_{ir}$ and $\bar{y}_{jr} \subset Y_{ir}$ if and only if $\bar{\bar{y}}_{js} \subset \bar{y}_{jr} \subset Y_{ir}$.

Let $\mathcal{L}(\mathcal{P}_{km})$ denote the face-lattice of $\mathcal{P}_{km}$. We note that the maps

$y_{jr} \in V(\mathcal{P}_{km})$ to $y^*_{jr} = Y_{jr} \in \mathcal{F}(\mathcal{P}_{km})$ and then to $Y^*_{jr} = \bar{\bar{y}}_{jr} \in V(\mathcal{P}_{km})$ are bijections with the property that $y_{ir} \in Y_{js}$ if and only if $Y^*_{js} \in y^*_{ir}$.

It is well known ([**3**], Ex 10.6) that this implies that there is an anti-isomorphism on $\mathcal{L}(\mathcal{P}_{km})$ that extends Φ. Specifically, if $\mathcal{G} \in \mathcal{L}(\mathcal{P})$ has the vertices $x_1, x_2 \ldots, x_n$ and is contained in exactly the facets $F_1, F_2, \ldots, F_v$, then $\mathcal{G}^* = [F^*_1, F^*_2, \ldots, F^*_v] = \cap^n_{j=1} x^*_j$. □

Corollary 1. *Let $\mathcal{Q}_{1m}$ be a self-dual $d-$polytope with the involutory anti-isomorphism given by ϕ_1 that sends y_{1r} to $\bar{y}_{1r}$; that is, $\bar{\bar{y}}_{1r} = y_{1r}$. Then P_{km} is involutory self-dual with $\Phi(y_{ir}) = y^*_{ir}$; moreover, $y_{ja} \in y^*_{ja}$ for some j and a if, and only if, $y_{1a} \in \bar{y}_{1a}$*

3. Examples of $\mathcal{P}_{km}$, for $d \geq 3$

In this section, we present examples of prismoidally constructed $\mathcal{P}_{km}$ in $\mathbb{R}^d$ that are self-dual. Specifically, we describe (in terms of vertices) facets of $\mathcal{P}_{km}$ for some small values of k, m and d.

1) Let $\mathcal{Q}_{k3} = [x_{11}, x_{12}, x_{13}, x_{21}, x_{22}, x_{23}, \dots, x_{k1}, x_{k2}, x_{k3}]$, $k \geq 2$, be a k-layered 3-prismoid, and $\mathcal{F}(\mathcal{Q}_{k3}) = \{[x_{11}, x_{12}, x_{13}], [x_{k1}, x_{k2}, x_{k3}]\} \cup \{G_{i1} = [x_{i2}, x_{i3}, x_{(i+1)2}, x_{(i+1)3}], G_{i2} = [x_{i1}, x_{i3}, x_{(i+1)1}, x_{(i+1)3}], G_{i3} = [x_{i1}, x_{i2}, x_{(i+1)1}, x_{(i+1)2}] | \; i = 1, \dots, k-1\}$.

Let $\mathcal{P}_{k3} = [x_{00}, \mathcal{Q}_{k3}]$ with x_{00} beyond exactly the facet $[x_{11}, x_{12}, x_{13}]$ of $\mathcal{Q}_{k3}$, and $F_{00} = [x_{k1}, x_{k2}, x_{k3}]$, $F_{k1} = [x_{00}, x_{12}, x_{13}]$, $F_{k2} = [x_{00}, x_{11}, x_{12}]$, $F_{k3} = [x_{00}, x_{11}, x_{13}]$ and $F_{(k-i)j} = G_{ij}$ for $i = k-1, k-2, \dots, 1$ and $j = 1, 2, 3$.

We have that $\mathcal{F}(\mathcal{P}_{k3}) = \{F_{00}, F_{11}, F_{12}, F_{13}, \dots, F_{k1}, F_{k2}, F_{k3}\}$ and $\mathcal{P}_{k3}$ is a self-dual 3-polytope under the mapping induced by $\Phi(x_{ij}) = F_{ij}$. We note that $x_{ij} \notin F_{ij} = x^*_{ij}$.

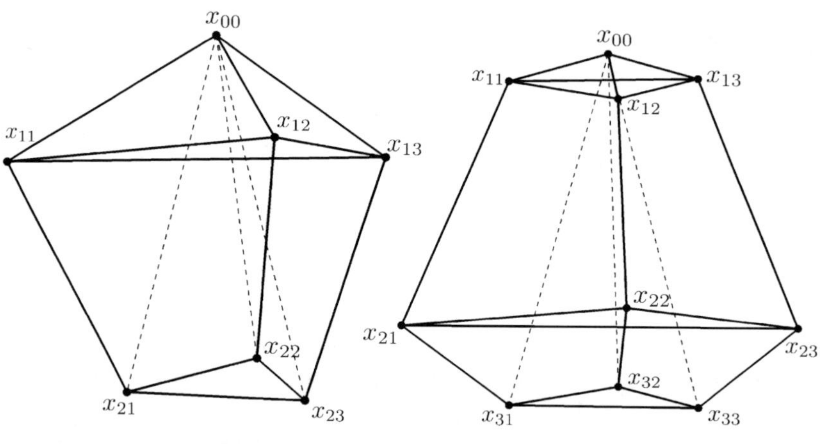

Figure 1. Self-dual $\mathcal{P}_{23}$ (right) and $\mathcal{P}_{33}$(left).

2) Let $\mathcal{Q}_{km} = [x_{11}, x_{12}, \dots, x_{1m}, x_{21}, x_{22}, \dots, x_{2m}, \dots, x_{k1}, x_{k2}, \dots, x_{km}]$, $k \geq [\frac{m}{2}]$, be a k-layered 3-prismoid, $m > d = 3$, $m = 2u+1$ where $\mathcal{Q}_{1m}$ is an m-gon and

$$\mathcal{F}(\mathcal{Q}_{km}) = \{[x_{11}, x_{12}, \dots, x_{1m}], [x_{k1}, x_{k2}, \dots, x_{km}]\}$$

$\cup\{G_{ij} = [x_{ij}, x_{i(j+1)}, x_{(i+1)j}, x_{(i+1)(j+1)}] | \; i = 1, \dots, k-1, j = 1, \dots, m, x_{u1} = x_{u,(m+1)}\}$.

Let $\mathcal{P}_{km} = [x_{00}, \mathcal{Q}_{km}]$ with x_{00} beyond exactly the facet $[x_{11}, x_{12}, \dots x_{1m}]$ of $\mathcal{Q}_{km}$. Then with $F_{00} = [x_{k1}, x_{k2}, \dots, x_{km}]$, $F_{kj} = [x_{00}, x_{1(j+[\frac{m}{2}])}, x_{1(j+[\frac{m}{2}]+1)}]$, and $F_{(k-i)j} = G_{i(j+[\frac{m}{2}])}$ for $i = k-1, k-2, \dots, 1$ and $j = 1, 2, \dots, m$. $\mathcal{F}(\mathcal{P}_{km}) = \{F_{00}, F_{11}, F_{12},$
$\dots, F_{1m}, \dots, F_{k1}, F_{k2}, \dots, F_{km}\}$.

Clearly $\mathcal{Q}_{1m} = [x_{11}, x_{12}, \dots, x_{1m}]$ with facets $\mathcal{F}(\mathcal{Q}_{1m}) = \{\bar{x}_{11}, \bar{x}_{12}, \dots, \bar{x}_{1m}\}$ and $\bar{x}_{1r} = [x_{1(r+u)}, x_{1(r+u+1)}]$, evaluated (mod $m+1$).

Then $\mathcal{P}_{km}$ is a self-dual 3-polytope under the mapping induced by $\Phi(x_{ij}) = F_{ij}$. We note that $x_{ij} \notin F_{ij} = x_{ij}^*$. Note that for m even, the same map yields an isomorphism but not an involutory one.

3) $m = d = 4$. Let $\mathcal{Q}_{k4} = [x_{11}, x_{12}, x_{13}, x_{14}, x_{21}, x_{22}, x_{23}, x_{24}, \ldots, x_{k1}, x_{k2}, x_{k3}, x_{k4}]$ be a k-layered 4-prismoid with $k \geq 2$. $\mathcal{Q}_{14}$ is the regular tetrahedron $[x_{11}, x_{12}, x_{13}, x_{14}]$, and $\mathcal{F}(\mathcal{Q}_{14}) = \{\bar{x}_{11}, \bar{x}_{12}, \bar{x}_{13}, \bar{x}_{14}\}$, with $\bar{x}_{1r} = [x_{1s}, x_{1t}, x_{1u}]$ for $\{r, s, t, u\} = \{1, 2, 3, 4\}$.

$$\mathcal{F}(\mathcal{Q}_{k4}) = \{[x_{11}, x_{12}, x_{13}, x_{14}], [x_{k1}, x_{k2}, x_{k3}, x_{k4}]\}$$

$$\cup\{G_{ir} = [x_{is}, x_{it}, x_{iu}, x_{(i+1)s}, x_{(i+1)t}, x_{(i+1)u}], \mid i = 1, \ldots, k-1, \}, \{r, s, t, u\} = \{1, 2, 3, 4\}\}.$$

Let $\mathcal{P}_{k4} = [x_{00}, \mathcal{Q}_{k4}]$ with x_{00} beyond exactly the facet $[x_{11}, x_{12}, x_{13}, x_{14}]$ of $\mathcal{Q}_{k4}$. $F_{00} = [x_{k1}, x_{k2}, x_{k3}, x_{k4}]$, $F_{kr} = [x_{00}, x_{1s}, x_{1t}, x_{1u}]$ and $F_{ir} = G_{(k-i)r}$ for $i = k-1, k-2, \ldots, 1$. $\mathcal{F}(\mathcal{P}_{k4}) = \{F_{00}, F_{11}, F_{12}, F_{13}, F_{14}, \ldots, F_{k1}, F_{k2}, F_{k3}, F_{k4}\}$. $\mathcal{P}_{k4}$ is an involutory self-dual 4-polytope with the same mapping as above.

4. Metric embeddings of self-dual polytopes

Throughout this section we will assume the following:

i) $\mathcal{Q}_{1m} = [y_{11}, y_{12}, \ldots y_{1m}]$ is an involutory self-dual $(d-1)-$polytope with anti-isomorphism given by ϕ_1 that sends y_{1r} to $\bar{y}_{1r}$ and $\bar{\bar{y}}_{1r} = y_{1r}$ for every $1 \leq r \leq m$.

ii) The vertices of the polytope $\mathcal{Q}_{1m}$ lie on the sphere $\mathbb{S}_t$ of radius $0 < t \leq 1$ with centre at the origin. Thus, we are assuming that $\|y_{1r}\| = t$ for every vertex $y_{1r} \in V(\mathcal{Q}_{1m})$.

iii) $\mathcal{Q}_{1m}$ is embedded metrically in $\mathbb{R}^{d-1}$; that is, the vertices of such polytope satisfy that the distance between any vertex, and any vertex of its dual facet is 1; that is, for every vertex $y_{1r} \in V(\mathcal{Q}_{1m})$, $\|y_{1r} - x_{1s}\| = 1$ if $x_{1s} \in V(\bar{y}_{1r})$.

Examples of such $\mathcal{Q}_{1m}$ are odd regular polygons, and regular $(d-1)$-simplexes.

We refer to the paper of Lovász [**8**] (and the references therein) for the definition and construction of strongly self-dual polytopes, and their relation to the chromatic number of the graph of diagonals.

Our goal is to show that i), ii) and iii) yield that there is a metric embedding of the $\mathcal{P}_{km}$ defined in Section 2.

In the following and as a simplification, we write y_{1r} and x_{1s} in $\bar{y}_{1r}$ for some $1 \leq r \leq m$ in $\mathbb{R}^{d-1}$ satisfying i),ii) and iii) as $y_{1r} = a = (a_1, a_2, \ldots, a_{d-1})$, and $x_{1s} = b = (b_1, b_2, \ldots, b_{d-1})$, and y_{1r} and x_{1s} in $\bar{y}_{1r}$ in $\mathbb{R}^d$ as $y_{1r} = (\lambda_1 a, p_1) = (\lambda a_1, \lambda_1 a_2, \ldots, \lambda_1 a_{d-1}, p_1)$ and $x_{1s} = (\lambda_1 b, p_1)$ with λ_1 and p_1 to be defined and determined next. We are interested here only in $\|a\| = \|b\| = t$ and $\|a - b\| = 1$.

We recall also that $\langle \mathcal{Q}_{im} \rangle$ for $\mathcal{P}_{km}$ are parallel distinct hyperplanes in $\mathbb{R}^d$, and we may assume that $\langle \mathcal{Q}_{im} \rangle : x_d = p_i$ with $p_1 > p_2 > \cdots > p_k = 0$ and the centres of the homothets that yield $\mathcal{Q}_{im}$ are on the x_d-axis. We observe next that the orthogonal projection of the vertices of $\mathcal{Q}_{im}$ on $\langle \mathcal{Q}_{1m} \rangle$ also lie on a sphere with centre at the origin, with say, radius λ_i. Hence, $y_{1r} = (\lambda_1 a, p_1) = (\lambda_1 a_1, \lambda_1 a_2, \ldots \lambda_1 a_{d-1}, p_1)$ and then $y_{ir} = (\lambda_i a, p_i) = (\lambda_i a_1, \lambda_i a_2, \ldots \lambda_i a_{d-1}, p_i)$. Similarly, $x_{1s} = (\lambda_1 b, p_1)$ and then $x_{is} = (\lambda_i b, p_i)$. Finally, as y_{00} is beyond only the facet $\mathcal{Q}_{1m}$ of $[\mathcal{Q}_{1m}, \mathcal{Q}_{12m}, \ldots \mathcal{Q}_{km}]$, we may assume that $y_{00} = (0, 0 \ldots, 0, q)$ with $q > p_1$.

Since $\mathcal{P}_{km}$ is self-dual via the anti-isomorphism Φ such that $\Phi(y_{00}) = Y_{00}$ and $\Phi(y_{ir}) = Y_{ir}$ from Theorem 2.1, we observe that there is a metrical embedding of $\mathcal{P}_{km}$ if for each $1 \le r \le m$ and associated s, the following equations have a solution:

(0) $\|y_{00} - y_{kr}\|^2 = 1 = (\lambda_k)^2\|a\|^2 + q^2$,

(j) $\|y_{jr} - x_{(k-j)s}\|^2 = 1 = \|y_{jr} - x_{(k-j+1)s}\|^2$ for $j = 1, 2, \dots, k-1$ and

(k) $\|y_{kr} - x_{1s}\|^2 = 1$.

Observe first that if $\lambda_k < 1$, then equation (0) has a real solution for q. Since $p_k = 0$, then, for equation (k) with $\lambda_k < 1$, and $\lambda_1 < 1$, $p_1^2 = 1 - \|\lambda_k a - \lambda_1 b\|^2$ has a real solution because $\|\lambda_k a - \lambda_1 b\|^2 < \lambda_1^2$ if $\lambda_k < \lambda_1$ and $\|\lambda_k a - \lambda_1 b\|^2 < \lambda_k^2$ if $\lambda_1 < \lambda_k$.

In the case of equation (j), we observe that:

If $k = 2n$, then

(n) $1 = \|y_{nr} - x_{ns}\|^2 = \|(\lambda_n a, p_n) - (\lambda_n b, p_n)\|^2 = \lambda_n^2\|a-b\|^2 = \lambda_n^2$,

and if $k = 2n+1$, then

(n+1) $1 = \|y_{(n+1)r} - x_{(n+1)s}\|^2 = \|(\lambda_{n+1}a, p_{n+1}) - (\lambda_{n+1}b, p_{n+1})\|^2 = \lambda_{n+1}^2\|a-b\|^2 = \lambda_{n+1}^2$.

Then we need only to check what happens in the rest of the cases. For $j = 2, \dots k-1$, we see that (j) is equivalent to the following two equations:

$(p_j - p_{k-j})^2 = 1 - [(\lambda_j a_1 - \lambda_{k-j} b_1)^2 + \cdots + (\lambda_j a_{d-1} - \lambda_{k-j} b_{d-1})^2] = \beta_j$

$(p_j - p_{k-j+1})^2 = 1 - [(\lambda_j a_1 - \lambda_{k-j+1} b_1)^2 + \cdots + (\lambda_j a_{d-1} - \lambda_{k-j+1} b_{d-1})^2] = \gamma_j$

Assume that $0 < \lambda_k < \lambda_1 < \lambda_{k-1} < \lambda_2 < \lambda_{k-2} \dots \lambda_j < \lambda_{k-j} < \cdots < \lambda_{[\frac{k+1}{2}]} = 1$, with possibly $\lambda_{k-j} = \lambda_j$. Then:

$$\beta_j > 1 - \lambda_{k-j}^2\|a-b\|^2 \ge 1 - \lambda_{k-j} > 0$$

and

$$\gamma_j > 1 - \lambda_j^2\|a-b\|^2 \ge 1 - \lambda_j > 0.$$

Therefore, $\sqrt{\beta_j}$ and $\sqrt{\gamma_j}$ have a real solution.

Using the value of p_1 in (k), we know now that:

for $(j = 1)$, $p_{k-1} = p_1 - \sqrt{\beta_1}$.

Now we just show by induction that if we assume $p_{k-m} = p_m - \sqrt{\beta_m}$ for $j = m$ then $p_{m+1} = \sqrt{\gamma_{m+1}} + p_{k-(m+1)+1} = \sqrt{\gamma_{m+1}} + p_{k-m}$ for $j = m+1$.

THEOREM 4.1. *If $\mathcal{Q}_{1m}$ is a convex $(d-1)$-polytope satisfying properties i), ii),iii), then there is $\mathcal{P}_{km} = [y_{00}, R_{km}]$, with $\mathcal{P}_{km}$ and Y_{ir} defined as in Section 2, and the property that for each y_{ir}, $\|y_{ir} - y\| = 1$ for any $y \in Y_{ir}$; furthermore if $k = 2$ then $\mathcal{P}_{km}$, has a metrical embedding.*

For example, in dimension 3, consider the regular tetrahedron $\mathcal{Q}_{14}$ with edge length 1 and coordinates:

$[x_{11}, x_{12}, x_{13}, x_{14}] = [(0, \frac{1}{\sqrt{3}}, -\frac{1}{2\sqrt{2}\sqrt{3}}), (-\frac{1}{2}, -\frac{1}{2\sqrt{3}}, -\frac{1}{2\sqrt{2}\sqrt{3}}), (\frac{1}{2}, -\frac{1}{2\sqrt{3}}, -\frac{1}{2\sqrt{2}\sqrt{3}}), (0, 0, \frac{\sqrt{3}}{2\sqrt{2}})]$.

Then in dimension 4, $\mathcal{P}_{k4}$ is presented in example 3) of Section 3.

For $k = 2$, $\mathcal{P}_{24}$ can be embedded metrically by solving three equations that yield the following solutions for $\lambda_2 = \frac{1}{2} < \lambda_1 = 1$, $q = \frac{1}{8}\sqrt{29}\sqrt{2}$, $p_2 = 0$ and $p_1 = \frac{1}{8}\sqrt{13}\sqrt{2}$. That is, the vertices of $\mathcal{P}_{24}$ have the following coordinates:

$[x_{11}, x_{12}, x_{13}, x_{14}, x_{21}, x_{22}, x_{23}, x_{24}, y_{00}]$ where $y_{00} = (0, 0, 0, \frac{1}{8}\sqrt{29}\sqrt{2})$, $x_{21} = \frac{1}{2}(0, \frac{1}{\sqrt{3}}, -\frac{1}{2\sqrt{2}\sqrt{3}}, 0)$, $x_{22} = \frac{1}{2}(-\frac{1}{2}, -\frac{1}{2\sqrt{3}}, -\frac{1}{2\sqrt{2}\sqrt{3}}, 0)$, $x_{23} = \frac{1}{2}(\frac{1}{2}, -\frac{1}{2\sqrt{3}}, -\frac{1}{2\sqrt{2}\sqrt{3}}, 0)$,

$x_{24} = \frac{1}{2}(0, 0, \frac{\sqrt{3}}{2\sqrt{2}}, 0)]$, $x_{11} = (0, \frac{1}{\sqrt{3}}, -\frac{1}{2\sqrt{2}\sqrt{3}}, \frac{1}{8}\sqrt{13\sqrt{2}})$, $x_{12} = (-\frac{1}{2}, -\frac{1}{2\sqrt{3}}, -\frac{1}{2\sqrt{2}\sqrt{3}}, \frac{1}{8}\sqrt{13\sqrt{2}})$, $x_{13} = (\frac{1}{2}, -\frac{1}{2\sqrt{3}}, -\frac{1}{2\sqrt{2}\sqrt{3}}, \frac{1}{8}\sqrt{13\sqrt{2}})$ and $x_{14} = (0, 0, \frac{\sqrt{3}}{2\sqrt{2}}, \frac{1}{8}\sqrt{13\sqrt{2}})$.

For $k = 3$, we may embed $\mathcal{P}_{34}$ metrically by solving four equations that yield the following solutions for $\lambda_3 = \frac{9}{20} < \lambda_1 = \frac{1}{2} < \lambda_2 = 1$, $q = \frac{1}{80}\sqrt{2957\sqrt{2}}$, $p_3 = 0$, $p_2 = \frac{1}{80}\sqrt{2}(\sqrt{2477} - \sqrt{1333}) \approx 0.2343$ and $p_1 = \frac{1}{80}\sqrt{2477\sqrt{2}} \approx 0.8798$.

That is, the vertices of $\mathcal{P}_{34}$ have the following coordinates:

$[x_{11}, x_{12}, x_{13}, x_{14}, x_{21}, x_{22}, x_{23}, x_{24}, x_{31}, x_{32}, x_{33}, x_{34}, y_{00}]$ where $y_{00} = (0, 0, 0, q)$, $x_{31} = \frac{9}{20}(0, \frac{1}{\sqrt{3}}, -\frac{1}{2\sqrt{2}\sqrt{3}}, 0)$, $x_{32} = \frac{9}{20}(-\frac{1}{2}, -\frac{1}{2\sqrt{3}}, -\frac{1}{2\sqrt{2}\sqrt{3}}, 0)$, $x_{33} = \frac{9}{20}(\frac{1}{2}, -\frac{1}{2\sqrt{3}}, -\frac{1}{2\sqrt{2}\sqrt{3}}, 0)$, $x_{34} = \frac{9}{20}(0, 0, \frac{\sqrt{3}}{2\sqrt{2}}, 0)]$, $x_{21} = (0, \frac{1}{\sqrt{3}}, -\frac{1}{2\sqrt{2}\sqrt{3}}, p_2)$, $x_{22} = (-\frac{1}{2}, -\frac{1}{2\sqrt{3}}, -\frac{1}{2\sqrt{2}\sqrt{3}}, p_2)$, $x_{23} = (\frac{1}{2}, -\frac{1}{2\sqrt{3}}, -\frac{1}{2\sqrt{2}\sqrt{3}}, p_2)$, $x_{24} = (0, 0, \frac{\sqrt{3}}{2\sqrt{2}}, p_2)]$, $x_{11} = \frac{1}{2}(0, \frac{1}{\sqrt{3}}, -\frac{1}{2\sqrt{2}\sqrt{3}}, p_1)$, $x_{12} = \frac{1}{2}(-\frac{1}{2}, -\frac{1}{2\sqrt{3}}, -\frac{1}{2\sqrt{2}\sqrt{3}}, p_1)$, $x_{13} = \frac{1}{2}(\frac{1}{2}, -\frac{1}{2\sqrt{3}}, -\frac{1}{2\sqrt{2}\sqrt{3}}, p_1)$, $x_{14} = \frac{1}{2}(0, 0, \frac{\sqrt{3}}{2\sqrt{2}}, p_1)$.

5. The boundary points of the Reuleaux 4-simplex

For a point $p_a \in \mathbb{R}^4$, we let S_a denote the 3-sphere in $\mathbb{R}^4$ with centre p_a and radius 1, and $B_a = [S_a]$.

Let $P = [p_1, p_2, p_3, p_4, p_5] \subset \mathbb{R}^4$ be a regular 4-simplex of edge-length 1, and $Y = \{p_1, \ldots, p_5\} = \{p_r, p_s, p_t, p_u, p_v\}$. P is clearly involutory self-dual with duality: p_v mapped to $[p_r, p_s, p_t, p_u]$ and $[p_u, p_v]$ mapped to $[p_r, p_s, p_t]$. We consider the ball-polytope

$$\Phi = B_1 \cap B_2 \cap B_3 \cap B_4 \cap B_5.$$

Then $bd(\Phi) \subset S_1 \cup S_2 \cup S_3 \cup S_4 \cup S_5$, and each of S_r, $S_r \cap S_s$ and $S_r \cap S_s \cap S_t$ is a supporting sphere of Φ. We note that:

1. $S_s \cap S_t$ is a 2-sphere that contains p_r, p_u, p_v and has radius $\frac{\sqrt{3}}{2} \approx 0.86602$, and
2. $S_r \cap S_s \cap S_t$ is a circle that contains p_u, p_v and has radius $\sqrt{\frac{2}{3}} \approx 0.81649$.

Let $p \in bd(\Phi)$. We will say that p is a k-point if it is contained in exactly k of $S_1, \ldots, S_5$. Then for $1 \leq k \leq 4$, Y is the set of 4-points of $bd(\Phi)$, and we consider the sets

$$S^*_{st} = \{p \in S_s \cup S_t |\ p \text{ is a } 2 - \text{point of } bd(\Phi)\}$$

and

$$\tilde{P}_{st} = \{p \in S_r \cup S_u \cap S_v |\ p \text{ is a } 3 - \text{point of } bd(\Phi)\}.$$

We observe from Figure 2, that $\tilde{P}_{uv}$ is a open arc with end-points p_u and p_v, on the circle $S_r \cap S_s \cap S_t$ of radius $\sqrt{\frac{2}{3}}$ and centre at c_{uv}. Next, m_{uv} is the mid-point of the arc $\tilde{P}_{uv}$ and $p_b \in (S_u \cap S_v) \cap int(B_r \cap B_s \cap B_t)$; that is, p_b is a 2-point of $bd(\Phi)$, and the distance $|p_b m_{uv}| > 1$. Accordingly, Φ is not of constant width 1.

From Figure 3, we observe that S^*_{st} is an open spherical convex set bounded by the arcs $\tilde{P}_{ru}, \tilde{P}_{rv}$ and $\tilde{P}_{uv}$.

LEMMA 5.1. *$|m_{uv} m_{rt}| > 1$ for the mid-points m_{uv} of $\tilde{P}_{uv}$ and m_{rt} of $\tilde{P}_{rt}$.*

PROOF. Observe that, without loss of generality, we may assume that $P = [p_1, p_2, p_3, p_4, p_5] \subset \mathbb{R}^4$ is the regular 4-simplex with the vertex coordinates $p_1 = (-\frac{1}{2}, \sqrt{\frac{5}{12}}, \frac{1}{\sqrt{12}}, 0)$, $p_2 = (\frac{1}{2}, \sqrt{\frac{5}{12}}, \frac{1}{\sqrt{12}}, 0)$, $p_3 = (0, 0, 0, -\frac{1}{2})$, $p_4 = (0, 0, 0, \frac{1}{2})$,

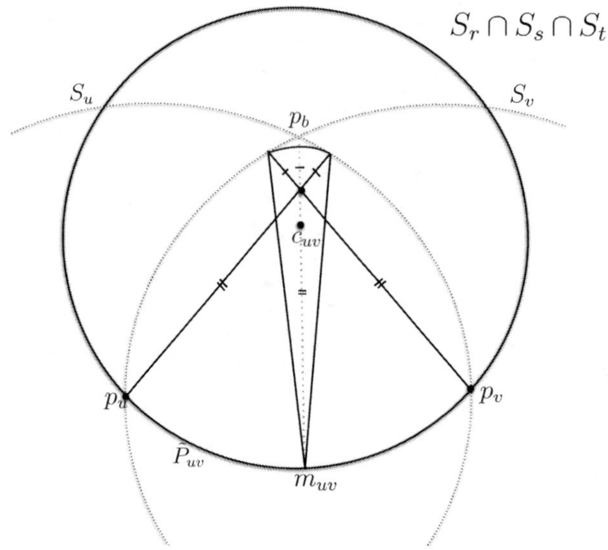

FIGURE 2. The circle $S_r \cap S_s \cap S_t$ on the boundary of Φ.

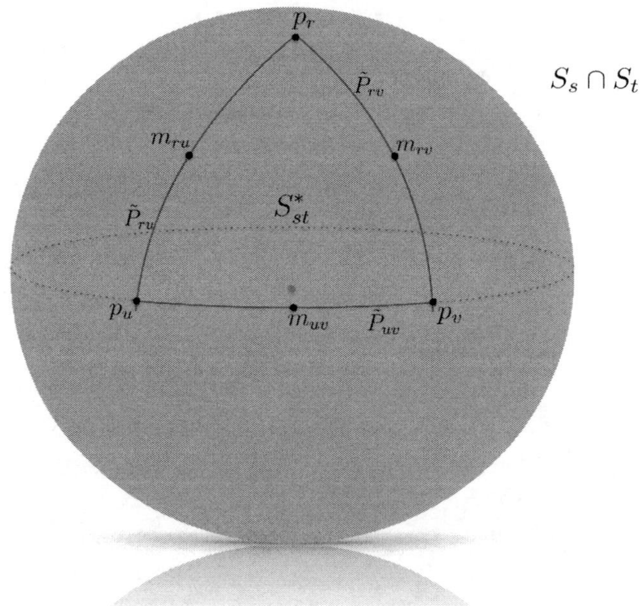

FIGURE 3. The 2-sphere $S_s \cap S_t$ on the boundary of Φ.

$p_5 = (0, 0, \frac{\sqrt{3}}{2}, 0)$. Due to its symmetries, it is sufficient to show that $|m_{12}m_{34}| > 1$, say.

We note that $\langle S_u \cap S_v \rangle$ is the hyperplane $H = \{q \in \mathbb{R}^4 | |qp_u| = |qp_v|\}$. Thus, the equations of interest are (denoted by $\langle \ \rangle : ax + by + cz + dw = e$)

$$\langle S_1 \cap S_2 \rangle \ : \ x = 0 \text{ and } \langle S_3 \cap S_4 \rangle : w = 0,$$

$\langle S_3 \cap S_4 \rangle \cap (S_1 \cap S_2) : y - \sqrt{\frac{5}{12}})^2 = \frac{3}{4}$, $x = 0$ and $w = 0$,
$\langle S_3 \cap S_4 \rangle \cap S_5 : x^2 + y^2 + z^2 - \sqrt{3}z = \frac{1}{4}$ and $w = 0$,
$\langle S_3 \cap S_4 \rangle : x^2 + y^2 + z^2 = \frac{3}{4}$ and $w = 0$,
$\langle S_3 \cap S_4 \cap S_5 \rangle : z = \frac{1}{\sqrt{12}}$ and $w = 0$.

Then for $m_{12} = (a, b, c, d) \in \langle S_1 \cap S_2 \rangle \cap \tilde{P}_{12} \subset \langle S_1 \cap S_2 \rangle \cap (S_3 \cap S_4 \cap S_5)$, we obtain that $a = 0$, $c = \frac{1}{\sqrt{12}}$, $d = 0$ and $b^2 = \frac{4}{9}$. It is clear that the y-coordinate of each point of $\tilde{P}_{12}$ is greater than $\sqrt{\frac{5}{12}}$, and so, $b = \sqrt{\frac{2}{3}}$ and $m_{12} = (0, \sqrt{\frac{2}{3}}, \frac{1}{\sqrt{12}}, 0)$.

Similarly from $m_{34} = (a, b, c, d) \in \langle S_3 \cap S_4 \rangle \cap \tilde{P}_{34} \subset \langle S_3 \cap S_4 \rangle \cap (S_1 \cap S_2 \cap S_5)$, we obtain that $a = 0$, $d = 0$, and $b = \frac{2}{\sqrt{5}}c$ and $m_{34} = (0, \frac{2}{\sqrt{5}}c, c, 0)$ with $c^2 - \frac{5\sqrt{3}}{9}c - \frac{5}{36} = 0$. Solving for c yields that $m_{34} = (\frac{\sqrt{5}-\sqrt{8}}{3\sqrt{3}})(0, 1, \sqrt{5}, 0)$. Then $|m_{12}m_{34}|^2 = \frac{1}{36}(88 - 16\sqrt{10}) = 1.03898\ldots$ and $|m_{12}m_{34}| = 1.01930\ldots$. □

In the case of the regular 3-simplex $Q = [q_1, q_2, q_3, q_4] = [q_i, q_j, q_k, q_l]$ with edges of length 1, the involutory self-duality maps q_i into $[q_j, q_k, q_l]$ and $[q_i, q_j]$ to $[q_k, q_l]$. In terms of the ball polytope $\Sigma = B_1 \cap B_2 \cap B_3 \cap B_4 \subset \mathbb{R}^3$, this means that $\Sigma_{ij} = (B_i \cap B_j) \cap bd(\Sigma)$ is mapped to $\Sigma_{kl} = (B_k \cap B_l) \cap bd(\Sigma)$. The Meissner-Montejano-Roldan operation is to perform surgery on either Σ_{ij} or Σ_{kl} for each choice of $\{i, j, k, l\} = \{1, 2, 3, 4\}$ in order to obtain a convex body Γ. Thus, for example, surgery on Σ_{12},Σ_{24},Σ_{34} and having Σ_{23}, Σ_{13} and Σ_{12} in the $bd(\Gamma)$.

If such operation carries over for the 4-simplex P, the involutory self-duality maps p_v to $[p_r, p_s, p_t, p_u]$, and $[p_u, p_v]$ to $[p_r, p_s, p_t]$. In terms of the ball-polytope $\Phi = B_1 \cap B_2 \cap B_3 \cap B_4 \cap B_5 \subset \mathbb{R}^4$, this means that the involutory self-duality maps S^*_{uv} to $\tilde{P}_{st}$. If one wants to generalise the Meissner-Montejano-Roldan operation for the 4-simplex in order to obtain a body of constant width $\Psi \in \mathbb{R}^4$ then the $bd(\Psi)$ will contain either S^*_{uv} or $\tilde{P}_{st}$ for each distinct p_u and p_v.

We observe next, that the following theorem holds.

THEOREM 5.2. *Let $\Psi \subset \mathbb{R}^4$ be a convex body that contains a regular 4-simplex with edges of length 1 and has constant width 1. Then $\Psi \subset \Phi$ and Ψ contains at most four $\tilde{P}_{uv}$'s and at most one S^*_{st}.*

PROOF. Assume $bd(\Psi)$ contains, say, $\tilde{P}_{12}$ and another $\tilde{P}_{uv}$. Then

$$\tilde{P}_{uv} \in \{\tilde{P}_{13}, \tilde{P}_{14}, \tilde{P}_{15}, \tilde{P}_{23}, \tilde{P}_{24}\tilde{P}_{25}\}$$

by Lemma 5.1, and $\tilde{P}_{uv} = \tilde{P}_{13}$, say. Then $bd(\Psi)$ may contain in addition either $\tilde{P}_{23}$ or $\tilde{P}_{14}$ and $\tilde{P}_{15}$. In the case of the former (latter), Ψ contains at most one (zero) S^*_{st}. □

The necessary condition for $\Psi \subset \mathbb{R}^4$ above implies that the construction proposed by Meissner in [**10**] does not follow directly in dimension 4.

It becomes interesting to investigate whether another involutory self-dual 4-polytope allows such a construction, or if there is another similar construction to perform on the ball polytope Φ to obtain a body of constant width.

References

[1] Dan Archdeacon and Seiya Negami, *The construction of self-dual projective polyhedra*, J. Combin. Theory Ser. B **59** (1993), no. 1, 122–131, DOI 10.1006/jctb.1993.1059. MR1234388

[2] Károly Bezdek, Zsolt Lángi, Márton Naszódi, and Peter Papez, *Ball-polyhedra*, Discrete Comput. Geom. **38** (2007), no. 2, 201–230, DOI 10.1007/s00454-007-1334-7. MR2343304

[3] Arne Brøndsted, *An introduction to convex polytopes*, Graduate Texts in Mathematics, vol. 90, Springer-Verlag, New York-Berlin, 1983. MR683612

[4] Branko Grünbaum, *Convex polytopes*, 2nd ed., Graduate Texts in Mathematics, vol. 221, Springer-Verlag, New York, 2003. Prepared and with a preface by Volker Kaibel, Victor Klee and Günter M. Ziegler. MR1976856

[5] B. Grünbaum and G. C. Sheppard, *Is selfduality involutory?*, Amer. Math. Monthly 95(1988), 729–733.

[6] Y. S. Kupitz, H. Martini, and M. A. Perles, *Ball polytopes and the Vázsonyi problem*, Acta Math. Hungar. **126** (2010), no. 1-2, 99–163, DOI 10.1007/s10474-009-9030-0. MR2593321

[7] Stanislav Jendroľ, *A noninvolutory self-duality*, Discrete Math. **74** (1989), no. 3, 325–326, DOI 10.1016/0012-365X(89)90144-1. MR992743

[8] L. Lovász, *Self-dual polytopes and the chromatic number of distance graphs on the sphere*, Acta Sci. Math. (Szeged) **45** (1983), no. 1-4, 317–323. MR708798

[9] Horst Martini, Luis Montejano, and Déborah Oliveros, *Bodies of constant width*, Birkhäuser/Springer, Cham, 2019. An introduction to convex geometry with applications. MR3930585

[10] E. Meissner and F. Schilling, *Drei Gipsmodelle von Flachen konstanter Breite*, Z. Math. Phys. 60 (1912), 92–94.

[11] L. Montejano and E. Roldán-Pensado, *Meissner polyhedra*, Acta Math. Hungar. **151** (2017), no. 2, 482–494, DOI 10.1007/s10474-017-0697-3. MR3620844

[12] Brigitte Servatius and Herman Servatius, *Self-dual graphs*, Discrete Math. **149** (1996), no. 1-3, 223–232, DOI 10.1016/0012-365X(94)00351-I. MR1375109

[13] Hans Raj Tiwary and Khaled Elbassioni, *Self-duality of polytopes and its relations to vertex enumeration and graph isomorphism*, Graphs Combin. **30** (2014), no. 3, 729–742, DOI 10.1007/s00373-013-1299-7. MR3195812

[14] I. M. Jaglom and V. G. Boltjanskiĭ, *Convex figures*, Translated by Paul J. Kelly and Lewis F. Walton, Holt, Rinehart and Winston, New York, 1960. MR0123962

[15] Günter M. Ziegler, *Lectures on polytopes*, Graduate Texts in Mathematics, vol. 152, Springer-Verlag, New York, 1995. MR1311028

Department of Mathematics and Statistics, University of Calgary, Canada
Email address: tbisztri@ucalgary.ca

Instituto de Matemáticas, Universidad Nacional Autónoma de México
Email address: dolivero@matem.unam.mx

Contemporary Mathematics
Volume **764**, 2021
https://doi.org/10.1090/conm/764/15329

On the ranks of string C-group representations for symplectic and orthogonal groups

Peter A. Brooksbank

ABSTRACT. We determine the ranks of string C-group representations of the groups $\mathrm{PSp}(4, \mathbb{F}_q) \cong \Omega(5, \mathbb{F}_q)$, and comment on those of higher-dimensional symplectic and orthogonal groups.

1. Introduction

Abstract polytopes are incidence structures that generalize classical geometric objects such as the Platonic solids. The study of these structures (and their associated symmetry groups) continues to be a fertile area of research. Abstract *regular* polytopes and their symmetry groups are fundamentally linked by the notion of a *string C-group*, namely a group G having a distinguished generating sequence $\rho_0, \ldots, \rho_{n-1}$ of involutions satisfying the following two conditions:

$$\forall\, 0 \leqslant i < j \leqslant n-1, \qquad j - i > 1 \implies [\rho_i, \rho_j] = 1; \tag{1.1}$$

$$\forall\, I, J \subseteq \{0, \ldots, n-1\}, \qquad \langle \rho_i \mid i \in I \rangle \cap \langle \rho_j \mid j \in J \rangle = \langle \rho_k \mid k \in I \cap J \rangle. \tag{1.2}$$

The first condition—the *string property*—asserts that G is the quotient of a Coxeter group whose diagram is a string (under the conventional assumption that commuting generators are not joined by an edge). Indeed, if we label each edge in a n-node string diagram by the order of the corresponding product $\rho_i\rho_{i+1}$, and define Γ to be the Coxeter group having this diagram, then G is a *smooth* quotient of Γ. The second condition—known as the *intersection property*—is satisfied by all Coxeter groups; hence, condition (1.2) requires that the quotient G inherit this property from its parent group Γ. The integer n is the *rank* of the string C-group G.

Each string C-group G has an associated abstract regular polytope $\mathcal{P}(G)$ whose symmetry group is G. Conversely, if G is the symmetry group of some abstract regular polytope $\mathcal{P}$, one can build a generating sequence of involutions in G satisfying (1.1) and (1.2). Thus, the study of abstract regular polytopes is equivalent to the study of string C-groups [**McS**, Section 2], and in this paper we adopt the group perspective. For a group G, let $\mathrm{rk}(G)$ denote the set of integers n such that G has a string C-group representation of rank n. We prove the following:

THEOREM 1.1. *Let $\mathbb{F}_q$ be the finite field with q elements and $G = \mathrm{PSp}(4, \mathbb{F}_q)$. Then $\mathrm{rk}(G) = \emptyset$ if $q = 3$, and $\mathrm{rk}(G) = \{3, 4, 5\}$ if $q \neq 3$.*

This work was partially supported by a grant from the Simons Foundation (#281435 to the author).

Group	Reference(s)
Alt(6), Alt(7)	[**Nu2**]
$\mathrm{PSL}(3,\mathbb{F}_q)$, $\mathrm{PSU}(3,\mathbb{F}_q)$, $\mathrm{PSp}(4,\mathbb{F}_3)$	[**Nu1**, **Nu3**, **Nu4**]
$\mathrm{PSL}(4,\mathbb{F}_q)$ (q even), $\mathrm{PSU}(4,\mathbb{F}_q)$ (q even)	[**Nu1**]
$\mathrm{PSU}(4,\mathbb{F}_3)$, $\mathrm{PSU}(5,\mathbb{F}_2)$	M. Macaj & G. Jones
M_{11}, M_{22}, M_{23}, McL	[**Ma**]

FIGURE 1. Simple groups with no rank 3 string C-group representation. The exceptions of Macaj and Jones, initially overlooked by Nuzhin, were communicated via D. Leemans.

G	Restrictions	$\mathrm{rk}(G)$	Reference(s)
$\mathrm{Sym}(n)$	$n \geqslant 5$	$\{3,\ldots,n-1\}$	[**FL1**]
$\mathrm{Alt}(n)$	$n \geqslant 12$	$\{3,\ldots,\lfloor\frac{n-1}{2}\rfloor\}$	[**FL2**]
$\mathrm{O}^{\pm}(2n,\mathbb{F}_{2^e})$	$e \geqslant 2$	$\{3,\ldots,2n\}$	[**BFL**]
$\mathrm{PSp}(2n,\mathbb{F}_{2^e})$	$e \geqslant 2$	$\{3,\ldots,2n+1\}$	[**BFL**]
$\mathrm{PSL}(4,\mathbb{F}_q)$	q odd	$\supseteq \{3,4\}$	[**BL1**]
$\mathrm{PSp}(4,\mathbb{F}_q)$	$q \neq 3$	$\{3,4,5\}$	Theorem 1.1

FIGURE 2. Infinite families of simple groups (and their variants) known to have string C-group representations of rank at least 4.

Theorem 1.1 may be seen as contributing to an ongoing effort to determine the ranks of string C-group representations for families of finite simple groups.

A great deal is known about rank 3 representations: see [**Co1**, **Co2**, **Nu2**] for the alternating groups, [**Nu1**, **Nu3**, **Nu4**] for the simple groups of Lie type, and [**Ma**] for the sporadic simple groups. The upshot is that most finite simple groups have string C-group representations of rank 3; Figure 1 lists the exceptions.

The picture for ranks higher than 4 is far less complete. Some negative results have been proved for linear groups of low Lie rank: for example, none of $\mathrm{PSL}(3,\mathbb{F}_q)$, $\mathrm{PSU}(3,\mathbb{F}_q)$, $\mathrm{PSL}(4,\mathbb{F}_{2^e})$, $\mathrm{PSU}(4,\mathbb{F}_{2^e})$ have string C-group representations of any rank [**BV**, **BL1**, **BFL**], while $4 \in \mathrm{rk}(\mathrm{PSL}(2,\mathbb{F}_q))$ if, and only if, $q \in \{11,19\}$ [**LS**]. Figure 2 summarizes the positive results for infinite families of simple groups and their variants.

To prove Theorem 1.1 we use the Klein correspondence to move from the 4-dimensional projective representation of $\mathrm{PSp}(4,\mathbb{F}_q)$ to the 5-dimensional linear representation of its isomorphic orthogonal group $\Omega(5,\mathbb{F}_q)$. This was also the approach in [**BFL**], where it was shown that $\mathrm{PSp}(2n,\mathbb{F}_{2^e}) \cong \Omega(2n+1,\mathbb{F}_{2^e})$ has string C-group representations of rank $2n+1$. (Note, in odd characteristic this isomorphism holds only when $n = 2$.) That construction may be adapted in odd characteristic to generate $\mathrm{O}(5,\mathbb{F}_q)$ as a string C-group of rank 5, but we must work harder to obtain such a representation for its simple subgroup $\Omega(5,\mathbb{F}_q)$ of index 4. Having done so, we apply a recently discovered technique [**BL2**] to reduce the rank of this representation and obtain the desired constructions of ranks 3 and 4.

2. Orthogonal groups and their geometries

Let $\mathbb{F}_q$ be the field wth $q = p^e$ elements, and V a d-dimensional $\mathbb{F}_q$-space. Although we shall be primarily interested in the case $d = 5$, we work for a while with arbitrary d. Let $\varphi\colon V \to \mathbb{F}_q$ be a quadratic form on V, and

$$\forall u, v \in V, \qquad (u,v) = \varphi(u+v) - \varphi(u) - \varphi(v) \tag{2.1}$$

its associated symmetric bilinear form. If q is odd, φ can be recovered from $(\ ,\ )$ using the equation $\varphi(v) = (v,v)/2$.

To each $U \leqslant V$ we associate a subspace $U^\perp = \{v \in V \mid (v,U) = 0\}$, and say $V^\perp$ is the *radical* of V. If d is even, or d and $|\mathbb{F}_q|$ are both odd, we insist that $V^\perp = 0$. If d is odd and $|\mathbb{F}_q|$ is even, $V^\perp$ is always nonzero; here, we insist that $V^\perp = \langle z\rangle$ with $\varphi(z) \neq 0$. A subspace U of V is *nonsingular* if the restriction of φ to U has these properties. In particular, a 1-space $\langle v\rangle$ is *nonsingular* if $\varphi(v) \neq 0$, and is otherwise *singular* (we extend this terminology to vectors).

The *orthogonal group* corresponding to φ is the group

$$\mathrm{O}(V) = \{g \in \mathrm{GL}(V) \mid \varphi(vg) = \varphi(v) \text{ for all } v \in V\} \tag{2.2}$$

of *isometries* of φ. Although we generally work with orthogonal groups and their underlying $\mathbb{F}_q$-spaces using standard (coordinate-free) linear transformation notation, it will from time to time be helpful to compute explicitly with matrices. If x is a matrix, x^{tr} denotes its transpose. Fixing an ordered basis $v_1, \ldots, v_d$ for V, we represent φ as a $d \times d$ upper-triangular matrix $\Phi := [[\varphi_{ij}]]$, where $\varphi_{ii} = \varphi(v_i)$, and $\varphi_{ij} = (v_i, v_j)$ for $1 \leqslant i < j \leqslant d$. Then $\Phi + \Phi^{\mathrm{tr}}$ is the matrix $[[(v_i, v_j)]]$ representing $(\ ,\ )$ relative to $v_1, \ldots, v_d$, and

$$\mathrm{O}(V) = \{g \in \mathrm{GL}(d, \mathbb{F}_q) \mid g\Phi g^{\mathrm{tr}} = \Phi\}. \tag{2.3}$$

Throughout this section we shall appeal to standard results from Taylor's text [**Tay**], to which we refer the reader for a thorough treatment of classical groups and their geometries.

2.1. Symmetries, and their geometric properties. The involutions best suited to generating string C-groups of high rank are those with ± 1–eigenspaces of highest possible dimension. Such involutions arise from elements of $\mathrm{O}(V)$ known as *symmetries* that are defined in terms of a nonsingular vector $u \in V$ as follows:

$$\forall v \in V, \qquad \sigma_u\colon v \mapsto v - \frac{(u,v)}{\varphi(u)}\,u. \tag{2.4}$$

Observe that σ_u is the identity on the $(d-1)$-space space $u^\perp$. If $|\mathbb{F}_q|$ is odd, $\sigma_u(u) = -u$, so σ_u is a *reflection* and has determinant -1. If $|\mathbb{F}_q|$ is even, σ_u is also the identity on $V/\langle u\rangle$, and hence is a *transvection*. Note, $\sigma_u = \sigma_{\lambda u}$ for all $\lambda \in \mathbb{F}_q^\times$, so symmetries correspond to nonsingular points of the projective geometry $\mathbb{P}(V)$ and we write $\sigma_u = \sigma_{\langle u\rangle}$. For $X \subseteq V$, define

$$\Sigma(X) = \langle \sigma_u \mid u \in X\rangle. \tag{2.5}$$

We typically frame our arguments in geometric terms, and the following elementary observation is particularly useful. For nonsingular points x, y in $\mathbb{P}(V)$,

$$[\sigma_x, \sigma_y] = 1 \iff y \text{ lies on } x^\perp. \tag{2.6}$$

The *support* of $h \in \mathrm{GL}(V)$ is the subspace $[V,h] := \{v - vh \mid v \in V\}$. If $H \leqslant \mathrm{GL}(V)$ is generated by S, then its support is $[V,H] := \sum_{h\in S}[V,h]$.

2.2. Generating with symmetries. It is well known that unless $V = \mathbb{F}_2^4$ the orthogonal group $\mathrm{O}(V)$ is generated by its symmetries; cf. [**Tay**, Theorem 11.39]. If $\dim V$ is odd and $|\mathbb{F}_q|$ is even—often considered the degenerate case—then $\mathrm{O}(V) \cong \mathrm{Sp}(V/V^{\perp})$ is the simple group we wish to work with. In the other cases $\mathrm{O}(V)$ has normal subgroups, which we now describe.

The map $g \mapsto \dim[V,g] \pmod 2$ is a group homomorphism $\pi\colon \mathrm{O}(V) \to \mathbb{Z}_2$ having kernel of index 2. In fact, if $g = \sigma_{u_1}\dots\sigma_{u_r}$, then $\dim[V,g] \pmod 2) = r \pmod 2$ and $\ker\pi$ is the subgroup of $\mathrm{O}(V)$ consisting of products of even numbers of symmetries. Next, the map $g \mapsto \varphi(u_1)\dots\varphi(u_r)(\mathbb{F}_q^\times)^2$ is a (well-defined) homomorphism $\theta\colon \mathrm{O}(V) \to \mathbb{F}_q^\times/(\mathbb{F}_q^\times)^2$, and $\theta(g)$ is called the *spinor norm* of g. For convenience, we denote the image of θ by $\{\bar 0, \bar 1\}$, where $\bar 0$ is the square class, and $\bar 1$ is the non-square class. Evidently, $\ker\theta$ has index 2 in $\mathrm{O}(V)$ if, and only if, $|\mathbb{F}_q|$ is odd, in which case

$$\ker\theta = \langle \sigma_u \mid u \in V \text{ is nonsingular, and } \varphi(u) \text{ is a square in } \mathbb{F}_q^\times\rangle. \tag{2.7}$$

Our principal focus is the subgroup

$$\Omega(V) = \begin{cases} \ker\pi \cap \ker\theta & \text{if } |\mathbb{F}_q| \text{ is odd} \\ \mathrm{O}(V) & \text{if } |\mathbb{F}_q| \text{ is even, and } \dim(V) \text{ is odd} \\ \ker\pi & \text{if } |\mathbb{F}_q| \text{ is even, and } \dim(V) \text{ is even} \end{cases} \tag{2.8}$$

of $\mathrm{O}(V)$. *We restrict now to the case when* $\dim V$ *is odd*, and consider generation of $\Omega(V)$ using symmetries.

When $|\mathbb{F}_q|$ is even, $\mathrm{O}(V) = \Omega(V)$ is generated by symmetries.

When $|\mathbb{F}_q|$ is odd, $\ker\pi = \mathrm{SO}(V)$, the subgroup of determinant 1 isometries of φ. Consider the map $\mathrm{O}(V) \to \mathbb{Z}_2 \times \{\bar 0, \bar 1\}$ sending $g \mapsto (g\pi, g\theta)$. The proper subgroups of the codomain give three maximal subgroups of $\mathrm{O}(G)$: $M_1 = \ker\theta$ is the preimage of $\langle(1,\bar 0)\rangle$, M_2 is the preimage of $\langle(1,\bar 1)\rangle$, and $\mathrm{SO}(V) = \ker\pi$ is the preimage of $\langle(0,\bar 1)\rangle$.

Consider $-1 \in \mathrm{O}(V)$. As $\det(-1) = -1$ and $M_1 \cap M_2 = \Omega(V) \leqslant \mathrm{SO}(V)$, it follows that -1 belongs to precisely one of the maximal subgroups M_i. As $\det(-\sigma_u) = (-1)(-1) = 1$ for each nonsingular $u \in V$, it follows that

$$\langle -\sigma_u \mid \varphi(u) \text{ is a square}\rangle, \qquad \langle -\sigma_u \mid \varphi(u) \text{ is a non-square}\rangle$$

are both subgroups of $\mathrm{SO}(V)$. Indeed, since $\dim V$ is odd,

$$\Omega(V) = \langle -\sigma_u \mid \theta(-1) = \varphi(u)\,(\mathbb{F}_q^\times)^2\rangle. \tag{2.9}$$

Of course, one can simply multiply φ by a suitable scalar to ensure that $\theta(-1) = \bar 0$, but we do not wish to place constraints on the specific φ that we work with.

For $X \subseteq V$, define subsets

$$X_{\square} = \langle u \in X \mid \varphi(u) \text{ is a square}\rangle \qquad X_{\boxtimes} = \langle u \in X \mid \varphi(u) \text{ is a non-square}\rangle \tag{2.10}$$

of X. We record the following result for easy reference.

LEMMA 2.1. *If (V,φ) is a 5-dimensional orthogonal $\mathbb{F}_q$-space, then either*
(i) $\theta(-1) = \bar 0$ and $\mathrm{PSp}(4,\mathbb{F}_q) \cong \Omega(V) = \langle -\sigma_u \mid u \in V_{\square}\rangle$, or
(ii) $\theta(-1) = \bar 1$ and $\mathrm{PSp}(4,\mathbb{F}_q) \cong \Omega(V) = \langle -\sigma_u \mid u \in V_{\boxtimes}\rangle$.

REMARK 2.2. This trick for generating $\Omega(V)$ with elements having an eigenspace of dimension $\dim V - 1$ works whenever $\dim V$ is odd. Indeed, by extending the methods in this paper it is anticipated that *for all finite fields* $\mathbb{F}_q$ of order at least 4, the simple group $\Omega(2n+1, \mathbb{F}_q)$ has a string C-group representation of rank $2n+1$. When $\dim V = 2n$ is even, on the other hand, $-\sigma_u$ still has determinant -1. Thus, it seems unlikely that $\Omega^{\pm}(2n, \mathbb{F}_q)$ has string C-group representations of rank $2n$.

The following version of [**BFL**, Proposition 3.3] is more restrictive in that it applies just to nonsingular subspaces of an orthogonal space V, but it also encompasses all finite fields. The restricted version is all we need here, and the proof is simpler than that of [**BFL**, Proposition 3.3].

PROPOSITION 2.3. *Let U, W be nonsingular subspaces of an orthogonal $\mathbb{F}_q$-space V such that $U \cap W$ is nonsingular. Then $\Sigma(U) \cap \Sigma(W) = \Sigma(U \cap W)$.*

PROOF. Let $G = \mathrm{O}(V)$. For $X \subseteq V$, $\mathrm{cent}_G(X) = \{g \in G \mid \forall x \in X,\ xg = x\}$ is the subgroup of G fixing X pointwise. As U is a nonsingular subspace of V, the stabilizer $\mathrm{stab}_G(U) = \{g \in G \mid Ug = U\}$ factorizes as a direct product

$$\mathrm{stab}_G(U) = \mathrm{cent}_G(U) \times \mathrm{cent}_G(U^{\perp}).$$

Furthermore, $\mathrm{cent}_G(U^{\perp})$ induces the full group $\mathrm{O}(U)$ of isometries on U; we say $\mathrm{cent}_G(U^{\perp})$ is a *natural embedding* of $\mathrm{O}(U)$ in G. In particular, $\mathrm{cent}_G(U^{\perp}) = \Sigma(U)$.

As W and $U \cap W$ are also nonsingular, we have $\Sigma(U \cap W) \leqslant \Sigma(U) \cap \Sigma(W)$. If $g \in \Sigma(U) \cap \Sigma(W) = \mathrm{cent}_G(U^{\perp}) \cap \mathrm{cent}_G(W^{\perp})$, then g is the identity on $U^{\perp}$ and $W^{\perp}$ and hence on $U^{\perp} + W^{\perp} = (U \cap W)^{\perp}$. As $U \cap W$ is nonsingular, it follows that $\mathrm{cent}_G((U \cap W)^{\perp}) = \Sigma(U \cap W)$, so the result follows. □

REMARK 2.4. Restricting the types of symmetries used to generate—working instead with the groups $\Sigma(X_{\square})$ and $\Sigma(X_{\boxtimes})$—yields equivalent results, namely

$$\Sigma(U_{\square}) \cap \Sigma(W_{\square}) = \Sigma((U \cap W)_{\square}), \text{ and } \quad \Sigma(U_{\boxtimes}) \cap \Sigma(W_{\boxtimes}) = \Sigma((U \cap W)_{\boxtimes}).$$

3. The rank 5 construction

The construction of a string C-group representation for $\mathrm{PSp}(4, \mathbb{F}_q) \cong \Omega(5, \mathbb{F}_q)$ is similar to the one given in [**BFL**] to prove that $\mathrm{O}(2m+1, \mathbb{F}_{2^e}) \cong \mathrm{PSp}(2m, \mathbb{F}_{2^e})$ is a string C-group of rank $2m+1$. We could of course appeal to [**BFL**] for the case $\mathbb{F}_{2^e}$ and focus just on $|\mathbb{F}_q|$ odd, but we prefer to give a (somewhat uniform) self-contained treatment for all finite fields. Before digging in, though, we record the following useful shortcut to string C-group verification.

LEMMA 3.1. *Let G be a group generated by involutions $\rho_0, \ldots, \rho_{n-1}$ such that*

(i) *$\langle \rho_0, \ldots, \rho_{n-2} \rangle$ and $\langle \rho_1, \ldots, \rho_{n-1} \rangle$ are both string C-groups, and*
(ii) *$\langle \rho_0, \ldots, \rho_{n-2} \rangle \cap \langle \rho_1, \ldots, \rho_{n-1} \rangle = \langle \rho_1, \ldots, \rho_{n-2} \rangle$.*

Then $(G\,; \{\rho_0, \ldots, \rho_{n-1}\})$ is a string C-group representation.

PROOF. See [**McS**, Proposition 2E16]. □

Let (V, φ) be an orthogonal $\mathbb{F}_q$-space, and $u_1, \ldots, u_d$ a basis of V consisting of nonsingular vectors. Consider a putative string C-group representation of a group generated by symmetries $\sigma_{u_1}, \ldots, \sigma_{u_d}$. From the commutator relation (2.6), we see that the string condition (1.1) translates to an orthogonality condition on their associated vectors $u_1, \ldots, u_d$, namely $(u_i, u_{i+k}) = 0$ if, and only if, $k > 1$. One can replace u_i with $u_i/(u_i, u_{i+1})$ for $i < d$ to ensure that $(u_i, u_{i+1}) = 1$ without changing the symmetries. Thus, φ is represented relative to $u_1, \ldots, u_d$ by a matrix

$$\Phi(\alpha_1, \ldots, \alpha_d) = \begin{bmatrix} \alpha_1 & 1 & & & \\ & \alpha_2 & 1 & & \\ & & \ddots & \ddots & \\ & & & \alpha_{d-1} & 1 \\ & & & & \alpha_d \end{bmatrix}. \tag{3.1}$$

Restricting to the case $d = 5$, our strategy is to choose the scalars α_i so that everything works nicely. For us, this will mean that the following conditions hold.

(a) For each $1 \leqslant i \leqslant 5$, the dihedral group $\langle \sigma_{u_i}, \sigma_{u_{i+1}} \rangle$ is equal to $\Sigma(\langle u_i, u_{i+1} \rangle)$ if $|\mathbb{F}_q|$ is even, or to one of $\Sigma(\langle u_i, u_{i+1} \rangle_{\square})$ or $\Sigma(\langle u_i, u_{i+1} \rangle_{\boxtimes})$ if $|\mathbb{F}_q|$ is odd. This means that each product $\sigma_{u_i}\sigma_{u_{i+1}}$ has order $q \pm 1$ if q is even, or $(q \pm 1)/2$ if q is odd. An easy calculation shows that the restriction of this product to $\langle u_i, u_{i+1} \rangle$ is represented by the matrix

$$h(\alpha_i, \alpha_{i+1}) = \begin{bmatrix} -1 & 1/\alpha_{i+1} \\ -1/\alpha_i & 1/(\alpha_i\alpha_{i+1}) - 1 \end{bmatrix},$$

(b) For each $1 \leqslant i < j \leqslant 5$, the subspace $U_{ij} = \langle u_i, \ldots u_j \rangle$ is nonsingular.

(c) If q is odd, then $\theta(\sigma_{u_i}) = \theta(-1)$ for each $1 \leqslant i \leqslant 5$.

To achieve this, we consider separately the cases $|\mathbb{F}_q|$ even and $|\mathbb{F}_q|$ odd.

$|\mathbb{F}_q|$ **is even.** Fix $\xi \in \mathbb{F}_q^\times$ such that $h(\xi, \xi)$ has order $q \pm 1$. For $\alpha \in \mathbb{F}_q^\times - \{\xi\}$, consider the quadratic form φ represented by the matrix

$$\Phi_{\text{even}}(\alpha) := \Phi(\xi, \xi, \alpha, \alpha, \xi). \tag{3.2}$$

For condition (a), we simply require that both $h(\alpha, \alpha)$ and $h(\xi, \alpha)$ have order $q \pm 1$. For condition (b), notice that

$$U_{ij}^{\perp} \cap U_{ij} = \begin{cases} 0 & \text{if } j - i \text{ is odd} \\ \langle u_i + u_{i+2} \rangle & \text{if } j = i + 2, \text{ and} \\ \langle u_1 + u_3 + u_5 \rangle & \text{if } i = 1 \text{ and } j = 5 \end{cases}$$

As $\varphi(u_i + u_{i+2}) = \xi + \alpha \neq 0$ for $1 \leqslant i \leqslant 3$, and $\varphi(u_1 + u_3 + u_5) = \alpha \neq 0$, it follows that U_{ij} is always nonsingular, as required. Thus, we define

$$\Gamma_{\text{even}}(\xi) := \{\alpha \in \mathbb{F}_q^\times - \{\xi\} \mid \text{ both } h(\alpha, \alpha) \text{ and } h(\xi, \alpha) \text{ have order } q \pm 1\}, \tag{3.3}$$

so that $\Phi_{\text{even}}(\alpha)$ satisfies conditions (a) and (b) for any $\alpha \in \Gamma_{\text{even}}(\xi)$.

$|\mathbb{F}_q|$ **is odd.** Fix $\xi \in \mathbb{F}_q^\times - \{\frac{1}{2}\}$ such that $\xi^2 - \frac{1}{2}$ is a nonzero square, and $h(1, \xi^2)$ has order $(q \pm 1)/2$. For $\alpha \in \mathbb{F}_q^\times$ consider the quadratic form φ represented by

$$\Phi_{\text{odd}}(\alpha) := \Phi(1, \xi^2, 1, \alpha^2, 1). \tag{3.4}$$

Condition (a) is satisfied so long as $h(1, \alpha^2)$ has order $(q \pm 1)/2$, so let A denote the set of squares in $\mathbb{F}_q^\times$ having this property. Choosing the diagonal entries of Φ to be squares ensures that all $\theta(\sigma_{u_i}) = \bar{0}$. Hence, for $1 \leqslant i \leqslant 4$ we have $\langle \sigma_{u_i}, \sigma_{u_{i+1}} \rangle = \Sigma(\langle u_i, u_{i+1} \rangle_{\square})$.

Computing determinants of submatrices of the matrix $\Phi_{\text{odd}}(\alpha) + \Phi_{\text{odd}}(\alpha)^{\text{tr}}$ representing the symmetric bilinear form associated to φ, we observe that condition

(b) holds provided α^2 is selected from the set

$$\Gamma_0(\xi) = A - \left\{ \frac{1}{2}, \frac{1}{4}, \frac{\xi^2}{4\xi^2-1}, \frac{2\xi^2-\frac{1}{4}}{4\xi^2-1}, \frac{\xi^2-\frac{1}{4}}{4\xi^2-2}, \frac{\xi^2-\frac{3}{8}}{2\xi^2-1} \right\} \tag{3.5}$$

To analyze condition (c), we compute an orthogonal basis $\{w_i \mid 1 \leqslant i \leqslant 5\}$ for V. For then, $-1 = \prod_i \sigma_{w_i}$, so that $\theta(-1) = \prod_i \varphi(w_i)\,(\mathbb{F}_q^\times)^2$. Put $U = \langle u_1, u_3, u_5\rangle$, so that $U^\perp = \langle -u_1 + 2u_2 - u_3, -u_3 + 2u_4 - u_5\rangle$. As

$$\varphi(-u_1 + 2u_2 - u_3) = 4\xi^2 - 2 \qquad \varphi(-u_3 + 2u_4 - u_5) = 4\alpha^2 - 2,$$

and $\xi^2, \alpha^2 \neq \frac{1}{2}$, it follows that $w_1 := -u_1 + 2u_2 - u_3$ and $-u_3 + 2u_4 - u_5$ are nonsingular. Notice $\varphi(w_1) = 4\xi^2 - 2 = 2^2(\xi^2 - \frac{1}{2})$ is a square, so $\theta(\sigma_{w_1}) = \bar{0}$. Also,

$$w_2 := -u_1 + 2u_2 + (1-4\xi^2)u_3 + (8\xi^2-4)u_4 + (2-4\xi^2)u_5 \in w_1^\perp \cap U^\perp,$$

so we compute

$$\varphi(w_2) = 16\alpha^2(2\xi^2-1)^2 - (32\xi^4 - 28\xi^2 + 6).$$

Define $m(\xi) = 16(2\xi^2-1)^2 \neq 0$, $b(\xi) = -(32\xi^4 - 28\xi^2 + 6)$, and

$$\Gamma_{\text{odd}}(\xi) = \{\alpha^2 \in \Gamma_0(\xi) \mid \alpha^2 m(\xi) + b(\xi) \text{ is a nonzero square}\}. \tag{3.6}$$

We have shown that $\Phi_{\text{odd}}(\alpha)$ satisfies conditions (a), (b) and (c) for any $\alpha \in \Gamma_{\text{odd}}(\xi)$. Note, we could have chosen ξ so that $\xi^2 - \frac{1}{2}$ is a non-square, in which case the condition on α in (3.6) would change to $\alpha^2 m(\xi) + b(\xi)$ a non-square. This increases the pool of scalars for a fixed $\mathbb{F}_q$, but our choice suffices to establish existence.

PROPOSITION 3.2. *Let $q \geqslant 4$, $\mathbb{F}_q$ be the field with q elements, and $\xi \in \mathbb{F}_q$ any element having the properties described separately above for q even and q odd. Let $\Gamma(\xi)$ be the set defined in (3.3) or (3.6) for q even or odd, respectively. Suppose $\Gamma(\xi) \neq \emptyset$, and fix $\alpha \in \Gamma(\xi)$. If φ is the quadratic form represented, relative to basis $\{u_i \mid 1 \leqslant i \leqslant 5\}$, by the matrix $\Phi_{\text{even}}(\alpha)$ or $\Phi_{\text{odd}}(\alpha)$, then*

$$(\,\Omega(V;\varphi)\,;\,\{\ \sigma_{u_i} \mid 1 \leqslant i \leqslant 5\}\,)$$

is a string C-group representation of $\Omega(V) \cong \mathrm{PSp}(4, \mathbb{F}_q)$ of Schläfli type $[p,p,p,p]$, where $p \in \{q-1, q+1\}$ if q is even, and $p \in \{(q-1)/2, (q+1)/2\}$ if q is odd.

PROOF. We first claim that, for any such choice of scalars ξ, α:

$$\text{for } 1 \leqslant i < j \leqslant 5, \quad \langle \sigma_{u_i}, \dots, \sigma_{u_j}\rangle = \begin{cases} \Sigma(\langle u_i, \dots, u_j\rangle) & \text{if } q \text{ is even} \\ \Sigma(\langle u_i, \dots, u_j\rangle_\square) & \text{if } q \text{ is odd} \end{cases}. \tag{3.7}$$

Recall that for $\alpha \in \Gamma(\xi)$, conditions (a), (b) and (c) are satisfied for the form φ represented by matrix $\Phi_{\text{even}}(\alpha)$ or $\Phi_{\text{odd}}(\alpha)$. In particular, condition (a) ensures that (3.7) holds whenever $j = i+1$. This is the base case of an induction on $j - i$. For q even, follow the argument in [**BFL**, Lemma 5.3]—although the form matrix is different, only conditions (a) and (b) matter. The q odd case is identical except that $\langle \sigma_{u_i}, \dots, \sigma_{u_j}\rangle = \Sigma(\langle u_i, \dots, u_j\rangle_\square)$ since we generate only with symmetries having square spinor norm; see Remark 2.4.

We next show, again using induction, that $\sigma_{u_1}, \dots, \sigma_{u_5}$ generates a string C-group. For $1 \leqslant i \leqslant 4$, the dihedral group $\langle \sigma_{u_i}, \sigma_{u_{i+1}}\rangle$ (with its defining generators) is clearly a string C-group. For $j - i > 1$, consider the group $\langle \sigma_{u_i}, \dots, \sigma_{u_j}\rangle$. By induction, both $\langle \sigma_{u_i}, \dots, \sigma_{u_{j-1}} \rangle$ and $\langle \sigma_{u_{i+1}}, \dots, \sigma_{u_j} \rangle$ are string C-groups. Furthermore, by Proposition 2.3 and the claim above,

$$\begin{aligned}\langle \sigma_{u_i}, \dots, \sigma_{u_{j-1}} \rangle \cap \langle \sigma_{u_{i+1}}, \dots, \sigma_{u_j} \rangle &= \Sigma(\langle u_i, \dots, u_{j-1} \rangle) \cap \Sigma(\langle u_{i+1}, \dots, u_j \rangle) \\ &= \Sigma(\langle u_i, \dots, u_{j-1} \rangle \cap \langle u_{i+1}, \dots, u_j \rangle) \\ &= \Sigma(\langle u_{i+1}, \dots, u_{j-1} \rangle) \\ &= \langle \sigma_{u_{i+1}}, \dots, \sigma_{u_{j-1}} \rangle.\end{aligned}$$

It now follows from Lemma 3.1 that $\langle \sigma_{u_i}, \dots, \sigma_{u_j} \rangle$ is a string C-group.

Finally, using the initial claim again, together with the discussion in Section 2.2, we have $\langle \sigma_{u_1}, \dots, \sigma_{u_5} \rangle = \ker \theta$. When q is even, $\mathrm{O}(V) = \ker \theta = \Omega(V)$, so $(\Omega(V)\,;\, \{\sigma_{u_1}, \dots, \sigma_{u_5}\})$ is a string C-group representation. When q is odd, the choices of ξ and α ensure that condition (c) holds, so that $\theta(-1) = \bar{0}$. Thus, by Lemma 2.1(i), $(\Omega(V)\,;\, \{-\sigma_{u_1}, \dots, -\sigma_{u_5}\})$ is a string C-group representation. □

4. Proof of Theorem 1.1

We first limit the rank of a string C-group representation for $G = \mathrm{PSp}(4, \mathbb{F}_q)$. Once again we approach this within the context of the isomorphic group $\Omega(V)$, where V is a 5-dimensional orthogonal space equipped with quadratic form φ.

LEMMA 4.1. *Let $\mathbb{F}_q$ be any finite field, V a 5-dimension $\mathbb{F}_q$ space equipped with quadratic form φ, and $n \geqslant 6$ an integer. If H is a subgroup of $\mathrm{O}(V)$ generated by a sequence of involutions $\rho_0, \dots, \rho_{n-1}$ satisfying* (1.1), *then $\rho_0, \dots, \rho_{n-1}$ violates the intersection condition* (1.2).

PROOF. A non-central involution of $\mathrm{O}(V)$ is one of $\pm\sigma_u$ or $\pm\sigma_u\sigma_w$ for $u, w \in V$ nonsingular and $w \in u^\perp$. Consider the non-commuting subgroups $L = \langle \rho_0, \rho_1, \rho_2 \rangle$ and $R = \langle \rho_3, \rho_4, \rho_5 \rangle$ of H (possible since $n \geqslant 6$). As each product $\rho_i\rho_{i+1}$ has order exceeding 2, L induces on $[V, L]$, of dimension at least 3, a group generated by non-commuting dihedral groups. Similarly R induces on $[V, R]$, of dimension at least 3, a second such group. As $\dim V = 5$, so $[V, L] \cap [V, R]$ is nontrivial. It follows that $L \cap R$ is also nontrivial, so condition (1.2) fails for the sequence $\rho_0, \dots, \rho_{n-1}$. □

Lemma 4.1 shows that $\mathrm{rk}(G) \subseteq \{3, 4, 5\}$. To complete the proof of Theorem 1.1 we establish the existence of a suitable rank 5 string C-group representation for $G \cong \Omega(V)$. This will show that $5 \in \mathrm{rk}(G)$. We will then apply the following *rank reduction* technique to show that $\{3, 4\} \subseteq \mathrm{rk}(G)$.

THEOREM 4.2. *Let $(G; \{\rho_0, \dots, \rho_{n-1}\})$ be an irreducible string C-group representation of rank $n \geqslant 4$. If $\rho_2\rho_3$ has odd order, then $(G; \{\rho_1, \rho_0\rho_2, \rho_3, \dots, \rho_{n-1}\})$ is a string C-group representation of rank $n - 1$.*

REMARK 4.3. This 'Petrie like' construction was developed in [**HL**] and first used as a rank reduction technique for the symmetric groups in [**FL1**]. The general technique, along with the criteria in Theorem 4.2, was established in the recent paper [**BL2**].

PROOF OF THEOREM 1.1. As the entire result can be verified by brute force for moderate values of q on a computer (see Section 5.4), for convenience we assume that $q \geqslant 4$ if q is even and that $q \geqslant 11$ if q is odd. Referring to Proposition 3.2, our first task is to establish the existence of a suitable scalar ξ, and to show that $\Gamma(\xi)$ is nonempty. We consider q even and odd separately.

Referring to (3.3) for q even, the only condition is that there are distinct scalars $\xi, \alpha \in \mathbb{F}_q^\times$ such that the matrices $h(\xi,\xi)$, $h(\alpha,\alpha)$, $h(\xi,\alpha)$ have order $q \pm 1$. For each $\xi, \alpha \in \mathbb{F}_q^\times - \{0,1\}$, all of these matrices have order dividing $q \pm 1$. The proportion of scalars defining generators of those cyclic groups behaves as $1/(c \log\log q)$ and there are at least two such scalars in $\mathbb{F}_q^\times$ for all $q \geqslant 4$.

Referring to (3.5) and (3.6) for q odd, while we halve the number of scalars we consider by only using squares, the proportion of these elements generating cyclic groups of the appropriate order is the same as in the even case. Upon restricting to $\Gamma_0(\xi)$, we discard up to 6 more scalars. Finally, consider the condition (as α^2 ranges over $\Gamma_0(\xi)$ for fixed ξ) that $\alpha^2 m(\xi) + b(\xi)$ is a square. If $b(\xi) = 0$, the condition always holds, but otherwise it will hold for roughly half of the choices of α^2. In particular, $\Gamma_{\text{odd}}(q)$ is nonempty so long as q is large enough (one can check on the computer that indeed $q \geqslant 11$ is large enough).

Thus, by Proposition 3.2, for q large enough there is a string C-group representation of rank 5 for $\Omega(V)$. When q is odd, moreover, it is possible to choose scalars so that the Schläfli type of the representation consists of any combination of $(q-1)/2$ and $(q+1)/2$. In particular, observing the residue class of q modulo 4, we can arrange for this to be a sequence of odd numbers. (Note, when q is even, our construction always produces Schläfli types of odd numbers $q \pm 1$.) Hence, we can now simply apply Theorem 4.2 twice to our string C-group to obtain new representations of ranks 4 and 3. This completes the proof. □

REMARK 4.4. Although our proof gives string C-group representations of rank 3 for $\text{PSp}(4, \mathbb{F}_q)$ via rank reduction, we remark that at least one other construction was already known. Indeed, as mentioned in the introduction, a complete determination of the simple groups of Lie type having string C-group representations of rank 3 may be extracted from the work of Nuzhin.

5. Concluding remarks

In this final section of the paper we make a number of remarks pertaining to the results of the foregoing sections.

5.1. Choosing scalars. We chose scalars $\xi, \alpha \in \mathbb{F}_q^\times$ in Section 3 and defined quadratic forms $\Phi_{\text{even}}(\alpha)$ and $\Phi_{\text{odd}}(\alpha)$ so as to make the verification of the corresponding string C-group representation as direct as possible. However, there is a lot of flexibility to choose other scalars in such a way that new (non-isomorphic) representations are obtained. For instance we do not absolutely require that all of the subspaces $\langle u_i, \dots, u_j \rangle$ be nonsingular; indeed, the construction in [**BFL**] does not insist on this.

5.2. Higher rank representations in higher dimensions. As we have already noted, the idea in [**BFL**] is to use symmetries to represent $\text{O}(d, \mathbb{F}_{2^e})$ as a string C-group of rank d. It was not fully appreciated at the time that the same technique can be used to build rank d representations for $\text{O}(d, \mathbb{F}_q)$ when $|\mathbb{F}_q|$ is odd. By restricting the scalars α_i in (3.1) to be either all squares or all nonsquares, one can further generate the two maximal subgroups of $\text{O}(d, \mathbb{F}_q)$ generated by symmetries. It seems likely, then, that our trick for generating $\Omega(5, \mathbb{F}_q)$ by negating symmetries extends to $\Omega(2n+1, \mathbb{F}_q)$, so that $2n+1 \in \text{rk}(\Omega(2n+1, \mathbb{F}_q))$

for all $q \geqslant 4$. However, we have no intuition to offer in regard to the highest rank string C-group representation of the quasisimple groups $\Omega^{\pm}(2n, \mathbb{F}_q)$.

5.3. Another rank 4 construction. In an earlier draft of this paper—before the emergence of the rank reduction technique—a direct construction of a string C-group representation of rank 4 for $\Omega(V)$ was discovered that utilized, to a greater extent, geometric properties of involutions. For completeness we provide this construction, but in the interest of brevity omit the verification.

Let $q \geqslant 5$ be an odd prime power, and $\mathbb{F}_q$ the field of q elements; an explicit rank 4 construction of $\mathrm{O}(5, \mathbb{F}_{2^e})$ was given in [**BFL**, Theorem 6.1]. The tuple

$$\begin{bmatrix} -1 & \cdot & \cdot & \cdot & \cdot \\ \cdot & 1 & \cdot & \cdot & \cdot \\ \cdot & \cdot & 1 & \cdot & \cdot \\ \cdot & \cdot & \cdot & 1 & \cdot \\ \cdot & \cdot & \cdot & \cdot & -1 \end{bmatrix}, \begin{bmatrix} f(\alpha) & g(\alpha) & \cdot & \cdot & \cdot \\ g(\alpha) & -f(\alpha) & \cdot & \cdot & \cdot \\ \cdot & \cdot & 1 & \cdot & \cdot \\ \cdot & \cdot & \cdot & -f(\alpha) & g(\alpha) \\ \cdot & \cdot & \cdot & g(\alpha) & f(\alpha) \end{bmatrix}, \begin{bmatrix} -1 & \cdot & \cdot & \cdot & \cdot \\ \cdot & f(\beta) & -g(\beta) & \cdot & \cdot \\ \cdot & -g(\beta) & -f(\beta) & \cdot & \cdot \\ \cdot & \cdot & \cdot & -1 & \cdot \\ \cdot & \cdot & \cdot & \cdot & -1 \end{bmatrix}, \begin{bmatrix} \cdot & \cdot & \cdot & \cdot & 1 \\ \cdot & \cdot & \cdot & 1 & \cdot \\ \cdot & \cdot & 1 & \cdot & \cdot \\ \cdot & 1 & \cdot & \cdot & \cdot \\ 1 & \cdot & \cdot & \cdot & \cdot \end{bmatrix},$$

where α and β come from (different) restricted sets of scalars, and

$$f(\alpha) = \frac{1-\alpha^2}{1+\alpha^2} \qquad g(\alpha) = \frac{-2\alpha}{1+\alpha^2}, \tag{5.1}$$

defines a string C-group representation for $\Omega(5, \mathbb{F}_q) \cong \mathrm{PSp}(4, \mathbb{F}_q)$ of rank 4. It is not isomorphic to the one given in the proof of Theorem 1.1, which applied reduction to the rank 5 representation: the first, second, and fourth generators all have 2-dimensional -1-eigenspace, whereas the earlier construction had three generators with 4-dimensional -1-eigenspace.

5.4. Software. A software package called SGGI has been implemented by the author in the MAGMA system [**BCP**] and is publicly available on GitHub. Among other things the package contains the explicit constructions in this paper, including the one in Section 5.3, and functions to carry out exhaustive searches for string C-group representations. These functions can be used to verify Theorem 1.1 for small values of q, but we caution that $\mathrm{PSp}(4, \mathbb{F}_{19})$ is nearing the limit of practicability for exhaustive searches. We note that other functions to compute with string C-groups, written by Dimitri Leemans, are distributed with MAGMA.

Acknowledgments

The author thanks J. T. Ferrara for the collection and synthesis of experimental data that was particularly helpful when designing the rank 4 construction in Section 5.3. He also thanks the anonymous referees for several helpful suggestions.

References

[BCP] Wieb Bosma, John Cannon, and Catherine Playoust, *The Magma algebra system. I. The user language*, J. Symbolic Comput. **24** (1997), no. 3-4, 235–265, DOI 10.1006/jsco.1996.0125. Computational algebra and number theory (London, 1993). MR1484478

[BFL] P. A. Brooksbank, J. T. Ferrara, and D. Leemans, *Orthogonal groups in characteristic 2 acting on polytopes of high rank*, Discrete Comput. Geom. **63** (2020), no. 3, 656–669, DOI 10.1007/s00454-019-00083-0. MR4074338

[BL1] P. A. Brooksbank and D. Leemans, *Polytopes of large rank for* $\mathrm{PSL}(4, \mathbb{F}_q)$, J. Algebra **452** (2016), 390–400, DOI 10.1016/j.jalgebra.2015.11.051. MR3461073

[BL2] Peter A. Brooksbank and Dimitri Leemans, *Rank reduction of string C-group representations*, Proc. Amer. Math. Soc. **147** (2019), no. 12, 5421–5426, DOI 10.1090/proc/14666. MR4021100

[BV] Peter A. Brooksbank and Deborah A. Vicinsky, *Three-dimensional classical groups acting on polytopes*, Discrete Comput. Geom. **44** (2010), no. 3, 654–659, DOI 10.1007/s00454-009-9212-0. MR2679061

[Co1] Marston D. E. Conder, *Generators for alternating and symmetric groups*, J. London Math. Soc. (2) **22** (1980), no. 1, 75–86, DOI 10.1112/jlms/s2-22.1.75. MR579811

[Co2] Marston D. E. Conder, *More on generators for alternating and symmetric groups*, Quart. J. Math. Oxford Ser. (2) **32** (1981), no. 126, 137–163, DOI 10.1093/qmath/32.2.137. MR615190

[FL1] Maria Elisa Fernandes and Dimitri Leemans, *Polytopes of high rank for the symmetric groups*, Adv. Math. **228** (2011), no. 6, 3207–3222, DOI 10.1016/j.aim.2011.08.006. MR2844941

[FL2] Maria Elisa Fernandes and Dimitri Leemans, *String C-group representations of alternating groups*, Ars Math. Contemp. **17** (2019), no. 1, 291–310, DOI 10.26493/1855-3974.1953.c53. MR4031526

[HL] Michael I. Hartley and Dimitri Leemans, *A new Petrie-like construction for abstract polytopes*, J. Combin. Theory Ser. A **115** (2008), no. 6, 997–1007, DOI 10.1016/j.jcta.2007.11.008. MR2423344

[LS] Dimitri Leemans and Egon Schulte, *Groups of type $L_2(q)$ acting on polytopes*, Adv. Geom. **7** (2007), no. 4, 529–539, DOI 10.1515/ADVGEOM.2007.031. MR2360900

[Ma] V. D. Mazurov, *On the generation of sporadic simple groups by three involutions, two of which commute* (Russian, with Russian summary), Sibirsk. Mat. Zh. **44** (2003), no. 1, 193–198, DOI 10.1023/A:1022028807652; English transl., Siberian Math. J. **44** (2003), no. 1, 160–164. MR1967616

[McS] Peter McMullen and Egon Schulte, *Abstract regular polytopes*, Encyclopedia of Mathematics and its Applications, vol. 92, Cambridge University Press, Cambridge, 2002. MR1965665

[MS] B. Monson and Egon Schulte, *Reflection groups and polytopes over finite fields. III*, Adv. in Appl. Math. **41** (2008), no. 1, 76–94, DOI 10.1016/j.aam.2007.07.001. MR2419764

[Nu1] Ya. N. Nuzhin, *Generating triples of involutions of Chevalley groups over a finite field of characteristic* 2 (Russian), Algebra i Logika **29** (1990), no. 2, 192–206, 261, DOI 10.1007/BF02001358; English transl., Algebra and Logic **29** (1990), no. 2, 134–143 (1991). MR1131150

[Nu2] Ya. N. Nuzhin, *Generating triples of involutions of alternating groups* (Russian), Mat. Zametki **51** (1992), no. 4, 91–95, 142, DOI 10.1007/BF01250552; English transl., Math. Notes **51** (1992), no. 3-4, 389–392. MR1172472

[Nu3] Ya. N. Nuzhin, *Generating triples of involutions of Lie-type groups over a finite field of odd characteristic. I* (Russian, with Russian summary), Algebra i Logika **36** (1997), no. 1, 77–96, 118, DOI 10.1007/BF02671953; English transl., Algebra and Logic **36** (1997), no. 1, 46–59. MR1454692

[Nu4] Ya. N. Nuzhin, *Generating triples of involutions of Lie-type groups over a finite field of odd characteristic. II* (Russian, with Russian summary), Algebra i Logika **36** (1997), no. 4, 422–440, 479; English transl., Algebra and Logic **36** (1997), no. 4, 245–256. MR1601503

[Tay] Donald E. Taylor, *The geometry of the classical groups*, Sigma Series in Pure Mathematics, vol. 9, Heldermann Verlag, Berlin, 1992. MR1189139

DEPARTMENT OF MATHEMATICS, BUCKNELL UNIVERSITY, LEWISBURG, PENNSYLVANIA 17837
Email address: pbrooksb@bucknell.edu

Contemporary Mathematics
Volume **764**, 2021
https://doi.org/10.1090/conm/764/15356

Perfect colorings of regular graphs

Joseph Ray Clarence Damasco and Dirk Frettlöh

Abstract. A vertex coloring of some graph is called *perfect* if each vertex of color i has exactly a_{ij} neighbors of color j. Being perfect imposes several restrictions on the color adjacency matrix (a_{ij}). We give a characterization of color adjacency matrices of perfect colorings of graphs, and in particular, connected graphs. Using this result we determine the lists of all color adjacency matrices corresponding to perfect colorings of 3-regular, 4-regular and 5-regular graphs with two, three and four colors. Finally, using these lists, we determine all perfect colorings of the edge graphs of the Platonic solids with two, three and four colors, respectively.

1. Introduction

Perfect colorings of graphs and related concepts have been studied in several contexts: algebraic graph theory, combinatorial designs, coding theory, finite geometry; and under several different names: equitable partitions, completely regular vertex sets, distance partitions, association schemes, etc. Some connections between these contexts are stated in Remark 1.3 below. For a broader overview see [**12**, **18**].

Throughout the paper let $G = (V, E)$ be a finite, undirected, simple, loop-free graph. A partition of V into disjoint nonempty sets $V_1, \ldots, V_m$ is called an *m-coloring* of G. Note that we do not require adjacent vertices to have different colors.

Definition 1.1. A coloring of the vertex set V of some graph $G = (V, E)$ with m colors is called *perfect* if (1) all colors are used, and (2) for all i, j the number of neighbors of colors j of any vertex v of color i is a constant a_{ij}. The matrix $A = (a_{ij})_{1 \le i,j \le m}$ is called the *color adjacency matrix* of the perfect coloring.

See Figures 1-5 in Appendix B for some examples of perfect colorings. Note that for $m = |V|$ the color adjacency matrix equals the adjacency matrix of G.

Remark 1.2. In some sources (e.g. [**16**, Sec. 9.3] and [**6**]) perfect colorings are called *equitable partitions*. However, it seems that the term "equitable partition" is used for two different concepts in graph theory: one is what we call perfect coloring

2010 *Mathematics Subject Classification.* Primary 05C15, 05C50.

The first author is grateful to the University of the Philippines System for financial support through its Faculty, REPS, and Administrative Staff Development Program.

The second author thanks the Research Center of Mathematical Modeling (RCM^2) at Bielefeld University for financial support.

above, the second is a coloring where every pair of adjacent vertices has different colors, and where the number of elements of any two color classes differs by at most one. See for instance [**17**], or [**13**] and references therein. Hence we will refer to the first concept by the term perfect coloring here.

REMARK 1.3. Perfect colorings — or very related concepts — have been studied in several contexts. The first source we know about is [**21**], where the color adjacency matrix was introduced to study spectral properties of certain graphs. In particular, the following result was shown in [**9**, Theorem 4.5], or [**16**, Theorem 9.3.3].

THEOREM 1.4. *Let M be the adjacency matrix of some graph G and let A be the color adjacency matrix of some perfect coloring of G. Then the characteristic polynomial of A divides the characteristic polynomial of M. In particular, each eigenvalue of A is an eigenvalue of M (with multiplicities).*

To name just a few more examples: Any subgroup of the automorphism group of a graph G induces a perfect coloring of G by considering the orbits of the group [**16**, Sec. 9.3]. However, not every perfect coloring arises from a graph automorphism. As another example, each *distance partition* (coloring the vertices w.r.t. their distance to some fixed vertex) of a distance regular graph [**14**] yields a perfect coloring. For more related work see [**12, 18, 19**].

Some concrete perfect colorings for small graphs were constructed for instance in [**2, 4, 5, 11, 15, 20**]. Here we generalize several results from these papers.

This paper is organized as follows: In Section 2, we give necessary and sufficient conditions for a matrix to be a color adjacency matrix of a perfect coloring of some graph, and in particular, some connected graph. In Section 3, we relate the cardinalities of the color classes of perfect colorings with two, three, and four colors to the entries of color adjacency matrices. Using these results we compute in Section 5 the lists of all color adjacency matrices of perfect colorings with two, three and four colors for k-regular connected graphs for $k \in \{3, 4, 5\}$, respectively, up to equivalence by permutations of colors. To our best knowledge the lists for three and four colors have not been published before.

The computations were carried out both in `sagemath` and in `scilab`. The implementation is described in Section 4.

As an application we determine in Section 6 all perfect colorings of the edge graphs of the Platonic solids using two, three and four colors, respectively. All perfect 2-colorings of the edge graphs of the Platonic solids have been determined in [**3**] already. The perfect 3-colorings of the edge graphs of the Platonic solids were studied in [**1**], but some cases were missed in the preprint version. To our knowledge the perfect 4-colorings of the edge graphs of the Platonic solids given in this paper are new.

2. Characterization of color adjacency matrices

We will find necessary and sufficient conditions under which a given nonnegative integer matrix corresponds to a perfect coloring of a graph as it seems that the relevant literature on perfect colorings of graphs lack such explicitly stated conditions. In the ensuing discussion, given a perfect coloring for a graph G with color adjacency matrix $A = (a_{ij}) \in \mathbb{N}^{m \times m}$, let v_i denote the number of vertices in the color class V_i, for $1 \leq i \leq m$.

THEOREM 2.1. *Suppose $A = (a_{ij}) \in \mathbb{N}^{m\times m}$. Then A is a color adjacency matrix for a perfect m-coloring of some graph $G = (V, E)$ if and only if the following hold:*

(1) *(Weak symmetry) For $1 \le i, j \le m$, $a_{ij} = 0$ if and only if $a_{ji} = 0$.*
(2) *(Consistency) For any nontrivial cycle $(n_1\, n_2\, \dots\, n_t)$ in the symmetric group S_m on the set $\{1, 2, \dots, m\}$,*

$$a_{n_1,n_2}a_{n_2,n_3}\cdots a_{n_{t-1},n_t}a_{n_t,n_1} = a_{n_2,n_1}a_{n_3,n_2}\cdots a_{n_t,n_{t-1}}a_{n_1,n_t}.$$

Moreover, there is a connected graph G with a perfect coloring corresponding to A if and only if A fulfills (1) and (2), and A is irreducible.

A symmetric matrix M is called irreducible if it is not conjugate via a permutation matrix to a block diagonal matrix having more than one block. (By "block diagonal matrix" we mean a square matrix having square matrices on its main diagonal, and all other entries being zero.) It is well-known that a directed graph G is connected if and only if its adjacency matrix is irreducible. A weaker statement is true here: if a graph G is connected then its color adjacency matrix is irreducible. (Because one can travel from any color to any other color.)

Before proving Theorem 2.1, we note that in any perfect coloring, the subgraph of G induced by the vertices of color i is an a_{ii}-regular graph. In addition, the edges between the vertices of color i and those of color j form the edge set of an (a_{ij}, a_{ji})-biregular graph, where by a (p, q)-biregular graph we mean a bipartite graph with bipartition (U, W) such that each vertex in U has degree p and each vertex in W has degree q.

We shall prove the sufficiency of conditions (1) and (2) of Theorem 2.1 constructively using Lemmas 2.2 and 2.3, which characterize regular and biregular graphs. All variables in these statements are nonnegative integers. We include a proof of Lemma 2.3 because we are not aware of a reference containing a proof of it.

LEMMA 2.2 ([**8**]). *There exists a k-regular graph with n vertices if and only if $n \ge k + 1$ and nk is even.*

Lemma 2.2 is a simple consequence of the Erdős-Gallai Theorem [**10**].

LEMMA 2.3. *There exists a (p, q)-biregular graph with bipartition (U, W) and $|U| = r$, $|W| = s$ if and only if $p \le s$, $q \le r$, and $pr = qs$.*

PROOF. That $p \le s$, $q \le r$, and $pr = qs$ are necessary follows from the definition of biregular graphs. To prove the converse, we assume p, $q \ne 0$ and construct a graph with the desired properties. Denote the vertices in U by $u_0, u_1, \dots, u_{r-1}$ and the vertices in W by $w_0, w_1, \dots, w_{s-1}$. We will use a greedy construction: join vertex u_0 with $w_0, w_1, \dots, w_{p-1 \bmod s}$, join vertex u_1 with $w_{p \bmod s}, w_{p+1 \bmod s}, \dots, w_{2p-1 \bmod s}$, and so on. That is, for each $0 \le a \le r-1$, join vertex u_a to $w_{ap+b \bmod s}$ for all $0 \le b \le p-1$. Because $pr = qs$, every vertex in U is adjacent to p vertices in W, and every vertex in W is adjacent to exactly q vertices in U. □

We now prove our characterization of color adjacency matrices.

PROOF OF THEOREM 2.1. The necessity of the first condition follows trivially from the symmetry of the vertex adjacency relation. Meanwhile, the necessity of the second condition when $t = 2$ is also trivial: $a_{n_1,n_2}a_{n_2,n_1} = a_{n_2,n_1}a_{n_1,n_2}$.

When $2 < t \leq m$, the identity arises from the fact that for any i and j, $a_{ij}v_i = a_{ji}v_j$. This is true by counting the number of edges between V_i and V_j in two ways. Thus, $a_{n_1,n_2}v_{n_1} = a_{n_2,n_1}v_{n_2}$, and therefore $a_{n_1,n_2}a_{n_2,n_3}v_{n_1} = a_{n_2,n_1}a_{n_2,n_3}v_{n_2} = a_{n_2,n_1}a_{n_3,n_2}v_{n_3}$. It follows by induction that

$$a_{n_1,n_2}a_{n_2,n_3}\cdots a_{n_{t-1},n_t}v_{n_1} = a_{n_2,n_1}a_{n_3,n_2}\cdots a_{n_t,n_{t-1}}v_{n_t}. \tag{2.1}$$

Combining this with $a_{n_1,n_t}v_{n_1} = a_{n_t,n_1}v_{n_t}$ gives the equation in condition (2).

Before establishing the converse, we note the equation arising from the induction in the following remark as it will be used later.

REMARK 2.4.

(1) For any nontrivial cycle $(n_1\, n_2\, \ldots\, n_t)$ in the symmetric group S_m on the set $\{1, 2, \ldots, m\}$,

$$a_{n_1,n_2}a_{n_2,n_3}\cdots a_{n_{t-1},n_t}v_{n_1} = a_{n_2,n_1}a_{n_3,n_2}\cdots a_{n_t,n_{t-1}}v_{n_t}.$$

(2) The consistency condition means that for any i and j, the different ways of relating v_i and v_j by products of the a_{n_r,n_s}'s must all agree.

To prove the converse, given a matrix $A \in \mathbb{N}^{m\times m}$ satisfying conditions (1) and (2), we construct a graph G with color adjacency matrix A, where the color classes are denoted by $V_1, V_2, \ldots, V_m$. By possibly conjugating via a permutation matrix, we assume that A is written as a block diagonal matrix with the largest number of blocks possible. (For instance, a block consisting of a diagonal matrix is interpreted as many blocks of size 1.)

Suppose first that there is only one block. This means that the matrix is not permutation-conjugate to a block diagonal matrix with more than one block, and that there is a path from any color to any other color. Hence, plugging the nonzero nondiagonal entries into Equation (2.1) we obtain the ratio of v_i to v_j for *every* $i \neq j$ There may be several ways of relating v_i and v_j by entries of A, but because of condition (2), we know all these ratios are consistent, and there is an ordered m-tuple of positive integers $(v'_1, v'_2, \ldots, v'_m)$ satisfying all required relations.

Moreover, there is a large enough multiple $(v_1, v_2, \ldots, v_m)$ of the m-tuple above such that for each i, $v_i \geq a_{ii} + 1$, $a_{ii}v_i$ is even, and $v_i \geq a_{ji}$ for $j \neq i$. Let V_i have v_i elements. By Lemmas 2.2 and 2.3, we may form an a_{ii}-regular graph using the vertices in V_i, and the edge set of an (a_{ij}, a_{ji})-biregular bipartite graph between distinct cells V_i and V_j. The resulting graph has the perfect coloring $(V_1, \ldots, V_m)$ with color adjacency matrix A. This graph may still be disconnected, but any component of this graph satisfies the adjacency relations described by A. Thus, we choose G to be one of the components of that auxiliary graph. Then G is a connected graph having a perfect m-coloring with color adjacency matrix A.

On the other hand, if A has multiple blocks, we perform the procedure above for each block of A, and let G be the union of the graphs formed for each block. Then G has a perfect m-coloring with color adjacency matrix A. Moreover, G is necessarily disconnected, as each vertex in G corresponding to one block of A is not adjacent to any vertex of G corresponding to any other block of A.

This completes the characterization of color adjacency matrices. □

REMARK 2.5. Given a realizable color adjacency matrix $A = (a_{ij})_{m\times m}$, let G' be the directed multigraph having the color classes V_i as vertices and A as adjacency matrix. We note that A is permutation-conjugate to a block diagonal matrix with more than one block if and only if G' is disconnected. In our ensuing computations to find candidate color adjacency matrices A for connected graphs G, we use the equivalent fact that G' must also have a spanning tree. Thus, there is an arrangement V_{n_1}, V_{n_2}, …, V_{n_m} of the color classes such that for $i \geq 1$, $V_{n_{i+1}}$ is connected to V_{n_j} for some $j \leq i$. In other words, there exists an $(m-1)$-tuple $(a_{n_1,n_2}, a_{n'_3,n_3}, \ldots, a_{n'_m,n_m})$ of nonzero entries such that $n_1 \neq n_2$, and if $m \geq 3$, $n'_i \in \{n_1, n_2, \ldots, n_{i-1}\}$ and $n_i \notin \{n_1, n_2, \ldots, n_{i-1}\}$ for $3 \leq i \leq m$.

3. Counting lemmas

For $m \in \{2, 3, 4\}$, the next lemmas count the number of vertices in each color class. Recall that there may be several ways to express v_i in terms of v_j and the a_{n_r,n_s}'s depending on the different ways how the induced graph G' (see above) might be connected. Let G'' be the simple graph induced by the multigraph G' above by identifying multiple edges and removing loops. By Cayley's tree formula the number of possible spanning trees of G'' equals m^{m-2} [**7**]. Hence there is only $2^0 = 1$ case to consider for $m = 2$, there are 3^1 cases for $m = 3$, and 4^2 cases for $m = 4$. The case $m = 2$ (Lemma 3.1) appears in [**3**]. Hence we sketch a proof only for the case when $m = 3$.

LEMMA 3.1. *Let $A = (a_{ij}) \in \mathbb{N}^{2\times 2}$ be a color adjacency matrix of some connected graph $G = (V, E)$. Then a_{12} and a_{21} are both nonzero, and if v_i denotes the number of vertices of color i, then*

$$v_1 = \frac{|V|}{1 + \frac{a_{12}}{a_{21}}}, v_2 = \frac{|V|}{\frac{a_{21}}{a_{12}} + 1}.$$

For $m = 3$, while the graph G' may have multiple spanning trees, there is only one up to isomorphism, namely, a path on three vertices. Dealing with the three cases determined by Cayley's formula may then be summarized in the following lemma.

LEMMA 3.2. *Let $A = (a_{ij}) \in \mathbb{N}^{3\times 3}$ be a color adjacency matrix of some connected graph $G = (V, E)$. Then there is a permutation $(n_1\, n_2\, n_3)$ of $(1\,2\,3)$ such that $a_{n_1,n_2} a_{n_1,n_3} \neq 0$. If v_i denotes the number of vertices of color i, then*

$$\begin{aligned} v_{n_1} &= \frac{|V|}{1 + \frac{a_{n_1,n_2}}{a_{n_2,n_1}} + \frac{a_{n_1,n_3}}{a_{n_3,n_1}}}, \\ v_{n_2} &= \frac{|V|}{\frac{a_{n_2,n_1}}{a_{n_1,n_2}} + 1 + \frac{a_{n_2,n_1} a_{n_1,n_3}}{a_{n_1,n_2} a_{n_3,n_1}}}, \\ v_{n_3} &= \frac{|V|}{\frac{a_{n_3,n_1}}{a_{n_1,n_3}} + \frac{a_{n_3,n_1} a_{n_1,n_2}}{a_{n_1,n_3} a_{n_2,n_1}} + 1}. \end{aligned}$$

The permutation referred to in Lemma 3.2 is determined by a spanning tree in G'. As for the values enumerated, the proof goes along the lines of counting the total number of vertices as $|V| = v_{n_1} + v_{n_2} + v_{n_3}$, considering that we have $a_{n_1,i} v_{n_1} = a_{i,n_1} v_i$, hence $|V| = v_{n_1} + \frac{a_{n_1,n_2}}{a_{n_2,n_1}} v_{n_1} + \frac{a_{n_1,n_3}}{a_{n_3,n_1}} v_{n_1}$. Here, a_{n_2,n_3} and a_{n_3,n_2} may equal zero. Hence they cannot necessarily be used to relate v_{n_2} and v_{n_3}. But, as in the counting procedure in the proof of Theorem 2.1, one obtains $a_{n_2,n_1} a_{n_1,n_3} v_{n_2} = a_{n_1,n_2} a_{n_3,n_1} v_{n_3}$, and consequently the expressions for v_{n_2} and v_{n_3} above.

Note that the permutation in Lemma 3.2 is not necessarily unique. But the consistency condition of Theorem 2.1 ensures that the values obtained are independent of the choice of permutation. We note that Proposition 2.1 in [**1**], which distinguished the mutually exclusive possibilities, follows from Theorem 2.1 and Lemma 3.2.

For the four color case, there are two possible spanning trees up to isomorphism, namely, a star graph with three leaves, or a path on four vertices. The sixteen cases from Cayley's tree formula break down into four star graphs as there are four choices for the central vertex, and twelve paths arising from the different ways of arranging four objects in a row, up to reversal of order.

LEMMA 3.3. *Let $A = (a_{ij}) \in \mathbb{N}^{4\times 4}$ be a color adjacency matrix of some connected graph $G = (V, E)$. Then there is a permutation (n_1, n_2, n_3, n_4) of $(1, 2, 3, 4)$ such that $a_{n_1,n_2}a_{n_1,n_3}a_{n_1,n_4} \neq 0$ or $a_{n_1,n_2}a_{n_2,n_3}a_{n_3,n_4} \neq 0$. Let v_i denote the number of vertices of color i.*

(1) *If $a_{n_1,n_2}a_{n_1,n_3}a_{n_1,n_4} \neq 0$, then*

$$\begin{aligned}
v_{n_1} &= \frac{|V|}{1 + \frac{a_{n_1,n_2}}{a_{n_2,n_1}} + \frac{a_{n_1,n_3}}{a_{n_3,n_1}} + \frac{a_{n_1,n_4}}{a_{n_4,n_1}}},\\
v_{n_2} &= \frac{|V|}{\frac{a_{n_2,n_1}}{a_{n_1,n_2}} + 1 + \frac{a_{n_2,n_1}a_{n_1,n_3}}{a_{n_1,n_2}a_{n_3,n_1}} + \frac{a_{n_2,n_1}a_{n_1,n_4}}{a_{n_1,n_2}a_{n_4,n_1}}},\\
v_{n_3} &= \frac{|V|}{\frac{a_{n_3,n_1}}{a_{n_1,n_3}} + \frac{a_{n_3,n_1}a_{n_1,n_2}}{a_{n_1,n_3}a_{n_2,n_1}} + 1 + \frac{a_{n_3,n_1}a_{n_1,n_4}}{a_{n_1,n_3}a_{n_4,n_1}}},\\
v_{n_4} &= \frac{|V|}{\frac{a_{n_4,n_1}}{a_{n_1,n_4}} + \frac{a_{n_4,n_1}a_{n_1,n_2}}{a_{n_1,n_4}a_{n_2,n_1}} + \frac{a_{n_4,n_1}a_{n_1,n_3}}{a_{n_1,n_4}a_{n_3,n_1}} + 1}.
\end{aligned}$$

(2) *If $a_{n_1,n_2}a_{n_2,n_3}a_{n_3,n_4} \neq 0$, then*

$$\begin{aligned}
v_{n_1} &= \frac{|V|}{1 + \frac{a_{n_1,n_2}}{a_{n_2,n_1}} + \frac{a_{n_1,n_2}a_{n_2,n_3}}{a_{n_2,n_1}a_{n_3,n_2}} + \frac{a_{n_1,n_2}a_{n_2,n_3}a_{n_3,n_4}}{a_{n_2,n_1}a_{n_3,n_2}a_{n_4,n_3}}},\\
v_{n_2} &= \frac{|V|}{\frac{a_{n_2,n_1}}{a_{n_1,n_2}} + 1 + \frac{a_{n_2,n_3}}{a_{n_3,n_2}} + \frac{a_{n_2,n_3}a_{n_3,n_4}}{a_{n_3,n_2}a_{n_4,n_3}}},\\
v_{n_3} &= \frac{|V|}{\frac{a_{n_3,n_2}a_{n_2,n_1}}{a_{n_2,n_3}a_{n_1,n_2}} + \frac{a_{n_3,n_2}}{a_{n_2,n_3}} + 1 + \frac{a_{n_3,n_4}}{a_{n_4,n_3}}},\\
v_{n_4} &= \frac{|V|}{\frac{a_{n_4,n_3}a_{n_3,n_2}a_{n_2,n_1}}{a_{n_3,n_4}a_{n_2,n_3}a_{n_1,n_2}} + \frac{a_{n_4,n_3}a_{n_3,n_2}}{a_{n_3,n_4}a_{n_2,n_3}} + \frac{a_{n_4,n_3}}{a_{n_3,n_4}} + 1}.
\end{aligned}$$

The proof of this lemma is in complete analogy to the proof of Lemma 3.2: We count $|V|$ by $|V| = v_{n_1} + v_{n_2} + v_{n_3} + v_{n_4}$. Then we express for instance v_{n_1} in terms of $v_{n_2}, v_{n_3}, v_{n_4}$ using the appropriate (nonzero) a_{ij}'s, depending on the possible spanning tree for G''.

4. Implementation

Let G be a k-regular connected graph and $A \in \mathbb{N}^{m\times m}$ be a color adjacency matrix for a perfect m-coloring of G. Clearly each row sum of A equals k. Note that for each row there are $\binom{k+m-1}{m-1}$ different ways to distribute the entries such that the row sum equals k. Hence there are $\binom{k+m-1}{m-1}^m$ matrices to consider altogether.

The conditions above yield the following procedure to enumerate all color adjacency matrices for connected regular graphs. We need $m(m-1)$ nested loops to go through all matrices $A = (a_{ij}) \in \mathbb{N}^{m \times m}$ with constant row sum k.

(1) Check for $i \neq j$ whether it is true that "$a_{ij} = 0$ if and only if $a_{ji} = 0$" (weak symmetry condition of Theorem 2.1).
(2) Ensure connectedness by applying Remark 2.5. For $m = 2$, this means a_{12} must be nonzero; for $m = 3$ and $m = 4$, we find the right permutation such that the product in the condition of the corresponding lemma is nonzero.
(3) Check whether the consistency condition of Theorem 2.1 is satisfied by going through all relevant products. Several of these products may be zero, but connectedness implies that there is a way to relate any v_r and v_s by products of nonzero a_{ij}'s as in Remark 2.4.

    Performing the preceding steps for given m and k yields all color adjacency matrices for perfect m-colorings of connected k-regular graphs. The following steps are merely for removal of matrices that essentially the same partitions, just with the colors permuted. Without loss of generality, we also adopt the convention that $v_i \leq v_{i+1}$ for $i < m$.
(4) We identify a suitable case in Lemma 3.1, 3.2, or 3.3 that A satisfies. Again, when using Lemma 3.2 or 3.3, by the consistency condition, it is enough to consider only one spanning tree of G'. Observe that for each i, the value of $\frac{v_i}{|V|}$ depends only on the entries a_{ij}. We check whether these expressions are in nondecreasing order when arranged according to increasing i. For the two-color case, it suffices to check if $a_{12} \leq a_{21}$.
(5) Finally, we identify matrices if they are conjugate via a permutation matrix.

We summarize our procedure in Algorithm 1.

The tests above were implemented both in `scilab` and `sagemath` [**22**]. The `sagemath` worksheets are available for download [**23**]. There are three worksheets, one for each number of colors. The worksheets are organized in sections, one for each degree k of regularity ($k \in \{3, 4, 5\}$). The comments in the code indicate the different cases and tests. After executing all cells in all sections in the worksheet the list `l` contains all color adjacency matrices passing the tests (1.)-(5.) for the respective value of k. Each section contains further code to determine all perfect colorings of Platonic graphs, see Section 6.

The worksheets for two and three colors will need at most a few minutes computing time on an ordinary laptop or desktop computer. The worksheets for four colors need several hours of computation on a modern laptop. The most time-consuming part is step 5. Therefore we also provide a sage data file and a pdf file containing all color adjacency matrices for download [**23**]. One can download the sage data file (for instance `4col-list.sage`), store them in some folder (for instance `/home/user/sage`) and load the content into any sage worksheet using `open('/home/user/sage/4col-list.sage')`, for instance. After executing the above command, the list `l43` contains all color adjacency matrices for perfect 4-colorings of 3-regular graphs, `l44` contains all color adjacency matrices for perfect 4-colorings of 4-regular graphs, and `l45` the corresponding list for perfect 4-colorings

Algorithm 1 Generate color adjacency matrices of perfect m-colorings of connected k-regular graphs

procedure CAM(m, k)
    L' = empty list
    for $A \in \mathbb{N}^{m \times m}$ with constant row sum k **do**
        if $a_{ij} = 0$ if and only if $a_{ji} = 0$ for $i \neq j$ **then**
            if there exists $(m-1)$-tuple satisfying Remark 2.5 **then**
                if for every nontrivial cycle in S_m, the equation in condition (2) of Theorem 2.1 is satisfied **then**
                    if $\frac{v_i}{|V|} \leq \frac{v_{i+1}}{|V|}$ for each $i < m$ in Lemma 3.1, 3.2, or 3.3 **then**
                        Add A to L'
                    end if
                end if
            end if
        end if
    end for
    L = list containing first element of L'
    for $A \in L'$, starting with second element **do**
        if for each permutation matrix P, PAP^{-1} is not equal to any element of L **then**
            Add A to L
        end if
    end for
    return L
end procedure

of 5-regular graphs. These lists can then be processed further, as seen in the examples in Section 6.

5. Color adjacency matrices of k-regular graphs

Using these criteria all color adjacency matrices A for perfect 2-colorings of k-regular graphs with $k \in \{3, 4, 5\}$ are only the ones listed in the Table 1.

k	A
3	$\begin{pmatrix} 0&3\\1&2 \end{pmatrix}, \begin{pmatrix} 0&3\\2&1 \end{pmatrix}, \begin{pmatrix} 0&3\\3&0 \end{pmatrix}, \begin{pmatrix} 1&2\\1&2 \end{pmatrix}, \begin{pmatrix} 1&2\\2&1 \end{pmatrix}, \begin{pmatrix} 2&1\\1&2 \end{pmatrix}$
4	$\begin{pmatrix} 0&4\\1&3 \end{pmatrix}, \begin{pmatrix} 0&4\\2&2 \end{pmatrix}, \begin{pmatrix} 0&4\\3&1 \end{pmatrix}, \begin{pmatrix} 0&4\\4&0 \end{pmatrix}, \begin{pmatrix} 1&3\\1&3 \end{pmatrix}, \begin{pmatrix} 1&3\\2&2 \end{pmatrix}, \begin{pmatrix} 1&3\\3&1 \end{pmatrix}, \begin{pmatrix} 2&2\\1&3 \end{pmatrix}, \begin{pmatrix} 2&2\\2&2 \end{pmatrix}, \begin{pmatrix} 3&1\\1&3 \end{pmatrix}$
5	$\begin{pmatrix} 0&5\\1&4 \end{pmatrix}, \begin{pmatrix} 0&5\\2&3 \end{pmatrix}, \begin{pmatrix} 0&5\\3&2 \end{pmatrix}, \begin{pmatrix} 0&5\\4&1 \end{pmatrix}, \begin{pmatrix} 0&5\\5&0 \end{pmatrix}, \begin{pmatrix} 1&4\\1&4 \end{pmatrix}, \begin{pmatrix} 1&4\\2&3 \end{pmatrix}, \begin{pmatrix} 1&4\\3&2 \end{pmatrix}, \begin{pmatrix} 1&4\\4&1 \end{pmatrix}, \begin{pmatrix} 2&3\\1&4 \end{pmatrix},$ $\begin{pmatrix} 2&3\\2&3 \end{pmatrix}, \begin{pmatrix} 2&3\\3&2 \end{pmatrix}, \begin{pmatrix} 3&2\\1&4 \end{pmatrix}, \begin{pmatrix} 3&2\\2&3 \end{pmatrix}, \begin{pmatrix} 4&1\\1&4 \end{pmatrix}$

TABLE 1. All color adjacency matrices A for k-regular graphs with two colors.

All color adjacency matrices A for perfect 3-colorings of k-regular graphs with $k \in \{3, 4, 5\}$ are given in Appendix A. There are 18 possible matrices for 3-regular graphs, 64 for 4-regular graphs, and 153 for 5-regular graphs.

The lists of all color adjacency matrices A for perfect 4-colorings of k-regular graphs with $k \in \{3, 4, 5\}$ are quite long: there are 72 matrices for 3-regular graphs,

485 for 4-regular graphs, and 2042 for 5-regular graphs. They are available online at [**23**] in two forms: as a list in `pdf`, and as a loadable `sage` data file, see Section 4. Table 2 below compares the number of all matrices in $\mathbb{N}^{m\times m}$ with all row sums equal to k with the number of all color adjacency matrices for perfect colorings for 4-colorings of k-regular graphs with $k \in \{3,4,5\}$.

$m \setminus k$	3	4	5
2	6 of 16	10 of 25	15 of 36
3	18 of 1000	64 of 3375	153 of 9261
4	72 of 16 000	485 of 1 500 625	2042 of 9 834 496

TABLE 2. A comparison of the number of all color adjacency matrices for perfect colorings of connected graphs (passing the tests (1.)-(5.)) with the number of matrices in $\mathbb{N}^{m\times m}$ with all row sums equal to k.

6. Perfect colorings of Platonic graphs

Theorem 1.4 may now be used as a further necessary criterion for possible color adjacency matrices for a particular graph G. We illustrate this with the Platonic graphs (i.e., the edge graphs of the Platonic solids). The eigenvalues of these graphs are given in Table 3. These values can be found for instance in [**9**]. An entry a^n means that a is an eigenvalue of algebraic multiplicity n.

G	eigenvalues
tetrahedron	$-1^3, 3$
cube	$-3, -1^3, 1^3, 3$
octahedron	$-2^2, 0^3, 4$
dodecahedron	$-\sqrt{5}^3, -2^4, 0^4, 1^5, \sqrt{5}^3, 3$
icosahedron	$-\sqrt{5}^3, -1^5, \sqrt{5}^3, 5$

TABLE 3. The eigenvalues of the Platonic graphs. A superscript denotes the multiplicity of the respective eigenvalue.

It follows from Theorem 1.4 that to determine all perfect colorings of the Platonic graphs with two colors one can check which of the matrices in Table 1 have eigenvalues in the respective spectrum of the Platonic graphs. This is the actual test we implemented in `sagemath`. One could refine it in order to include counting the multiplicities, but we found by inspection that for these graphs the latter condition does not exclude further matrices.

6.1. The perfect 2-colorings of Platonic graphs. By the methods described above, we obtained a list of all color adjacency matrices for perfect 2-colorings of k-regular graphs for $k \in \{3,4,5\}$, see Table 1. For each matrix in each of these lists we now check whether the corresponding expressions for v_i in Lemma 3.1 are integers, and whether the eigenvalues of the matrix are eigenvalues of the Platonic graph under consideration. For example, since the cube graph is 3-regular, we check for all six matrices in the first row of Table 1 whether the expressions in

Lemma 3.1 are all integers, and whether all eigenvalues of the matrix are contained in $\{-3, -1, 1, 3\}$. In this manner we obtained the following candidates for color adjacency matrices for 2-colorings of the Platonic graphs, respectively.

(1) Tetrahedron: $\begin{pmatrix} 0 & 3 \\ 1 & 2 \end{pmatrix}, \begin{pmatrix} 1 & 2 \\ 2 & 1 \end{pmatrix}$
(2) Cube: $\begin{pmatrix} 0 & 3 \\ 1 & 2 \end{pmatrix}, \begin{pmatrix} 0 & 3 \\ 3 & 0 \end{pmatrix}, \begin{pmatrix} 1 & 2 \\ 2 & 1 \end{pmatrix}, \begin{pmatrix} 2 & 1 \\ 1 & 2 \end{pmatrix}$
(3) Octahedron: $\begin{pmatrix} 0 & 4 \\ 2 & 2 \end{pmatrix}, \begin{pmatrix} 1 & 3 \\ 3 & 1 \end{pmatrix}^{\dagger}, \begin{pmatrix} 2 & 2 \\ 2 & 2 \end{pmatrix}$
(4) Dodecahedron: $\begin{pmatrix} 0 & 3 \\ 2 & 1 \end{pmatrix}, \begin{pmatrix} 2 & 1 \\ 1 & 2 \end{pmatrix}$
(5) Icosahedron: $\begin{pmatrix} 0 & 5 \\ 1 & 4 \end{pmatrix}, \begin{pmatrix} 1 & 4 \\ 2 & 3 \end{pmatrix}, \begin{pmatrix} 2 & 3 \\ 3 & 2 \end{pmatrix}$

For the tetrahedron, the cube, the dodecahedron, and the icosahedron, all possible color adjacency matrices in the list above actually correspond to perfect 2-colorings. These colorings are shown in Figures 1, 2, 4 and 5. For the octahedron there are only two perfect 2-colorings, shown in Figure 3. In this case, one of the matrices above does not correspond to a perfect 2-coloring of the octahedral graph: the matrix marked with † can be checked to be impossible in a straightforward manner by attempting to color the vertices of an octahedral graph according to these color adjacencies. One may also argue combinatorially: if this matrix is a color adjacency matrix for a connected graph G, then G must have at least 8 vertices. This is because $a_{12} = a_{21} = 3$ imply $v_1, v_2 \geq 3$ and $a_{11} = a_{22} = 1$ imply v_1 and v_2 must be even. Thus, G cannot be the octahedral graph. In any case, the list confirms the results in [**3**].

6.2. The perfect 3-colorings of Platonic graphs. Applying the analogous procedure, and with Lemma 3.2 rather than Lemma 3.1, we obtain a list of all color adjacency matrices for 3-colorings of the Platonic graphs, respectively. In this case, all candidates are valid color adjacency matrices for perfect colorings of Platonic graphs.

(1) Tetrahedron: $\begin{pmatrix} 0 & 1 & 2 \\ 1 & 0 & 2 \\ 1 & 1 & 1 \end{pmatrix}$
(2) Cube: $\begin{pmatrix} 0 & 1 & 2 \\ 1 & 0 & 2 \\ 1 & 1 & 1 \end{pmatrix}, \begin{pmatrix} 1 & 0 & 2 \\ 0 & 1 & 2 \\ 1 & 1 & 1 \end{pmatrix}$
(3) Octahedron: $\begin{pmatrix} 0 & 0 & 4 \\ 0 & 0 & 4 \\ 1 & 1 & 2 \end{pmatrix}, \begin{pmatrix} 0 & 2 & 2 \\ 2 & 0 & 2 \\ 2 & 2 & 0 \end{pmatrix}, \begin{pmatrix} 0 & 2 & 2 \\ 2 & 1 & 1 \\ 2 & 1 & 1 \end{pmatrix}$
(4) Dodecahedron: $\begin{pmatrix} 0 & 0 & 3 \\ 0 & 0 & 3 \\ 1 & 1 & 1 \end{pmatrix}, \begin{pmatrix} 0 & 3 & 0 \\ 1 & 0 & 2 \\ 0 & 1 & 2 \end{pmatrix}, \begin{pmatrix} 1 & 0 & 2 \\ 0 & 1 & 2 \\ 1 & 2 & 0 \end{pmatrix}$
(5) Icosahedron: $\begin{pmatrix} 0 & 1 & 4 \\ 1 & 0 & 4 \\ 1 & 1 & 3 \end{pmatrix}, \begin{pmatrix} 0 & 2 & 3 \\ 1 & 1 & 3 \\ 1 & 2 & 2 \end{pmatrix}, \begin{pmatrix} 1 & 2 & 2 \\ 2 & 1 & 2 \\ 2 & 2 & 1 \end{pmatrix}$

The perfect colorings corresponding to the color adjacency matrices above are shown in Figures 1-5. This list corrects a preprint version of [**1**] by providing the three cases missing there, namely $\begin{pmatrix} 0 & 2 & 2 \\ 2 & 0 & 2 \\ 2 & 2 & 0 \end{pmatrix}$ for the octahedral graph and $\begin{pmatrix} 0 & 3 & 0 \\ 1 & 0 & 2 \\ 0 & 1 & 2 \end{pmatrix}$ and $\begin{pmatrix} 1 & 0 & 2 \\ 0 & 1 & 2 \\ 1 & 2 & 0 \end{pmatrix}$ for the dodecahedral graph. The final version of [**1**] is correct.

6.3. The perfect 4-colorings of Platonic graphs. We obtained in a similar manner the following candidates for color adjacency matrices for 4-colorings of the Platonic graphs, respectively.

(1) Tetrahedron: $\begin{pmatrix} 0 & 1 & 1 & 1 \\ 1 & 0 & 1 & 1 \\ 1 & 1 & 0 & 1 \\ 1 & 1 & 1 & 0 \end{pmatrix}$
(2) Cube: $\begin{pmatrix} 0 & 0 & 0 & 3 \\ 0 & 0 & 3 & 0 \\ 0 & 1 & 0 & 2 \\ 1 & 0 & 2 & 0 \end{pmatrix}, \begin{pmatrix} 0 & 0 & 1 & 2 \\ 0 & 0 & 2 & 1 \\ 1 & 2 & 0 & 0 \\ 2 & 1 & 0 & 0 \end{pmatrix}, \begin{pmatrix} 0 & 1 & 1 & 1 \\ 1 & 0 & 1 & 1 \\ 1 & 1 & 0 & 1 \\ 1 & 1 & 1 & 0 \end{pmatrix}, \begin{pmatrix} 0 & 1 & 1 & 1 \\ 1 & 0 & 1 & 1 \\ 1 & 1 & 1 & 0 \\ 1 & 1 & 0 & 1 \end{pmatrix}, \begin{pmatrix} 1 & 0 & 1 & 1 \\ 0 & 1 & 1 & 1 \\ 1 & 1 & 1 & 0 \\ 1 & 1 & 0 & 1 \end{pmatrix}$
(3) Octahedron: $\begin{pmatrix} 0 & 0 & 2 & 2 \\ 0 & 0 & 2 & 2 \\ 1 & 1 & 0 & 2 \\ 1 & 1 & 2 & 0 \end{pmatrix}, \begin{pmatrix} 0 & 0 & 2 & 2 \\ 0 & 0 & 2 & 2 \\ 1 & 1 & 1 & 1 \\ 1 & 1 & 1 & 1 \end{pmatrix}$

(4) Dodecahedron: $\begin{pmatrix}0&0&0&3\\0&0&2&1\\0&2&0&1\\1&1&1&0\end{pmatrix}$, $\begin{pmatrix}0&0&0&3\\0&1&1&1\\0&1&1&1\\1&1&1&0\end{pmatrix}$, $\begin{pmatrix}0&0&1&2\\0&0&1&2\\1&1&1&0\\1&1&0&1\end{pmatrix}$, $\begin{pmatrix}0&0&1&2\\0&2&0&1\\1&0&2&0\\2&1&0&0\end{pmatrix}$, $\begin{pmatrix}1&0&0&2\\0&1&0&2\\0&0&1&2\\1&1&1&0\end{pmatrix}$

(5) Icosahedron: $\begin{pmatrix}0&0&0&5\\0&0&5&0\\0&1&2&2\\1&0&2&2\end{pmatrix}$, $\begin{pmatrix}0&1&1&3\\1&0&1&3\\1&1&0&3\\1&1&1&2\end{pmatrix}$, $\begin{pmatrix}0&1&2&2\\1&0&2&2\\1&1&1&2\\1&1&2&1\end{pmatrix}$, $\begin{pmatrix}0&1&2&2\\1&2&0&2\\2&0&2&1\\2&2&1&0\end{pmatrix}$

All of these candidates have corresponding perfect colorings, and these are shown in Figures 1-5, respectively.

7. Further questions

The results and methods above give rise to several questions.

(1) Using the lists of realizable color adjacency matrices generated in Section 5, one may also try to determine the perfect colorings of other regular graphs starting with special classes of graphs, say the Archimedean graphs. These graphs are all regular with valency at most 5.

(2) The matrix marked † in Section 6.1 could have been excluded from the list by adding conditions checking if the order of the color class V_i is even if the corresponding diagonal entry is odd. Then, the procedure becomes sufficient to enumerate the realizable color adjacency matrices for Platonic graphs. It would be interesting to understand why this is so, and to characterize all regular graphs for which this modified method is sufficient.

(3) Recall that not all perfect colorings correspond to orbit partitions. We then ask if there are conditions under which a given realizable color adjacency matrix corresponds to an orbit partition of a graph. For this question it might be instructive to start with graphs possessing high degrees of symmetry, vertex transitivity, and edge transitivity.

Appendix A: All color adjacency matrices for 3-colorings

All color adjacency matrices A for perfect 3-colorings of k-regular graphs with $k \in \{3, 4, 5\}$:

<u>3-regular graphs:</u>

$\begin{pmatrix}0&0&3\\0&0&3\\1&1&1\end{pmatrix}$ $\begin{pmatrix}0&0&3\\0&0&3\\1&2&0\end{pmatrix}$ $\begin{pmatrix}0&0&3\\0&1&2\\1&1&1\end{pmatrix}$ $\begin{pmatrix}0&0&3\\0&1&2\\1&2&0\end{pmatrix}$ $\begin{pmatrix}0&0&3\\0&2&1\\1&1&1\end{pmatrix}$ $\begin{pmatrix}0&0&3\\0&2&1\\2&1&0\end{pmatrix}$ $\begin{pmatrix}0&1&2\\1&0&2\\1&1&1\end{pmatrix}$ $\begin{pmatrix}0&1&2\\1&1&1\\2&1&0\end{pmatrix}$ $\begin{pmatrix}0&1&2\\1&2&0\\1&0&2\end{pmatrix}$ $\begin{pmatrix}0&1&2\\1&2&0\\2&0&1\end{pmatrix}$
$\begin{pmatrix}0&3&0\\1&0&2\\0&1&2\end{pmatrix}$ $\begin{pmatrix}1&0&2\\0&0&3\\1&2&0\end{pmatrix}$ $\begin{pmatrix}1&0&2\\0&1&2\\1&1&1\end{pmatrix}$ $\begin{pmatrix}1&0&2\\0&1&2\\1&2&0\end{pmatrix}$ $\begin{pmatrix}1&0&2\\0&2&1\\1&1&1\end{pmatrix}$ $\begin{pmatrix}1&1&1\\1&1&1\\1&1&1\end{pmatrix}$ $\begin{pmatrix}1&1&1\\1&2&0\\1&0&2\end{pmatrix}$ $\begin{pmatrix}1&2&0\\1&0&2\\0&1&2\end{pmatrix}$

<u>4-regular graphs:</u>

$\begin{pmatrix}0&0&4\\0&0&4\\1&1&2\end{pmatrix}$ $\begin{pmatrix}0&0&4\\0&0&4\\1&2&1\end{pmatrix}$ $\begin{pmatrix}0&0&4\\0&0&4\\1&3&0\end{pmatrix}$ $\begin{pmatrix}0&0&4\\0&0&4\\2&2&0\end{pmatrix}$ $\begin{pmatrix}0&0&4\\0&1&3\\1&1&2\end{pmatrix}$ $\begin{pmatrix}0&0&4\\0&1&3\\1&2&1\end{pmatrix}$ $\begin{pmatrix}0&0&4\\0&1&3\\1&3&0\end{pmatrix}$ $\begin{pmatrix}0&0&4\\0&1&3\\2&2&0\end{pmatrix}$ $\begin{pmatrix}0&0&4\\0&2&2\\1&1&2\end{pmatrix}$ $\begin{pmatrix}0&0&4\\0&2&2\\1&2&1\end{pmatrix}$
$\begin{pmatrix}0&0&4\\0&2&2\\2&1&1\end{pmatrix}$ $\begin{pmatrix}0&0&4\\0&2&2\\2&2&0\end{pmatrix}$ $\begin{pmatrix}0&0&4\\0&3&1\\1&1&2\end{pmatrix}$ $\begin{pmatrix}0&0&4\\0&3&1\\2&1&1\end{pmatrix}$ $\begin{pmatrix}0&0&4\\0&3&1\\3&1&0\end{pmatrix}$ $\begin{pmatrix}0&1&3\\1&0&3\\1&1&2\end{pmatrix}$ $\begin{pmatrix}0&1&3\\1&0&3\\2&2&0\end{pmatrix}$ $\begin{pmatrix}0&1&3\\1&2&1\\3&1&0\end{pmatrix}$ $\begin{pmatrix}0&1&3\\1&3&0\\1&0&3\end{pmatrix}$ $\begin{pmatrix}0&1&3\\1&3&0\\2&0&2\end{pmatrix}$
$\begin{pmatrix}0&1&3\\1&3&0\\3&0&1\end{pmatrix}$ $\begin{pmatrix}0&2&2\\1&0&3\\1&3&0\end{pmatrix}$ $\begin{pmatrix}0&2&2\\1&1&2\\1&2&1\end{pmatrix}$ $\begin{pmatrix}0&2&2\\1&2&1\\1&1&2\end{pmatrix}$ $\begin{pmatrix}0&2&2\\1&3&0\\1&0&3\end{pmatrix}$ $\begin{pmatrix}0&2&2\\2&0&2\\1&1&2\end{pmatrix}$ $\begin{pmatrix}0&2&2\\2&0&2\\2&2&0\end{pmatrix}$ $\begin{pmatrix}0&2&2\\2&1&1\\2&1&1\end{pmatrix}$ $\begin{pmatrix}0&2&2\\2&2&0\\1&0&3\end{pmatrix}$ $\begin{pmatrix}0&2&2\\2&2&0\\2&0&2\end{pmatrix}$
$\begin{pmatrix}0&4&0\\1&0&3\\0&1&3\end{pmatrix}$ $\begin{pmatrix}0&4&0\\1&0&3\\0&2&2\end{pmatrix}$ $\begin{pmatrix}0&4&0\\1&1&2\\0&1&3\end{pmatrix}$ $\begin{pmatrix}0&4&0\\2&0&2\\0&1&3\end{pmatrix}$ $\begin{pmatrix}1&0&3\\0&0&4\\1&2&1\end{pmatrix}$ $\begin{pmatrix}1&0&3\\0&0&4\\1&3&0\end{pmatrix}$ $\begin{pmatrix}1&0&3\\0&1&3\\1&1&2\end{pmatrix}$ $\begin{pmatrix}1&0&3\\0&1&3\\1&2&1\end{pmatrix}$ $\begin{pmatrix}1&0&3\\0&1&3\\1&3&0\end{pmatrix}$ $\begin{pmatrix}1&0&3\\0&1&3\\2&2&0\end{pmatrix}$
$\begin{pmatrix}1&0&3\\0&2&2\\1&1&2\end{pmatrix}$ $\begin{pmatrix}1&0&3\\0&2&2\\1&2&1\end{pmatrix}$ $\begin{pmatrix}1&0&3\\0&2&2\\2&2&0\end{pmatrix}$ $\begin{pmatrix}1&0&3\\0&3&1\\1&1&2\end{pmatrix}$ $\begin{pmatrix}1&0&3\\0&3&1\\2&1&1\end{pmatrix}$ $\begin{pmatrix}1&1&2\\1&1&2\\1&1&2\end{pmatrix}$ $\begin{pmatrix}1&1&2\\1&2&1\\2&1&1\end{pmatrix}$ $\begin{pmatrix}1&1&2\\1&3&0\\1&0&3\end{pmatrix}$ $\begin{pmatrix}1&1&2\\1&3&0\\2&0&2\end{pmatrix}$ $\begin{pmatrix}1&3&0\\1&0&3\\0&1&3\end{pmatrix}$
$\begin{pmatrix}1&3&0\\1&0&3\\0&2&2\end{pmatrix}$ $\begin{pmatrix}1&3&0\\1&1&2\\0&1&3\end{pmatrix}$ $\begin{pmatrix}1&3&0\\2&0&2\\0&1&3\end{pmatrix}$ $\begin{pmatrix}2&0&2\\0&0&4\\1&3&0\end{pmatrix}$ $\begin{pmatrix}2&0&2\\0&1&3\\1&2&1\end{pmatrix}$ $\begin{pmatrix}2&0&2\\0&1&3\\1&3&0\end{pmatrix}$ $\begin{pmatrix}2&0&2\\0&2&2\\1&1&2\end{pmatrix}$ $\begin{pmatrix}2&0&2\\0&2&2\\1&2&1\end{pmatrix}$ $\begin{pmatrix}2&0&2\\0&3&1\\1&1&2\end{pmatrix}$ $\begin{pmatrix}2&1&1\\1&2&1\\1&1&2\end{pmatrix}$
$\begin{pmatrix}2&1&1\\1&3&0\\1&0&3\end{pmatrix}$ $\begin{pmatrix}2&2&0\\1&0&3\\0&1&3\end{pmatrix}$ $\begin{pmatrix}2&2&0\\1&0&3\\0&2&2\end{pmatrix}$ $\begin{pmatrix}2&2&0\\1&1&2\\0&1&3\end{pmatrix}$

5-regular graphs:

$$\begin{pmatrix}0&0&5\\0&0&5\\1&1&3\end{pmatrix} \begin{pmatrix}0&0&5\\0&0&5\\1&2&2\end{pmatrix} \begin{pmatrix}0&0&5\\0&0&5\\1&3&1\end{pmatrix} \begin{pmatrix}0&0&5\\0&0&5\\1&4&0\end{pmatrix} \begin{pmatrix}0&0&5\\0&0&5\\2&2&1\end{pmatrix} \begin{pmatrix}0&0&5\\0&0&5\\2&3&0\end{pmatrix} \begin{pmatrix}0&0&5\\0&1&4\\1&1&3\end{pmatrix} \begin{pmatrix}0&0&5\\0&1&4\\1&2&2\end{pmatrix} \begin{pmatrix}0&0&5\\0&1&4\\1&3&1\end{pmatrix} \begin{pmatrix}0&0&5\\0&1&4\\1&4&0\end{pmatrix}$$

$$\begin{pmatrix}0&0&5\\0&1&4\\2&2&1\end{pmatrix} \begin{pmatrix}0&0&5\\0&1&4\\2&3&0\end{pmatrix} \begin{pmatrix}0&0&5\\0&2&3\\1&1&3\end{pmatrix} \begin{pmatrix}0&0&5\\0&2&3\\1&2&2\end{pmatrix} \begin{pmatrix}0&0&5\\0&2&3\\1&3&1\end{pmatrix} \begin{pmatrix}0&0&5\\0&2&3\\2&2&1\end{pmatrix} \begin{pmatrix}0&0&5\\0&2&3\\2&3&0\end{pmatrix} \begin{pmatrix}0&0&5\\0&2&3\\3&2&0\end{pmatrix} \begin{pmatrix}0&0&5\\0&3&2\\1&1&3\end{pmatrix} \begin{pmatrix}0&0&5\\0&3&2\\1&2&2\end{pmatrix}$$

$$\begin{pmatrix}0&0&5\\0&3&2\\2&1&2\end{pmatrix} \begin{pmatrix}0&0&5\\0&3&2\\2&2&1\end{pmatrix} \begin{pmatrix}0&0&5\\0&3&2\\3&2&0\end{pmatrix} \begin{pmatrix}0&0&5\\0&4&1\\1&1&3\end{pmatrix} \begin{pmatrix}0&0&5\\0&4&1\\2&1&2\end{pmatrix} \begin{pmatrix}0&0&5\\0&4&1\\3&1&1\end{pmatrix} \begin{pmatrix}0&0&5\\0&4&1\\4&1&0\end{pmatrix} \begin{pmatrix}0&1&4\\1&0&4\\1&1&3\end{pmatrix} \begin{pmatrix}0&1&4\\1&0&4\\2&2&1\end{pmatrix} \begin{pmatrix}0&1&4\\1&2&2\\2&1&2\end{pmatrix}$$

$$\begin{pmatrix}0&1&4\\1&3&1\\4&1&0\end{pmatrix} \begin{pmatrix}0&1&4\\1&4&0\\1&0&4\end{pmatrix} \begin{pmatrix}0&1&4\\1&4&0\\2&0&3\end{pmatrix} \begin{pmatrix}0&1&4\\1&4&0\\3&0&2\end{pmatrix} \begin{pmatrix}0&1&4\\1&4&0\\4&0&1\end{pmatrix} \begin{pmatrix}0&2&3\\1&1&3\\1&2&2\end{pmatrix} \begin{pmatrix}0&2&3\\1&4&0\\1&0&4\end{pmatrix} \begin{pmatrix}0&2&3\\2&0&3\\1&1&3\end{pmatrix} \begin{pmatrix}0&2&3\\2&0&3\\2&2&1\end{pmatrix} \begin{pmatrix}0&2&3\\2&1&2\\3&2&0\end{pmatrix}$$

$$\begin{pmatrix}0&2&3\\2&2&1\\3&1&1\end{pmatrix} \begin{pmatrix}0&2&3\\2&3&0\\1&0&4\end{pmatrix} \begin{pmatrix}0&2&3\\2&3&0\\2&0&3\end{pmatrix} \begin{pmatrix}0&2&3\\2&3&0\\3&0&2\end{pmatrix} \begin{pmatrix}0&3&2\\2&3&0\\1&0&4\end{pmatrix} \begin{pmatrix}0&3&2\\3&0&2\\1&1&3\end{pmatrix} \begin{pmatrix}0&3&2\\3&2&0\\1&0&4\end{pmatrix} \begin{pmatrix}0&5&0\\1&0&4\\0&1&4\end{pmatrix} \begin{pmatrix}0&5&0\\1&0&4\\0&2&3\end{pmatrix} \begin{pmatrix}0&5&0\\1&0&4\\0&3&2\end{pmatrix}$$

$$\begin{pmatrix}0&5&0\\1&1&3\\0&1&4\end{pmatrix} \begin{pmatrix}0&5&0\\1&1&3\\0&2&3\end{pmatrix} \begin{pmatrix}0&5&0\\1&2&2\\0&1&4\end{pmatrix} \begin{pmatrix}0&5&0\\2&0&3\\0&1&4\end{pmatrix} \begin{pmatrix}0&5&0\\2&0&3\\0&2&3\end{pmatrix} \begin{pmatrix}0&5&0\\2&1&2\\0&1&4\end{pmatrix} \begin{pmatrix}0&5&0\\3&0&2\\0&1&4\end{pmatrix} \begin{pmatrix}1&0&4\\0&0&5\\1&2&2\end{pmatrix} \begin{pmatrix}1&0&4\\0&0&5\\1&3&1\end{pmatrix} \begin{pmatrix}1&0&4\\0&0&5\\1&4&0\end{pmatrix}$$

$$\begin{pmatrix}1&0&4\\0&0&5\\2&3&0\end{pmatrix} \begin{pmatrix}1&0&4\\0&1&4\\1&1&3\end{pmatrix} \begin{pmatrix}1&0&4\\0&1&4\\1&2&2\end{pmatrix} \begin{pmatrix}1&0&4\\0&1&4\\1&3&1\end{pmatrix} \begin{pmatrix}1&0&4\\0&1&4\\1&4&0\end{pmatrix} \begin{pmatrix}1&0&4\\0&1&4\\2&2&1\end{pmatrix} \begin{pmatrix}1&0&4\\0&1&4\\2&3&0\end{pmatrix} \begin{pmatrix}1&0&4\\0&2&3\\1&1&3\end{pmatrix} \begin{pmatrix}1&0&4\\0&2&3\\1&2&2\end{pmatrix} \begin{pmatrix}1&0&4\\0&2&3\\1&3&1\end{pmatrix}$$

$$\begin{pmatrix}1&0&4\\0&2&3\\2&2&1\end{pmatrix} \begin{pmatrix}1&0&4\\0&2&3\\2&3&0\end{pmatrix} \begin{pmatrix}1&0&4\\0&3&2\\1&1&3\end{pmatrix} \begin{pmatrix}1&0&4\\0&3&2\\1&2&2\end{pmatrix} \begin{pmatrix}1&0&4\\0&3&2\\2&1&2\end{pmatrix} \begin{pmatrix}1&0&4\\0&3&2\\2&2&1\end{pmatrix} \begin{pmatrix}1&0&4\\0&3&2\\3&2&0\end{pmatrix} \begin{pmatrix}1&0&4\\0&4&1\\1&1&3\end{pmatrix} \begin{pmatrix}1&0&4\\0&4&1\\2&1&2\end{pmatrix} \begin{pmatrix}1&0&4\\0&4&1\\3&1&1\end{pmatrix}$$

$$\begin{pmatrix}1&1&3\\1&1&3\\1&1&3\end{pmatrix} \begin{pmatrix}1&1&3\\1&1&3\\2&2&1\end{pmatrix} \begin{pmatrix}1&1&3\\1&3&1\\3&1&1\end{pmatrix} \begin{pmatrix}1&1&3\\1&4&0\\1&0&4\end{pmatrix} \begin{pmatrix}1&1&3\\1&4&0\\2&0&3\end{pmatrix} \begin{pmatrix}1&1&3\\1&4&0\\3&0&2\end{pmatrix} \begin{pmatrix}1&2&2\\1&0&4\\1&4&0\end{pmatrix} \begin{pmatrix}1&2&2\\1&1&3\\1&3&1\end{pmatrix} \begin{pmatrix}1&2&2\\1&2&2\\1&2&2\end{pmatrix} \begin{pmatrix}1&2&2\\1&3&1\\1&1&3\end{pmatrix}$$

$$\begin{pmatrix}1&2&2\\1&4&0\\1&0&4\end{pmatrix} \begin{pmatrix}1&2&2\\2&1&2\\1&1&3\end{pmatrix} \begin{pmatrix}1&2&2\\2&1&2\\2&2&1\end{pmatrix} \begin{pmatrix}1&2&2\\2&2&1\\2&1&2\end{pmatrix} \begin{pmatrix}1&2&2\\2&3&0\\1&0&4\end{pmatrix} \begin{pmatrix}1&2&2\\2&3&0\\2&0&3\end{pmatrix} \begin{pmatrix}1&4&0\\1&0&4\\0&1&4\end{pmatrix} \begin{pmatrix}1&4&0\\1&0&4\\0&2&3\end{pmatrix} \begin{pmatrix}1&4&0\\1&0&4\\0&3&2\end{pmatrix} \begin{pmatrix}1&4&0\\1&1&3\\0&1&4\end{pmatrix}$$

$$\begin{pmatrix}1&4&0\\1&1&3\\0&2&3\end{pmatrix} \begin{pmatrix}1&4&0\\1&2&2\\0&1&4\end{pmatrix} \begin{pmatrix}1&4&0\\2&0&3\\0&1&4\end{pmatrix} \begin{pmatrix}1&4&0\\2&0&3\\0&2&3\end{pmatrix} \begin{pmatrix}1&4&0\\2&1&2\\0&1&4\end{pmatrix} \begin{pmatrix}1&4&0\\3&0&2\\0&1&4\end{pmatrix} \begin{pmatrix}2&0&3\\0&0&5\\1&2&2\end{pmatrix} \begin{pmatrix}2&0&3\\0&0&5\\1&3&1\end{pmatrix} \begin{pmatrix}2&0&3\\0&0&5\\1&4&0\end{pmatrix} \begin{pmatrix}2&0&3\\0&1&4\\1&2&2\end{pmatrix}$$

$$\begin{pmatrix}2&0&3\\0&1&4\\1&3&1\end{pmatrix} \begin{pmatrix}2&0&3\\0&1&4\\1&4&0\end{pmatrix} \begin{pmatrix}2&0&3\\0&1&4\\2&3&0\end{pmatrix} \begin{pmatrix}2&0&3\\0&2&3\\1&1&3\end{pmatrix} \begin{pmatrix}2&0&3\\0&2&3\\1&2&2\end{pmatrix} \begin{pmatrix}2&0&3\\0&2&3\\1&3&1\end{pmatrix} \begin{pmatrix}2&0&3\\0&2&3\\2&2&1\end{pmatrix} \begin{pmatrix}2&0&3\\0&2&3\\2&3&0\end{pmatrix} \begin{pmatrix}2&0&3\\0&3&2\\1&1&3\end{pmatrix} \begin{pmatrix}2&0&3\\0&3&2\\1&2&2\end{pmatrix}$$

$$\begin{pmatrix}2&0&3\\0&3&2\\2&2&1\end{pmatrix} \begin{pmatrix}2&0&3\\0&4&1\\1&1&3\end{pmatrix} \begin{pmatrix}2&0&3\\0&4&1\\2&1&2\end{pmatrix} \begin{pmatrix}2&1&2\\1&2&2\\1&1&3\end{pmatrix} \begin{pmatrix}2&1&2\\1&3&1\\2&1&2\end{pmatrix} \begin{pmatrix}2&1&2\\1&4&0\\1&0&4\end{pmatrix} \begin{pmatrix}2&1&2\\1&4&0\\2&0&3\end{pmatrix} \begin{pmatrix}2&3&0\\1&0&4\\0&1&4\end{pmatrix} \begin{pmatrix}2&3&0\\1&0&4\\0&2&3\end{pmatrix} \begin{pmatrix}2&3&0\\1&0&4\\0&3&2\end{pmatrix}$$

$$\begin{pmatrix}2&3&0\\1&1&3\\0&1&4\end{pmatrix} \begin{pmatrix}2&3&0\\1&1&3\\0&2&3\end{pmatrix} \begin{pmatrix}2&3&0\\1&2&2\\0&1&4\end{pmatrix} \begin{pmatrix}2&3&0\\2&0&3\\0&1&4\end{pmatrix} \begin{pmatrix}2&3&0\\2&0&3\\0&2&3\end{pmatrix} \begin{pmatrix}2&3&0\\2&1&2\\0&1&4\end{pmatrix} \begin{pmatrix}3&0&2\\0&0&5\\1&3&1\end{pmatrix} \begin{pmatrix}3&0&2\\0&0&5\\1&4&0\end{pmatrix} \begin{pmatrix}3&0&2\\0&1&4\\1&3&1\end{pmatrix} \begin{pmatrix}3&0&2\\0&1&4\\1&4&0\end{pmatrix}$$

$$\begin{pmatrix}3&0&2\\0&2&3\\1&2&2\end{pmatrix} \begin{pmatrix}3&0&2\\0&2&3\\1&3&1\end{pmatrix} \begin{pmatrix}3&0&2\\0&3&2\\1&1&3\end{pmatrix} \begin{pmatrix}3&0&2\\0&3&2\\1&2&2\end{pmatrix} \begin{pmatrix}3&0&2\\0&4&1\\1&1&3\end{pmatrix} \begin{pmatrix}3&1&1\\1&3&1\\1&1&3\end{pmatrix} \begin{pmatrix}3&1&1\\1&4&0\\1&0&4\end{pmatrix} \begin{pmatrix}3&2&0\\1&0&4\\0&1&4\end{pmatrix} \begin{pmatrix}3&2&0\\1&0&4\\0&2&3\end{pmatrix} \begin{pmatrix}3&2&0\\1&0&4\\0&3&2\end{pmatrix}$$

$$\begin{pmatrix}3&2&0\\1&1&3\\0&1&4\end{pmatrix} \begin{pmatrix}3&2&0\\1&1&3\\0&2&3\end{pmatrix} \begin{pmatrix}3&2&0\\1&2&2\\0&1&4\end{pmatrix}$$

Appendix B: Perfect colorings of the Platonic graphs

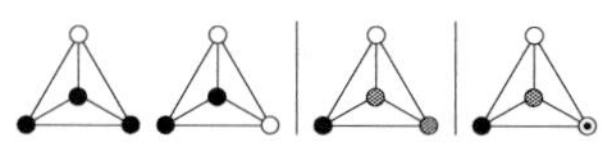

FIGURE 1. The perfect 2-, 3- and 4-colorings of the tetrahedral graph. Here and in the following figures white is color 1, black is color 2, grey is color 3, and color 4 is represented by a dotted vertex.

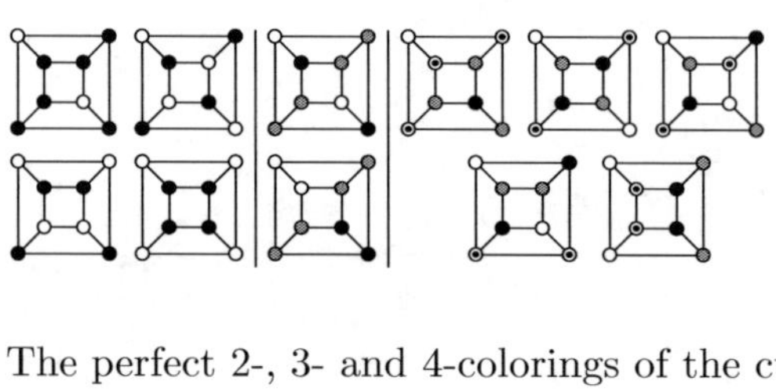

FIGURE 2. The perfect 2-, 3- and 4-colorings of the cube graph.

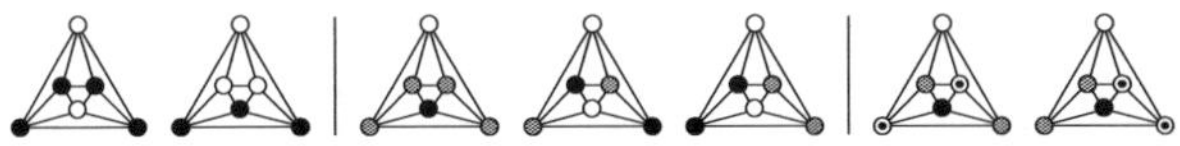

FIGURE 3. The perfect 2-, 3- and 4-colorings of the octahedral graph.

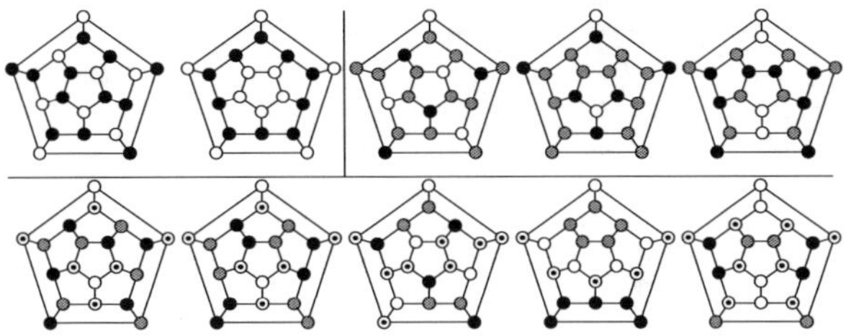

FIGURE 4. The perfect 2-, 3- and 4-colorings of the dodecahedral graph.

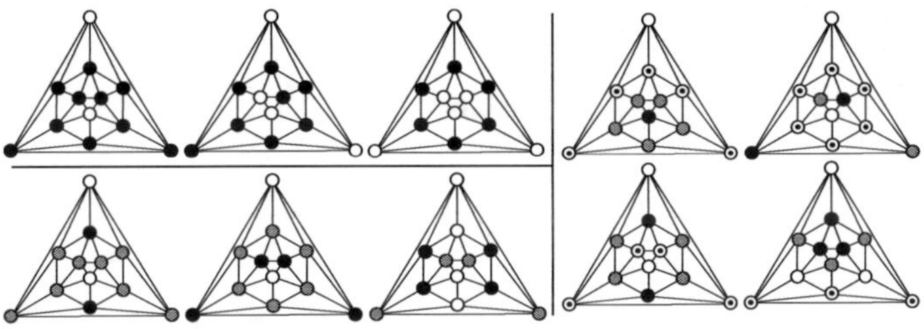

FIGURE 5. The perfect 2-, 3- and 4-colorings of the icosahedral graph.

Acknowledgments

We are grateful to Ferdinand Ihringer and to Michael Stiebitz for valuable information, and to the two anonymous referees for valuable remarks. Special thanks to Caya and Lasse Schubert for providing some of the 3- and 4-colorings of the cube and the dodecahedron.

References

[1] M. Alaeiyan and A. Mehrabani, *Perfect 3-colorings of the platonic graph*, Iran. J. Sci. Technol. Trans. A Sci. **43** (2019), no. 4, 1863–1871, DOI 10.1007/s40995-018-0646-1. MR3978600

[2] Mohammad Hadi Alaeiyan and Amirabbas Abedi, *Perfect 3-colorings of the Johnson graphs $J(4,2)$, $J(5,2)$, $J(6,2)$ and Petersen graph*, Ars Combin. **140** (2018), 199–213. MR3822000

[3] M. H. Alaeiyan and H. Karami, *Perfect 2-colorings of the Platonic graphs*, Internat. J. Nonlinear Analysis and Appl. 8 (2017) 29-35.

[4] Mehdi Alaeiyan and Ayoob Mehrabani, *Perfect 3-colorings of the cubic graphs of order 10*, Electron. J. Graph Theory Appl. (EJGTA) **5** (2017), no. 2, 194–206, DOI 10.5614/ejgta.2017.5.2.3. MR3716894

[5] S. V. Avgustinovich and I. Yu. Mogil′nykh, *Perfect 2-colorings of the Johnson graphs $J(8,3)$ and $J(8,4)$* (Russian, with English and Russian summaries), Diskretn. Anal. Issled. Oper. **17** (2010), no. 2, 3–19, 100. MR2682086

[6] Wayne Barrett, Amanda Francis, and Benjamin Webb, *Equitable decompositions of graphs with symmetries*, Linear Algebra Appl. **513** (2017), 409–434, DOI 10.1016/j.laa.2016.10.017. MR3573808

[7] A. Cayley, *A theorem on trees*, Quart. J. Pure Appl. Math. 23 (1889) 376-378.

[8] G. Chartrand and L. Lesniak, *Graphs & digraphs*, 4th ed., Chapman & Hall/CRC, Boca Raton, FL, 2005. MR2107429

[9] Dragoš M. Cvetković, Michael Doob, and Horst Sachs, *Spectra of graphs*, Pure and Applied Mathematics, vol. 87, Academic Press, Inc. [Harcourt Brace Jovanovich, Publishers], New York-London, 1980. Theory and application. MR572262

[10] P. Erdős and T. Gallai, *Gráfok előírt fokszámú pontokkal*, Matematikai Lapok (1960) 264-274.

[11] D. G. Fon-Der-Flaas, *Perfect 2-colorings of a hypercube* (Russian, with Russian summary), Sibirsk. Mat. Zh. **48** (2007), no. 4, 923–930, DOI 10.1007/s11202-007-0075-4; English transl., Siberian Math. J. **48** (2007), no. 4, 740–745. MR2355384

[12] Yuval Filmus and Ferdinand Ihringer, *Boolean degree 1 functions on some classical association schemes*, J. Combin. Theory Ser. A **162** (2019), 241–270, DOI 10.1016/j.jcta.2018.11.006. MR3874601

[13] Hanna Furmańczyk, *Equitable coloring of graphs*, Graph colorings, Contemp. Math., vol. 352, Amer. Math. Soc., Providence, RI, 2004, pp. 35–53, DOI 10.1090/conm/352/03. MR2076998

[14] C. D. Godsil, *Bounding the diameter of distance-regular graphs*, Combinatorica **8** (1988), no. 4, 333–343, DOI 10.1007/BF02189090. MR981891

[15] Alexander L. Gavrilyuk and Sergey V. Goryainov, *On perfect 2-colorings of Johnson graphs $J(v, 3)$*, J. Combin. Des. **21** (2013), no. 6, 232–252, DOI 10.1002/jcd.21327. MR3150904

[16] Chris Godsil and Gordon Royle, *Algebraic graph theory*, Graduate Texts in Mathematics, vol. 207, Springer-Verlag, New York, 2001. MR1829620

[17] A. Hajnal and E. Szemerédi, *Proof of a conjecture of P. Erdős*, Combinatorial theory and its applications, II (Proc. Colloq., Balatonfüred, 1969), North-Holland, Amsterdam, 1970, pp. 601–623. MR0297607

[18] F. Ihringer, *Translating terminology: equitable partitions and related concepts*, blog entry: `https://ratiobound.wordpress.com/2018/12/02/` (accessed 7. Jan. 2019).

[19] Denis S. Krotov, *On calculation of the interweight distribution of an equitable partition*, J. Algebraic Combin. **40** (2014), no. 2, 373–386, DOI 10.1007/s10801-013-0492-3. MR3239290

[20] I. Yu. Mogil′nykh, *On the regularity of perfect colorings of the Johnson graph in two colors* (Russian, with Russian summary), Problemy Peredachi Informatsii **43** (2007), no. 4, 37–44, DOI 10.1134/S0032946007040035; English transl., Probl. Inf. Transm. **43** (2007), no. 4, 303–309. MR2406142

[21] H. Sachs, *Über Teiler, Faktoren und charakteristische Polynome von Graphen. I* (German), Wiss. Z. Tech. Hochsch. Ilmenau **12** (1966), 7–12. MR194361

[22] The Sage Developer, *SageMath, the Sage Mathematics Software System*, (Version 7.4, 2017).

[23] Online: `https://www.math.uni-bielefeld.de/~frettloe/perfect-col-graph/`

Institute of Mathematics, College of Science, University of the Philippines Diliman, Quezon City, Philippines
Email address: `jrcgdamasco@math.upd.edu.ph`

Bielefeld University, Postfach 100131, 33501 Bielefeld, Germany
Email address: `dirk.frettloeh@udo.edu`
URL: `https://math.uni-bielefeld.de/~frettloe`

Contemporary Mathematics
Volume **764**, 2021
https://doi.org/10.1090/conm/764/15332

Tverberg theorems over discrete sets of points

J. A. De Loera, T. A. Hogan, F. Meunier, and N. H. Mustafa

ABSTRACT. This paper discusses Tverberg-type theorems with coordinate constraints (i.e., versions of these theorems where all points lie within a subset $S \subset \mathbb{R}^d$ and the intersection of convex hulls is required to have a non-empty intersection with S). We determine the m-Tverberg number, when $m \geq 3$, of any discrete subset S of $\mathbb{R}^2$ (a generalization of an unpublished result of J.-P. Doignon). We also present improvements on the upper bounds for the Tverberg numbers of $\mathbb{Z}^3$ and $\mathbb{Z}^j \times \mathbb{R}^k$ and an integer version of the well-known positive-fraction selection lemma of J. Pach.

Introduction

Consider n points in $\mathbb{R}^d$ and a positive integer $m \geq 2$. If $n \geq (m-1)(d+1)+1$, the points can always be partitioned into m subsets whose convex hulls contain a common point. This is the celebrated theorem of Tverberg [**Tve66**], which has been the topic of many generalizations and variations since it was first proved in 1966 [**BS18, DLGMM19**]. In this paper we focus on new versions of Tverberg-type theorems where some of the coordinates of the points are restricted to discrete subsets of a Euclidean space. The associated discrete Tverberg numbers are much harder to compute than their classical real-version counterparts (see for instance the complexity discussion of [**Onn91**]).

We begin our work remembering the following unpublished Tverberg-type result of Doignon. Consider n points with coordinates in $\mathbb{Z}^2$ and a positive integer $m \geq 3$. If $n \geq 4m-3$, then the points can be partitioned into m subsets whose convex hulls contain a common point in $\mathbb{Z}^2$. According to Eckhoff [**Eck00**] this result was stated by Doignon in a conference.

A partition of points where the intersection of the convex hulls contains at least one lattice point is called an *integer m-Tverberg partition* and such a common point is an *integer Tverberg point* for that partition. Regarding the case $m = 2$, the integer 2-Tverberg partitions are called *integer Radon partitions* and the integer Tverberg points are called *integer Radon points.* Any configuration of at least six

2010 *Mathematics Subject Classification.* Primary 52A35, 52C05, 52C07.

This work was partially supported by NSF grant DMS-1440140, while the first and third authors were in residence at the Mathematical Sciences Research Institute in Berkeley, California, during the Fall 2017 semester.

The first and second authors were also partially supported by NSF grants DMS-1522158 and DMS-1818169.

The fourth author was supported by ANR grant ADDS (ANR-19-CE48-0005).

points in $\mathbb{Z}^2$ admits an integer Radon partition. This was proved by Doignon in his PhD thesis [**Doi75**] and later discovered independently by Onn [**Onn91**]. All these values for $\mathbb{Z}^2$ are optimal as shown by the following examples. The 5-point configuration $\{(0,0),(0,1),(2,0),(1,2),(3,2)\}$, exhibited by Onn in the cited paper, has no integer Radon partition. To address the optimality when $m \geq 3$, consider the set $\{(i,i),(i,-i+1) \colon i = -m+2,-m+3,\ldots,m-2,m-1\}$. (According to Eckhoff [**Eck00**], this set was proposed by Doignon during the aforementioned conference.) This set has $4m-4$ points and a moment of reflection might convince the reader that it has no integer m-Tverberg partition.

More generally, one can define the *Tverberg number* $\mathrm{Tv}(S,m)$ for any subset S of $\mathbb{R}^d$ and an integer $m \geq 2$ as the smallest positive integer n with the following property: Any multiset of n points in S admits a partition into m subsets $A_1, A_2, \ldots, A_m$ with

$$\left(\bigcap_{i=1}^{m} \mathrm{conv}(A_i)\right) \cap S \neq \varnothing.$$

Here, by "partition of a multiset", we mean that each element of a multiset A is contained in a number of sub-multisets $A_1, \ldots, A_m$ so that the sum of its multiplicities in the A_i is equal to its multiplicity in A. If no such number exists, we say that $\mathrm{Tv}(S,m) = \infty$. Note that Doignon's theorem, together with the discussion that follows, allows us to state

$$\mathrm{Tv}(\mathbb{Z}^2, m) = \begin{cases} 6 & \text{if } m = 2, \\ 4m-3 & \text{otherwise.} \end{cases}$$

Our contributions. All our results deal with discrete versions of Tverberg's theorem. This latter theorem, in its classical version, and its variations have many applications, e.g., to the computation of simplicial data depth that is important in statistics [**RH99**], to data classification algorithms [**DLH19**], and to the study and computation of centerpoints whose applications in integer optimization are well-documented in [**BO16**]; see also the surveys [**BS18, DLGMM19**] and the references therein. We can thus expect that our results extend the range of these applications. Moreover, they also form a contribution to the theory of abstract convexity; see the book by van de Vel for an introduction to this theory [**vdV93**].

Our first main result generalizes Doignon's theorem. We determine the exact m-Tverberg number (when m is at least three) for any discrete subset S of $\mathbb{R}^2$, as considered in [**DLHRS17**]. Before stating this result we recall the *Helly number* $\mathrm{H}(S)$ of a discrete subset S of $\mathbb{R}^d$ as the smallest positive integer h with the following property: Suppose $\mathcal{F}$ is a finite family of convex sets in $\mathbb{R}^d$, and that $\bigcap \mathcal{G}$ (the intersection of all elements in $\mathcal{G}$) intersects S in at least one point for every subfamily $\mathcal{G}$ of $\mathcal{F}$ having at most h members. Then $\bigcap \mathcal{F}$ intersects S in at least one point. If no such integer exists, we say that $\mathrm{H}(S) = \infty$.

Then we have the following theorem. (The theorem is stated for S with finite Helly number, as any $S \subset \mathbb{R}^d$ with $\mathrm{H}(S) = \infty$ has $\mathrm{Tv}(S,m) = \infty$ for all $m \geq 2$ [**Lev51**].)

THEOREM 1. *Suppose S is a discrete subset of $\mathbb{R}^2$ with $\mathrm{H}(S) < \infty$. Then for all $m \geq 2$, we have*

$$\mathrm{Tv}(S,m) = \mathrm{H}(S)(m-1)+1,$$

except for the case $m = 2$, $\mathrm{H}(S) = 4$, *for which we have*

$$5 \leq \mathrm{Tv}(S, 2) \leq 6$$

and both values are possible.

In particular we present a proof of Doignon's theorem, the special case of Theorem 1 where $S = \mathbb{Z}^2$ (noting that $\mathrm{H}(\mathbb{Z}^d) = 2^d$ as established in [**Doi73**]).

REMARK. Theorem 1 shows that, except for an exceptional case, the Tverberg number of a planar set can be written as a function of its Helly number (see [**Ave13, AGS+17**] and the references there). For the case $\mathrm{H}(S) = 4$, the bounds on $\mathrm{Tv}(S, 2)$ given above cannot be improved. For example, $S' = \{(0,0), (0,1), (1,0), (1,1)\}$ and $\mathbb{Z}^2$ both have Helly number four, but $\mathrm{Tv}(\mathbb{Z}^2, 2) = 6$, while the pigeonhole principle implies that $\mathrm{Tv}(S', 2) = 5$.

Our second main result improves the upper bound on the integer Tverberg numbers for the three-dimensional case $S = \mathbb{Z}^3$.

THEOREM 2. *The following inequality holds for all* $m \geq 2$*:*

$$\mathrm{Tv}(\mathbb{Z}^3, m) \leq 24m - 31.$$

Our third main result is an inequality that will be used to derive improved bounds on S-Tverberg numbers when S is a product of a Euclidean space with some subset S' of a Euclidean space.

THEOREM 3. *Let* $S' \subset \mathbb{R}^j$. *Then for all integers* $k \geq 1$ *and all* $m \geq 2$, *we have*

$$\mathrm{Tv}(S' \times \mathbb{R}^k, m) \leq \mathrm{Tv}(S', \mathrm{Tv}(\mathbb{R}^k, m)).$$

For example, choosing $S' = \mathbb{Z}^j$ leads to the "mixed integer" case. Then Theorem 3 implies that for all positive integers j, k and all $m \geq 2$, we have

$$\mathrm{Tv}(\mathbb{Z}^j \times \mathbb{R}^k, m) \leq \mathrm{Tv}(\mathbb{Z}^j, \mathrm{Tv}(\mathbb{R}^k, m)).$$

Moreover, we will use Theorem 3 to obtain the following bound:

$$2^j(m-1)(k+1) + 1 \leq \mathrm{Tv}(\mathbb{Z}^j \times \mathbb{R}^k, m) \leq j2^j(m-1)(k+1) + 1. \tag{1}$$

Our fourth main result is a generalization of *Pach's positive-fraction selection lemma* [**Pac98**] (see [**KKP+15**] for related bounds). Here is Pach's result: Given an integer d, there exists a constant c_d such that for any set P of n points in $\mathbb{R}^d$, there exist a point $\boldsymbol{q} \in \mathbb{R}^d$, and $d+1$ disjoint subsets of P, say $P_1, \ldots, P_{d+1}$, such that $|P_i| \geq c_d \cdot n$ for all i and every simplex defined by a transversal of $P_1, \ldots, P_{d+1}$ contains $\boldsymbol{q}$. (By "transversal", we mean a set containing exactly one element from each P_i.)

Unfortunately the point $\boldsymbol{q}$ need not be an integer point; furthermore, the proof uses the so-called "second selection lemma" that currently does not exist for integer points (see Pach [**Pac98**] and Matoušek [**Mat02**, Chapter 9]). In Section 4, we strengthen the above theorem, such that, as a consequence, the theorem now extends to the integer case—indeed, to any scenario where one has points of high half-space depth in the following sense:

Given a finite set P of points in $\mathbb{R}^d$ and a point $\boldsymbol{q} \in \mathbb{R}^d$, we say that $\boldsymbol{q}$ is *of half-space depth* t *with respect to* P if any half-space containing $\boldsymbol{q}$ contains at least t points of P (when the context is clear, we will simply say that $\boldsymbol{q}$ is of *depth* t). Then here is our theorem.

THEOREM 4. *For any integer $d \geq 1$ and real number $\alpha \in (0,1]$, there exists a constant $c_{d,\alpha}$ such that the following holds. For any set P of n points in $\mathbb{R}^d$ and any point $\boldsymbol{q} \in \mathbb{R}^d$ of half-space depth at least $\alpha \cdot n$, there exist $d+1$ disjoint subsets of P, say $P_1, \ldots, P_{d+1}$, such that*

- $|P_i| \geq c_{d,\alpha} \cdot n$ *for* $i = 1, \ldots, d+1$, *and*
- *every simplex defined by a transversal of* $P_1, \ldots, P_{d+1}$ *contains* $\boldsymbol{q}$.

REMARK. Our proof yields a constant $c_{d,\alpha}$ whose value is exponentially decreasing with the dimension d.

Note that the existence of integer points of high half-space depth (Lemma 1) together with Theorem 4 implies the following integer version of the positive-fraction selection lemma.

COROLLARY 1. *Let P be a set of $n \geq d+1$ points in $\mathbb{Z}^d$. Then there exist a point $\boldsymbol{q} \in \mathbb{Z}^d$, and $d+1$ disjoint subsets of P, say $P_1, \ldots, P_{d+1}$, such that $|P_i| \geq c_{d,2^{-d}} \cdot n$ for all $i = 1, \ldots, d+1$, and the simplex defined by every transversal of $P_1, \ldots, P_{d+1}$ contains $\boldsymbol{q}$.*

REMARK. In particular, this implies that $\boldsymbol{q}$ belongs to many *distinct* Tverberg partitions—at least $\left(\lceil c_{d,2^{-d}} \cdot n \rceil !\right)^d$ *distinct* Tverberg partitions, with each such Tverberg partition containing $\lceil c_{d,2^{-d}} \cdot n \rceil$ sets. To see this, assume without loss of generality that the P_i are equally sized with cardinality equal to the lower bound given in our theorem, say $|P_i| = k$. Then consider them as columns of a matrix A. Each row of A is a transversal of the P_i and contains $\boldsymbol{q}$ in its convex hull. Thus, regardless of the ordering of the points in each column, we get a Tverberg partition of P into k subsets where each subset is a row of A. Matrices obtained by permuting independently the points in each column provide a same partition only if the same permutation is applied to each column, and thus there are at least $(k!)^d$ distinct Tverberg partitions in total.

Related results and organization of the paper. The problem of computing the Tverberg number for $\mathbb{Z}^d$ with $d \geq 3$ seems to be challenging. It has been identified as an interesting problem since the 1970's [**GS79**] and yet the following inequalities are almost all that is known about this problem: for the general case, De Loera et al. [**DLHRS17**] proved

$$2^d(m-1)+1 \leq \mathrm{Tv}(\mathbb{Z}^d, m) \leq d2^d(m-1)+1, \quad \text{for } d \geq 1 \text{ and } m \geq 2. \tag{2}$$

Two special cases get better bounds:

$$\mathrm{Tv}(\mathbb{Z}^3, 2) \leq 17 \qquad \text{and} \qquad 5 \cdot 2^{d-2}+1 \leq \mathrm{Tv}(\mathbb{Z}^d, 2) \quad \text{for } d \geq 1. \tag{3}$$

The left-hand side inequality is due to Bezdek and Blokhuis [**BB03**] and the right-hand side was proved by Doignon in his PhD thesis (and rediscovered by Onn).

Previously established bounds for the "mixed integer" case include the bounds for the Radon number (2-Tverberg number) found by Averkov and Weismantel [**AW12**].

$$2^j(k+1)+1 \leq \mathrm{Tv}(\mathbb{Z}^j \times \mathbb{R}^k, 2) \leq (j+k)2^j(k+1) - j - k + 2.$$

Later, De Loera et al. [**DLHRS17**] gave the following general bound for all mixed integer Tverberg numbers:

$$\mathrm{Tv}(\mathbb{Z}^j \times \mathbb{R}^k, m) \leq (j+k)2^j(m-1)(k+1)+1.$$

Note that (1) above is a simultaneous improvement of both of these if $k \geq 1$ or if $k = 0$ and $j \leq 1$.

Previous bounds and related work on more general S-Tverberg numbers can also be found in [**DLHRS17**], including the following bound for any discrete $S \subset \mathbb{R}^d$ and $m \geq 2$:

$$\mathrm{Tv}(S, m) \leq \mathrm{H}(S)(m-1)d + 1.$$

The following lemma about integer points of high half-space depth is used throughout the paper. See [**BO16**] for a proof and related results.

LEMMA 1. *Consider a multiset A of points in $\mathbb{Z}^d$. If $|A| \geq 2^d(m-1)+1$ (counting multiplicities), then there is a point $\boldsymbol{q} \in \mathbb{Z}^d$ of half-space depth m in A.*

The paper is organized as follows. In Section 1, we prove Theorem 1 using a somewhat similar strategy to Birch's proof of the planar case of the original Tverberg theorem [**Bir59**]. In Section 2, we prove Theorem 2 using techniques reminiscent of those in [**DLHRS17**]. In Section 3, we prove Theorem 3 and collect some consequences of the main theorems presented above, including (1). Finally, in Section 4, we prove Theorem 4 by proving a new lemma and adapting the methods of Pach in [**Pac98**].

1. Tverberg numbers over discrete subsets of $\mathbb{R}^2$: Proof of Theorem 1

We start with the proof of the special case $S = \mathbb{Z}^2$ (where $\mathrm{H}(\mathbb{Z}^2) = 4$) because it nicely illustrates the techniques of the more general proof of Theorem 1.

1.1. Proof of the special case $S = \mathbb{Z}^2$. The lower bound for the theorem is given by Inequality (2) given in the introduction. The upper bound will follow easily from the following two lemmas, the first covering the case $m \geq 3$ and the second the case $m = 2$.

LEMMA 2. *Consider a multiset A of points in $\mathbb{R}^2$ with $|A| \geq 4m-3$ and $m \geq 3$. If $\boldsymbol{p} \notin A$ is a point of depth m, then there is an m-Tverberg partition of A with $\boldsymbol{p}$ as Tverberg point.*

LEMMA 3. *Consider a multiset A of points in $\mathbb{R}^2$ with $|A| \geq 6$. If $\boldsymbol{p} \notin A$ is a point of depth two, then there is a Radon partition of A with $\boldsymbol{p}$ as Radon point.*

PROOF OF THEOREM 1 WHEN $S = \mathbb{Z}^2$. The inequality $\mathrm{Tv}(\mathbb{Z}^2, m) \geq 4m - 3$ is given by Inequality (2). The proof consists thus in establishing the upper bound.

We start with the case $m \geq 3$. Consider a multiset A of at least $4m-3$ points in $\mathbb{Z}^2$. By Lemma 1, A has an integer point $\boldsymbol{p}$ of depth m. If $\boldsymbol{p}$ is an element of A with multiplicity μ, then take the singletons $\{\boldsymbol{p}\}$ as μ of the sets in the Tverberg partition. Then $\boldsymbol{p}$ is a point of depth $m-\mu$ of the remaining $4m-\mu-3$ points. If $\mu \geq m$, we are done, and if $\mu = m-1$, the point $\boldsymbol{p}$ is in the convex hull of the remaining points and we take them to be the last set in the desired partition. If $\mu \leq m-3$, according to Lemma 2, there is an $(m-\mu)$-Tverberg partition of the remaining points with $\boldsymbol{p}$ as Tverberg point. There is thus an m-Tverberg partition of A with $\boldsymbol{p}$ as Tverberg point. The case $\mu = m-2$ is treated similarly with the help of Lemma 3 in place of Lemma 2.

For the case $m = 2$, we proceed similarly, except that we start with a set A of 6 points in $\mathbb{Z}^2$, in order to be able to apply Lemma 3. □

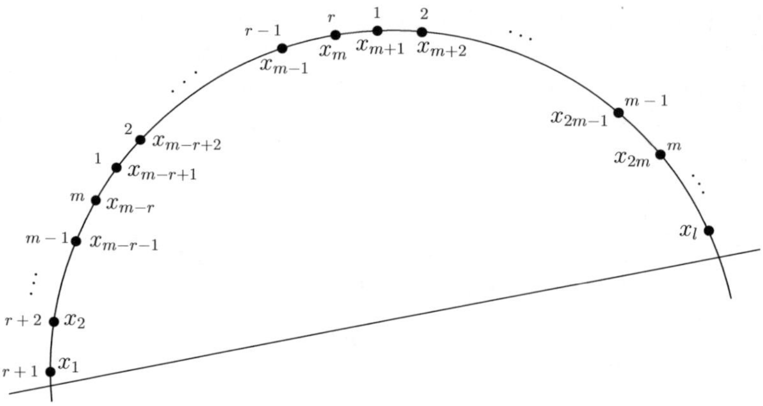

FIGURE 1. Labeling of the points in the half-plane H_-.

PROOF OF LEMMA 2. Since $\boldsymbol{p}$ is not in A, up to a radial projection, we can assume that the points of A are arranged in a circle around $\boldsymbol{p}$. This is without loss of generality since points in A contain $\boldsymbol{p}$ in their convex hull if and only if their radial projections from $\boldsymbol{p}$ do so.

Define q and r to be respectively the quotient and the remainder of the Euclidean division of $|A|$ by m. Define moreover e to be $\lceil \frac{r}{q} \rceil$.

Suppose first that $\boldsymbol{p}$ is a point of depth $m+e$. Since $qe \geq r$, we can choose k_i with $i \in [q]$, and $0 \leq k_i \leq e$, such that $k_1 + k_2 + \cdots + k_q = r$. Then we arbitrarily select a first point in A, and label clockwise the points with elements in $[m]$ according to the following pattern:

$$1, 2, \ldots, m, 1, 2, \ldots, k_1, 1, 2, \ldots, m, 1, 2, \ldots, k_2, \ldots, 1, 2, \ldots, m, 1, 2, \ldots, k_q.$$

Each half-plane delimited by a line passing through $\boldsymbol{p}$ contains at least $m+e$ consecutive points in this pattern and thus has at least one point with each of the m different labels. Partitioning the points so that each subset consists of all points with a fixed label, we therefore obtain an m-Tverberg partition with $\boldsymbol{p}$ as Tverberg point.

Suppose now that $\boldsymbol{p}$ is not a point of depth $m+e$. There is thus a closed half-plane H_+, delimited by a line passing through $\boldsymbol{p}$, with $|H_+ \cap A| < m+e$. The complementary closed half-plane to H_+, which we denote by H_-, is such that $|H_- \cap A| > 4m - 3 - (m+e)$. Define ℓ to be $|H_- \cap A|$. Since $e \leq \frac{m}{3}$, we have $\ell \geq 2m$. Denote the points in $H_- \cap A$ by $\boldsymbol{x}_1, \ldots, \boldsymbol{x}_\ell$, where the indices are increasing when we move clockwise. We label $\boldsymbol{x}_i$ with $r+i$ from $\boldsymbol{x}_1$ to $\boldsymbol{x}_{m-r}$, and then label $\boldsymbol{x}_{m-r+j}$ with j from $\boldsymbol{x}_{m-r+1}$ to $\boldsymbol{x}_m$. We then continue labeling the points of A, still moving clockwise, using labels $1, 2, \ldots, m, \ldots, 1, 2, \ldots m, 1, 2, \ldots r$. See Figure 1 for an illustration of the labeling scheme.

The labeling pattern is such that any sequence of m consecutive points either has all m labels, or contains the two consecutive points $\boldsymbol{x}_m$ and $\boldsymbol{x}_{m+1}$. Let us prove that any closed half-plane H delimited by a line passing through $\boldsymbol{p}$ contains at least one point with each label. Once this is proved, the conclusion will be immediate by taking as subsets of points those with same labels, as above.

If such an H does not simultaneously contain $\boldsymbol{x}_m$ and $\boldsymbol{x}_{m+1}$, then H contains at least one point with each label. Consider thus a closed half-plane H delimited by a line passing through $\boldsymbol{p}$ and containing $\boldsymbol{x}_m$ and $\boldsymbol{x}_{m+1}$. Note that according to Farkas' lemma ([**Sch03**] Theorem 5.3) , $\boldsymbol{x}_{m+1}$ cannot be separated from $\boldsymbol{x}_1$ and $\boldsymbol{x}_\ell$ by a line passing through $\boldsymbol{p}$, since they are all in H_-. This means that either H contains $\boldsymbol{x}_1, \boldsymbol{x}_2, \ldots, \boldsymbol{x}_{m+1}$, or H contains $\boldsymbol{x}_{m+1}, \boldsymbol{x}_{m+2}, \ldots, \boldsymbol{x}_\ell$. In any case, H contains a point with each label. □

PROOF OF LEMMA 3. As before, we assume that the points in A are arranged on a circle centered at $\boldsymbol{p}$. If $|A|$ is even, it clearly suffices to label the points in order, alternating between 1 and 2. We may therefore assume that $|A|$ is odd, and thus $|A| \geq 7$. If $\boldsymbol{p}$ is a point of depth three, it suffices to label the points alternating labels between 1 and 2, except with two consecutive points labeled 1. If $|A|$ is odd but $\boldsymbol{p}$ is not a point of depth three, then $|A| \geq 7$ and there is a half-plane H_+ containing $\boldsymbol{p}$ with $|H_+ \cap A| = 2$. The complementary half-plane H_- has $|H_- \cap A| \geq 5$ and we follow a similar strategy as in the second half of Lemma 2. Namely, we denote the points in $H_- \cap A$ by $\boldsymbol{x}_1, \ldots, \boldsymbol{x}_\ell$, where the indices are increasing when we move clockwise. Then we label $\boldsymbol{x}_1$ with 2, $\boldsymbol{x}_2$ with 1, $\boldsymbol{x}_3$ with 1, and $\boldsymbol{x}_4$ with 2. We continue this pattern for $\alpha \geq 5$, labeling $\boldsymbol{x}_\alpha$ with 1 if α is odd, and $\boldsymbol{x}_\alpha$ with 2 if α is even. For the remaining points in A we continue labeling clockwise, alternating between the labels 1 and 2.

The labeling pattern is such that any sequence of 2 consecutive points either has both labels, or contains the two consecutive points $\boldsymbol{x}_2$ and $\boldsymbol{x}_3$. As in Lemma 2 it suffices to show that any closed half-plane H delimited by a line passing through $\boldsymbol{p}$ contains at least one point with each label.

If such an H does not simultaneously contain $\boldsymbol{x}_2$ and $\boldsymbol{x}_3$, then H contains at least one point with each label. Consider thus a closed half-plane H delimited by a line passing through $\boldsymbol{p}$ and containing $\boldsymbol{x}_2$ and $\boldsymbol{x}_3$. Note that according to Farkas' lemma, $\boldsymbol{x}_3$ cannot be separated from $\boldsymbol{x}_1$ and $\boldsymbol{x}_4$ by a line passing through $\boldsymbol{p}$, since they are all in H_-. This means that either H contains $\boldsymbol{x}_1, \boldsymbol{x}_2, \boldsymbol{x}_3$, or H contains $\boldsymbol{x}_3$ and $\boldsymbol{x}_4$. In any case, H contains a point with each label. □

1.2. Proof of the general case. The proof of the general case is split into three lemmas addressing the lower bound, the upper bound for $\mathrm{H}(S) \geq 4$, and the upper bound for $\mathrm{H}(S) \leq 3$, respectively.

LEMMA 4. *For any discrete set $S \subset \mathbb{R}^2$ with finite Helly number $\mathrm{H}(S)$ and for all $m \geq 2$, we have $\mathrm{Tv}(S, m) \geq \mathrm{H}(S)(m-1)+1$.*

PROOF. It suffices to exhibit a subset $R \subseteq S$, of cardinality $|R| = \mathrm{H}(S)(m-1)$, with the property that no point in S is of half-space depth m with respect to R. By Lemma 2.6 in [**ADLS17**], there exists a set R' of $\mathrm{H}(S)$ points in S in convex position with the property that $\operatorname{conv}(R') \cap S = R'$. Let R be the multiset given by taking each point in R' with multiplicity $m-1$, so $|R| = \mathrm{H}(S)(m-1)$. No points of $S - R$ are in $\operatorname{conv}(R)$. Since R' was taken to be in convex position, for any point in R, there exists a line such that one side of that line has at most $m-1$ points in R. Thus S cannot contain a point of half-space depth m with respect to R. □

LEMMA 5. *For any discrete set $S \subset \mathbb{R}^2$ with finite Helly number $\mathrm{H}(S) \geq 4$ and for all $m \geq 2$, we have $\mathrm{Tv}(S, m) \leq \mathrm{H}(S)(m-1)+1$, except when $m = 2$ and $\mathrm{H}(S) = 4$ simultaneously, in which case we only have $\mathrm{Tv}(S, 2) \leq 6$.*

PROOF. The proof of Lemma 5 is the same as the proof of Theorem 1 for $S = \mathbb{Z}^2$, except that we use the following result (Theorem 2 in [**BO16**] with μ

being the uniform probability measure on A) in place of Lemma 1: For any discrete subset S of a Euclidean space with finite Helly number $\mathrm{H}(S)$, and any set $A \subseteq S$ with $|A| \geq \mathrm{H}(S)(m-1)+1$, there exists a point $\boldsymbol{p} \in S$ that is of depth m with respect to A. $\square$

LEMMA 6. *For a discrete set $S \subset \mathbb{R}^2$ with finite Helly number $\mathrm{H}(S) \leq 3$ and for all $m \geq 2$, we have $\mathrm{Tv}(S,m) \leq \mathrm{H}(S)(m-1)+1$.*

PROOF. The case $\mathrm{H}(S) = 1$ implies that S consists of a single point, so the result trivially follows. If $\mathrm{H}(S) = 2$, it must be that all points in S are collinear (as any set containing a non-degenerate triangle has Helly number at least 3), and thus we can take the median of any set with at least $2(m-1)+1$ points in S as the desired m-Tverberg point. Thus for the remainder of the proof we assume that $\mathrm{H}(S) = 3$.

Given any set A of $\mathrm{H}(S)(m-1)+1 = 3m-2$ points in S, there exists an m-Tverberg partition, say $\mathcal{P}$ by the classical Tverberg theorem. We denote by $K_1, \ldots, K_m$ the m convex hulls of the subsets in $\mathcal{P}$. As $\bigcap_{1 \leq i \leq m} K_i$ is a nonempty polygon, say Q, (possibly just a point or line segment) we pick an arbitrary vertex $\boldsymbol{q}$ of Q.

It suffices to show that $\boldsymbol{q} \in S$. We can assume that $\boldsymbol{q}$ is not a vertex of any K_i, since otherwise $\boldsymbol{q} \in A \subseteq S$.

Since $\boldsymbol{q}$ is a vertex of Q, it must be contained in a one dimensional face F_1 of at least one K_i. Since $\boldsymbol{q}$ is not a vertex of any K_i, in fact $\boldsymbol{q}$ is in the relative interior of F_1. For $\boldsymbol{q}$ to be a vertex of Q, it must also be in another one dimensional face, say F_2, of some other K_i, such that F_1 is not parallel to F_2. Moreover, $\boldsymbol{q}$ must be in the relative interior of F_2, and we also have $F_1 \cap F_2 = \{\boldsymbol{q}\}$.

Denote by $\{\boldsymbol{a}, \boldsymbol{b}\}$ and $\{\boldsymbol{c}, \boldsymbol{d}\}$ the vertices of F_1 and F_2 respectively. We have that $\boldsymbol{a}, \boldsymbol{b}, \boldsymbol{c}, \boldsymbol{d} \in S$ are the vertices of a convex quadrilateral with diagonals intersecting at $\boldsymbol{q}$, by the assumption that F_1 and F_2 are non parallel. Out of the four triangles $\mathrm{conv}(\{\boldsymbol{a},\boldsymbol{b},\boldsymbol{c}\}), \mathrm{conv}(\{\boldsymbol{a},\boldsymbol{b},\boldsymbol{d}\}), \mathrm{conv}(\{\boldsymbol{a},\boldsymbol{c},\boldsymbol{d}\}), \mathrm{conv}(\{\boldsymbol{b},\boldsymbol{c},\boldsymbol{d}\})$, any three have at least one vertex in common, and therefore intersect in S. Since $\mathrm{H}(S) = 3$, the four triangles therefore all intersect in S. This intersection point is $\boldsymbol{q}$, the point where the diagonals of the quadrilateral intersect. $\square$

2. Tverberg numbers over $\mathbb{Z}^3$: Proof of Theorem 2

As in Section 1, the proof of Theorem 2 will follow from some lemmas.

We state the following lemma without proof; it is a consequence, upon close inspection of the argument, of the proof of the main theorem in the already mentioned paper by Bezdek and Blokhuis [**BB03**].

LEMMA 7. *Consider a multiset A of at least 17 points in $\mathbb{R}^3$ and a point $\boldsymbol{p}$ of depth three in A. There is a bipartition of A into two subsets whose convex hulls contain $\boldsymbol{p}$.*

The next lemma will be proved later in this section.

LEMMA 8. *Consider a multiset A of points in $\mathbb{R}^3$ with $|A| \geq 4m+9$ and $m \geq 2$. If $\mathbf{p} \notin A$ is a point of depth $3m-3$, then there is an m-Tverberg partition of A with $\mathbf{p}$ as Tverberg point.*

The proof of Theorem 2 follows then from these two lemmas.

PROOF OF THEOREM 2. Consider a multiset A of $24m-31$ points in $\mathbb{Z}^3$. For the case $m=2$, note that Lemma 1 yields a point of depth three, and we can then apply Lemma 8 to obtain the result. Assume that $m \geq 3$. Applying Lemma 1, A has an integer point $\boldsymbol{p}$ of depth $3m-3$. If $\boldsymbol{p}$ is an element of A with multiplicity μ, then take the singletons $\{\boldsymbol{p}\}$ as μ of the sets in the Tverberg partition.

If $\mu \geq m$, we are done. If $\mu = m-1$, the point $\boldsymbol{p}$ is still in the convex hull of points in A, and thus we are done. And if $\mu \leq m-2$, the point $\boldsymbol{p}$ is still a point of depth $3m-\mu-3 \geq 3(m-\mu)-3$ of the remaining $24m-\mu-31 \geq 24(m-\mu)-31$ points. Thus, we may apply Lemma 8 to get an $(m-\mu)$-Tverberg partition of the remaining points, with $\boldsymbol{p}$ as Tverberg point, and conclude the result. □

PROOF OF LEMMA 8. Since $\boldsymbol{p}$ is not an element of A, after radially projecting the points in A, we can assume without loss of generality that the points of A are located on a sphere centered at $\boldsymbol{p}$, as in the proof of Lemma 2.

We claim that there exist pairwise disjoint subsets $X_1, X_2, \ldots, X_{m-2}$ of A, each having $\boldsymbol{p}$ in its convex hull and each being of cardinality at most 4. (Here "pairwise disjoint" means that each element of A is present in a number of X_i's that does not exceed its multiplicity in A.) We proceed by contradiction. Suppose that we can find at most $s < m-2$ such subsets X_i's. Then, by Carathéodory's theorem, $\boldsymbol{p}$ is not in the convex hull of the remaining points in A. Therefore there is a half-space H_+ delimited by a plane containing $\boldsymbol{p}$ such that $H_+ \cap A \subseteq \bigcup_{i=1}^{s} X_i$. On the other hand, since each X_i contains $\boldsymbol{p}$ in its convex hull (and we can assume the X_i are minimal with respect to containing $\boldsymbol{p}$), we have $|H_+ \cap X_i| \leq 3$ for all $i \in [s]$. Therefore $|H_+ \cap A| \leq |H_+ \cap (\bigcup_{i=1}^{s} X_i)| \leq 3s < 3(m-2)$, which is a contradiction since $\boldsymbol{p}$ is a point of depth $3m-3$ in A. There are thus $m-2$ disjoint subsets $X_1, X_2, \ldots, X_{m-2}$ as claimed.

Let X denote $\bigcup_{i=1}^{m-2} X_i$. Consider an arbitrary half-space H_+ delimited by a plane containing $\boldsymbol{p}$. Since $|H_+ \cap X_i| \leq 3$ for all i, we have $|H_+ \cap X| \leq 3(m-2)$. Furthermore $|H_+ \cap A| \geq 3m-3$, so $|H_+ \cap (A \setminus X)| \geq 3$. Since H_+ is arbitrary, $\boldsymbol{p}$ is a point of depth 3 of $A \setminus X$. Also, $|A \setminus X| \geq |A| - 4(m-2) \geq 17$, so Lemma 7 implies that $A \setminus X$ can be partitioned into two sets whose convex hulls contain $\boldsymbol{p}$. With the subsets X_i, we have therefore an m-Tverberg partition of A, with $\boldsymbol{p}$ as Tverberg point. □

3. Tverberg numbers over $S' \times \mathbb{R}^k$: Proof of Theorem 3

In this section, we prove Theorem 3. We adapt an approach by Mulzer and Werner [**MW13**, Lemma 2.3] and show how the results of our paper can be combined to improve known bounds and to determine new exact values for the Tverberg number in the mixed integer case, as well as better bounds for certain S-Tverberg numbers.

PROOF OF THEOREM 3. Let $t = \mathrm{Tv}(\mathbb{R}^k, m) = (m-1)(k+1)+1$. Choose a multiset A in $S' \times \mathbb{R}^k$ with $|A| \geq \mathrm{Tv}(S', t)$. It suffices to prove that A can be partitioned into m subsets whose convex hulls contain a common point in $S' \times \mathbb{R}^k$.

Let A' be the multiset projection of A onto S' (so that $|A'| = |A|$). Since $|A'| \geq \mathrm{Tv}(S', t)$, there is a partition of A' into t submultisets $Q_1', \ldots, Q_t'$ whose convex hulls contain a common point $\boldsymbol{q}$ in S'. The Q_i' are the projections onto S' of t disjoint subsets $Q_i \subseteq A$ forming a partition of A. For each $i \in [t]$, we can find a point $\boldsymbol{q}_i \in \mathrm{conv}(Q_i)$ projecting onto $\boldsymbol{q}$.

The t points $\boldsymbol{q}_1, \ldots, \boldsymbol{q}_t$ belong to $\{\boldsymbol{q}\} \times \mathbb{R}^k$. As $t = \mathrm{Tv}(\mathbb{R}^k, m)$, there exists a partition of $[t]$ into $I_1, \ldots, I_m$ and a point $\boldsymbol{p} \in \{\boldsymbol{q}\} \times \mathbb{R}^k$ such that $\boldsymbol{p} \in \operatorname{conv}\left(\bigcup_{i \in I_\ell} \boldsymbol{q}_i\right)$ for all $\ell \in [m]$. For each $\ell \in [m]$, define A_ℓ to be $\bigcup_{i \in I_\ell} Q_i$. We have, for each $\ell \in [m]$

$$\boldsymbol{p} \in \operatorname{conv}\left(\bigcup_{i \in I_\ell} \boldsymbol{q}_i\right) \subseteq \operatorname{conv}\left(\bigcup_{i \in I_\ell} \operatorname{conv}(Q_i)\right) = \operatorname{conv}(A_\ell)$$

and the A_ℓ form the desired partition. □

Here are the new bounds and exact values we get:

(1) $\mathrm{Tv}(\mathbb{Z} \times \mathbb{R}^k, m) = 2(m-1)(k+1)+1$.
(2) $\mathrm{Tv}(\mathbb{Z}^2 \times \mathbb{R}^k, m) = 4(m-1)(k+1)+1$.
(3) $\mathrm{Tv}(\mathbb{Z}^3 \times \mathbb{R}^k, m) \leq 24(m-1)(k+1)-7$.
(4) $2^j(m-1)(k+1)+1 \leq \mathrm{Tv}(\mathbb{Z}^j \times \mathbb{R}^k, m) \leq j2^j(m-1)(k+1)+1$.
(5) If $S' \subset \mathbb{R}^2$ has finite Helly number $\mathrm{H}(S')$, then

$$\mathrm{Tv}(S' \times \mathbb{R}^k, m) \leq \mathrm{H}(S')(m-1)(k+1)+1.$$

The lower bound in (4) is obtained by repeated applications of Proposition 1 below. The upper bounds follow from Theorem 3, combined with the fact that $\mathrm{Tv}(\mathbb{Z}, m) = 2m-1$, Theorem 1 for $S = \mathbb{Z}^2$, Theorem 2, the upper bound in Equation (2), and Theorem 1 respectively.

PROPOSITION 1. *Let j and k be two non-negative integers. Then we have*

$$\mathrm{Tv}(\mathbb{Z}^{j+1} \times \mathbb{R}^k, m) \geq 2\,\mathrm{Tv}(\mathbb{Z}^j \times \mathbb{R}^k, m) - 1.$$

We prove Proposition 1 by following the idea of the proof of Proposition 2.1 in [**Onn91**].

PROOF OF PROPOSITION 1. Assume toward a contradiction that

$$\mathrm{Tv}(\mathbb{Z}^{j+1} \times \mathbb{R}^k, m) \leq 2\,\mathrm{Tv}(\mathbb{Z}^j \times \mathbb{R}^k, m) - 2.$$

Choose A to be a set of $\mathrm{Tv}(\mathbb{Z}^j \times \mathbb{R}^k, m) - 1$ points in $\mathbb{Z}^j \times \mathbb{R}^k$ with no m-Tverberg partition. Let $A_i = \{(i, \boldsymbol{a}) \colon \boldsymbol{a} \in A\}$, for $i \in \{0,1\}$. Since $A_0 \cup A_1 \subset \mathbb{Z}^{j+1} \times \mathbb{R}^k$ has cardinality $2\,\mathrm{Tv}(\mathbb{Z}^j \times \mathbb{R}^k, m) - 2$, there exists an m-Tverberg partition $Y_1, Y_2, \ldots, Y_m$ of $A_0 \cup A_1$ with $\boldsymbol{p} \in \bigcap_{i \in [m]} \operatorname{conv}(Y_i)$. Furthermore $\boldsymbol{p}$ is in $\mathbb{Z}^{j+1} \times \mathbb{R}^k$. That implies either $\boldsymbol{p} \in \operatorname{conv}(A_0)$ or $\boldsymbol{p} \in \operatorname{conv}(A_1)$. In either case A_0 or A_1 has an m-Tverberg partition, a contradiction with our choice of A. □

4. A generalized fraction selection lemma: Proof of Theorem 4

Our proof relies on the simplicial partition theorem of Matoušek, used in a similar manner as in [**MR17**], which states the following.

THEOREM 5 ([**Mat92**]; see also [**Cha00**]). *Given an integer $d \geq 1$ and a parameter r, there exists a constant $c_d \geq 1$ such that for any set P of n points in $\mathbb{R}^d$, there exists an integer s and a partition $\{P_1, \ldots, P_s\}$ of P such that*

- *for each $i = 1, \ldots, s$, $\frac{n}{r} \leq |P_i| \leq \frac{2n}{r}$, and*
- *any hyperplane intersects the convex hull of less than $c_d \cdot r^{1-\frac{1}{d}}$ sets of the partition.*

The constant c_d is independent of P and depends only on d.

We now prove the following key lemma.

LEMMA 9. *For any integer $d \geq 1$, there exists a constant c_d such that the following holds. For any set P of n points in $\mathbb{R}^d$ and a real number $\alpha \in (0,1]$, there exists a partition $\mathcal{P} = \{P_1, \ldots, P_r\}$, $r = \left\lceil \left(\frac{4c_d}{\alpha}\right)^d \right\rceil$, of P such that*

- $\frac{n}{2r} \leq |P_i| \leq \frac{2n}{r}$ *for each $i = 1, \ldots, r$, and*
- *the convex hull of any transversal Q of $\mathcal{P}$ contains all points in $\mathbb{R}^d$ of half-space depth at least $\alpha \cdot n$.*

PROOF. Apply the simplicial partition theorem (Theorem 5) to P with $r = \left\lceil \left(\frac{4c_d}{\alpha}\right)^d \right\rceil$, and let the resulting partition be $\{P'_1, \ldots, P'_s\}$. Note that as $\frac{n}{r} \leq |P'_i| \leq \frac{2n}{r}$ for each $i = 1, \ldots, s$, we have $\frac{r}{2} \leq s \leq r$. Now partition arbitrarily each of the $r - s$ most numerous sets in $\{P'_1, \ldots, P'_s\}$ into two equal parts, and let the resulting partition be $\{P_1, \ldots, P_r\}$. Clearly each set of this partition has size in the interval $\left[\frac{n}{2r}, \frac{2n}{r}\right]$. This proves the first part. Note also that each hyperplane intersects the convex hull of at most twice as many sets, i.e., less than $2c_d \cdot r^{1-\frac{1}{d}}$ sets of the partition $\{P_1, \ldots, P_r\}$.

To see the second part, let $\boldsymbol{c}$ be any point of half-space depth at least $\alpha \cdot n$, and Q any transversal of $\mathcal{P}$. For contradiction, assume that $\boldsymbol{c} \notin \operatorname{conv}(Q)$. Then there exists a hyperplane H containing $\boldsymbol{c}$ in one of its two open half-spaces, say H^-, and containing $\operatorname{conv}(Q)$ in the half-space H^+. We will show that then there exists an index $i \in \{1, \ldots, r\}$ such that $P_i \subseteq H^-$. But then $P_i \cap Q = \varnothing$, a contradiction to the fact that Q is a transversal of $\mathcal{P}$.

It remains to show the existence of a set $P_i \in \mathcal{P}$ such that $P_i \subseteq H^-$. Towards this, we bound $|P \cap H^-|$. Each point of P lying in H^- belongs to a set $P' \in \mathcal{P}$ such that either

- $P' \subseteq H^-$, in which case we are done, or
- $P' \not\subseteq H^-$. As H^- contains at least one point of P', we must have $\operatorname{conv}(P') \cap H \neq \varnothing$. As argued earlier, there are less than $2c_d \cdot r^{1-\frac{1}{d}}$ such sets.

Thus we have

$$|P \cap H^-| < 2c_d \cdot r^{1-\frac{1}{d}} \cdot \frac{2n}{r} = \frac{4c_d \cdot n}{\left\lceil \left(\frac{4c_d}{\alpha}\right)^d \right\rceil^{\frac{1}{d}}} \leq \alpha \cdot n. \tag{4}$$

On the other hand, as $\boldsymbol{c}$ has half-space depth at least $\alpha \cdot n$ and $\boldsymbol{c} \in H^-$, we have $|P \cap H^-| \geq \alpha n$, a contradiction to inequality (4). □

REMARK. In particular, for $r = \left\lceil \left(\frac{4c_d}{\alpha}\right)^d \right\rceil$, there exist at least $\left(\frac{n}{2r}\right)^r$ r-sized subsets, each of whose convex hull contains all integer points of depth at least $\alpha \cdot n$.

PROOF OF THEOREM 4. Given the point set P in $\mathbb{R}^d$ and a point $\boldsymbol{q} \in \mathbb{R}^d$ of half-space depth $\alpha \cdot n$, apply Lemma 9 with P and α to get a partition consisting of $r \geq d+1$ sets, where $r = \left\lceil \left(\frac{4c_d}{\alpha}\right)^d \right\rceil$. By discarding at most $\frac{n}{2}$ points of P, we can derive a partition on the remaining points of P, say $\mathcal{P} = \{P_1, \ldots, P_r\}$, such that the P_i's are equal-sized disjoint subsets of P, i.e., $|P_i| = \frac{n}{2r}$ for all $i = 1, \ldots, r$. Furthermore, every transversal of $\mathcal{P}$ contains all points in $\mathbb{R}^d$ of half-space depth at least αn, and thus $\boldsymbol{q}$.

For each transversal Q of $\mathcal{P}$, the point $\boldsymbol{q}$ lies in the convex hull of Q, and by Carathéodory's theorem, there exists a $(d+1)$-sized subset of Q whose convex hull also contains $\boldsymbol{q}$. By the pigeonhole principle, there must exist $d+1$ sets of $\mathcal{P}$, say the sets $P_1, \ldots, P_{d+1}$, such that at least

$$\frac{\left(\frac{n}{2r}\right)^r}{\binom{r}{d+1}\left(\frac{n}{2r}\right)^{r-(d+1)}} \geq \frac{\left(\frac{n}{2r}\right)^{d+1}}{\left(\frac{er}{d+1}\right)^{d+1}} = \frac{1}{\left(\frac{er}{d+1}\right)^{d+1}} \cdot \prod_{i=1}^{d+1} |P_i|. \tag{5}$$

distinct transversals of $\{P_1, \ldots, P_{d+1}\}$ contain $\boldsymbol{q}$.

The rest of the proof follows the one of Pach [**Pac98**]. In brief, we view the P_i's as parts of a $(d+1)$-partite hypergraph with vertices corresponding to points in P and a hyperedge corresponding to each transversal of $\mathcal{P}$ containing $\boldsymbol{q}$. As there are $\Omega\left(n^{d+1}\right)$ such transversals by inequality (5), we apply a weak form of the hypergraph version of Szemerédi's regularity lemma (see [**Mat92**] Theorem 9.4.1) to derive the existence of constant-fraction sized subsets $P_1' \subseteq P_1, \ldots, P_{d+1}' \subseteq P_{d+1}$ such that the following is true, for some constant c_d':

> for *any* $P_1'' \subseteq P_1', \ldots, P_{d+1}'' \subseteq P_{d+1}'$, with $|P_i''| \geq c_d' \cdot |P_i'|$ for $i = 1, \ldots, d+1$, we have the property that there exists at least one transversal of $\left\{P_1'', \ldots, P_{d+1}''\right\}$ whose convex hull contains $\boldsymbol{q}$.

Then the so-called same-type lemma ([**BV98**] Theorem 2) applied to $\left\{P_1', \ldots, P_{d+1}', \{\boldsymbol{q}\}\right\}$ gives constant-fraction sized subsets $X_1 \subseteq P_1', \ldots, X_{d+1} \subseteq P_{d+1}'$ such that either the convex hull of *each* transversal of $\{X_1, \ldots, X_{d+1}\}$ contains $\boldsymbol{q}$ or none of them do.

We can set up the parameters for the same-type lemma and the weak regularity lemma such that $|X_i| \geq c_d' \cdot |P_i'|$, for all $i = 1, \ldots, d+1$. Then the weak regularity lemma implies that there exists at least one transversal of $\{X_1, \ldots, X_{d+1}\}$ whose convex hull contains $\boldsymbol{q}$. This implies that the convex hull of *each* transversal of $\{X_1, \ldots, X_{d+1}\}$ contains $\boldsymbol{q}$. These are the required subsets.

The size of each X_i is a constant-fraction of n, say $|X_i| \geq c_{d,\alpha} \cdot n$, where the constant $c_{d,\alpha}$ depends on the constants in inequality (5), in the weak regularity lemma and in the same-type lemma. All of these depend only on α and d. □

Acknowledgments

We would like to thank the anonymous referees for their excellent detailed comments. We appreciate their thoroughness and careful reading. Last, but not least, we are grateful for the information and comments we received from N. Amenta, G. Averkov, A. Basu, J.P. Doignon, P. Soberón, D. Oliveros, and S. Onn. In particular we thank Pablo Soberón for noticing that the same methods used in an earlier version could be extended to yield Theorem 1.

References

[ADLS17] N. Amenta, J. A. De Loera, and P. Soberón, *Helly's theorem: new variations and applications*, Algebraic and geometric methods in discrete mathematics, Contemp. Math., vol. 685, Amer. Math. Soc., Providence, RI, 2017, pp. 55–95, DOI 10.1090/conm/685. MR3625571

[AGS+17] G. Averkov, B. González Merino, I. Paschke, M. Schymura, and S. Weltge, *Tight bounds on discrete quantitative Helly numbers*, Adv. in Appl. Math. **89** (2017), 76–101, DOI 10.1016/j.aam.2017.04.003. MR3655733

[Ave13] G. Averkov, *On maximal S-free sets and the Helly number for the family of S-convex sets*, SIAM J. Discrete Math. **27** (2013), no. 3, 1610–1624, DOI 10.1137/110850463. MR3106473

[AW12] G. Averkov and R. Weismantel, *Transversal numbers over subsets of linear spaces*, Adv. Geom. **12** (2012), no. 1, 19–28, DOI 10.1515/advgeom.2011.028. MR2911157

[BB03] K. Bezdek and A. Blokhuis, *The Radon number of the three-dimensional integer lattice*, Discrete Comput. Geom. **30** (2003), no. 2, 181–184, DOI 10.1007/s00454-003-0003-8. U.S.-Hungarian Workshops on Discrete Geometry and Convexity (Budapest, 1999/Auburn, AL, 2000). MR2007959

[Bir59] B. J. Birch, *On 3N points in a plane*, Proc. Cambridge Philos. Soc. **55** (1959), 289–293, DOI 10.1017/s0305004100034071. MR109315

[BO16] A. Basu and T. Oertel, *Centerpoints: a link between optimization and convex geometry*, Integer programming and combinatorial optimization, Lecture Notes in Comput. Sci., vol. 9682, Springer, [Cham], 2016, pp. 14–25, DOI 10.1007/978-3-319-33461-5_2. MR3534718

[BS18] I. Bárány and P. Soberón, *Tverberg's theorem is 50 years old: a survey*, Bull. Amer. Math. Soc. (N.S.) **55** (2018), no. 4, 459–492, DOI 10.1090/bull/1634. MR3854075

[BV98] I. Bárány and P. Valtr, *A positive fraction Erdős-Szekeres theorem*, Discrete Comput. Geom. **19** (1998), no. 3, Special Issue, 335–342, DOI 10.1007/PL00009350. Dedicated to the memory of Paul Erdős. MR1608874

[Cha00] B. Chazelle, *The discrepancy method*, Cambridge University Press, Cambridge, 2000. Randomness and complexity. MR1779341

[DLGMM19] J. A. De Loera, X. Goaoc, F. Meunier, and N. H. Mustafa, *The discrete yet ubiquitous theorems of Carathéodory, Helly, Sperner, Tucker, and Tverberg*, Bull. Amer. Math. Soc. (N.S.) **56** (2019), no. 3, 415–511, DOI 10.1090/bull/1653. MR3974609

[DLH19] J. A. De Loera and T. A. Hogan, *Stochastic Tverberg theorems and their applications in multi-class logistic regression, data separability, and centerpoints of data*, 2019, `arXiv:1907.09698`.

[DLHRS17] J. A. De Loera, R. N. La Haye, D. Rolnick, and P. Soberón, *Quantitative Tverberg theorems over lattices and other discrete sets*, Discrete Comput. Geom. **58** (2017), no. 2, 435–448, DOI 10.1007/s00454-016-9858-3. MR3679944

[Doi73] J. P. Doignon, *Convexity in cristallographical lattices*, J. Geom. **3** (1973), 71–85, DOI 10.1007/BF01949705. MR387090

[Doi75] J. P. Doignon, *Segments et ensembles convexes*, Ph.D. thesis, Université Libre de Bruxelles, 1975.

[Eck00] J. Eckhoff, *The partition conjecture*, Discrete Math. **221** (2000), no. 1-3, 61–78, DOI 10.1016/S0012-365X(99)00386-6. Selected papers in honor of Ludwig Danzer. MR1778908

[GS79] P. M. Gruber and R. Schneider, *Problems in geometric convexity*, Contributions to geometry (Proc. Geom. Sympos., Siegen, 1978), Birkhäuser, Basel-Boston, Mass., 1979, pp. 255–278. MR568503

[KKP+15] R. Karasev, J. Kynčl, P. Paták, Z. Patáková, and M. Tancer, *Bounds for Pach's selection theorem and for the minimum solid angle in a simplex*, Discrete Comput. Geom. **54** (2015), no. 3, 610–636, DOI 10.1007/s00454-015-9720-z. MR3392968

[Lev51] F. W. Levi, *On Helly's theorem and the axioms of convexity.* A, J. Indian Math. Soc. (N.S.) **15** (1951), 65–76. MR43487

[Mat92] J. Matoušek, *Efficient partition trees*, Discrete Comput. Geom. **8** (1992), no. 3, 315–334, DOI 10.1007/BF02293051. ACM Symposium on Computational Geometry (North Conway, NH, 1991). MR1174360

[Mat02] J. Matoušek, *Lectures in Discrete Geometry*, Springer-Verlag, New York, NY, 2002.

[MR17] N. H. Mustafa and S. Ray, *ε-Mnets: hitting geometric set systems with subsets*, Discrete Comput. Geom. **57** (2017), no. 3, 625–640, DOI 10.1007/s00454-016-9845-8. MR3614776

[MW13] W. Mulzer and D. Werner, *Approximating Tverberg points in linear time for any fixed dimension*, Discrete Comput. Geom. **50** (2013), no. 2, 520–535, DOI 10.1007/s00454-013-9528-7. MR3090531

[Onn91] S. Onn, *On the geometry and computational complexity of Radon partitions in the integer lattice*, SIAM J. Discrete Math. **4** (1991), no. 3, 436–446, DOI 10.1137/0404039. MR1105949

[Pac98] J. Pach, *A Tverberg-type result on multicolored simplices*, Comput. Geom. **10** (1998), no. 2, 71–76, DOI 10.1016/S0925-7721(97)00022-9. MR1614605

[RH99] P. J. Rousseeuw and M. Hubert, *Regression depth*, J. Amer. Statist. Assoc. **94** (1999), no. 446, 388–433, DOI 10.2307/2670155. With discussion and a reply by the authors and Stefan Van Aelst. MR1702314

[Sch03] A. Schrijver, *Combinatorial optimization. Polyhedra and efficiency. Vol. A*, Algorithms and Combinatorics, vol. 24, Springer-Verlag, Berlin, 2003. Paths, flows, matchings; Chapters 1–38. MR1956924

[Tve66] H. Tverberg, *A generalization of Radon's theorem*, J. London Math. Soc. **41** (1966), 123–128, DOI 10.1112/jlms/s1-41.1.123. MR187147

[vdV93] M. L. J. van de Vel, *Theory of convex structures*, North-Holland Mathematical Library, vol. 50, North-Holland Publishing Co., Amsterdam, 1993. MR1234493

Department of Mathematics, University of California, Davis, California
Email address: deloera@math.ucdavis.edu

Department of Mathematics, University of California, Davis, California
Email address: tahogan@math.ucdavis.edu

Université Paris-Est, CERMICS, École Nationale des Ponts et Chaussées, Marne-la-Vallée, France
Email address: frederic.meunier@enpc.fr

Université Paris-Est, Laboratoire d'Informatique Gaspard-Monge, ESIEE Paris, Marne-la-Vallée, France
Email address: mustafan@esiee.fr

Contemporary Mathematics
Volume **764**, 2021
https://doi.org/10.1090/conm/764/15336

The vertices of primitive zonotopes

Antoine Deza, Lionel Pournin, and Rado Rakotonarivo

ABSTRACT. Primitive zonotopes arise naturally in various research areas, such as discrete geometry, combinatorial optimization, and theoretical physics. We provide geometric and combinatorial properties for these polytopes that allow us to estimate the size of their vertex sets. In particular, we show that the logarithm of the complexity of convex matroid optimization is quadratic, and we improve the bounds on the number of generalized retarded functions from quantum field theory. We also give a sharp asymptotic estimate for the number of vertices of a primitive zonotope that, in terms of Minkowski sums, is an intermediate between the permutohedra of types A and B.

1. Introduction

For any positive integers d and p, denote by $\mathcal{G}_q(d,p)$ the set of the points g in $\mathbb{Z}^d\backslash\{0\}$ whose greater common divisor of coordinates is equal to 1, whose last non-zero coordinate is positive, and whose q-norm satisfies $\|g\|_q \leq p$. Consider the Minkowski sum $H_q(d,p)$ of the segments incident to 0 on one end and to a point in $\mathcal{G}_q(d,p)$ on the other. The resulting polytopes, introduced in [**4, 5**], are called *primitive zonotopes*. The elements of $\mathcal{G}_q(d,p)$ will be referred to as the *generators* of $H_q(d,p)$. In [**4, 5**], the first non-zero coordinate of the generators of $H_q(d,p)$ is positive instead of the last. However, the polytopes resulting from these two definitions are translates of one another, and the convention we take here will simplify the exposition. A second family of primitive zonotopes, denoted by $H_q^+(d,p)$, is introduced in [**4, 5**]. The set $\mathcal{G}_q^+(d,p)$ of their generators is made up of the points in $\mathcal{G}_q(d,p)$ whose all coordinates are non-negative. As above, $H_q^+(d,p)$ is the Minkowski sum of the segments between 0 and a generator. Observe that primitive zonotopes can be equivalently defined as the set of all the linear combinations of their generators with coefficients in the unit segment $[0,1]$, or as the convex hull of all the possible subsums of their generators.

We estimate the number of vertices of primitive zonotopes. Our results follow from geometric and combinatorial properties of these polytopes that we establish

2010 *Mathematics Subject Classification.* Primary 52B20, 52B05; Secondary 05B35, 81T28.
Key words and phrases. Zonotopes, Minkowski sums, convex matroid optimization, generalized retarded functions, maximal unbalanced families, thermal field theory.
The first author was partially supported by the Natural Sciences and Engineering Research Council of Canada Discovery Grant program (RGPIN-2015-06163).
The second and third authors were partially supported by the ANR project SoS (Structures on Surfaces), grant number ANR-17-CE40-0033, and by the PHC project number 42703TD.

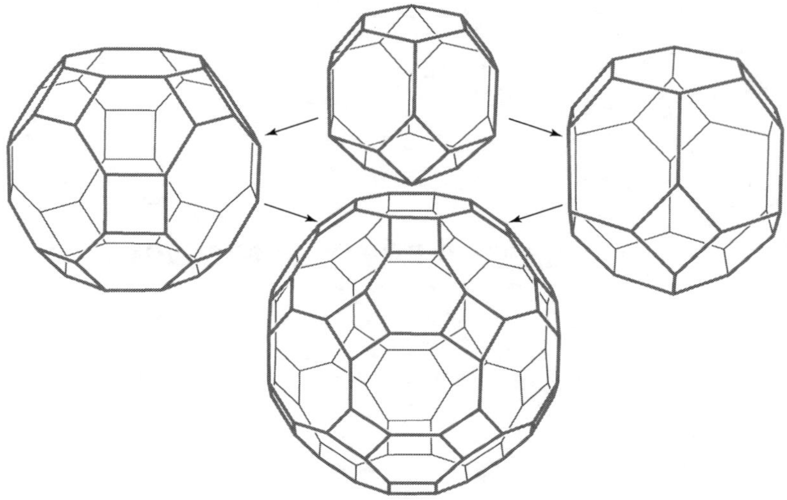

FIGURE 1. The primitive zonotopes $H_1^+(3,2)$ (top), $H_\infty^+(3,1)$ (right), $H_1(3,2)$ (left), and $H_\infty(3,1)$ (bottom) ordered by the inclusion of their sets of generators.

in Section 2. Denote by $a_q(d,p)$ the number of vertices of $H_q^+(d,p)$ whose none of the coordinates is equal to 0. Further denote $a_q(0,p) = 1$ as a convention. The first result of Section 2 is the following expression for the number $f_0\big(H_q^+(d,p)\big)$ of vertices of the primitive zonotope $H_q^+(d,p)$.

THEOREM 1.1.

$$f_0\big(H_q^+(d,p)\big) = \sum_{i=0}^{d} \binom{d}{i} a_q(i,p).$$

While the proof of Theorem 1.1 is rather straightforward, we shall see that it admits several interesting consequences. The remainder of Section 2 is devoted to studying the geometry of the primitive zonotopes $H_1(d,2)$, $H_1^+(d,2)$, $H_\infty(d,1)$, and $H_\infty^+(d,2)$, whose coordinates of generators belong to $\{-1,0,1\}$. These primitive zonotopes, depicted in Fig. 1 when $d=3$, are of particular interest in various research areas and exhibit additional structural properties. For instance, slicing $H_1^+(d,2)$ with the hyperplanes of $\mathbb{R}^d$ wherein the last coordinate is a fixed integer results in the Minkowski sums of $H_1^+(d-1,2)$ with the $(d-1)$-dimensional hypersimplices. In Section 3, we derive from Theorem 1.1 an implicit expression for the number of vertices of $H_1^+(d,2)$ that allows for a sharp asymptotic estimate.

THEOREM 1.2.

$$f_0\big(H_1^+(d,2)\big) \sim \frac{d!}{(\ln 2)^{d+1}}.$$

In terms of Minkowski sums, $H_1^+(d,2)$ can be thought of as an intermediate between the permutohedra of types A and B. Indeed, as mentioned in [**4**], $H_1^+(d,2)$ is the Minkowski sum of the permutohedron of type A with the hypercube $[0,1]^d$. Moreover, the primitive zonotope $H_1(d,2)$ is homothetic to the permutohedron of type B [**4**] and since $\mathcal{G}_1^+(d,2)$ is a subset of $\mathcal{G}_1(d,2)$, it can be obtained as the Minkowski sum of $H_1^+(d,2)$ with a zonotope. This is reflected in the estimate given

d	$f_0(H^+_\infty(d,1))$	$a_\infty(d,1)$
1	2	1
2	6	3
3	32	19
4	370	271
5	11 292	9 711
6	1 066 044	1 003 281
7	347 326 352	340 089 233
8	419 172 756 930	416 423 387 255

TABLE 1

by Theorem 1.2, that lies between the number of vertices of the permutohedron of type A ($d!$) and that of the permutohedron of type B ($d!2^d$).

As shown in [**4**], the worst case complexity of d-criteria, p-bounded convex matroid optimization (see [**11**–**13**]) is equal to the number of vertices of the primitive zonotope $H_\infty(d,p)$. It is shown in [**4**, **11**] that

$$d!2^d \leq f_0(H_\infty(d,1)) \leq O\left(3^{d(d-1)}\right).$$

We improve the lower bound in Section 4 and the upper bound in Section 5.

THEOREM 1.3.

$$\prod_{i=0}^{d-1}(3^i+1) \leq f_0(H_\infty(d,1)) \leq 2\left(3^{d-1}+1\right)^{d-1}.$$

The number of vertices of $H^+_\infty(d,1)$ appears in several contexts [**2**, **3**, **7**, **9**, **14**]. It is, for instance, the number of generalized retarded functions on $d+1$ variables in quantum field theory [**7**] and the number of maximal unbalanced families of subsets of $\{1,2,...,d+1\}$ in combinatorics [**2**]. The values of $f_0(H^+_\infty(d,1))$ have been computed up to $d=8$ [**7**, **9**, **15**], and can be found in the Online Encyclopedia of Integer Sequences. We report them in Table 1 as well as the corresponding values of $a_\infty(d,1)$, obtained from Theorem 1.1. It is shown in [**2**] that

$$\prod_{i=0}^{d-1}(2^i+1) \leq f_0\left(H^+_\infty(d,1)\right) < 2^{d^2}.$$

We will refine both of these bounds. While the improvement is not significant, this illustrates the benefits of looking at the problem in terms of primitive zonotopes. Our lower bound, established in Section 4, is another consequence of Theorem 1.1 and our upper bound, proven in Section 5, is obtained by identifying large regions of the hypercube $[0,2^{d-1}]^d$ that do not contain any vertex of $H^+_\infty(d,1)$.

THEOREM 1.4. *For any* $d \geq 3$,

$$6\prod_{i=1}^{d-2}(2^{i+1}+i) \leq f_0\left(H^+_\infty(d,1)\right) \leq 2(d+4)2^{(d-1)(d-2)}.$$

Recently, Gutekunst, Mészáros, and Petersen [**8**] have further significantly improved the lower bound on $f_0(H^+_\infty(d,1))$ and shown that the upper bound is the right asymptotic estimate. Note that the number of vertices of $H^+_\infty(d,1)$ is the sum of the Betti numbers of its dual hyperplane arrangement. The first two non-trivial of these Betti numbers have been recently determined by Lukas Kühne [**10**].

2. Geometric and combinatorial properties

We denote by x_1 to x_d the coordinates of a point x in $\mathbb{R}^d$. Moreover, if $i < d$, we will think of $\mathbb{R}^i$ as the subspace of $\mathbb{R}^d$ spanned by the first i coordinates.

PROPOSITION 2.1. *The intersection of $H_q^+(d,p)$ with a facet of the cone $[0,+\infty[^d$ is isometric to $H_q^+(d-1,p)$ by a permutation of the coordinates.*

PROOF. By definition, the intersection of $H_q^+(d,p)$ with the cone $[0,+\infty[^{d-1}$ is precisely $H_q^+(d-1,p)$. As shown in [**4**], $H_q^+(d,p)$ is invariant under any permutation of the coordinates and the desired result holds. □

PROOF OF THEOREM 1.1. Consider an i-dimensional face F of $[0,+\infty[^d$. Using Proposition 2.1 recursively, one obtains that the intersection of $H_q^+(d,p)$ with F can be recovered from $H_q^+(i,p)$ by a permutation of the coordinates. Here, we will take the convention that $H_q^+(0,p)$ is equal to $\{0\}$. As a consequence, the number of vertices of $H_q^+(d,p)$ contained in F, but not in any face of $[0,+\infty[^d$ of dimension less than i, is exactly $a_q(i,p)$. In particular, the face complex of $[0,+\infty[^d$ induces a partition of the vertex set of $H_q^+(d,p)$ into subsets of size $a_q(i,p)$, where i ranges from 0 to d. In this partition, the number of subsets of size $a_q(i,p)$ is equal to the number of i-dimensional faces of the cone $[0,+\infty[^d$. Since this cone has $\binom{d}{i}$ faces of dimension i, we obtain the desired result. □

Consider the intersection of a primitive zonotope with the hyperplane $S(d,h)$ of $\mathbb{R}^d$ made up of all the points x such that x_d is equal to an integer h. We will characterize this intersection as a Minkowski sum for the families of primitive zonotopes whose set of generators is a subset of $\{-1,0,1\}^d$.

PROPOSITION 2.2. *Consider a primitive zonotope Z. If the set of the generators of Z is a subset of $\{-1,0,1\}^d$ then, for any integer h, the polytope obtained as the intersection of Z and $S(d,h)$ shares all of its vertices with Z.*

PROOF. First consider a vertex v of $Z \cap S(d,h)$ and assume that v is not a vertex of Z. In this case, v must be the intersection of the hyperplane $S(d,h)$ with an edge of Z whose vertices a and b satisfy $a_d < h$ and $b_d > h$. In particular, $b_d - a_d \geq 2$. By the definition of primitive zonotopes, $b-a$ is a generator of Z, and therefore, the set of the generators of Z cannot be a subset of $\{-1,0,1\}^d$. □

There are four families of primitive zonotopes whose set of generators is a subset of $\{-1,0,1\}$: $H_1(d,2)$, $H_1^+(d,2)$, $H_q(d,1)$, and $H_q^+(d,1)$. Note that the latter two families are distinct only when $q=\infty$ and, when they coincide, they are equal to the hypercube $[0,1]^d$. In the remainder of the section, we consider any of these four families and denote by $H(d)$ its d-dimensional member. We will characterize the polytopes obtained by slicing $H(d)$ with the hyperplane $S(d,h)$ as the Minkowski sums of $H(d-1)$ with well-defined polytopes where, as a convention, $H(0)$ is taken equal to $\{0\}$. Denote by $\kappa(H(d))$ the largest possible value for the last coordinate of a vertex of $H(d)$. In other words, $\kappa(H(d))$ is the sum of the last coordinates of the generators of $H(d)$. For instance, $\kappa\big(H_1^+(d,2)\big)$ is equal to d, $\kappa(H_\infty^+(d,1))$ to 2^{d-1}, and $\kappa(H_\infty(d,1))$ to 3^{d-1} [4]. Further note that $\kappa\big(H_q^+(d,p)\big)$ is the smallest integer r such that $H_q^+(d,p)$ is contained in the hypercube $[0,r]^d$.

LEMMA 2.3. *For any integer h such that $0 < h \leq \kappa(H(d))$, the intersection of $H(d)$ with $S(d,h)$ is the Minkowski sum of $H(d-1)$ with the convex hull of the sums of exactly h generators of $H(d)$ whose last coordinate is equal to* 1.

PROOF. Denote by $\mathcal{G}(d)$ the set of the generators of $H(d)$. Recall that $\mathbb{R}^{d-1}$ is identified with the hyperplane of $\mathbb{R}^d$ spanned by the first $d-1$ coordinates. In particular, the generators of $H(d-1)$ coincide with the generators of $H(d)$ whose last coordinate is equal to 0. Therefore, we think of $\mathcal{G}(d-1)$ as a subset of $\mathcal{G}(d)$. As mentioned above, $\mathcal{G}(d)$ is a subset of $\{-1,0,1\}^d$. By the definition of primitive zonotopes, the last non-zero coordinate of any generator of $H(d)$ is positive. Hence, $\mathcal{G}(d)\backslash\mathcal{G}(d-1)$ is exactly the set of the points in $\mathcal{G}(d)$ whose last coordinate is equal to 1. Now pick an integer h such that $0 \leq h \leq \kappa(H(d))$ and denote by P the convex hull of the sums of exactly h elements of $\mathcal{G}(d)\backslash\mathcal{G}(d-1)$.

Recall that the primitive zonotope $H(d)$ is the convex hull of all the possible subsums of its generators. It therefore follows from Proposition 2.2 that the intersection of $H(d)$ with $S(d,h)$ is the convex hull of all the possible subsums of elements of $\mathcal{G}(d)$ such that exactly h of them belong to $\mathcal{G}(d)\backslash\mathcal{G}(d-1)$. In such a sum, the terms from $\mathcal{G}(d-1)$ sum to a point in $H(d-1)$, and the terms from $\mathcal{G}(d)\backslash\mathcal{G}(d-1)$ sum to a point in P. As a consequence, the intersection of $H(d)$ with $S(d,h)$ is a subset of $H(d-1)+P$. Inversely, $H(d-1)+P$ is the convex hull of all the sums whose terms are any number of points from $\mathcal{G}(d-1)$ and exactly h points from $\mathcal{G}(d)\backslash\mathcal{G}(d-1)$. Since any such sum is a point in the intersection $H(d)\cap S(d,h)$, the Minkowski sum of $H(d-1)$ with P is a subset of that intersection. □

Recall that the $(d-1)$-dimensional standard hypersimplices are the convex hulls of the vertices of the hypercube $[0,1]^d$ whose coordinates sum to a fixed integer h such that $0 < h < d$. Therefore, by Lemma 2.3, the intersections $H_1^+(d,2)\cap S(d,h)$ are, up to translation, the Minkowski sums of $H_1^+(d-1,2)$ with the orthogonal projection on $\mathbb{R}^{d-1}$ of the $(d-1)$-dimensional standard hypersimplices.

3. An asymptotic estimate for the number of vertices of $H_1^+(d,2)$

We first establish, as a consequence of Lemma 2.3, the following result on the placement of the vertices of $H_1^+(d,2)$.

LEMMA 3.1. *Every vertex of $H_1^+(d,2)$ belongs to a facet of $[0,d]^d$.*

PROOF. We proceed by induction on d. Note that $H_1^+(1,2) = [0,1]$ and that $H_1^+(2,2)$ is the hexagon obtained as the convex hull of all the lattice points in the square $[0,2]^2$ except for two opposite vertices of this square. Hence, the lemma holds when d is equal to 1 or 2. Now assume that $d \geq 3$ and consider a vertex x of $H_1^+(d,2)$. For any positive integer i less than d, denote by g^i the generator of $H_1^+(d,2)$ whose two non-zero coordinates are g_i^i and g_d^i. Further denote by g^0 the point in $\mathcal{G}_1^+(d,2)$ whose last coordinate is equal to 1, and whose all other coordinates are equal to 0. By Lemma 2.3, there exists a vertex y of $H_1^+(d-1,2)$ satisfying

$$x = y + \sum_{i\in\mathcal{I}} g^i,$$

where $\mathcal{I}$ is a subset of exactly x_d elements of $\{0, 1, ..., d-1\}$. By induction, y admits a coordinate equal to 0 or a coordinate equal to $d-1$. Let us first study the latter case. We can assume without loss of generality that $y_1 = d-1$. If $\mathcal{I} = \{0\}$, then $x = y + g^0$. In this case, consider the triangle with vertices $y + g^1$, $y + g^2$ and g^0. This triangle is contained in $H_1^+(d, 2)$ and, since $y \neq 0$, the point $y + g^0$ belongs to its relative interior. Hence, x cannot be a vertex of $H_1^+(d, 2)$. Now assume that $\mathcal{I} \neq \{0\}$. Assume, in addition, that $y_1 \neq d$. In this case, $\mathcal{I}$ does not contain 1. Yet, it must contain a positive integer and we assume without loss of generality that 2 belongs to $\mathcal{I}$. By symmetry, the point y' obtained by exchanging the first and second coordinates of y is a vertex of $H_1^+(d, 2)$ [**4**] and the point

$$x' = y' + \sum_{i \in \mathcal{I}} g^i$$

necessarily belongs to $H_1^+(d, 2)$. Observe that $x - g^2 + g^1$ also belongs to $H_1^+(d, 2)$. By construction, x is in the relative interior of the segment with extremities x' and $x - g^2 + g^1$ and it cannot be a vertex of $H_1^+(d, 2)$, a contradiction. This shows that 1 belongs to $\mathcal{I}$ and, as a consequence, that x_d is equal to d.

Now assume that one of the coordinates of y, say y_j, is equal to 0. In this case, x_j is equal to 0 or to 1. Since $H_1^+(d, 2)$ is centrally-symmetric with respect to the center of the hypercube $[0, d]^d$ [**4**], the symmetric x' of x with respect to the center of that hypercube is a vertex of $H_1^+(d, 2)$. Therefore, by Lemma 2.3, there exists a vertex y' of $H_1^+(d-1, 2)$ such that

$$x' = y' + \sum_{i \in \mathcal{I}'} g^i,$$

where $\mathcal{I}'$ is a subset of $\{0, 1, ..., d-1\}$. By symmetry, x'_j is equal to $d-1$ or to d. Therefore, y'_j must be equal to $d-1$. As shown above, in this case x'_j must be equal to d and, by symmetry, x belongs to a facet of the hypercube $[0, d]^d$. □

THEOREM 3.2.

$$f_0\big(H_1^+(d, 2)\big) = 2a_1(d, 2).$$

PROOF. Consider a vertex x of $H_1^+(d, 2)$. It follows from Lemma 3.1 that some coordinate of x must be equal to 0 or to d. By proposition 2.1 and Lemma 3.1, if a coordinate of x is equal to 0 then none of its coordinates can be greater than $d-1$. Therefore, the vertices of $H_1^+(d, 2)$ with at least one coordinate equal to 0 and the vertices of $H_1^+(d, 2)$ with at least one coordinate equal to d form a partition of the vertex set of $H_1^+(d, 2)$. Since $H_1^+(d, 2)$ is centrally-symmetric with respect to the center of the hypercube $[0, d]^d$, the number of vertices of $H_1^+(d, 2)$ is equal to twice the number of its vertices with at least one coordinate equal to d, or equivalently, to twice the number of its vertices whose all coordinates are positive. □

The following result is a consequence of Theorems 1.1 and 3.2.

COROLLARY 3.3.

$$a_1(d, 2) = \sum_{i=0}^{d-1} \binom{d}{i} a_1(i, 2).$$

Recall that, as a convention $a_1(0, 2)$ is equal to 1. In this case, the recursive expression provided by Corollary 3.3 results in a well known integer sequence, the Fubini numbers. Coincidentally, $a_1(d, 2)$ is therefore also equal to the number of non-empty faces of the $(d-1)$-dimensional permutohedron. This observation and Theorem 3.2 immediately provide the following statement.

COROLLARY 3.4. *The number of vertices of the primitive zonotope $H_1^+(d,2)$ is equal to twice the d-th Fubini number.*

The following asymptotic estimate is proven in [**1**].

$$a_1(d,2) \sim \frac{d!}{2(\ln 2)^{d+1}}.$$

Theorem 1.2 is obtained from this estimate and from Theorem 3.2.

4. Lower bounds on the number of vertices of $H_\infty(d,1)$ and $H_\infty^+(d,1)$

The lower bound on the number of vertices of $H_\infty(d,1)$ provided by Theorem 1.3 is a rather straightforward consequence of Lemma 2.3.

THEOREM 4.1.

$$f_0(H_\infty(d,1)) \geq \prod_{i=0}^{d-1}(3^i+1)\,.$$

PROOF. It is shown in [**4**] that $\kappa(H_\infty(d,1))$ is equal to 3^{d-1}. Since $\mathcal{G}_\infty(d,1)$ is a subset of $\{-1,0,1\}^d$, it follows from Lemma 2.3 that, for any integer h such that $0<h\leq 3^{d-1}$, the intersection of $H_\infty(d,1)$ with $S(d,h)$ has at least $f_0(H_\infty(d-1,1))$ vertices. Indeed, the Minkowski sum of two polytopes has at least as many vertices as any of them. Moreover, according to Proposition 2.2, the vertex sets of the intersections $H_\infty(d,1)\cap S(d,h)$, when h ranges from 0 to 3^{d-1}, form a partition of the vertex set of $H_\infty(d,1)$. As a consequence,

$$f_0(H_\infty(d,1)) \geq (3^{d-1}+1)f_0(H_\infty(d-1,1)).$$

Since $H_\infty(1,1)$ has two vertices, we obtain the desired inequality. □

As a consequence, the order of the logarithm to base 3 of d-criteria, 1-bounded convex matroid optimization is quadratic in d.

We now turn our attention to $H_\infty^+(d,1)$. As shown in [**4**], $\kappa(H_\infty^+(d,1))$ is equal to 2^{d-1} and we can use the same argument as in the proof of Theorem 4.1 in order to recover the lower bound on $f_0(H_\infty^+(d,1))$ from [**2**]. In order to improve on this, we derive a lower bound on $a_\infty(d,1)$ from Lemma 2.3.

THEOREM 4.2. *If $d\geq 2$, then*

$$a_\infty(d,1) \geq 2^{d-2}\big[f_0\big(H_\infty^+(d-1,1)\big)+a_\infty(d-1,1)\big]. \tag{4.1}$$

PROOF. Recall that $\kappa(H_\infty^+(d,1))=2^{d-1}$. Hence, by Lemma 2.3, any vertex x of $H_\infty^+(d,1)$ belongs to the hypercube

$$\big[0,2^{d-2}+x_d\big]^{d-1}\times\{x_d\}.$$

Since $H_\infty^+(d,1)$ is centrally-symmetric with respect to the center of $[0,2^{d-1}]^d$, a vertex of $H_\infty^+(d,1)$ whose last coordinate is greater than 2^{d-2} only has positive coordinates. Since the Minkowski sum of two polytopes has at least as many vertices as either of them, it follows from Proposition 2.2 and Lemma 2.3 that the number of vertices of $H_\infty^+(d,1)$ whose last coordinate is greater than 2^{d-2} is at least

$$2^{d-2}f_0\big(H_\infty^+(d-1,1)\big).$$

This quantity is the first term in the right-hand side of (4.1). Now, let h be an integer such that $0 < h \leq 2^{d-2}$. We will prove that $S(d,h)$ contains at least $a_\infty(d-1,1)$ vertices of $H_\infty^+(d,1)$ whose all coordinates are positive.

According to Lemma 2.3,

$$H_\infty^+(d,1) \cap S(d,h) = H_\infty^+(d-1,1) + Q, \tag{4.2}$$

where Q is a polytopes contained in the positive orthant $[0,+\infty[^d$. Observe that there exists an injection ϕ from the vertex set of $H_\infty^+(d-1,1)$ into the vertex set of the Minkowski sum $H_\infty^+(d-1,1)+Q$ that sends every vertex of $H_\infty^+(d-1,1)$ to its sum with a vertex of Q (see for instance Lemma 2.3 from [**6**]). In particular, if x is a vertex of $H_\infty^+(d-1,1)$ whose all coordinates are positive, then all the coordinates of $\phi(x)$ are also necessarily positive. Since ϕ is an injection, $H_\infty^+(d-1,1)+Q$ admits at least $a_\infty(d-1,1)$ vertices whose all coordinates are positive. By (4.2) and Proposition 2.2, all the vertices of $H_\infty^+(d-1,1)+Q$ are vertices of $H_\infty^+(d,1)$. Hence, the number of vertices of $H_\infty^+(d,1)$ whose all coordinates are positive and whose last coordinate does not exceed 2^{d-2} must be at least

$$2^{d-2}a_\infty(d-1,1).$$

This quantity is the second term in the right-hand side of (4.1). □

We now establish the lower bound stated by Theorem 1.4.

THEOREM 4.3. *For all* $d \geq 3$,

$$f_0\big(H_\infty^+(d,1)\big) \geq 6\prod_{i=1}^{d-2}\big(2^{i+1}+i\big).$$

PROOF. One can check using the values of $f_0(H_\infty^+(d,1))$ reported in Table 1 that the theorem holds when d is equal to 3 or 4.

We will prove that, for all $d \geq 5$,

$$a_\infty(d,1) \geq 6\prod_{i=1}^{d-2}\big(2^{i+1}+i\big). \tag{4.3}$$

Since $f_0(H_\infty^+(d,1)) \geq a_\infty(d,1)$, the theorem will follow. We proceed by induction on d. First observe that (4.3) holds when d is equal to 5 or to 6, as can be checked using the values of $a_\infty(5,1)$ and $a_\infty(6,1)$ reported in Table 1.

Now assume that $d \geq 7$. By Theorem 1.1,

$$f_0\big(H_\infty^+(d-1,1)\big) \geq a_\infty(d-1,1)+(d-1)a_\infty(d-2,1).$$

Combining this with (4.1), we obtain

$$a_\infty(d,1) \geq 2^{d-1}a_\infty(d-1,1)+2^{d-2}(d-1)a_\infty(d-2,1).$$

Observe that $2^{d-2} \geq (d-2)(d-3)$. Therefore,

$$a_\infty(d,1) \geq 2^{d-1}a_\infty(d-1,1)+(d-2)\big[2^{d-2}+d-3\big]a_\infty(d-2,1). \tag{4.4}$$

By induction, $a_\infty(d-1,1)$ and $a_\infty(d-2,1)$ can be bounded below using (4.3). Combining these bounds with (4.4) completes the proof. □

5. Upper bounds on the number of vertices of $H_\infty(d,1)$ and $H_\infty^+(d,1)$

Recall that the primitive zonotope $H_\infty^+(d,1)$ is contained in the hypercube $[0,2^{d-1}]^d$. In particular, the number of vertices of $H_\infty^+(d,1)$ is at most the number of lattice points in this hypercube. Since at most two vertices can differ only in the last coordinate, this bound can be improved into twice the number of lattice points in the hypercube $[0,2^{d-1}]^{d-1}$. Therefore, we obtain the inequality

$$f_0\big(H_\infty^+(d,1)\big) \le 2\big(2^{d-1}+1\big)^{d-1}, \tag{5.1}$$

that improves the upper bound of 2^{d^2} from [**2**]. The number of vertices of $H_\infty(d,1)$ can be bounded above using the same argument. Indeed, this polytope is contained, up to translation, in the hypercube $[0,3^{d-1}]^d$. Therefore, the number of its vertices is at most twice the number of lattice points in $[0,3^{d-1}]^{d-1}$. This results in the upper bound stated by Theorem 1.3.

THEOREM 5.1.

$$f_0(H_\infty(d,1)) \le 2\big(3^{d-1}+1\big)^{d-1}.$$

The upper bound provided by Theorem 1.4 essentially divides by 2^d the right-hand side of (5.1). Our strategy consists in identifying large portions of the hypercube $[0,2^{d-1}]^d$ disjoint from $H_\infty^+(d,1)$.

LEMMA 5.2. *If x is a vertex of $H_\infty^+(d,1)$ and $i\neq j$, then $|x_i-x_j|\le 2^{d-2}$.*

PROOF. Consider a vertex x of $H_\infty^+(d,1)$. By symmetry, we can assume that $x_i\ge x_j$. Observe that $\mathcal{G}_\infty^+(d,1)=\{0,1\}^d$. Hence, it follows from the definition of $H_\infty^+(d,1)$ that there is a subset $\mathcal{A}$ of $\{0,1\}^d$ whose sum of elements is equal to x. Let $\mathcal{B}$ denote the elements x in $\mathcal{A}$ such that $x_i=1$ and $x_j=0$. Further denote by $\mathcal{C}$ the complement of $\mathcal{B}$ in $\mathcal{A}$. The following holds.

$$x_i-x_j=\sum_{g\in\mathcal{B}}(g_i-g_j)+\sum_{g\in\mathcal{C}}(g_i-g_j).$$

Note that g_i-g_j is equal to 1 when $g\subset\mathcal{B}$ and to 0 or to -1 when $g\in\mathcal{C}$. Hence, x_i-x_j is, at most, the number of elements of $\mathcal{B}$. Since there are 2^{d-2} points g in $\{0,1\}^d$ such that $g_i=1$ and $g_j=0$, the lemma is proven. $\square$

We are now ready to complete the proof of Theorem 1.4. The upper bound stated by this theorem can be roughly estimated as the number of lattice points in $2d$ copies of the $(d-1)$-dimensional hypercube $[0,2^{d-2}]^{d-1}$.

THEOREM 5.3.

$$f_0\big(H_\infty^+(d,1)\big)\le 2(d+4)2^{(d-1)(d-2)}.$$

PROOF. Observe that the theorem holds when d is equal to 1. We therefore assume in the remainder of the proof that $d\ge 2$. Denote by u the lattice vector in $\mathbb{R}^d$ whose all coordinates are equal to 1 and by Q the union of the facets of the cone $[0,+\infty[^d$. Now consider a point x in $\mathbb{N}^d$, and its projection on Q along u, which we denote by $\pi(x)$. In other words, $\pi(x)$ is the unique point in Q such that $x-\pi(x)=ku$ for some non-negative integer k. It follows from Lemma 5.2 that, if x is a vertex of $H_\infty^+(d,1)$, then $\pi(x)$ is in the intersection of Q with the hypercube $[0,2^{d-2}]^d$. By convexity, a point in this intersection cannot be the image by π of more than two vertices of $H_\infty^+(d,1)$. Therefore, $f_0(H_\infty^+(d,1))$ is bounded above by twice the number of lattice points in $Q \cap [0, 2^{d-2}]^d$; that is,

$$f_0\big(H_\infty^+(d,1)\big) \le 2\sum_{i=0}^{d-1}\binom{d}{i}2^{i(d-2)}.$$

Factoring the largest term in the right-hand side of this inequality yields

$$f_0\big(H_\infty^+(d,1)\big) \le 2d2^{(d-1)(d-2)}\left[1+\frac{1}{d}\sum_{i=0}^{d-2}\binom{d}{i}2^{(i-d+1)(d-2)}\right].$$

Since $(i-d+1)(d-2) \le 2-d$ when $i \le d-2$, we obtain

$$f_0\big(H_\infty^+(d,1)\big) \le 2d2^{(d-1)(d-2)}\left[1+\frac{2^{2-d}}{d}\sum_{i=0}^{d-2}\binom{d}{i}\right].$$

Bounding above the sum of binomial coefficients in the right-hand side by 2^d and then rearranging the terms provide the desired result. □

Acknowledgments

The authors thank the anonymous referees for providing valuable comments and suggestions.

References

[1] Jean-Pierre Barthélémy, *An asymptotic equivalent for the number of total preorders on a finite set*, Discrete Math. **29** (1980), no. 3, 311–313, DOI 10.1016/0012-365X(80)90159-4. MR560774

[2] Louis J. Billera, Justin Tatch Moore, Costandino Dufort Moraites, Yipu Wang, and Kameryn Williams, *Maximal unbalanced families*, preprint, arXiv:1209.2309, 2012.

[3] Jonathan Block and Shmuel Weinberger, *Aperiodic tilings, positive scalar curvature and amenability of spaces*, J. Amer. Math. Soc. **5** (1992), no. 4, 907–918, DOI 10.2307/2152713. MR1145337

[4] Antoine Deza, George Manoussakis, and Shmuel Onn, *Primitive zonotopes*, Discrete Comput. Geom. **60** (2018), no. 1, 27–39, DOI 10.1007/s00454-017-9873-z. MR3807347

[5] Antoine Deza, George Manoussakis, and Shmuel Onn, *Small primitive zonotopes*, Discrete geometry and symmetry, Springer Proc. Math. Stat., vol. 234, Springer, Cham, 2018, pp. 87–107, DOI 10.1007/978-3-319-78434-2_5. MR3816872

[6] Antoine Deza and Lionel Pournin, *Diameter, decomposability, and Minkowski sums of polytopes*, Canad. Math. Bull. **62** (2019), no. 4, 741–755, DOI 10.4153/s0008439518000668. MR4028484

[7] Tim S. Evans, *What is being calculated with thermal field theory?*, Particle Physics and Cosmology, Proceedings of the Ninth Lake Louise Winter Institute, World Scientific, 1995, 343–352.

[8] Samuel C. Gutekunst, Karola Mészáros, and T. Kyle Petersen, *Root cones and the resonance arrangement*, preprint, arXiv:1903.06595, 2019.

[9] Hidehiko Kamiya, Akimichi Takemura, and Hiroaki Terao, *Ranking patterns of unfolding models of codimension one*, Adv. in Appl. Math. **47** (2011), no. 2, 379–400, DOI 10.1016/j.aam.2010.11.002. MR2803809

[10] Lukas Kühne, *The universality of the resonance arrangement and its Betti numbers*, preprint, arXiv:2008.10553, 2020.

[11] Michal Melamed and Shmuel Onn, *Convex integer optimization by constantly many linear counterparts*, Linear Algebra Appl. **447** (2014), 88–109, DOI 10.1016/j.laa.2014.01.007. MR3200209

[12] Shmuel Onn, *Nonlinear discrete optimization, an algorithmic theory*, Zurich Lectures in Advanced Mathematics, European Mathematical Society (EMS), Zürich, 2010. MR2724387

[13] Shmuel Onn and Uriel G. Rothblum, *Convex combinatorial optimization*, Discrete Comput. Geom. **32** (2004), no. 4, 549–566, DOI 10.1007/s00454-004-1138-y. MR2096748
[14] Justin Tatch Moore, *Amenability and Ramsey theory*, Fund. Math. **220** (2013), no. 3, 263–280, DOI 10.4064/fm220-3-6. MR3040674
[15] Michel A. van Eijck, *Thermal field theory and finite-temperature renormalisation group*, PhD thesis, University of Amsterdam, 1995.

Advanced Optimization Laboratory, McMaster University, Hamilton, Canada
Email address: deza@mcmaster.ca

LIPN, Université Paris 13, Villetaneuse, France
Email address: lionel.pournin@univ-paris13.fr

LIPN, Université Paris 13, Villetaneuse, France
Email address: rakotonarivo@lipn.univ-paris13.fr

Contemporary Mathematics
Volume **764**, 2021
https://doi.org/10.1090/conm/764/15330

Barycenters of points in polytope skeleta

Michael Gene Dobbins and Florian Frick

ABSTRACT. The first author showed that for a given point p in an nk-polytope P there are n points in the k-faces of P, whose barycenter is p. Here we completely classify n-tuples of dimensions $(k_1, \dots, k_n)$ that sum to nk such that there are n points from faces of these prescribed dimensions whose barycenter is p. We show that we can increase the dimension of P by r, if we allow r of the points to be in $(k+1)$-faces. While we can force points with a prescribed barycenter into faces of dimensions k and $k+1$, we show that the gap in dimensions of these faces can never exceed one. We also investigate the weighted analogue of this question, where a convex combination with predetermined coefficients of n points in k-faces of an nk-polytope is supposed to equal a given target point. While weights that are not all equal may be prescribed for certain values of n and k, any coefficient vector that yields a point different from the barycenter cannot be prescribed for fixed n and sufficiently large k.

1. Introduction

Given an abelian group $(G, +)$ with identity element 0 and an integer $n \geq 2$, zero-sum problems aim to find sufficient conditions on sequences $x_1, \dots, x_n$ of n elements of G to sum to zero, $x_1 + \dots + x_n = 0$. The seminal result for this problem area is a theorem of Erdős, Ginzburg, and Ziv [**4**]: any multiset $A \subset \mathbb{Z}/n$ of size $2n - 1$ contains n elements (counted with multiplicity) $x_1, \dots, x_n$ such that $x_1 + \dots + x_n = 0$. Various generalizations have been established, such as Reiher's proof of the Kemnitz conjecture [**6**].

While for finite G these problems fall into the realm of combinatorial number theory, they become geometric if G itself has geometry. Here we will study zero-sum problems in Euclidean space, $G = \mathbb{R}^d$, which seek sufficient conditions for a set $A \subset \mathbb{R}^d$ to contains n vectors that sum to zero, or equivalently, n vectors with the origin at their barycenter. For instance, it was first shown in a 13 author paper [**1**] that if A is the 1-skeleton of a 3-polytope that contains the origin, then there are $x_1, x_2, x_3 \in A$ with $x_1 + x_2 + x_3 = 0$. Tokuyama conjectured that the 1-skeleton of a polytope in $\mathbb{R}^n$ that contains the origin contains n vectors that sum to zero. More generally, the first author established that the k-skeleton of an nk-polytope P with $0 \in P$ contains n vectors that sum to zero [**3**]. A simplified proof was then given by Blagojević, the second author, and Ziegler [**2**].

The proofs in [**2**, **3**] depend on methods from equivariant topology, and thus crucially make use of the inherent symmetries of the problem. In particular, for a

2010 *Mathematics Subject Classification.* Primary 51M04, 51M20, 52B11.

polytope of dimension $d < nk$ with $d > n(k-1)$, the proofs do not generalize to force $nk-d$ of the x_i into faces of dimension $k-1$. Our first main result will extend slightly beyond the symmetric case and establish precisely that; see Theorem 2.2:

THEOREM 1.1. *Let P be a d-polytope with $0 \in P$. Let n, k, and r be non-negative integers, where $r \leq n-1$. Then for $d = nk+r$ there are points $x_1, \dots, x_{n-r}$ in k-faces of P, and $x_{n-r+1}, \dots, x_n$ in $(k+1)$-faces of P such that $x_1+\dots+x_n = 0$.*

While this might seem like a slight improvement, we show that this result is in fact optimal and classifies n-tuples of dimensions $(k_1, \dots, k_n)$ that sum to nk such that there are n points from faces of these prescribed dimensions whose barycenter is at the origin. We show that dimensions cannot be further decreased (Proposition 2.3(b)), and that the dimensions of faces that the x_i are constrained to cannot differ by more than one (Proposition 2.3(a)).

Since $\mathbb{R}^d$ is a vector space this gives us another chance to break symmetries and study unbalanced zero-sum problems: Given coefficients $\lambda_1, \dots, \lambda_n > 0$, find sufficient conditions on a set $A \subset \mathbb{R}^d$ to contain n vectors $x_1, \dots, x_n \in A$ with $\lambda_1 x_1 + \dots + \lambda_n x_n = 0$. Again, earlier proofs cannot easily be adapted to this asymmetric situation. In fact, we can use Theorem 1.1 to establish results for the case of unbalanced coefficients; see Theorem 3.1. We also show that if $0 = \lambda_1 x_1 + \dots + \lambda_n x_n$ for x_i in the k-faces of an nk-polytope, then the λ_i are almost equal for large k; see Corollary 3.4.

2. Inhomogeneous skeleta

Let $P^{(k)} \subset \mathbb{R}^{nk}$ denote the k-skeleton of a polytope P, that is, the collection of all faces of dimension at most k, and suppose that P contains the origin, then $P^{(k)}$ contains n vectors that sum to zero:

THEOREM 2.1 (Dobbins [**3**]). *Let $P \subset \mathbb{R}^{nk}$ be a polytope with $0 \in P$. Then there are $x_1, \dots, x_n \in P^{(k)}$ such that $x_1 + \dots + x_n = 0$.*

Equivalently, for any given point $p \in P$, where P is an nk-polytope, there are points $x_1, \dots, x_n \in P^{(k)}$ with their barycenter $\frac{1}{n}x_1 + \dots + \frac{1}{n}x_n$ at p. In the sequel, we will use arbitrary target points p in in the polytope, and not only the origin.

Let $\mathcal{P}(d; k_1, \dots, k_n)$ be the predicate: "For any polytope P of dimension at most d, and for any target point $p \in P$, there exist points $x_1, \dots, x_n$ such that x_i is in a k_i-face of P and the target point p is the barycenter of the points $x_1, \dots, x_n$." With this notation Theorem 1.1 can be rephrased as:

THEOREM 2.2. *Let n, k, and r be non-negative integers, where $r \leq n-1$. Then for $d = nk + r$ the statement*

$$\mathcal{P}(d; \underbrace{k, \dots, k}_{n-r}, \underbrace{k+1, \dots, k+1}_{r})$$

is true.

The proof of Theorem 2.2 will be postponed for now. In the theorem above, the dimension of any k-face cannot be decreased to $k-1$, even if all other x_i may be chosen from faces of arbitrary dimension, including the interior of P itself. We also have to show that we cannot decrease the dimension of any $(k+1)$-face to k (again allowing more freedom for the other x_i). We collect these two results here:

PROPOSITION 2.3. (a) *For integers* $n \geq 1$, $k \geq 0$, *and* $d \geq nk$ *the statement* $\mathcal{P}(d; k-1, \underbrace{d, \dots, d}_{n-1})$ *is false.*

(b) *For non-negatrive integers* n, k, *and* r *with* $r \leq n-1$, *and* $d = nk + r$ *the statement* $\mathcal{P}(d; \underbrace{k, \dots, k}_{n-r+1}, \underbrace{d, \dots, d}_{r-1})$ *is false.*

Before proving Theorem 2.2 and Proposition 2.3, we first need an additional lemma. In the following denote by Δ^d the regular d-dimensional simplex

$$\{(x_1, \dots, x_{d+1}) \in \mathbb{R}^{d+1} | x_i \geq 0, \sum_i x_i = 1\}.$$

We denote the standard basis of $\mathbb{R}^{d+1}$ by $e_1, \dots, e_{d+1}$, and the dual basis by $e_1^*, \dots, e_{d+1}^*$.

LEMMA 2.4. *There do not exist points* $x_1, \dots, x_n$ *in the regular* d*-simplex* Δ^d, $d < n$, *where* $x_1, \dots, x_{n-d+1}$ *are vertices, and the barycenter* $\frac{1}{n}\sum_i x_i$ *is equal to*

$$p = \left(\tfrac{d-1/2}{dn}, \dots, \tfrac{d-1/2}{dn}, \tfrac{n-d+1/2}{n}\right) \in \Delta^d.$$

PROOF. Suppose there are points $x_1, \dots, x_n \in \Delta^d$ with barycenter p, and that $x_1, \dots, x_{n-d+1}$ are vertices of Δ^d. We cannot have $x_i = e_{d+1}$ for all $i \leq n-d+1$, since that would give

$$e_{d+1}^*(p) = e_{d+1}^*\left(\tfrac{1}{n}x_1 + \dots + \tfrac{1}{n}x_n\right) \geq e_{d+1}^*\left(\tfrac{1}{n}x_1 + \dots + \tfrac{1}{n}x_{n-d+1}\right) = \tfrac{n-d+1}{n},$$

but $e_{d+1}^*(p) = \frac{n-d+1/2}{n} < \frac{n-d+1}{n}$. Therefore, at least one of $x_1, \dots, x_{n-d+1}$ is a vertex of Δ^d other than the vertex e_{d+1}. We may assume that $x_1 = e_1$, which gives

$$e_1^*(p) \geq e_1^*\left(\tfrac{1}{n}x_1\right) = \tfrac{1}{n},$$

but $e_1^*(p) = \frac{d-1/2}{dn} < \frac{1}{n}$, which is again a contradiction. Thus, no such points $x_1, \dots, x_n$ exist. □

PROOF OF THEOREM 2.2. For a given d-polytope P with $d = nk + r$ and $0 \leq r \leq n-1$, let $Q = P \times \Delta^s$ where $s = n - r$, and let

$$x = \left(\tfrac{s-1/2}{sn}, \dots, \tfrac{s-1/2}{sn}, \tfrac{n-s+1/2}{n}\right) \in \Delta^s.$$

Since Q is an $n(k+1)$-polytope, by Theorem 2.1 there are points $y_1, \dots, y_n$ in the $(k+1)$-faces of Q that have $(0, x) \in P \times \Delta^s$ as their barycenter. Let $x_i \in \mathbb{R}^{nk+r}$ be the first component of y_i in the product $Q = P \times \Delta$, and let $\tilde{y}_i$ be the second component of y_i. Then, $x_1, \dots, x_n$ are in the $(k+1)$-faces of P and sum to zero.

Suppose that at least $r+1$ of the points x_i are not in a k-face of P. Then at least $r + 1 = n - s + 1$ of the points $\tilde{y}_i$ are vertices of Δ^s, but that contradicts Lemma 2.4, since x is the barycenter of $\{\tilde{y}_1, \dots, \tilde{y}_n\}$. Thus, at least $n - r$ of the points x_i are in k-faces of P. □

PROOF OF PROPOSITION 2.3. (a) Let $P = \Delta^d$ and let

$$p = \left(\tfrac{1}{d+1}, \dots, \tfrac{1}{d+1}\right).$$

Suppose there are points $x_1, \dots, x_n \in \Delta^d$ with barycenter p, and that one of the points is in a $(k-1)$-face of Δ^d. We may assume $x_1 \in \operatorname{conv}\{e_1, \dots, e_k\}$. Let $\phi = e_1^* + \cdots + e_k^*$. Then, we have

$$\phi(p) = \phi\left(\tfrac{1}{n}x_1 + \cdots + \tfrac{1}{n}x_n\right) \geq \phi\left(\tfrac{1}{n}x_1\right) = \tfrac{1}{n},$$

but this is a contradiction, since $\phi(p) = \frac{k}{d+1} \leq \frac{k}{nk+1} < \frac{1}{n}$. Thus, no such points $x_1, \dots, x_n$ exist.

(b) Let

$$P = \underbrace{\Delta^n \times \cdots \times \Delta^n}_{k} \times \Delta^r$$

$$\begin{aligned} p &= p_1 \times \cdots \times p_{k+1} \\ &= \Big(\underbrace{\tfrac{1}{n+1}, \dots, \tfrac{1}{n+1}}_{k(n+1)}, \underbrace{\tfrac{r-1/2}{rn}, \dots, \tfrac{r-1/2}{rn}}_{r}, \tfrac{n-r+1/2}{n}\Big) \end{aligned}$$

where $p_1, \dots, p_k \in \Delta^n$ are each the target point from part (a) when $k = 1, r = 0$, and $p_{k+1} \in \Delta^r$ is the target point from Lemma 2.4.

Suppose there are points $x_1, \dots, x_n \in P$ with barycenter p, and that the points $x_1, \dots, x_{n-r+1}$ are in k-faces of P. Let $x_{i,j}$ be the j-th component of x_i from the product defining P above, and let $k_{i,j}$ be the dimension of the minimal face of the j-th simplex of the product containing $x_{i,j}$. That is, for $j \in \{1, \dots, k\}$, $x_{i,j}$ is given by the j-th block of $n+1$ consecutive coordinates of x_i,

$$x_{i,j} = (e^*_{(n+1)j-n}(x_i), \dots, e^*_{(n+1)j}(x_i)) \in \Delta^n,$$

and $x_{i,k+1}$ is given by the last $r+1$ coordinates of x_i,

$$x_{i,k+1} = (e^*_{(n+1)k+1}(x_i), \dots, e^*_{(n+1)k+r+1}(x_i)) \in \Delta^r.$$

Since the barycenter of $\{x_{1,j}, \dots, x_{n,j}\}$ for $j \in \{1, \dots, k\}$ is

$$p_j = \left(\tfrac{1}{n+1}, \dots, \tfrac{1}{n+1}\right),$$

by part (a), none of the points $x_{i,j}$ are vertices of Δ^n. Hence, $k_{i,j} \geq 1$ for $j \in \{1, \dots, k\}$, so $k_{i,1} + \cdots + k_{i,k} \geq k$. Since x_i is in a k-face for $i \in \{1, \dots, n-r+1\}$, we must have $k_{i,1} + \cdots + k_{i,k+1} \leq k$, so $k_{i,k+1} = 0$. That is, $n-r+1$ of the points $x_{i,k+1}$ must be vertices of Δ^r, but that is impossible since the barycenter of $\{x_{1,k+1}, \dots, x_{n,k+1}\}$ is

$$p_{k+1} = \left(\tfrac{r-1/2}{rn}, \dots, \tfrac{r-1/2}{rn}, \tfrac{n-r+1/2}{n}\right),$$

and by Lemma 2.4 at most $n-r$ of the points $x_{i,k+1}$ can be vertices of Δ^r. Thus, no such points $x_1, \dots, x_n$ exist. □

3. Unbalanced weights

Given positive integers n and k, what are all n-tuples $(\lambda_1, \dots, \lambda_n)$ of coefficients $\lambda_i > 0$ normalized to $\lambda_1 + \cdots + \lambda_n = 1$ such that for any nk-polytope $P \subset \mathbb{R}^{nk}$ with $0 \in P$ there are $x_1, \dots, x_n \in P^{(k)}$ with $\lambda_1 x_1 + \cdots + \lambda_n x_n = 0$? We denote the set of all such coefficients with $\lambda_1 \geq \lambda_2 \geq \cdots \geq \lambda_n$ by $\Lambda(n,k)$. The set $\Lambda(n,k)$ is nonempty since it contains $(\frac{1}{n}, \dots, \frac{1}{n})$ by Theorem 2.1. By taking P to be closer and closer approximations of the unit ball in $\mathbb{R}^{2k}$, we see that $\Lambda(2,k) = \{(\frac{1}{2}, \frac{1}{2})\}$. We will show that $\Lambda(n,1)$ may contain more than one element for $n > 2$.

THEOREM 3.1. *Let s, t, and k be positive integers, and let $n = sk + t(k+1)$. Then the coefficient vector $(\lambda_1, \ldots, \lambda_n)$ with $\lambda_1 = \cdots = \lambda_{sk} = \frac{1}{(s+t)k}$ and $\lambda_{sk+1} = \cdots = \lambda_n = \frac{1}{(s+t)(k+1)}$ is contained in $\Lambda(n, 1)$.*

PROOF. Let P be an n-polytope with $0 \in P$. By Theorem 2.2 there are points $x_1, \ldots, x_s$ in k-faces of P and points $x_{s+1}, \ldots, x_{s+t}$ in $(k+1)$-faces of P, such that $\sum x_i = 0$. Each x_i with $i \in \{1, \ldots, s\}$ can be written as $x_i = \frac{1}{k}\sum_j y_j^{(i)}$ for $y_1^{(i)}, \ldots, y_k^{(i)} \in P^{(1)}$. Similarly, each x_i with $i \in \{s+1, \ldots, s+t\}$ can be written as $x_i = \frac{1}{k+1}\sum_j y_j^{(i)}$ for $y_1^{(i)}, \ldots, y_{k+1}^{(i)} \in P^{(1)}$. Putting this together, we obtain

$$0 = \sum_{i=1}^{s} \frac{1}{k} \sum_j y_j^{(i)} + \sum_{i=s+1}^{s+t} \frac{1}{k+1} \sum_j y_j^{(i)}.$$

Normalizing these coefficients to sum up to one, we get the result. □

This shows that unbalanced weights may be prescribed for certain parameters. Our last goal is to show that asymptotically, that is, for fixed n and large k, unbalanced weights may not be prescribed. More precisely, given n positive real numbers $\lambda_1, \ldots, \lambda_n$ with $\lambda_1 + \cdots + \lambda_n = 1$ that are not all equal to $\frac{1}{n}$, there is an integer k and an nk-polytope P with $0 \in P$ such that $\lambda_1 x_1 + \cdots + \lambda_n x_n \neq 0$ for all $x_1, \ldots, x_n \in P^{(k)}$. We will need the following simple lemma:

LEMMA 3.2. *Let $k \geq 1$ and $n \geq 2$ be integers. Let $d = nk$ and $x_1, x_2 \in \Delta^d$, where x_1 is contained in a k-face of Δ^d. Suppose $\lambda_1 x_1 + \lambda_2 x_2 = (\frac{1}{d+1}, \ldots, \frac{1}{d+1})$ for $\lambda_1, \lambda_2 \geq 0$ with $\lambda_1 + \lambda_2 = 1$. Then $\lambda_1 \leq \frac{k+1}{d+1}$.*

PROOF. Let $p = (\frac{1}{d+1}, \ldots, \frac{1}{d+1})$. We may assume $x_1 \in \operatorname{conv}\{e_1, \ldots, e_{k+1}\}$. Let $\phi = e_1^* + \cdots + e_{k+1}^*$. Then, we have

$$\tfrac{k+1}{nk+1} = \tfrac{k+1}{d+1} = \phi(p) = \phi\left(\lambda_1 x_1 + \lambda_2 x_2\right) \geq \phi\left(\lambda_1 x_1\right) = \lambda_1.$$

□

THEOREM 3.3. *Let $n > 0$ and $k > 0$ be integers. If $(\lambda_1, \ldots, \lambda_n) \in \Lambda(n, k)$ then $\lambda_i \leq \frac{k+1}{nk+1}$ for all i.*

PROOF. Let $d = nk$. Translate the standard simplex Δ^d such that its barycenter is at the origin. Now suppose

$$0 = \lambda_1 x_1 + \cdots + \lambda_n x_n,$$

where the x_i are in k-faces of Δ^d, and the coefficients λ_i are nonnegative and satisfy $\lambda_1 + \cdots + \lambda_n = 1$. We also assume $\lambda_1 \geq \lambda_2 \geq \cdots \geq \lambda_n$. Then $\lambda_2 + \cdots + \lambda_n > 0$ and thus

$$0 = \lambda_1 x_1 + (\lambda_2 + \cdots + \lambda_n)\left(\tfrac{\lambda_2}{\lambda_2 + \cdots + \lambda_n} x_2 + \cdots + \tfrac{\lambda_n}{\lambda_2 + \cdots + \lambda_n} x_n\right).$$

Since $\frac{\lambda_2}{\lambda_2 + \cdots + \lambda_n} x_2 + \cdots + \frac{\lambda_n}{\lambda_2 + \cdots + \lambda_n} x_n$ is a convex combination of points in Δ^d, it is itself a point in Δ^d. Then by Lemma 3.2, we have that $\lambda_1 \leq \frac{k+1}{d+1} = \frac{k+1}{nk+1}$. □

It is a simple consequence that $\bigcap_k \Lambda(n, k) = \{(\frac{1}{n}, \ldots, \frac{1}{n})\}$, or in different words:

COROLLARY 3.4. *Let $\lambda = (\lambda_1, \ldots, \lambda_n)$ be some unbalanced coefficient vector, that is, $\lambda_i \geq 0$ for all i, $\lambda_1 + \cdots + \lambda_n = 1$, and $\lambda \neq (\frac{1}{n}, \ldots, \frac{1}{n})$. Then for k sufficiently large $\lambda \notin \Lambda(n, k)$.*

4. Final remarks

Our results are only for $A \subset \mathbb{R}^d$ that are skeleta of polytopes, but what about other kinds of sets? For instance, a problem that arises naturally in protein folding is to determine the Carathéodory number of an orbit of SO(3) acting on a higher dimensional space [**5**]. Can we find sufficient conditions for zero-sums to exist in sets of this type? Related to this, can our results be improved if we assume that A itself has some symmetry?

Acknowledgments

We would like to thank two referees for their help in improving the exposition, and for pointing us to [**5**]. The "Final remarks" section is based on comments of one referee.

References

[1] Luis Barba, Jean Lou De Carufel, Otfried Cheong, Michael Gene Dobbins, Rudolf Fleischer, Akitoshi Kawamura, Matias Korman, Yoshio Okamoto, János Pach, Yuan Tang, Takeshi Tokuyama, Sander Verdonschot, and Tianhao Wang, *Weight balancing on boundaries and skeletons*, Proc. 30th Symp. Comp. Geom., Kyoto, Japan, 2014, pp. 436–443.

[2] Pavle V. M. Blagojević, Florian Frick, and Günter M. Ziegler, *Barycenters of polytope skeleta and counterexamples to the topological Tverberg conjecture, via constraints*, J. Eur. Math. Soc. (JEMS) **21** (2019), no. 7, 2107–2116, DOI 10.4171/JEMS/881. MR3959859

[3] Michael Gene Dobbins, *A point in a nd-polytope is the barycenter of n points in its d-faces*, Invent. Math. **199** (2015), no. 1, 287–292, DOI 10.1007/s00222-014-0523-2. MR3294963

[4] P. Erdös, A. Ginzburg, and A. Ziv, *Theorem in the additive number theory*, Bull. Res. Council Israel Sect. F **10F** (1961), no. 1, 41–43. MR3618568

[5] Marco Longinetti, Luca Sgheri, and Frank Sottile, *Convex hulls of orbits and orientations of a moving protein domain*, Discrete Comput. Geom. **43** (2010), no. 1, 54–77, DOI 10.1007/s00454-008-9076-8. MR2575320

[6] Christian Reiher, *On Kemnitz' conjecture concerning lattice-points in the plane*, Ramanujan J. **13** (2007), no. 1-3, 333–337, DOI 10.1007/s11139-006-0256-y. MR2281170

Department of Mathematics, Binghamton University, Binghamton, New York 13902
Email address: mdobbins@binghamton.edu

Department of Mathematical Sciences, Carnegie Mellon University, Pittsburgh, Pennsylvania 15213
Email address: frick@cmu.edu

Contemporary Mathematics
Volume **764**, 2021
https://doi.org/10.1090/conm/764/15331

Two families of locally toroidal regular 4-hypertopes arising from toroids

Maria Elisa Fernandes, Dimitri Leemans, Claudio Alexandre Piedade, and Asia Ivić Weiss

ABSTRACT. In this paper we present two infinite families of locally toroidal hypertopes of rank 4 that are constructed from regular toroids of types $\{4,3,4\}_{(s,s,0)}$ and $\{3,3,4,3\}_{(s,0,0,0)}$. The Coxeter diagram of the first of the two families is star-shaped and the diagram of the other is a square. In both cases the toroidal residues are regular toroidal maps of type $\{3,6\}$.

1. Introduction

A regular abstract polytope is usually defined as a flag-transitive poset whose elements are faces [**9**]. An alternative but equivalent way of defining a regular abstract polytope is as an incidence geometry with linear diagram that is (1) thin, (2) residually-connected, and (3) flag-transitive. In [**5**] the authors decided to consider incidence geometries satisfying (1), (2), and (3) and whose diagram is not necessarily linear. A similar generalization from maps to hypermaps exists which is the reason why these combinatorial structures are called hypertopes.

It is well known that abstract regular polytopes are in one-to-one correspondence with string C-groups. That is, given a string C-group one can construct a regular polytope from it [**9**]. As the group of (type-preserving) automorphisms of a regular hypertope is a C-group one may expect that such a correspondence would exist in case of hypertopes as well. However, this is not the case as was shown for example in [**5**, Example 4.5] where a C-group can not be used to construct a regular hypertope. The conditions on a C-group for the construction to work are given in the Theorems of Section 2. In this paper we also give another example illustrating that one must be careful in drawing on analogy between polytopes and hypertopes.

Lists of finite universal locally toroidal regular hypertopes with toroidal residues of type $\{3,6\}$ were computed from groups in [**6**] using Tits' coset geometries. The idea was to use a family of groups with a given Coxeter diagram which were then parameterized using the parameters of their toroidal residues. Among the examples given in Table 8 of [**6**], we find a surprising phenomenon on the 2nd and 3rd lines: the locally toroidal hypertopes do not have the expected residues. That is, the

This work is supported by The Center for Research and Development in Mathematics and Applications (CIDMA) throught the Portuguese Foundation for Science and Technology (FCT-Fundação para a Ciência e a Tecnologia), references UIDB/04106/2020 and UIDP/04106/2020, and by NSERC.

hypertopes had the expected diagrams but the residues did not correspond to the given parameters (this will be explained in detail in Section 5). For that reason in Section 3 we revisit the family of hypertopes introduced in Section 7 of [**6**]. The diagram of the family can be seen as a hexagonal extension of the diagram of a tetrahedron having two toroidal non-isomorphic residues of type $\{3,6\}$. We show that the rank 3 residues for the family are correct. In Section 4 we deal the family of finite universal locally toroidal 4-hypertopes introduced in Section 8 of [**6**]. For this family, having a square diagram with labels 6 on non-adjacent edges, we also confirm the correctness of the residues. In both cases we establish that the construction yields regular hypertopes. Before dealing with these two families of hypertopes we introduce some preliminary background on regular hypertopes that we will make use of in subsequent sections. For background on toroidal maps and more generally toroids, which is also essential, we refer to [**9**].

2. Preliminaries

In this section, we start by giving the definition of a *regular hypertope* introduced in [**5**] as an incidence geometry that is thin, residually-connected, and flag-transitive. The precise definition is below but more details and general theory of these concepts can be found for instance in [**3**].

An *incidence system* is a 4-tuple $\Gamma := (X, *, t, I)$ satisfying the following conditions.

- X is the set of elements of Γ;
- I is the set of *types of* Γ (whose cardinality is called the *rank of* Γ);
- $t : X \to I$ is a type function, associating to each element $x \in X$ a type $t(x) \in I$ (an element x is said to be of *type* i, or an i*-element*, whenever $t(x) = i$, for $i \in I$); and
- $*$ is a binary relation in X called *incidence*, which is reflexive, symmetric and such that, for all $x, y \in X$, if $x * y$ and $t(x) = t(y)$, then $x = y$.

The *incidence graph* of Γ is a graph whose vertices are the elements of X and where two vertices are connected whenever they are incident in Γ. The type function determines an $|I|$-partition on the set of elements of the incidence graph. Observe that in general it does not induce a partial-order (which it does in the case of polytopes).

A *flag* F is a subset of X in which every two elements are incident. For a flag F the set $t(F) := \{t(x) \,|\, x \in F\}$ is called *the type of* F. When $t(F) = I$, F is called a *chamber*. An element x is incident to a flag F, denoted by $x * F$, when x is incident to all elements of F.

An incidence system Γ is a *geometry* (or an *incidence geometry*) if every flag of Γ is contained in a chamber, that is if all maximal flags of Γ are chambers. If a geometry Γ of rank 2 with $I = \{i, j\}$ is such that each of its i-elements are incident with each of its j-elements, then Γ is called a *generalized digon*. The *residue* of a flag F of an incidence geometry Γ is the incidence geometry $\Gamma_F := (X_F, *_F, t_F, I_F)$ where $X_F := \{x \in X : x * F, x \notin F\}$, $I_F := I \setminus t(F)$, and where t_F and $*_F$ are restrictions of t and $*$ to X_F and I_F respectively. An incidence system Γ is *connected* if its incidence graph is connected. Moreover, Γ is *residually connected* when Γ is connected and each residue of Γ of rank at least two is also connected. An incidence system Γ is *thin* when every residue of rank one of Γ contains exactly two elements. If an incidence geometry is thin, then given a chamber C there exists exactly one chamber differing from C in its i-element which we denote by C^i. We also say that C and C^i are *i-adjacent*. A *hypertope*, as defined in [**5**], is a thin incidence geometry which is residually connected.

In summary, a hypertope generalizes a concept of an abstract polytope. The elements, instead of forming a partially-ordered set, form an n-partite incidence graph. Such a graph defines a polytope if the following conditions are satisfied.

- The maximal cliques (called chambers) all have size n (including one element of each type).
- Every $(n-1)$-clique can be augmented to form an n-clique in exactly two ways (corresponding to the thinness condition for polytope).
- The graph is residually connected, meaning that it is connected and that for every k-clique with $1 \leq k \leq n-2$, the subgraph of elements that is incident to every element in the clique is also connected.

An *automorphism* of an incidence system $\Gamma := (X, *, t, I)$ is a permutation α of X inducing a permutation on I such that $t(x) = t(y)$ implies $t(\alpha(x)) = t(\alpha(y))$ for all $x, y \in X$, and preserving incidence (i.e. $x * y$ if and only if $\alpha(x) * \alpha(y)$). The set of automorphisms of Γ is a group denoted by $Aut(\Gamma)$. An automorphism α is called *type-preserving* if it induces the identity permutation on I (that is, α maps each element to an element of the same type, or more precisely $t(\alpha(x)) = t(x)$ for every $x \in X$). The set of type-preserving automorphisms of Γ is a subgroup of $Aut(\Gamma)$ denoted by $Aut_I(\Gamma)$.

The group of type-preserving automorphisms of an incidence geometry Γ acts faithfully on the set of its chambers. Moreover, if Γ is a hypertope this action is semi-regular. In fact, if $g \in Aut_I(\Gamma)$ fixes one chamber C, it also fixes its i-adjacent chamber C^i. Since Γ is residually connected, g must be the identity. We say that Γ is *chamber-transitive* if the action of $Aut_I(\Gamma)$ on the chambers is transitive, and in that case the action of Γ on the set of chambers is regular. For that reason Γ is then called a *regular hypertope*. Observe that, when Γ is a geometry, chamber-transitivity implies *flag-transitivity*; that is, for each $J \subseteq I$, there is a unique orbit on the flags of type J under the action of $Aut_I(\Gamma)$.

Let $\Gamma := (X, *, t, I)$ be a regular hypertope and let C be a (base) chamber of Γ. For each $i \in I$ there exists exactly one automorphism ρ_i sending C to C^i. The group generated by $\{\rho_i \,|\, i \in I\}$ is a C-group [**5**], which means that the following two conditions are satisfied.

- ρ_i is an involution for each $i \in I$.
- $\langle \rho_j \,|\, j \in J\rangle \cap \langle \rho_k \,|\, k \in K\rangle = \langle \rho_i \,|\, i \in J \cap K\rangle$ for J, $K \subseteq I$ (intersection property).

The *Coxeter diagram* of a C-group G is a graph with $|I|$ vertices corresponding to the generators of G and with an edge $\{i, j\}$ whenever the order p_{ij} of $\rho_i\rho_j$ is greater than 2. The edge is endowed with the label p_{ij} when $p_{ij} > 3$. The type-preserving automorphism group of an abstract polytope is a *string C-group*, that is, a C-group having a linear Coxeter diagram. The Coxeter diagram of the type-preserving automorphism group of a regular hypertope is isomorphic to its Buekenhout diagram defined as follows.

- The vertices of the diagram are the types.

- There is an edge between two types i and j with label p_{ij} when the residues of type $\{i, j\}$ are not generalized digons (observe that all rank 2 residues of a regular hypertope of the same type are isomorphic).

As is the case with Coxeter diagram, the edges of the Buekenhout diagram have labels p_{ij} only when $p_{ij} > 3$.

The poset of an abstract regular polytope can be built from a string C-group [**9**]. Analogously, it is also possible to construct a regular hypertope from a group, and particularly from a C-group, using the following proposition.

PROPOSITION 2.1 (Tits Algorithm, [**11**]). *Let n be a positive integer and $I := \{1, \ldots, n\}$. Let G be a group together with a family of subgroups $(G_i)_{i\in I}$, X the set consisting of all cosets $G_i g$ with $g \in G$ and $i \in I$, and $t : X \to I$ defined by $t(G_i g) = i$. Define an incidence relation $*$ on $X \times X$ by:*

$$G_i g_1 * G_j g_2 \text{ iff } G_i g_1 \cap G_j g_2 \neq \emptyset.$$

Then the 4-tuple $\Gamma := (X, *, t, I)$ is an incidence system having a chamber. Moreover, the group G acts by right multiplication as an automorphism group on Γ. Finally, the group G is transitive on the flags of rank less than 3.

The incidence system constructed using the proposition above will be denoted by $\Gamma(G; (G_i)_{i\in I})$. The subgroups $(G_i)_{i\in I}$, called the *maximal parabolic subgroups*, are the stabilizers of an element of type i. If $G = \langle \rho_i \,|\, i \in I\rangle$ is the type-preserving automorphism group of a regular hypertope Γ and $G_i := \langle \rho_j \,|\, j \neq i\rangle$ for each $i \in I$, the coset geometry $\Gamma(G; (G_i)_{i\in I})$ is isomorphic to Γ. Therefore, to get examples of regular hypertopes we may consider coset geometries from C-groups. Contrary to what happens in the case of regular polytopes [**1**], this construction might not give a regular hypertope, and indeed it might not give a geometry at all (see [**7**]). In rank three we have the following result.

PROPOSITION 2.2. [**5**, Proposition 4.3] *If $G := \langle \rho_0, \rho_1, \rho_2\rangle$ is a C-group of rank 3, and $\Gamma(G; (G_i)_{i\in\{0,1,2\}})$ with $G_i = \langle \rho_j \,|\, j \neq i\rangle$ is thin, then Γ is a regular hypertope.*

Unfortunately thinness does not suffice in higher ranks, but in that case we can use the following result.

THEOREM 2.3. [**5**, Theorem 4.6] *Let $G = \langle \rho_i \,|\, i \in I\rangle$ be a C-group and let $\Gamma := \Gamma(G; (G_i)_{i\in I})$ where $G_i := \langle \rho_j \,|\, j \neq i\rangle$ for all $i \in I$. If G is flag-transitive on Γ, then Γ is a regular hypertope.*

We recall a group-theoretical result of Tits that will be used to prove flag-transitivity.

LEMMA 2.4. [**12**] *Let K, H, Q be three subgroups of a group G. The following conditions are equivalent.*

(1) *$KH \cap KQ = K(H \cap Q)$.*
(2) *$(K \cap H)(K \cap Q) = K \cap (HQ)$.*
(3) *If the three cosets Kx, Hy and Qz have pairwise non-empty intersections, then $Kx \cap Hy \cap Qz \neq \emptyset$.*

To prove that an incidence system of rank 4 (the case we deal with in this paper) with $I = \{0, 1, 2, 3\}$ is flag-transitive it is sufficient to prove one of the three conditions of Lemma 2.4 for every subset of I mentioned in the following theorem.

THEOREM 2.5. [**3**, Theorem 1.8.10 (iii)] *Let $\Gamma(G,(G_i)_{i\in I})$ be the coset incidence system of G over $(G_i)_{i\in I}$. Then Γ is flag-transitive if and only if for each subset J of I of size three, the group G is transitive on the set of flags of type J, and for each $i \in I$ the subgroup G_i is flag-transitive on $\Gamma(G_i,(G_{i,j})_{j\in I\setminus\{i\}})$. If one (whence both) these properties hold, then Γ is a geometry.*

We denote by $G_{i,j}$ the subgroup of G generated by all the generators of G except ρ_i and ρ_j. When all maximal parabolic subgroups (with their respective generators) are C-groups the following proposition gives the conditions on their intersections for the group generated by all involutions to be a C-group.

PROPOSITION 2.6. [**4**, Proposition 6.1] *Let G be a group generated by n involutions $\rho_0,\ldots,\rho_{n-1}$. Suppose that G_i is a C-group for every $i \in \{0,\ldots,n-1\}$. Then G is a C-group if and only if $G_i \cap G_j = G_{i,j}$ for all $0 \le i,\, j \le n-1$.*

A rank 4 hypertope is *locally toroidal* if all its rank 3 residues are either spherical or toroidal with at least one of them being toroidal. When the defining relations of the type-preserving automorphism group of a locally toroidal hypertope are those corresponding to the Coxeter diagram with the only additional relations determining the toroidal residues, we say that the locally toroidal hypertope is *universal*.

In what follows we give two infinite families of finite universal regular locally toroidal hypertopes constructed from the two families of toroids $\{4,3,4\}_{(s,s,0)}$ and $\{3,3,4,3\}_{(s,0,0,0)}$.

3. A family of hypertopes arising from $\{4,3,4\}_{(s,s,0)}$

In Section 7 of [**6**] the authors suggested the existence of an infinite family of universal locally toroidal regular hypertopes with Coxeter diagram that is a *hexagonal extension of the diagram of a tetrahedron*, meaning that the diagram is as follows.

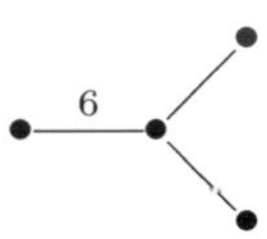

The family of groups found yields an incidence system with toroidal residues $\{3,6\}_{(2,0)}$ and $\{3,6\}_{(s,0)}$ $(s \ge 3)$. The authors also observed that each incidence system in the family, with the automorphism group G, can be obtained from the cubic toroid $\{4,3,4\}_{(s,s,0)}$ by Petrie operations and doubling of the fundamental region so that G is isomorphic to $[4,3,4]_{(s,s,0)}$ of order $48s^3$. However, the result of these transformations does not guarantee that G is a C-group nor that its coset geometry is flag-transitive. Nevertheless, this is the case as we now prove with the following theorem.

THEOREM 3.1. *Let G be the group with the following presentation*

$$G := \langle \rho_0,\rho_1,\rho_2,\rho_3 \mid \rho_0^2=\rho_1^2=\rho_2^2=\rho_3^2=(\rho_0\rho_1)^3=(\rho_0\rho_2)^2=(\rho_0\rho_3)^2= (\rho_1\rho_2)^6=(\rho_1\rho_3)^3=(\rho_2\rho_3)^2=(\rho_0(\rho_1\rho_2)^2\rho_1)^s=(\rho_3(\rho_1\rho_2)^2\rho_1)^2\rangle.$$

Then $\Gamma(G;(G_i)_{\{0,1,2,3\}})$ is a universal locally toroidal regular hypertope with toroidal residues $\{3,6\}_{(2,0)}$ and $\{3,6\}_{(s,0)}$.

Let $G := \langle \alpha_0,\ldots,\alpha_{r-1}\rangle$ where α_i are involutions. The *permutation representation graph* of $(G,\{\alpha_0,\ldots,\alpha_{r-1}\})$ *of degree* n is a graph on n vertices, with edges

$\{a,b\}$ labelled i whenever $a\alpha_i = b$ for some α_i. In [**6**] the authors gave a permutation representation graph for G of degree $4s$. We say that a permutation representation graph is a *CPR graph* if the involutions satisfy the intersection property. Hence our notion of CPR graph is a generalization to C-groups with nonlinear diagram of the concept of CPR graph given in [**10**] for linear diagrams. In what follows we recall the result of [**6**]. Observe that the intersection property is obtained from the intersections of the respective orbits of the maximal parabolic subgroups on the vertices of the graph.

PROPOSITION 3.2. *Let $s \geq 3$. The following graphs are CPR graphs, of degree $4s$, of the group G given in Theorem 3.1.*

s even

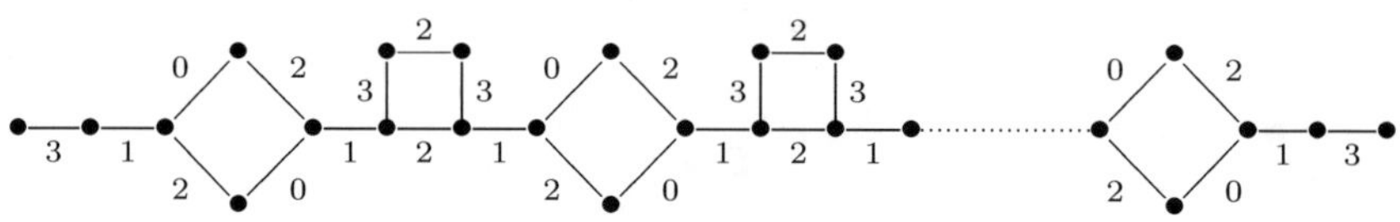

s odd

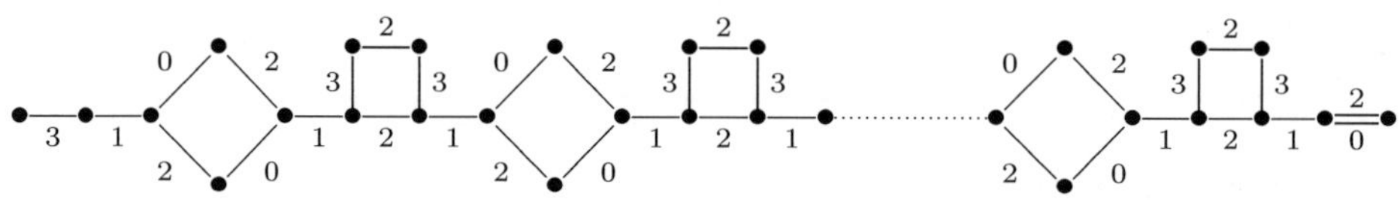

A CPR graph gives a faithful permutation representation of a C-group. The proposition above describes one CPR graph with $4s$ vertices for the C-group G. It is most probably the minimal degree of that group, having in mind the results obtained in [**8**].

By Theorem 2.3, to complete the proof of Theorem 3.1 we only need to prove flag-transitivity.

PROPOSITION 3.3. *If G is the group of Theorem 3.1 and $G_i := \langle \rho_j \mid j \in I \setminus \{i\} \rangle$ with $i \in I := \{0,1,2,3\}$, then $\Gamma(G;(G_i)_{i\in I})$ is flag-transitive.*

PROOF. To prove that G is flag-transitive on Γ, by Lemma 2.4 it is sufficient to prove the following equality for every possible subset $\{i,j,k\}$ of I with three distinct elements.

$$(G_i \cap G_j)(G_i \cap G_k) = G_i \cap (G_j G_k)$$

In any of these cases, the inclusion $(G_i \cap G_j)(G_i \cap G_k) \subseteq G_i \cap (G_j G_k)$ is trivial. Let us prove the other inclusion for each case separately.

CASE. $\{i,j,k\} = \{0,1,2\}$ Since $G_1 = \{id_G, \rho_0, \rho_2, \rho_3, \rho_0\rho_2, \rho_0\rho_3, \rho_2\rho_3, \rho_0\rho_2\rho_3\}$ and $G_1G_2 = \bigcup_{g\in G_1} gG_2$, it follows that $G_1G_2 = G_2 \cup \rho_2 G_2$ and thus $G_0 \cap G_1G_2 = G_{0,2} \cup (G_0 \cap \rho_2 G_2)$. As $\rho_2 \in G_0$, we get $G_0 \cap \rho_2 G_2 = \rho_2(G_0 \cap G_2) = \rho_2 G_{0,2} \subseteq G_{0,1}G_{0,2}$. It is now clear that $G_0 \cap G_1G_2 = G_{0,1}G_{0,2}$, as wanted.

The cases where $\{i,j,k\}$ are equal to $\{0,1,3\}$ and $\{2,1,3\}$ follow a similar proof.

CASE. $\{i,j,k\} = \{0,2,3\}$ Now we have $G_2G_3 = G_3 \cup \rho_3 G_3 \cup \rho_1\rho_3 G_3 \cup \rho_0\rho_1\rho_3 G_3$ and therefore

$$G_0 \cap G_2G_3 = G_{0,3} \cup (G_0 \cap \rho_3 G_3) \cup (G_0 \cap \rho_1\rho_3 G_3) \cup (G_0 \cap \rho_0\rho_1\rho_3 G_3).$$

Since $G_0 \cap \rho_3 G_3 \subseteq \rho_3 G_{0,3}$ and $G_0 \cap \rho_1\rho_3 G_3 \subseteq \rho_1\rho_3 G_{0,3}$, we get $G_0 \cap G_2 G_3 \subseteq G_{0,3} \cup \rho_3 G_{0,3} \cup \rho_1\rho_3 G_{0,3} \cup (G_0 \cap \rho_0\rho_1\rho_3 G_3)$. The only difficulty now is to prove that $G_0 \cap \rho_0\rho_1\rho_3 G_3 \subseteq G_{0,2}G_{0,3}$. Let $\alpha \in G_0 \cap \rho_0\rho_1\rho_3 G_3$ and consider the following numbering of the left side vertices of one of the CPR-graphs given in Proposition 3.2.

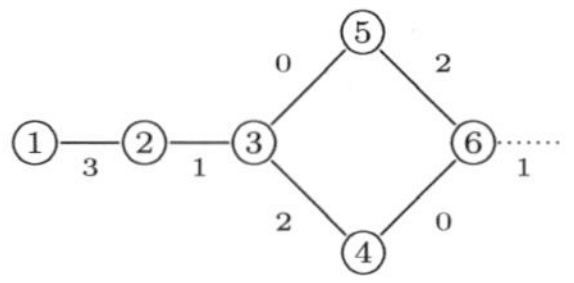

For every $\alpha \in \rho_0\rho_1\rho_3 G_3$, α acts on the CPR graph by sending vertex 5 to vertex 1. On the other hand 1 and 5 are in different orbits of G_0 and hence $G_0 \cap \rho_0\rho_1\rho_3 G_3 = \emptyset$. With this we conclude that $G_0 \cap G_2 G_3 \subseteq G_{0,2}G_{0,3}$ and therefore Γ is flag-transitive. □

4. A family of hypertopes arising from $\{3,3,4,3\}_{(s,0,0,0)}$

In Section 8 of **[6]** it was suggested that there exists an infinite family of universal finite locally toroidal regular hypertopes whose Coxeter diagram is a 4-circuit of type $(3,6,3,6)$, as pictured below, with four toroidal rank 3 residues being $\{3,6\}_{(s,s)}$, $\{3,6\}_{(1,1)}$, $\{3,6\}_{(2,0)}$, and $\{3,6\}_{(2,0)}$.

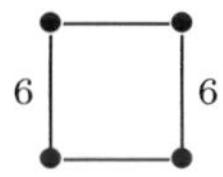

Moreover, the authors observed that this family could be obtained from toroids of type $\{3,3,4,3\}_{(s,0,0,0)}$. In the following proposition, we use this observation given in **[6]** to derive a group generated by involutions (ggi) whose maximal parabolic subgroups are automorphism groups of toroidal maps specified above.

PROPOSITION 4.1. *Let $s \geq 2$ and $H = \langle \rho_0, \rho_1, \rho_2, \rho_3, \rho_4 \rangle$ be the automorphism group of the regular toroid $\{3,3,4,3\}_{(s,0,0,0)}$. The group $G = \langle \tau_0, \tau_1, \tau_2, \tau_3 \rangle$ with $\tau_0 := \rho_1$, $\tau_1 := \rho_2$, $\tau_2 := \rho_1\rho_3$, $\tau_3 := \rho_0\rho_4$ is a ggi such that G_0, G_1, G_2 and G_3 are the groups of the toroidal maps $\{3,6\}_{(s,s)}$, $\{3,6\}_{(1,1)}$, $\{3,6\}_{(2,0)}$, and $\{3,6\}_{(2,0)}$ respectively. Moreover, the groups G and H are isomorphic.*

PROOF. Let H be the automorphism group of the rank 5 toroid $\{3,3,4,3\}_{(s,0,0,0)}$. The group H is obtained as the quotient of the affine Coxeter group $[3,3,4,3]$ having the Coxeter diagram

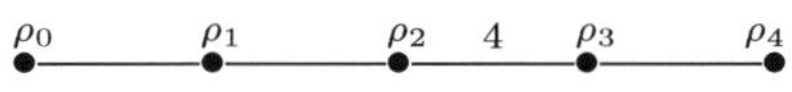

by adding the relation $(\rho_0\rho_1(\rho_2\rho_3\rho_4)^3)^{2s} = id_G$. Consider the group $G := \langle \tau_0, \tau_1, \tau_2, \tau_3 \rangle$ generated by the involutions

$$\tau_0 := \rho_1,\ \tau_1 := \rho_2,\ \tau_2 := \rho_1\rho_3,\ \tau_3 := \rho_0\rho_4.$$

It was shown in **[6]** that $G \cong H$ and that G has the following Coxeter diagram (as a ggi).

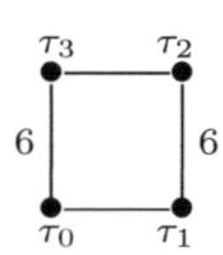

We now prove that the toroidal residues G_i are as stated in the proposition. From the diagram we see that every G_i determines a regular toroidal map of type $\{3,6\}$, and hence either a map $\{3,6\}_{(k,0)}$ with group of order $12k^2$ or a map $\{3,6\}_{(k,k)}$ with group of order $36k^2$. We now determine each group G_i for $i \in \{0, 1, 2, 3\}$.

The group $G_1 = \langle \tau_2, \tau_3, \tau_0 \rangle = \langle \rho_1\rho_3, \rho_0\rho_4, \rho_1 \rangle$. As $\rho_4 = (\rho_1\rho_0\rho_4)^3$ we conclude that $G_1 \cong H_2 \cong D_3 \times D_3$. Thus G_1 must be the group of the toroidal map $\{3,6\}_{(1,1)}$.

Now consider $G_2 = \langle \tau_1, \tau_0, \tau_3 \rangle = \langle \rho_2, \rho_1, \rho_0\rho_4 \rangle$. As $\rho_4 = (\rho_1\rho_0\rho_4)^3$ we conclude that $G_2 \cong H_3 \cong C_2 \times S_4$. Hence G_2 has order 48 and is the group of $\{3,6\}_{(2,0)}$.

Similarly, as $G_3 = \langle \tau_0, \tau_1, \tau_2 \rangle = \langle \rho_1, \rho_2, \rho_1\rho_3 \rangle$, $G_3 \cong H_{0,4} \cong C_2 \times S_4$. Hence G_3 has order 48 and is the group of $\{3,6\}_{(2,0)}$.

Let us now consider the group G_0. We have $G_0 = \langle \tau_3, \tau_2, \tau_1 \rangle = \langle \rho_0\rho_4, \rho_1\rho_3, \rho_2 \rangle$. To prove that G_0 is the group of $\{3,6\}_{(s,s)}$ we will prove that $(\tau_3(\tau_2\tau_1)^2)^2$ has order s and that $\tau_3(\tau_2\tau_1)^2\tau_2$ has order $3s$. Indeed, as $\{3,6\}_{(s,0)}$ lies in $\{3,6\}_{(s,s)}$, we have to guarantee that the group is not smaller than expected.

We may take the vertex-set of the toroid $\{3,3,4,3\}$ to be $\mathbb{Z}^4 \cup (\mathbb{Z}^4 + (\frac{1}{2}, \frac{1}{2}, -\frac{1}{2}, -\frac{1}{2}))$, the set of points of the euclidian 4-space whose cartesian coordinates are all integers or all halves of odd integers [**9**]. As a space-form, the toroid $\{3,3,4,3\}_{(s,0,0,0)}$ is the euclidian 4-space factorized by $\Lambda_{(s,0,0,0)} = s\Lambda_{(1,0,0,0)}$ whose basis is $\{e_1, e_2, e_3, \frac{1}{2}(e_1 + e_2 + e_3 + e_4)\}$ (where e_i, $i \in \{1, \dots, 4\}$ are the vectors of the canonical basis).

Let us now prove that $v := \tau_3(\tau_2\tau_1)^2\tau_2 (= \rho_0\rho_4(\rho_1\rho_3\rho_2)^2\rho_1\rho_3)$ is a translation of order $3s$. Consider the involution $\sigma := \rho_1\rho_2\rho_3\rho_2\rho_1$ and the hyperplane reflexions R_i (for $i \in \{0, \dots, 4\}$) and S as defined on page 171 of [**9**].

$$
\begin{aligned}
(x_1,x_2,x_3,x_4)R_0 &= (1-x_1,x_2,x_3,x_4)\\
(x_1,x_2,x_3,x_4)R_1 &= (x_1,x_2,x_3,x_4) - \tfrac{1}{2}(x_1-x_2-x_3-x_4)(1,-1,-1,-1)\\
(x_1,x_2,x_3,x_4)R_2 &= (x_1,x_2,x_3,-x_4)\\
(x_1,x_2,x_3,x_4)R_3 &= (x_1,x_2,x_4,x_3)\\
(x_1,x_2,x_3,x_4)R_4 &= (x_1,x_3,x_2,x_4)\\
(x_1,x_2,x_3,x_4)S &= (x_2,x_1,x_3,x_4)
\end{aligned}
$$

Then $v = \rho_0\rho_4\rho_3\rho_2\sigma\rho_3$ and its action on the euclidean 4-space is given by $V = R_0R_4R_3R_2SR_3$. We find that $xV = (x_3, 1-x_1, -x_2, x_4)$ and $(x_1,x_2,x_3,x_4)V^3 = (-1,1,-1,0) + (x_1,x_2,x_3,x_4)$. Consequently, V is a translation of order $3s$. Therefore G_0 is the group of either $\{3,6\}_{(s,s)}$ or of $\{3,6\}_{(3s,0)}$.

Let us now prove that $(\tau_3(\tau_2\tau_1)^2)^2 = (\rho_0\rho_4(\rho_1\rho_3\rho_2)^2)^2$ has order s, or equivalently that $U = (R_0R_4(R_1R_3R_2)^2)^2$ is a translation of order s. Let $P = R_0R_4(R_1R_3R_2)^2$. Then $U = P^2$. The action of P on the euclidean 4-space is described by the following equality.

$$
(x_1,x_2,x_3,x_4)P = \left(\frac{1}{2}, \frac{1}{2}, -\frac{1}{2}, -\frac{1}{2}\right) + \frac{1}{2}\begin{bmatrix} -1 & -1 & 1 & 1\\ -1 & 1 & 1 & -1\\ 1 & 1 & 1 & 1\\ 1 & -1 & 1 & -1 \end{bmatrix}\begin{bmatrix} x_1\\ x_2\\ x_3\\ x_4 \end{bmatrix}
$$

Let M be the 4×4 matrix above. As $M^2 = 4I_4$ (where I_4 denotes the 4 by 4 identity matrix) and $(\frac{1}{2}, \frac{1}{2}, -\frac{1}{2}, -\frac{1}{2})M = (-2,0,0,0)$, we get that $U (= P^2)$ is a translation defined by $(x_1,x_2,x_3,x_4)U = (-\frac{1}{2}, \frac{1}{2}, -\frac{1}{2}, -\frac{1}{2}) + (x_1,x_2,x_3,x_4)$ that has order s. With this we have shown that G_0 is the group of $\{3,6\}_{(s,s)}$. $\square$

PROPOSITION 4.2. *The group* $G = \langle \tau_0, \tau_1, \tau_2, \tau_3 \rangle$ *of Proposition* 4.1 *is a C-group.*

PROOF. Let H be the automorphism group of the rank 5 toroid $\{3,3,4,3\}_{(s,0,0,0)}$ and G as defined in the previous proof. We have shown in Proposition 4.1 that $G_1 = H_2$, $G_2 = H_3$ and $G_3 = H_{0,4}$. Since H is a C-group, $G_i \cap G_j = G_{i,j}$ whenever $\{i,j\} \subset \{1,2,3\}$. By Proposition 2.6 we only need to prove that $G_0 \cap G_k = G_{0,k}$ for every $k \in \{1,2,3\}$.

Let $G_0 = \langle \tau_1, \tau_2, \tau_3 \rangle = \langle \rho_2, \rho_1\rho_3, \rho_0\rho_4 \rangle$. For each k, the fact that $G_0 \cap G_k \supseteq G_{0,k}$ is trivial. To prove the other inclusion we consider each case in turn.

CASE. $k = 1$ We know that $G_1 = \langle \rho_1\rho_3, \rho_0\rho_4, \rho_1 \rangle = \langle \rho_3, \rho_0, \rho_4, \rho_1 \rangle = H_2$ and that $G_{0,1} \cong D_3$. With this, we have the inclusion chain $D_3 \cong G_{0,1} \subseteq G_0 \cap G_1 \subseteq G_1 \cong D_3 \times D_3$. Suppose that $G_0 \cap G_1 \not\cong D_3$. Then there exists $\alpha \in G_0 \cap G_1$ centralizing $G_{0,1}$. But there is no such element in $[3,6]_{(1,1)}$.

CASE. $k = 2$ We have $G_{0,2} \cong \langle \rho_2, \rho_0\rho_4 \rangle \cong C_2 \times C_2$ and $G_2 \cong H_3 \cong C_2 \times S_4$, and suppose that $G_0 \cap G_2 \neq G_{0,2}$. Looking at the subgroup chain of $C_2 \times S_4$ we conclude that $G_0 \cap G_2$ is either S_4, A_4, or D_4. First suppose $G_0 \cap G_2 \cong S_4$. This is only possible if $G_0 \cap G_2 = \langle \rho_0, \rho_1, \rho_2 \rangle$ which implies that both ρ_0 and ρ_1 are elements of G_0. In that case, $G_0 = \langle \rho_0, \rho_1, \rho_2, \rho_3, \rho_4 \rangle = G$, a contradiction. Now assume that $G_0 \cap G_2 \cong A_4$. Then, as A_4 does not contain a central involution, $G_0 \cap G_2 \leq \langle \rho_0, \rho_1, \rho_2 \rangle$ containing $\rho_0\rho_4$, again a contradicition. Finally, suppose that $G_{0,2} \cong D_4$, then, as $\rho_2(\rho_0\rho_4)$, ρ_2 and $\rho_0\rho_4$ are involutions that commute, two of them must be in the same conjugacy class in D_4. But then, as in addition $G_{0,2} \leq \langle \rho_4 \rangle \times \langle \rho_0, \rho_1, \rho_2 \rangle = H_3$, one get $\rho_4 \in \langle \rho_0, \rho_1, \rho_2 \rangle$ or ρ_0 and $\rho_0\rho_2$ are conjugated in $\langle \rho_0, \rho_1, \rho_2 \rangle$ (the group of a tetrahedron), in any case we get a contradiction. Consequently $G_{0,2} = G_0 \cap G_2$.

CASE. $k = 3$ In this case we have $G_{0,3} \cong \langle \rho_2, \rho_1\rho_3 \rangle \cong D_6$ and $G_3 = \langle \rho_1, \rho_2, \rho_1\rho_3 \rangle = H_{0,4} \cong C_2 \times S_4$. Suppose that $G_{0,3} \neq G_0 \cap G_3$. Then $G_0 \cap G_3 \cong S_4$. Taking this into account, adding ρ_1 to $G_0 \cap G_3$ we get G_3. But this would only be possible if ρ_1 commuted with both ρ_2 and $\rho_1\rho_3$, a contradiction.

Hence, G is a C-group. □

THEOREM 4.3. *Let* $G = \langle \tau_0, \tau_1, \tau_2, \tau_3 \rangle$ *be as defined in Proposition* 4.1. *The incidence system given by* $\Gamma(G; (G_i)_{i \in I})$ *with* $G_i := \langle \tau_j | j \in I \setminus \{i\} \rangle$ *for all* $i \in I := \{0,1,2,3\}$ *is a finite universal locally toroidal regular hypertope whose Coxeter diagram is a 4-circuit of type* $(3,6,3,6)$ *having the rank* 3 *residues* $\{3,6\}_{(s,s)}$, $\{3,6\}_{(1,1)}$, $\{3,6\}_{(2,0)}$, *and* $\{3,6\}_{(2,0)}$ *(with type-preserving automorphism groups* G_0, G_1, G_2 *and* G_3, *respectively).*

PROOF. From Theorem 2.3, Propositions 4.1 and 4.2 we only need to prove that Γ is flag-transitive. For that, by Lemma 2.4, we need to prove one of the two equivalent equalities $G_{i,j}G_{i,k} = G_i \cap (G_jG_k)$ or $G_iG_{j,k} = G_iG_j \cap G_iG_k$ for $\{i,j,k\} \subset \{0,1,2,3\}$. In any of the cases, the direct inclusion is trivial. We now prove the other inclusion for each case separately.

As in Proposition 4.1 let $H := \langle \rho_0, \rho_1, \rho_2, \rho_3, \rho_4 \rangle$ be the group of the regular polytope $\{3,3,4,3\}_{(s,0,0,0)}$.

CASE. $\{i,j,k\} = \{1,2,3\}$ Since H is a string C-group of a regular polytope and since $G_1 = H_2$, $G_2 = H_3$, and $G_3 = H_{0,4}$, the equality is satisfied.

CASE. $\{i,j,k\} = \{0,1,2\}$ In this case $G_1G_2 = G_2 \cup \rho_3G_2 \cup \rho_4\rho_3G_2$. Hence, $G_0 \cap (G_1G_2) = G_{0,2} \cup (G_0 \cap \rho_3G_2) \cup (G_0 \cap \rho_4\rho_3G_2)$. It is easy to see that $G_0 \cap \rho_3G_2 = \rho_1\rho_3G_0 \cap \rho_1\rho_3G_2 = \rho_1\rho_3G_{0,2}$ and that $G_0 \cap \rho_4\rho_3G_2 = G_0 \cap \rho_4\rho_0\rho_1\rho_3G_2 = \rho_4\rho_0\rho_1\rho_3(G_{0,2})$. Since both $\rho_3\rho_1$ and $\rho_3\rho_1\rho_4\rho_0 \in G_{0,1}$, we conclude that $G_0 \cap (G_1G_2) \subseteq G_{0,1}G_{0,2}$, as wanted.

CASE. $\{i,j,k\} = \{0,1,3\}$ We have $G_0G_1 = G_0 \cup G_0\rho_0 \cup G_0\rho_1 \cup G_0\rho_0\rho_1 \cup G_0\rho_0\rho_3 \cup G_0\rho_0\rho_1\rho_0$ and $G_0G_3 = G_0 \cup G_0\rho_1 \cup G_0\rho_1\rho_2 \cup G_0\rho_1\rho_2\rho_3$. We claim that the cardinality of the following set C of cosets is 8 (meaning that all cosets are pairwise distinct).

$$C := \{G_0,\ G_0\rho_0,\ G_0\rho_0\rho_1,\ G_0\rho_1\rho_0, G_0\rho_0\rho_1\rho_0,\ G_0\rho_1,\ G_0\rho_1\rho_2,\ G_0\rho_1\rho_2\rho_3\}.$$

As $G_0 \cap \langle \rho_0, \rho_1\rangle \subseteq G_0 \cap G_1 \cap \langle \rho_0, \rho_1\rangle = G_{0,1} \cap \langle \rho_0, \rho_1\rangle$, and the right-hand intersection is trivial, the first six cosets must be distinct. To prove that the remaining cosets are also distinct some more calculations are needed. Let us prove that $G_0\rho_0 \neq G_0\rho_1\rho_2\rho_3$. Assume that $\rho_0\rho_3\rho_2\rho_1 \in G_0$. Since $\rho_1\rho_3 \in G_0$ and ρ_3 commutes with both ρ_0 and ρ_1, we have that $\rho_3\rho_1(\rho_0\rho_3\rho_2\rho_1) = \rho_1\rho_0\rho_2\rho_1 \in G_0$. As $\rho_1\rho_0\rho_2\rho_1$ also belongs to G_2 it must belong to $G_0 \cap G_2 = \langle \rho_2, \rho_0\rho_4\rangle = \{id_G, \rho_2, \rho_0\rho_4, \rho_2\rho_4\rho_0\}$. We can easily see that any case leads to a contradiction: if $\rho_1\rho_0\rho_2\rho_1 = id_G$, then $\rho_0\rho_2 = id_G$; if $\rho_1\rho_0\rho_2\rho_1 = \rho_2$, then $\rho_1\rho_0\rho_2 = \rho_2\rho_1$, but $\rho_1\rho_0\rho_2$ has order 4 and $\rho_2\rho_1$ has order 3; if $\rho_1\rho_0\rho_2\rho_1 = \rho_0\rho_4$, then $\rho_4 = \rho_0\rho_1\rho_0\rho_2\rho_1$; at last if $\rho_1\rho_0\rho_2\rho_1 = \rho_2\rho_0\rho_4$, then $\rho_4 = \rho_0\rho_2\rho_1\rho_0\rho_2\rho_1$.

Similar calculations made us conclude that $|C| = 8$. From this we get $G_0G_1 \cap G_0G_3 = G_0 \cup G_0\rho_1 \subseteq G_0G_{1,3}$, as desired.

CASE. $\{i,j,k\} = \{0,2,3\}$ In this case $G_0G_2 = G_0 \cup G_0\rho_0 \cup G_0\rho_1 \cup G_0\rho_0\rho_1 \cup G_0\rho_1\rho_0 \cup G_0\rho_1\rho_2 \cup G_0\rho_0\rho_1\rho_0 \cup G_0\rho_0\rho_1\rho_2 \cup G_0\rho_1\rho_0\rho_2 \cup G_0\rho_0\rho_1\rho_0\rho_2 \cup G_0\rho_1\rho_0\rho_2\rho_1 \cup G_0\rho_0\rho_1\rho_0\rho_2\rho_1$ and $G_0G_3 = G_0 \cup G_0\rho_1 \cup G_0\rho_1\rho_2 \cup G_0\rho_1\rho_2\rho_3$. We claim that all the cosets of these partition decompositions of G_0G_2 and G_0G_3 are distinct when the representatives of the coset are different. We omit the calculations as the arguments are similar to those used in the above cases and can easily be checked. Then, $G_0G_2 \cap G_0G_3 = G_0 \cup G_0\rho_1 \cup G_0\rho_1\rho_2$ and consequently $G_0G_2 \cap G_0G_3 \subseteq G_0(G_2 \cap G_3)$, as desired.

With this we conclude that Γ is flag-transitive and therefore a regular hypertope. □

5. Final remarks

In the last section we proved the existence of a family of universal locally toroidal regular rank 4 hypertopes as suggested in the first line of Table 8 of [**6**]. This table lists universal locally toroidal hypertopes with a 4-circuit diagram of type $(3,6,3,6)$ that were obtained using MAGMA [**2**]. The authors used Tits algorithm to construct regular (and also chiral) incidence geometries from groups. The defining relations for the groups are the ones corresponding to the 4-circuit diagram of type $(3,6,3,6)$ together with in each case four additional relations which determine the maps $\{3,6\}_{\mathbf{s}}$, $\{3,6\}_{\mathbf{t}}$, $\{3,6\}_{\mathbf{u}}$, and $\{3,6\}_{\mathbf{v}}$.

The authors carefully checked that for each group, with parameters specified in lines 2 to 12, the construction leads to a hypertope. However, without writing it down, it was assumed that the toroidal rank 3 residues for each hypertope in the table is determined by the values of the eight parameters $\mathbf{s} = (a,b)$, $\mathbf{t} = (c,d)$,

$\mathbf{u} = (e, f)$, $\mathbf{v} = (g, h)$. To our great surprise, we now realize that in lines 2 and 3 this is not the case and some of the toroidal residues are in fact smaller than expected. Indeed, in both cases we get one toroidal residue with smaller group as described below.

The type-preserving automorphism group of the hypertope of line 2 of Table 8 of [**6**] is the group $G := G_{(\mathbf{s},\mathbf{t},\mathbf{u},\mathbf{v})}$ with $(\mathbf{s},\mathbf{t},\mathbf{u},\mathbf{v}) = ((2,0),\ (2,0),\ (1,1),\ (3,0))$ with the following presentation (in [**6**] the authors considered the rotational subgroup instead).

$$\begin{aligned} G &= \langle \rho_0, \rho_1, \rho_2, \rho_3 | \rho_0^2 = \rho_1^2 = \rho_2^2 = \rho_3^2 = (\rho_0\rho_1)^3 = (\rho_0\rho_2)^6 = (\rho_0\rho_3)^2 = \\ &= (\rho_1\rho_2)^2 = (\rho_1\rho_3)^6 = (\rho_2\rho_3)^3 = (\rho_2(\rho_3\rho_1)^2\rho_3)^3 = (\rho_3(\rho_2\rho_0)^2\rho_2)^2 = \\ &= (\rho_0(\rho_1\rho_3)^2)^2 = (\rho_1(\rho_0\rho_2)^2\rho_0)^2 = id_G \rangle \end{aligned}$$

The maximal parabolic subgroups G_1, G_2, and G_3 are the type-preserving automorphism groups of $\{3,6\}_{(2,0)}$, $\{3,6\}_{(1,1)}$, and $\{3,6\}_{(2,0)}$ respectively. However, on closer inspection G_0 turns out to be the group of a toroidal regular map $\{3,6\}_{(1,1)}$, that is covered by $\{3,6\}_{(3,0)}$. We note that changing the presentation of G given above by replacing the relator $(\rho_2(\rho_3\rho_1)^2\rho_3)^3$ by $(\rho_2(\rho_3\rho_1)^2))^2$ implies $(\rho_1\rho_3)^2 = id_G$ and the resulting hypertope is in fact the universal rank 4 locally toroidal regular polytope $\{\{3,6\}_{(2,0)}, \{6,3\}_{(2,0)}\}$ of type $\{3,6,3\}$ with 240 chambers (see [**13**] and also Table 11E1 of [**9**]) and not the hypertope determined by the group above.

The same happens to the hypertope in line 3 of Table 8 of [**6**] that was constructed from the rotational subgroup of the C-group $G := G_{(\mathbf{s},\mathbf{t},\mathbf{u},\mathbf{v})}$ with $(\mathbf{s},\mathbf{t},\mathbf{u},\mathbf{v}) = ((2,0),\ (2,0),\ (1,1),\ (6,0))$ with the following presentation.

$$\begin{aligned} G &= \langle \rho_0, \rho_1, \rho_2, \rho_3 | \rho_0^2 = \rho_1^2 = \rho_2^2 = \rho_3^2 = (\rho_0\rho_1)^3 = (\rho_0\rho_2)^6 = (\rho_0\rho_3)^2 = \\ &= (\rho_1\rho_2)^2 = (\rho_1\rho_3)^6 = (\rho_2\rho_3)^3 = (\rho_2(\rho_3\rho_1)^2\rho_3)^6 = (\rho_3(\rho_2\rho_0)^2\rho_2)^2 = \\ &= (\rho_0(\rho_1\rho_3)^2)^2 = (\rho_1(\rho_0\rho_2)^2\rho_0)^2 = id_G \rangle \end{aligned}$$

In this case, G_0 is the group of the toroidal map $\{3,6\}_{(2,2)}$ and not $\{3,6\}_{(6,0)}$ which covers it. It can easily be checked that replacing the relator $(\rho_2(\rho_3\rho_1)^2\rho_3)^6$ by $(\rho_2(\rho_3\rho_1)^2)^4$ yields the group of the same locally toroidal regular polytope as above.

It should also be noted that the ordering of the residues (arising from the Coxeter diagram) matters. In Section 4 (and the line 1 in the table above) we have a family of regular hypertopes with residues $\{3,6\}_{(s,s)}$, $\{3,6\}_{(1,1)}$, $\{3,6\}_{(2,0)}$, and $\{3,6\}_{(2,0)}$, but the same maps corresponding to other order of residues lead to different hypertopes or may not lead to hypertopes at all.

Acknowledgments

The authors would like to thank the anonymous referees for their comments and suggestions for improvement.

References

[1] Michael Aschbacher, *Flag structures on Tits geometries*, Geom. Dedicata **14** (1983), no. 1, 21–32, DOI 10.1007/BF00182268. MR701748

[2] Wieb Bosma, John Cannon, and Catherine Playoust, *The Magma algebra system. I. The user language*, J. Symbolic Comput. **24** (1997), no. 3-4, 235–265, DOI 10.1006/jsco.1996.0125. Computational algebra and number theory (London, 1993). MR1484478

[3] Francis Buekenhout and Arjeh M. Cohen, *Diagram geometry*, Ergebnisse der Mathematik und ihrer Grenzgebiete. 3. Folge. A Series of Modern Surveys in Mathematics [Results in Mathematics and Related Areas. 3rd Series. A Series of Modern Surveys in Mathematics], vol. 57, Springer, Heidelberg, 2013. Related to classical groups and buildings. MR3014979

[4] Maria Elisa Fernandes and Dimitri Leemans, *C-groups of high rank for the symmetric groups*, J. Algebra **508** (2018), 196–218, DOI 10.1016/j.jalgebra.2018.04.031. MR3810293

[5] Maria Elisa Fernandes, Dimitri Leemans, and Asia Ivić Weiss, *Highly symmetric hypertopes*, Aequationes Math. **90** (2016), no. 5, 1045–1067, DOI 10.1007/s00010-016-0431-1. MR3547707

[6] Maria Elisa Fernandes, Dimitri Leemans, and Asia Ivić Weiss, *Hexagonal extensions of toroidal maps and hypermaps*, Discrete geometry and symmetry, Springer Proc. Math. Stat., vol. 234, Springer, Cham, 2018, pp. 147–170, DOI 10.1007/978-3-319-78434-2_8. MR3816875

[7] Maria Elisa Fernandes, Dimitri Leemans, and Asia Ivić Weiss, *An Exploration of Locally Spherical Regular Hypertopes*, Discrete Comput. Geom. **64** (2020), no. 2, 519–534, DOI 10.1007/s00454-020-00209-9. MR4131559

[8] Maria Elisa Fernandes and Claudio Alexandre Piedade, *Faithful permutation representations of toroidal regular maps*, J. Algebraic Combin. **52** (2020), no. 3, 317–337, DOI 10.1007/s10801-019-00904-8. MR4154605

[9] Peter McMullen and Egon Schulte, *Abstract regular polytopes*, Encyclopedia of Mathematics and its Applications, vol. 92, Cambridge University Press, Cambridge, 2002. MR1965665

[10] Daniel Pellicer, *CPR graphs and regular polytopes*, European J. Combin. **29** (2008), no. 1, 59–71, DOI 10.1016/j.ejc.2007.01.001. MR2368614

[11] J. Tits, *Groupes et géométries de Coxeter*, Notes polycopiées I.H.E.S, Bures sur Yvette, 1961, 26 pages; Heritage of Mathematics, Jacques Tits, Oeuvres Collected Works Volume 1, 803–817, 2013.

[12] Jacques Tits, *Buildings of spherical type and finite BN-pairs*, Lecture Notes in Mathematics, Vol. 386, Springer-Verlag, Berlin-New York, 1974. MR0470099

[13] Asia Ivić Weiss, *An infinite graph of girth* 12, Trans. Amer. Math. Soc. **283** (1984), no. 2, 575–588, DOI 10.2307/1999147. MR737885

Center for Research and Development in Mathematics and Applications, Department of Mathematics, University of Aveiro, Portugal
Email address: maria.elisa@ua.pt

Université Libre de Bruxelles, Département de Mathématique, C.P.216 Algèbre et Combinatoire, Bld du Triomphe, 1050 Bruxelles, Belgium
Email address: dleemans@ulb.ac.be

Center for Research and Development in Mathematics and Applications, Department of Mathematics, University of Aveiro, Portugal
Email address: claudio.a.piedade@ua.pt

Department of Mathematics and Statistics, York University, Toronto, Ontario M3J 1P3, Canada
Email address: weiss@mathstat.yorku.ca

Contemporary Mathematics
Volume **764**, 2021
https://doi.org/10.1090/conm/764/15333

Self-polar polytopes

Alathea Jensen

This paper is dedicated to Jim Lawrence.

Abstract. Self-polar polytopes are convex polytopes that are equal to an orthogonal transformation of their polar sets. These polytopes were first studied by Lovász as a means of establishing the chromatic number of distance graphs on spheres. We investigate the existence, construction, and facial structure of self-polar polytopes, as well as the place of these polytopes within the broader set of self-dual polytopes.

1. Introduction

Convex polytopes are fundamental objects in the field of discrete geometry that have been studied since ancient times. They arise naturally as the feasible sets of systems of linear inequalities, and are also valued for their ability to encode complex combinatorial information of various sorts.

The faces of a polytope are its vertices, edges, facets, and so on, and together they can be arranged in a lattice, partially ordered by inclusion, which is known as the face lattice. The face lattice is also the combinatorial type of the polytope, so that many different polytopes which are not equal as sets in real space may nevertheless have the same combinatorial type.

A great deal has been written about the ways in which a combinatorial type can be realized, including characterizations of realization spaces [**12**], discussions of whether certain combinatorial types are realizable with rational coordinates [**15**], and procedures for determining whether a given lattice is the face lattice of a polytope [**3**].

All polytopes have a dual polytope whose face lattice is the dual of the original's face lattice, and some polytopes are also self-dual. Much study has also been given to self-dual polytopes, including the enumeration of types in low dimensions [**5**], the discovery that self-duality is not necessarily involutory [**1**], and the classification of self-dualities into internal and external types [**4**].

No one has yet devoted any major study, however, to the topic of realizations of self-dual polytopes. In this article, we examine this topic from the point of view of self-polar polytopes. Self-polar is a term we have coined to describe any set that is an orthogonal transformation of its own polar set. Thus, self-polar polytopes

2010 *Mathematics Subject Classification.* Primary 52B05, 52B12, 52B15; Secondary 05C69, 06D50.

Key words and phrases. Polytopes, self dual polytopes, dual polytopes, self dual, polar.

are a subset of self-dual polytopes. They include as a subset the strongly self-dual polytopes of Lovász [**11**], whose work inspired this study.

This article will investigate the basic properties of self-polar polytopes: their existence, construction, facial structure, symmetries, and applications. We will focus in particular on polytopes that are equal to the negative of their polar sets. Our ultimate question is whether all self-dual polytopes are self-polar.

We begin with the necessary definitions and preliminary information in Section 2, then discuss some properties of the orthogonal transformations in Section 3. In Section 4, we discuss self-polar polytopes in two and three dimensions, and in Section 5 we describe ways to build self-polar polytopes in higher dimensions from ones in lower dimensions, and vice versa. In Section 6, we characterize the number of vertices of negatively self-polar polytopes. In Section 7, we describe a way to construct a self-polar polytope from a polytope that is contained in an orthogonal transformation of its polar set, as well as a way to add vertices to a self-polar polytope while maintaining self-polarity. Finally, in Section 8, we summarize our findings and propose future work.

2. Definitions and preliminaries

We assume the reader's familiarity with many basic properties of convex polytopes, including vertex and halfspace definitions, faces, face lattices, duality, and self-duality. Standard references for these properties and for all the lemmas in this section are the books of Grünbaum [**6**] and Ziegler [**16**].

Regarding polytopes, we will only note that by a *polytope*, we mean a convex polytope; that is, a subset of real space $P \subset \mathbb{R}^d$ which can be described as the convex hull of a finite set of points in $\mathbb{R}^d$.

The *polar* of a set $A \subseteq \mathbb{R}^d$ is denoted A° where

$$A^\circ = \{x \in \mathbb{R}^d : \langle x, a \rangle \leq 1 \text{ for all } a \in A\}$$

As the polar operation forms the basis for the majority of the research in this article, we will list some of its properties in Lemma 2.1. Before listing the polar's properties, however, we need to define a notation that will be of much utility throughout this article.

For a set $A \subseteq \mathbb{R}^d$, we will use $[A]$ to denote the closure of the convex hull of A with the origin; that is:

$$[A] = \text{closure}(\text{conv}(A \cup \{o\}))$$

We will refer to this operation as the *polar closure*, and we will say that A is *closed with respect to the polar*. The polar closure operation is indeed a closure operation because Euclidean closure, convex hull, and union with zero are all themselves closure operations.

Basic properties of the polar operation are described by the following lemma.

LEMMA 2.1. *For any $A, B \subseteq \mathbb{R}^d$,*

(1) $A^\circ = [A^\circ]$.
(2) $A^\circ = [A]^\circ$.
(3) $A^{\circ\circ} = [A]$.
(4) $A^{\circ\circ\circ} = A^\circ$.
(5) $A \subset B \implies B^\circ \subset A^\circ$
(6) $(A \cup B)^\circ = A^\circ \cap B^\circ$

(7) $(A \cap B)^\circ = [A^\circ \cup B^\circ]$

The interaction of the polar operation with linear transformations of $\mathbb{R}^d$ is described by the following lemma. By an orthogonal matrix, we mean a matrix M such that $M^T = M^{-1}$.

LEMMA 2.2. *For any $A \subseteq \mathbb{R}^d$ and an invertible matrix $M \in \mathbb{R}^{d\times d}$, $(MA)^\circ = M^{-T}A^\circ$. In particular, if M is orthogonal, then $(MA)^\circ = MA^\circ$.*

Let $\mathcal{P}_0^d$ denote the set of all d-dimensional polytopes in $\mathbb{R}^d$ that contain the origin in their interior. Note that $P = [P]$ for all $P \in \mathcal{P}_0^d$, which is precisely why this set of polytopes is of interest.

It should be clear that every polytope is realizable as an element of $\mathcal{P}_0^d$ for some value of d. To obtain such a realization for a given polytope, we simply restrict the ambient space to the polytope's affine span to make it full-dimensional, and then translate the polytope so that the origin is in its interior.

Now we note two further properties of the polar as it applies to polytopes.

LEMMA 2.3. *The polar operation is an involution on $\mathcal{P}_0^d$; that is, for any $P \in \mathcal{P}_0^d$, $P^\circ \in \mathcal{P}_0^d$ and $P^{\circ\circ} = P$.*

LEMMA 2.4. *For any polytope $P \in \mathcal{P}_0^d$, P° and P are dual polytopes.*

3. Self-polarity

Since polar polytopes are realizations of dual polytopes, and there are many polytopes that are self-dual, it is natural to wonder whether there are any polytopes that are self-polar. It turns out that the answer to this question depends on what we mean by "self-polar". If we mean $P = P^\circ$, then the answer is the following.

THEOREM 3.1. *The only set $A \subseteq \mathbb{R}^d$ for which $A = A^\circ$ is the unit ball, $A = \{x \in \mathbb{R}^d : |x| \leq 1\}$.*

PROOF. Let B denote the unit ball; that is, $B = \{x \in \mathbb{R}^d : |x| \leq 1\}$. It should be clear that $B = B^\circ$ from the definition of the polar operation. Now suppose we have some other $A \subseteq \mathbb{R}^d$ such that $A = A^\circ$. For all $x \in A$, we also have $x \in A^\circ$, so we must have $\langle x, x\rangle \leq 1 \implies |x| \leq 1$. Hence $A \subseteq B = B^\circ$. Then using Lemma 2.1(5) on $A \subseteq B^\circ$, we get $B \subseteq A^\circ = A$. Hence $A = B$. □

In light of this, we must use a slightly more relaxed definition of self-polarity in order to find anything interesting to work with.

We define a set $A \subset \mathbb{R}^d$ as *self-polar* provided there exists some orthogonal transformation U of $\mathbb{R}^d$ such that $A = UA^\circ$. By orthogonal transformation, we mean any nonsingular linear transformation U such that $U^T = U^{-1}$. In general, these are rotations and reflections. In the case that $A = -A^\circ$, we say that A is *negatively self-polar.*

We will refer to such an orthogonal transformation as a *self-polarity map* of the set A. In the case of a polytope P, note that every self-polarity map induces a dual automorphism of the face lattice via the mapping from each face of P to its dual face in the polar polytope P°, to the image of that dual face in UP° after the orthogonal transformation, which is again some face of P. Hence a given polytope can have at most as many distinct self-polarity maps as its face lattice has distinct dual automorphisms.

We will soon begin constructing many examples of self-polar polytopes, so it is worthwhile to have a simple way of verifying the property of self-polarity. The next theorem fulfills that goal.

THEOREM 3.2. *For a polytope $P \in \mathcal{P}_0^d$ and an orthogonal transformation U, $P = UP^\circ$ if and only if both of the following are true.*

(1) *For all vertices v, w of P, $\langle Uv, w\rangle \leq 1$.*
(2) *For each facet F of P, there is a vertex v of P such that $\langle Uv, w\rangle = 1$ for all vertices $w \in \operatorname{vert}(F)$.*

PROOF. The first condition guarantees that $P \subseteq UP^\circ$. As for the second condition, each vertex v of UP° is generated by a facet F of P, so that $\langle v, Uw\rangle = 1$ for all vertices $w \in \operatorname{vert}(F)$. Thus the second condition guarantees that we must have $\operatorname{vert}(UP^\circ) \subseteq P$. Hence $P = UP^\circ$. □

Before proceeding to show examples of self-polar polytopes and explore their properties, it is interesting to note that we can say something about the properties of the self-polarity maps of polytopes even without knowing anything about the polytopes themselves. First, recall that a matrix M is *periodic* if there is some $n \in \mathbb{N}$ such that M^n is the identity. The *period of a matrix* M is the minimum $n \in \mathbb{N}$ such that M^n is the identity.

THEOREM 3.3. *Let $P = UP^\circ$ for some polytope $P \in \mathcal{P}_0^d$ and some orthogonal map U. Then $U^2P = P$ and U has an even period.*

PROOF. Assume the conditions of the theorem statement and then apply the polar operation to both sides of $P = UP^\circ$ to obtain $P^\circ = (UP^\circ)^\circ$. Using Lemmas 2.2 and 2.1(3), this becomes $P^\circ = UP^{\circ\circ} = UP$. Then, substituting UP for P° in $P = UP^\circ$, we get $P = U^2P$. Hence U^2 is a symmetry of P.

Any symmetry of a polytope must have a finite period. This is so because a symmetry maps vertices to vertices. Each $U^2, U^4, U^6, \ldots$ induces a permutation on the vertices of P, and there are a finite number of such permutations. Hence there exist some $i, j \in \mathbb{N}$ with $i < j$ such that U^{2i} and U^{2j} induce the same permutation of the vertices of P. But since the vertices of P, regarded as vectors, span the space $\mathbb{R}^d$, we must have $U^{2i} = U^{2j}$, which implies $I = U^{2(j-i)}$. Thus U is periodic.

Now suppose that U has an odd period, so that there is some $k \in \mathbb{N}$ such that $U^{2k-1} = I$. Then $U^{2k-1}P = P$. But we also know that $U^{2k}P = P$, so together these imply $U^{2k-1}P = U^{2k}P$, which implies $P = UP$. Then substituting into $P^\circ = UP$ we get $P^\circ = P$, which by Theorem 3.1 implies that P is a unit ball, contradicting our assumption that P is a polytope. Hence U has an even period. □

4. In low dimensions

4.1. Two dimensions. In two dimensions, it is well-known that every polygon is self-dual. If we start with a regular polygon, centered at the origin, we obtain as its polar a dilated and rotated copy of the same polygon. Let us refer to the distance between origin and vertex in such a polygon as its *radius*. When we increase the radius of such a polygon, the radius of its polar decreases, and vice versa. See Figure 1 for examples. Hence there is always some radius for which the polar radius is the same as the original radius.

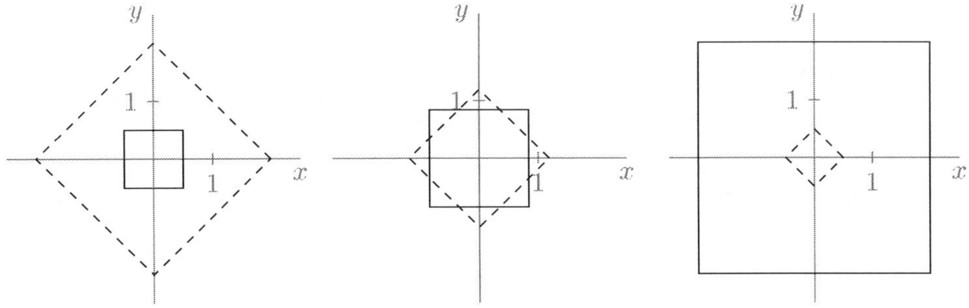

FIGURE 1. Examples of a regular, origin-centered square (solid) and its polar set (dashed) with three different radii.

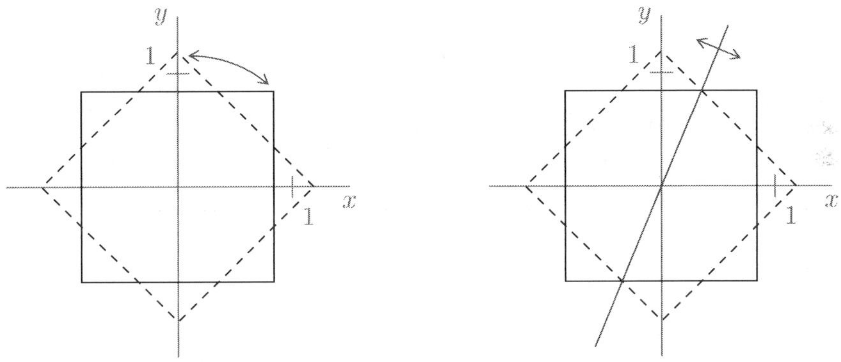

FIGURE 2. Examples of a self-polarity by a rotation of 45° (left panel) and by reflection across the red line (right panel).

When their radii are the same, a regular, origin-centered polygon and its polar can be made identical by a suitable rotation. We can also make the two polygons identical by a reflection, if we let the axis of symmetry connect the origin and any one of the points of intersection between the boundary of the polygon and the boundary of its polar. See Figure 2 for an example.

Since the dihedral group for an n-gon has $2n$ elements, there are evidently only $2n$ possible dual automorphisms of the n-gon's face lattice. We will show that any n-gon is realizable in such a way that it is self-polar by $2n$ distinct orthogonal transformations, each of which realizes a different one of the possible dual automorphisms.

THEOREM 4.1. *For $n \in \mathbb{N}$, an n-sided polygon is realizable as self-polar by $2n$ distinct orthogonal transformations, n of which are reflections and the other n of which are rotations.*

PROOF. Consider a regular polygon P with n sides, centered at the origin, and given by the vertices $\{v_1, v_2, \dots, v_n\}$ where

$$v_i := \left(\sqrt{\sec\left(\frac{\pi}{n}\right)}\cos\left(\frac{2\pi}{n}i\right), \sqrt{\sec\left(\frac{\pi}{n}\right)}\sin\left(\frac{2\pi}{n}i\right)\right)$$

Let U be a counterclockwise rotation of π/n. Note that U is an orthogonal transformation. Then each Uv_i is given by

$$Uv_i := \left(\sqrt{\sec\left(\frac{\pi}{n}\right)}\cos\left(\frac{2\pi}{n}i + \frac{\pi}{n}\right), \sqrt{\sec\left(\frac{\pi}{n}\right)}\sin\left(\frac{2\pi}{n}i + \frac{\pi}{n}\right)\right)$$

Consider that the maximum dot product between any v_i and Uv_j is between vertices that are as close together as possible, hence pairs of the form $\langle v_i, Uv_i\rangle$ or $\langle v_i, Uv_{i-1}\rangle$. Using trigonometric identities, we these dot products are

$$\begin{aligned}\langle v_i, Uv_i\rangle &= \sec\left(\frac{\pi}{n}\right)\left(\cos\left(\frac{2\pi}{n}i\right)\cos\left(\frac{2\pi}{n}i + \frac{\pi}{n}\right) + \sin\left(\frac{2\pi}{n}i\right)\sin\left(\frac{2\pi}{n}i + \frac{\pi}{n}\right)\right)\\ &= \sec\left(\frac{\pi}{n}\right)\cos\left(\frac{\pi}{n}\right)\\ &= 1\end{aligned}$$

and

$$\begin{aligned}\langle v_i, Uv_{i-1}\rangle &= \sec\left(\frac{\pi}{n}\right)\left(\cos\left(\frac{2\pi}{n}i\right)\cos\left(\frac{2\pi}{n}i - \frac{\pi}{n}\right) + \sin\left(\frac{2\pi}{n}i\right)\sin\left(\frac{2\pi}{n}i - \frac{\pi}{n}\right)\right)\\ &= \sec\left(\frac{\pi}{n}\right)\cos\left(\frac{\pi}{n}\right)\\ &= 1\end{aligned}$$

Hence the first condition of Theorem 3.2 is fulfilled. As for the second condition, the facets of P are of the form $\operatorname{conv}(v_{i-1}, v_i)$. Both these vertices have a dot product of 1 with Uv_{i-1}. Hence $P = UP^\circ$.

Now consider that U^2 is a symmetry of P, because P is regular with n sides and U is a rotation by π/n. This means $U^{2j}P = P$ for all $j \in \mathbb{N}$, and so $P = UP^\circ \implies P = U(U^{2j}P)^\circ = U^{2j+1}P^\circ$. Only $U, U^3, U^5, \ldots, U^{2n-1}$ are distinct, however, because $U^{2n} = I$. These are the n distinct rotations mentioned in the theorem statement.

Let R be reflection across the x-axis. Then there are n other symmetries of P, given by $R, U^2R, U^4R, U^6R, \ldots, U^{2n-2}R$, and, just as before, $P = UP^\circ \implies P = U(U^{2j}RP)^\circ = U^{2j+1}RP^\circ$. These are the n distinct reflections mentioned in the theorem statement. □

Negatively self-polar polytopes are of particular interest due to their application to sphere-coloring. From the last theorem, we can already say that odd polygons are realizable as negatively self-polar.

COROLLARY 4.2. *Polygons with an odd number of sides are realizable as negatively self-polar.*

PROOF. For $\mathbb{R}^2$, the negative transformation $-I$ is equivalent to a rotation of π. If P is as described in the proof of Theorem 4.1 and has n sides, and U is a rotation by π/n, then $U, U^3, U^5, \ldots, U^{2n-1}$ are self-polarity maps of P. If n is odd, then U^n must also be in this list. Since U^n is a rotation by π, we conclude that $P = -P^\circ$. □

Now let us consider even polygons. First of all, if our goal is to construct a negatively self-polar even polygon, it ought to be clear that a regular even polygon, centered at the origin, cannot be negatively self-polar. This is so because for any such realization, call it P, we have $P = -P$ and so $P^\circ = -P^\circ$, and so $P = -P^\circ$ would imply $P = P^\circ$, which is impossible except in the case of the unit ball.

It is not necessary to use a regular polygon to achieve self-polarity, however. This opens the door to tremendous number of possible realizations, suggesting perhaps a linear algebra-based approach. However, it is not necessary to turn to such techniques, because the reason for the impossibility of negatively self-polar even polygons is combinatorial, as we shall show.

Recall that any self-polarity map gives rise to a dual automorphism on the face lattice of the polytope. For reflections, such a map is an involution. We will show that for an even polygon, it is impossible to produce an involutory dual automorphism without mapping some facet to one of the vertices that it contains.

This is significant because when this is the case, it is impossible to realize the dual automorphism as the negative self-polarity map. This impossibility is true of any polytope, not only the two-dimensional sort, so we will pause to state it as a lemma before applying it to the case of even polygons.

LEMMA 4.3. *A dual automorphism ϕ on the face lattice of a polytope cannot be realized as the negative self-polarity map when there exists a vertex v for which $\phi(v)$ contains v.*

PROOF. In a negatively self-polar polytope, each vertex v is paired with a facet f which lies in the boundary of the half-space $\{x \in \mathbb{R}^d : \langle -v, x\rangle \leq 1\}$. If v is in fact an element of this facet, then $\langle -v, v\rangle = 1$. But $\langle -v, v\rangle = 1$ implies $|v|^2 = -1$, which is impossible. So v cannot be contained in f. Hence any dual automorphism ϕ for which $\phi(v)$ contains v cannot be realized by a negative self-polarity map. □

THEOREM 4.4. *No even polygon is negatively self-polar.*

PROOF. We begin by noting that an involutory dual automorphism partners each vertex with a facet, in such a way that if vertex v is in facet f, then the partner vertex of f is contained in the partner facet of v.

Now suppose that we have an even polygon with our vertices labeled 0, 1, 2, ..., $n-1$, going in order around the polygon, and we have likewise labeled the facets by ordered pairs of the vertices they contain, so that facet $(i, i+1)$ contains vertices i and $i+1$, and the addition operation in the expression $i+1$ is understood to be modulo n.

Now suppose we have a pairing of vertices and facets that corresponds to an involutory dual automorphism. Suppose that vertex 0 has facet $(i, i+1)$ as its partner. Then vertex i is paired with a facet that contains 0, either (Case 1) facet $(0, 1)$ or (Case 2) facet $(n-1, 0)$.

In Case 1, 0 is paired with $(i, i+1)$ and i with $(0, 1)$. Then 1 is paired with a facet that contains i, but there is only one of those that isn't paired yet: $(i-1, i)$. In fact, from this point on, every pairing is necessitated by the pairings that we have already looked at. In the order that we can deduce them, and including those previously decided, they are:

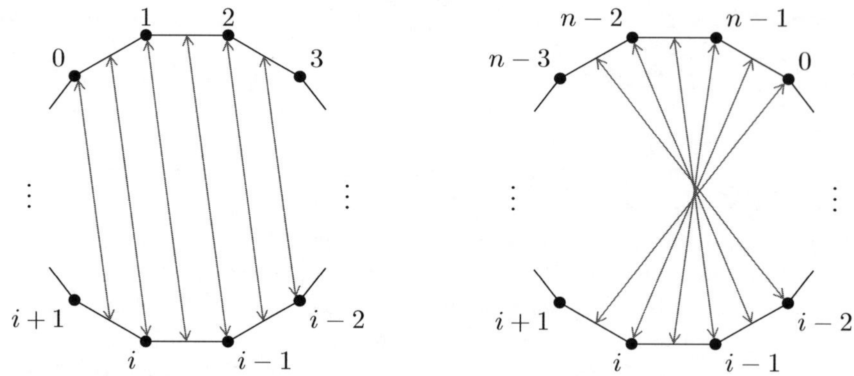

FIGURE 3. Diagram of pairings in Case 1 (left) and Case 2 (right) of Theorem 4.4.

$$\begin{aligned} 0 &\leftrightarrow (i, i+1) \\ i &\leftrightarrow (0, 1) \\ 1 &\leftrightarrow (i-1, i) \\ i-1 &\leftrightarrow (1, 2) \\ 2 &\leftrightarrow (i-2, i-1) \\ i-2 &\leftrightarrow (2, 3) \\ &\vdots \end{aligned}$$

See Figure 3 (left) for a diagram of the pairings.

We see that there are two alternating patterns in which vertices $0, 1, 2, \ldots$ are paired with facets $(i, i+1), (i-1, i), (i-2, i-1), \ldots$ respectively, and vertices $i, i-1, i-2, \ldots$ are paired with facets $(0, 1), (1, 2), (2, 3), \ldots$ respectively. In both patterns, vertex k is paired with facet $(i-k, i-k+1)$ (using addition modulo n), hence the patterns are consistent with one another.

Consider now the situation when i is even and $k = \frac{i}{2}$. Then vertex k is paired with facet $(k, k+1)$. On the other hand, if i is odd, then when $k = \frac{i+1}{2}$, vertex k is paired with facet $(k-1, k)$. In either case, the pattern of pairings forces the existence of a vertex paired with one of the facets containing it, which, by Lemma 4.3, means that this dual automorphism cannot be realized by a negative self-polarity.

Now for Case 2, in which 0 is paired with $(i, i+1)$ and i with $(n-1, 0)$. A similar situation arises here:

$$\begin{aligned} 0 &\leftrightarrow (i, i+1) \\ i &\leftrightarrow (n-1, 0) \\ n-1 &\leftrightarrow (i-1, i) \\ i-1 &\leftrightarrow (n-2, n-1) \\ n-2 &\leftrightarrow (i-2, i-1) \\ i-2 &\leftrightarrow (n-3, n-2) \\ &\vdots \end{aligned}$$

See Figure 3 (right) for a diagram of the pairings.

We see that there are two alternating patterns in which vertices $0, n-1, n-2, \ldots$ are paired with facets $(i, i+1), (i-1, i), (i-2, i-1), \ldots$ respectively, and vertices $i, i-1, i-2, \ldots$ are paired with facets $(n-1, 0), (n-2, n-1), (n-3, n-2), \ldots$ respectively. In the first pattern, vertex k is paired with facet $(k+i, k+i+1)$, while in the second pattern, vertex k is paired with facet $(k-i-1, k-i)$, where all of these expressions are modulo n.

These two patterns must be consistent with one another, of course. The first pattern pairs vertex 0 with facet $(i, i+1)$, while the second pairs vertex 0 with facet $(n-i-1, n-1)$. This is the same facet, so we have $i = n-i-1$, which implies $n = 2i+1$, meaning that n is odd, contrary to our assumption. □

4.2. Three dimensions. Having completed our investigation in two dimensions, we proceed to three dimensions. Here, we begin with the happy discovery that, as with two dimensions, every self-dual three-dimensional polytope has a self-polar realization.

First we need a lemma.

LEMMA 4.5. *Let the affine hull of a face f of a polytope $P \in \mathcal{P}_0^d$ be tangent to an origin-centered sphere of radius r at point x. Then the affine hull of the corresponding dual face g of P° is tangent to an origin-centered sphere of radius $1/r$ at point x/r^2.*

PROOF. Let x' be the point of tangency for $\operatorname{aff}(f)$. Then f is orthogonal to x', so every $x \in f$ can be expressed as $x = x' + w$ where w is some vector orthogonal to x'.

Now consider $y' = x'/r^2$. We know y' is in $\operatorname{aff}(g)$ because for any $x \in f$ there is some w orthogonal to x' so that we have $x = x' + w$, which gives us

$$\langle x, y' \rangle = \left\langle x' + w, \frac{x'}{r^2} \right\rangle = \frac{1}{r^2}\left(\langle x', x' \rangle + \langle w, x' \rangle\right) = \frac{1}{r^2}(r^2 + 0) = 1$$

Now consider that any vector which lies in g is orthogonal to y'. This is because any vector lying in g is of the form $y_1 - y_2$ for some $y_1, y_2 \in g$, and we have

$$\begin{aligned} \langle y', y_1 - y_2 \rangle = \left\langle \frac{x'}{r^2}, y_1 - y_2 \right\rangle &= \frac{1}{r^2}\langle x', y_1 - y_2 \rangle \\ &= \frac{1}{r^2}\left(\langle x', y_1 \rangle - \langle x', y_2 \rangle\right) = \frac{1}{r^2}(1 - 1) = 0 \end{aligned}$$

Since g is orthogonal to y', and $|y'| = 1/r$, the lemma's statement follows. □

THEOREM 4.6. *Every self-dual three-dimensional polytope has a self-polar realization.*

PROOF. The Koebe-Andreev-Thurston theorem implies that each 3-connected planar graph is realizable a 3-polytope which has all edges tangent to the unit sphere and that the edge-tangent realization for which the barycenter of the tangency points is the center of the sphere is unique up to orthogonal transformations. See [**13**] for more details and a proof.

For a given combinatorial type, let us call this realization P. From Lemma 4.5, the edges of P° will also be tangent to the unit sphere, with their tangency points in the same locations as those of P, hence having their barycenter at the origin. If the polar P° is of the same combinatorial type as P, then its realization in this form is unique up to orthogonal transformation, hence we have $P = UP^\circ$ for some orthogonal transformation U which maps edge-tangency points onto edge-tangency points. □

It is worthwhile to note at this point that self-polarities need not be involutory, as the following example will make clear.

EXAMPLE 4.7. The classic example of a self-dual polytope without an involutory dual automorphism is Jendroľ's polytope [**9**]. Jendroľ's polytope is self-polar, however, when realized as shown in Figure 4.

In this realization, edges aa', bb', ff', and gg' are parallel to the z-axis. Projection of the polytope onto the (x, y)-coordinates forms a regular, origin-centered decagon with each vertex at a distance of $\sqrt{\sec(\pi/10)}$ from the origin, as described in the proof of Theorem 4.1. Vertices a and f have z-coordinate 1, while vertices a' and f' have z-coordinate -1. Similarly, vertices b and g have z-coordinate $2(1 - \cos(\pi/5))$, while vertices b' and g' have z-coordinate $-2(1 - \cos(\pi/5))$. The rest of the vertices have z-coordinate 0. This realization of the polytope is self-polar via 90° rotation about the z-axis.

Now that we know all self-dual three-dimensional polytopes are realizable as self-polar, we would like to know more specifically which three-dimensional polytopes are realizable as negatively self-polar. It is not very difficult to construct a few examples that demonstrate we can have (almost) any number of vertices we like.

THEOREM 4.8. *For $n = 4$ and $n \geq 6$, there exists a 3-dimensional polytope $P = -P^\circ$ with n vertices.*

PROOF. The proof is by construction of three different types of polytopes: those with an even number of vertices, those with the number of vertices congruent to $3 \mod 4$, and those with the number of vertices congruent to $1 \mod 4$.

For the first type, those with an even number of vertices, we construct a pyramid over an odd polygon. For $k \in \mathbb{N}$, let P be the convex hull of $(0, 0, 1)$ and

$$\left(\sqrt{2\sec\left(\frac{\pi}{2k+1}\right)}\cos\left(\frac{2\pi i}{2k+1}\right), \sqrt{2\sec\left(\frac{\pi}{2k+1}\right)}\sin\left(\frac{2\pi i}{2k+1}\right), -1\right)$$

for $i = 1, 2, 3, ..., 2k + 1$. Then $P = -P^\circ$ and has $2k + 2$ vertices.

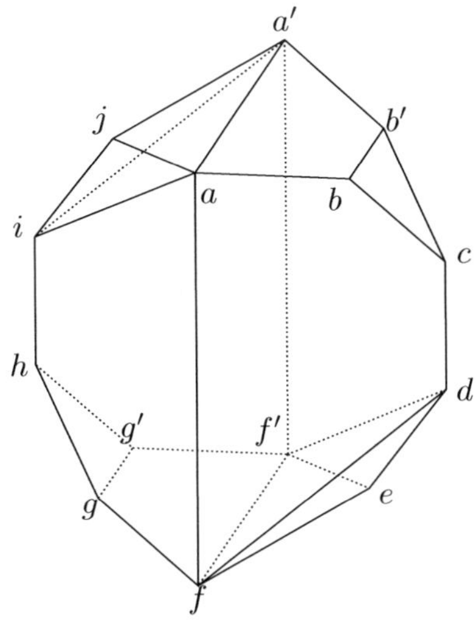

FIGURE 4. Self-polar realization of Jendroľ's polytope.

For the second type, we construct a pyramid over an odd polygon, glued to a prism over the same polygon. For $k \in \mathbb{N}$, let P be the convex hull of $(0,0,1)$ and

$$\left(\sqrt{\sec\left(\frac{\pi}{2k+1}\right)}\cos\left(\frac{2\pi i}{2k+1}\right), \sqrt{\sec\left(\frac{\pi}{2k+1}\right)}\sin\left(\frac{2\pi i}{2k+1}\right), 0\right)$$

for $i = 1, 2, 3, ..., 2k+1$, and

$$\left(\sqrt{\sec\left(\frac{\pi}{2k+1}\right)}\cos\left(\frac{2\pi i}{2k+1}\right), \sqrt{\sec\left(\frac{\pi}{2k+1}\right)}\sin\left(\frac{2\pi i}{2k+1}\right), -1\right)$$

for $i = 1, 2, 3, ..., 2k+1$. Then $P = -P^\circ$ and has $4k+3$ vertices.

For the third type, we construct a pyramid over an even polygon with deep truncations at each vertex of the base. For $k \in \mathbb{N}$ and $k \geq 2$, let P be the convex hull of $\left(0, 0, \cot\left(\frac{\pi}{2k}\right)\right)$ and

$$\left(\cos\left(\frac{\pi i}{k}\right), \sin\left(\frac{\pi i}{k}\right), 0\right)$$

for $i = 1, 2, 3, ..., 2k$, and

$$\left(\sec\left(\frac{\pi}{2k}\right)\cos\left(\frac{\pi i}{k} + \frac{\pi}{2k}\right), \sec\left(\frac{\pi}{2k}\right)\sin\left(\frac{\pi i}{k} + \frac{\pi}{2k}\right), -\tan\left(\frac{\pi}{2k}\right)\right)$$

for $i = 1, 2, 3, ..., 2k$. Then $P = -P^\circ$ and has $4k+1$ vertices. □

Shown in Figure 5 is an example with $k = 2$ for each of the constructions.

As for five vertices, which was missing from Theorem 4.8, there are only two different combinatorial types of polytopes with 5 vertices in $\mathbb{R}^3$, and, of those, only the square pyramid is self-dual. It turns out that the square pyramid cannot be realized as negatively self-polar because its base cannot be realized as negatively self-polar. The proof of this fact appears a bit later, however, in Theorem 5.5.

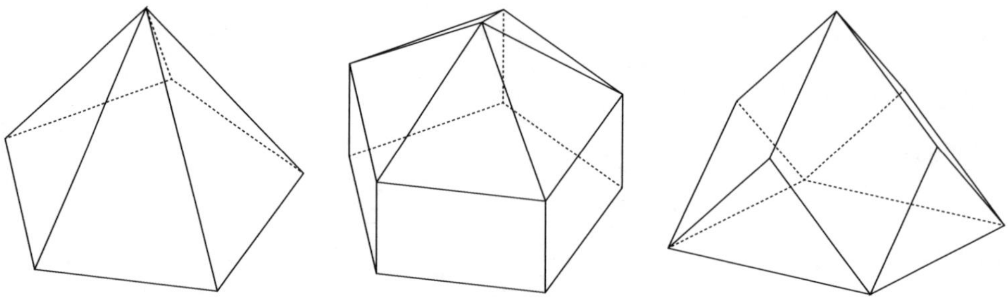

FIGURE 5. Examples of the constructions in Theorem 4.8 with $k = 2$.

Now that we know how many vertices are possible for negatively self-polar three-dimensional polytopes, we also know the possible f-vectors (f_0, f_1, f_2).

COROLLARY 4.9. *If $P = -P^\circ$ for $P \in \mathcal{P}_0^3$, then $f(P) = (n, 2n-2, n)$ for some $n \geq 4$ and $n \neq 5$.*

PROOF. The Euler-Poincaré formula tells us that $-1 + f_0 - f_1 + f_2 - 1 = 0$, hence $f_1 = f_0 + f_2 - 2$. We must have $f_0 = f_2$ for any self-dual three-dimensional polytope, so this becomes $f_1 = 2f_0 - 2$. We know the possible values of n from Theorem 4.8 and 5.5. □

5. In higher and lower dimensions

It was simple to construct concrete examples of self-polar polytopes in two dimensions, and not very difficult in three, but in higher dimensions, the path forward is less clear. We were able to show that all 3-dimensional self-dual polytopes are realizable as self-polar by exploiting the Koebe-Andreev-Thurston theorem, but the same proof technique will not work in higher dimensions because there is no higher-dimensional analog of the Koebe-Andreev-Thurston theorem. Indeed, it has been shown by Schulte [**14**] that for any dimension $d \geq 4$ and $k \in \{0, 1, \ldots, d-1\}$, there are infinitely many polytopes of dimension d which cannot be realized with their k-faces tangent to a sphere.

We can nevertheless construct many examples of self-polar polytopes in dimension 4 and greater by beginning with self-polar bases and building over them in higher dimensions.

5.1. Pyramids. We begin with a pyramidal construction that allows us to construct self-polar pyramids in higher dimensions over self-polar bases.

THEOREM 5.1. *For a polytope $P \in \mathcal{P}_0^d$ that is self-polar by orthogonal transformation U, a pyramid over P is realizable as self-polar by the transformation W, where*

$$W := \begin{pmatrix} U & 0 \\ 0 & -1 \end{pmatrix}$$

PROOF. First we will construct a realization of the pyramid. Let $Q \subseteq \mathbb{R}^{d+1}$ be the pyramid formed by appending $a \in \mathbb{R} \setminus \{0\}$ as the last coordinate to $\sqrt{1+a^2}P$, and taking the convex hull with $\left(0, ..., 0, -\frac{1}{a}\right)$. For $x = (x_1, x_2, \ldots, x_d) \in \mathbb{R}^d$ and $y \in \mathbb{R}$, we will use the notation $(x; y)$ to stand for $(x_1, x_2, \ldots, x_d, y) \in \mathbb{R}^{d+1}$.

Now we will use Theorem 3.2 to show that Q is self-polar by W. For the first condition, we need to identify the vertices of Q, which are $(0, ..., 0, -\frac{1}{a})$ and $(\sqrt{1+a^2}v; a)$ for $v \in \text{vert}(P)$. Let v, w be any two vertices of P. Then we have

(5.1)
$$\begin{aligned}\left\langle\left(\sqrt{1+a^2}v;a\right), W\left(\sqrt{1+a^2}w;a\right)\right\rangle &= \left\langle\left(\sqrt{1+a^2}v;a\right), \left(\sqrt{1+a^2}Uw;-a\right)\right\rangle \\ &= \left\langle\sqrt{1+a^2}v, \sqrt{1+a^2}Uw\right\rangle - a^2 \\ &= (1+a^2)\langle v, Uw\rangle - a^2 \\ &\leq (1+a^2)(1) - a^2 \\ &= 1\end{aligned}$$

As for the apex vertex of Q, we have

$$\begin{aligned}\left\langle\left(\sqrt{1+a^2}v;a\right), W\left(0, ..., 0, -\frac{1}{a}\right)\right\rangle &= \left\langle\left(\sqrt{1+a^2}v;a\right), \left(0, ..., 0, \frac{1}{a}\right)\right\rangle \\ &= a \cdot \frac{1}{a} \\ &= 1\end{aligned} \tag{5.2}$$

Hence not only do the vertices of Q fulfill the first condition of Theorem 3.2, but from Equation 5.2 we can see that we have also fulfilled the second condition for the facet of Q that is the scaled and translated copy of P.

As for the other facets of Q, each facet F of Q corresponds to a facet $\hat{F}$ of P in such a way that F has as its vertices $(0, ..., 0, -\frac{1}{a})$ and $(\sqrt{1+a^2}v; a)$ for $v \in \text{vert}(\hat{F})$. Since P is self-polar by U, we know that for each facet $\hat{F}$ of P there is a vertex w of P for which $\langle v, Uw\rangle = 1$ for all $v \in \text{vert}(\hat{F})$. Then Equation 5.1 applies, but with equality rather than inequality, and Equation 5.2 applies with w substituted for v. Thus we have fulfilled the second condition of Theorem 3.2 for all facets of Q. □

5.2. Joins. For polytopes $P \in \mathcal{P}_0^{d_1}$ and $Q \in \mathcal{P}_0^{d_2}$, the *join* of P and Q, denoted $P * Q$, is the combinatorial type of a $d_1 + d_2 + 1$ polytope which can be realized by embedding P and Q into orthogonal linear subspaces of $\mathbb{R}^{d_1+d_2+1}$, then translating the two subspaces apart along the one-dimensional subspace of $\mathbb{R}^{d_1+d_2+1}$ that is orthogonal to both, and finally taking the convex hull.

For every face f of P and g of Q, the join $f * g$ is a face of $P * Q$, and conversely, every face of $P * Q$ is the join of a face from P with a face from Q. Here, we are including P and Q as well as the empty set as faces. (Note that $f * \emptyset = f$.) Just as with P and Q themselves, the dimension of a face $f * g$ of $P * Q$ is $\dim(f) + \dim(g) + 1$. [**8**]

Since a pyramid is the join of a polytope with a point, it seems natural to suppose that, as with pyramids, joins of self-polar polytopes might be self-polar. This is indeed the case.

THEOREM 5.2. *For polytopes $P_1 \in \mathcal{P}_0^{d_1}$ and $P_2 \in \mathcal{P}_0^{d_2}$ that are self-polar by orthogonal transformations U_1 and U_2 respectively, the join of P_1 and P_2 is realizable* *as self-polar by the transformation*

$$W := \begin{pmatrix} U_1 & 0 & 0 \\ 0 & U_2 & 0 \\ 0 & 0 & -1 \end{pmatrix}$$

PROOF. First we construct a realization of the join. Let $Q \subseteq \mathbb{R}^{d_1+d_2+1}$ be the join formed by the following process. To each point of $\sqrt{1+a^2}P_1$, we append d_2 zeros and then $a \in \mathbb{R} \setminus \{0\}$ as the last coordinate. To each point of $\sqrt{1+\frac{1}{a^2}}P_2$, we insert d_1 zeros before the first coordinate, and then append $-\frac{1}{a}$ as the last coordinate. Finally, we take the convex hull.

For $x = (x_1, x_2, \dots, x_{d_1}) \in \mathbb{R}^{d_1}$, $y = (y_1, y_2, \dots, y_{d_2}) \in \mathbb{R}^{d_2}$, and $z \in \mathbb{R}$, we will use the notation $(x; y; z)$ to stand for $(x_1, x_2, \dots, x_{d_1}, y_1, y_2, \dots, y_{d_2}, z) \in \mathbb{R}^{d_1+d_2+1}$.

Now we will use Theorem 3.2 to show that Q is self-polar by W. For the first condition, we need to confirm that any two vertices of Q have a dot product of ≤ 1. Clearly the vertices of Q are of two types: vertices of the embedded copy of P_1 and vertices of the embedded copy of P_2. Let v, w be any two vertices of P_1. Then we have

$$\begin{aligned}(5.3) \quad &\left\langle \left(\sqrt{1+a^2}v; 0, \dots, 0; a\right), W\left(\sqrt{1+a^2}w; 0, \dots, 0; a\right)\right\rangle \\ &= \left\langle \left(\sqrt{1+a^2}v; 0, \dots, 0; a\right), \left(\sqrt{1+a^2}U_1 w; 0, \dots, 0; -a\right)\right\rangle \\ &= \left\langle \sqrt{1+a^2}v, \sqrt{1+a^2}U_1 w\right\rangle - a^2 \\ &= (1+a^2)\langle v, U_1 w\rangle - a^2 \\ &\leq (1+a^2)(1) - a^2 \\ &= 1\end{aligned}$$

On the other hand, let v, w be any two vertices of P_2. Then we have

$$\begin{aligned}(5.4) \quad &\left\langle \left(0, \dots, 0; \sqrt{1+\frac{1}{a^2}}v; -\frac{1}{a}\right), W\left(0, \dots, 0; \sqrt{1+\frac{1}{a^2}}w; -\frac{1}{a}\right)\right\rangle \\ &= \left\langle \left(0, \dots, 0; \sqrt{1+\frac{1}{a^2}}v; -\frac{1}{a}\right), \left(0, \dots, 0; \sqrt{1+\frac{1}{a^2}}U_2 w; \frac{1}{a}\right)\right\rangle \\ &= \left\langle \sqrt{1+\frac{1}{a^2}}v, \sqrt{1+\frac{1}{a^2}}U_2 w\right\rangle - \frac{1}{a^2} \\ &= \left(1+\frac{1}{a^2}\right)\langle v, U_2 w\rangle - \frac{1}{a^2} \\ &\leq \left(1+\frac{1}{a^2}\right)(1) - \frac{1}{a^2} \\ &= 1\end{aligned}$$

Finally, for a vertex v of P_1 and a vertex w of P_2, we have

$$\begin{aligned}(5.5)\quad &\left\langle\left(\sqrt{1+\frac{1}{a^2}}v;0,\ldots,0;a\right),W\left(0,\ldots,0;\sqrt{1+\frac{1}{a^2}}w;-\frac{1}{a}\right)\right\rangle\\ &=\left\langle\left(\sqrt{1+\frac{1}{a^2}}v;0,\ldots,0;a\right),\left(0,\ldots,0;\sqrt{1+\frac{1}{a^2}}U_2w;\frac{1}{a}\right)\right\rangle\\ &=a\cdot\frac{1}{a}\\ &=1\end{aligned}$$

Now for the second condition of Theorem 3.2, we need to identify the facets of Q. Since the dimension of each facet is d_1+d_2, and every face of Q is the join of a face from P_1 with a face from P_2, each facet of Q is either the join of P_1 with a facet of P_2, or the join of a facet of P_1 with P_2.

Let F be a facet of P_1, and let F' be the corresponding facet of Q which is the join of F with P_2. Since P_1 is self-polar by U_1, there is some vertex w of P_1 such that $\langle v, U_1 w\rangle = 1$ for all $v\in \mathrm{vert}(F)$. Then Equation 5.3 applies to all such v, with equality rather than inequality. As for the vertices of F' which come from P_2 rather than from P_1, Equation 5.5 applies to them.

Now for the other case, let F be a facet of P_2, and let F' be the corresponding facet of Q which is the join of F with P_1. Since P_2 is self-polar by U_2, there is some vertex w of P_2 such that $\langle v, U_2 w\rangle = 1$ for all $v \in \mathrm{vert}(F)$. Then Equation 5.4 applies to all such v, with equality rather than inequality. As for the vertices of F' which come from P_1 rather than from P_2, Equation 5.5 applies to them. □

5.3. Sections and projections. Now that we have shown how to construct self-polar polytopes in higher dimensions from those in lower dimensions, it is natural to consider whether we can also construct self-polar polytopes in lower dimensions from those in higher dimensions.

First, we need to establish how the polar operation works on sections and projections.

LEMMA 5.3. *For a polytope $P\in\mathcal{P}_0^d$ and a linear subspace H of $\mathbb{R}^d$, the polar of the orthogonal projection of P onto H is the intersection of the polar of P with H:*

$$(\mathrm{proj}_H(P))^\circ = P^\circ \cap H$$

PROOF. Note that $\mathrm{proj}_H(P)$ is given by AA^TP, where the columns of A are an orthonormal basis of H. Note also that we are taking the polar in H only, not in $\mathbb{R}^d$. Since our projection is orthogonal, we have

$$\begin{aligned}(\mathrm{proj}_H(P))^\circ &= \{y\in H : \langle y, AA^Tx\rangle \le 1 \text{ for all } x\in P\}\\ &= \{y\in H : \langle AA^Ty, x\rangle \le 1 \text{ for all } x\in P\}\\ &= \{y\in H : \langle y, x\rangle \le 1 \text{ for all } x\in P\}\\ &= P^\circ\cap H\end{aligned}$$

□

Now we can state conditions for the existence of a lower-dimensional self-polar projection or cross-section.

THEOREM 5.4. *Let $P = UP^\circ$ for $P \in \mathcal{P}_0^d$ and orthogonal transformation U. For a linear subspace H of $\mathbb{R}^d$, let $UH = H$ and $\operatorname{proj}_H(P) = P \cap H$. Then $P \cap H$ is self-polar by the restriction of U to H.*

PROOF. We know from Lemma 5.3 that $(\operatorname{proj}_H(P))^\circ = P^\circ \cap H$, so by assumption we have $(P \cap H)^\circ = P^\circ \cap H$. Multiplying both sides by U, we get $U(P \cap H)^\circ = UP^\circ \cap UH$. Since $P = UP^\circ$ and $UH = H$, we obtain $U(P \cap H)^\circ = P \cap H$. □

Another consequence of Lemma 5.3 is that it gives us a condition for the existence of a self-polar pyramid for the negative transformation.

THEOREM 5.5. *Let $P \in \mathcal{P}_0^d$ be a pyramid with base Q. If P is negatively self-polar, then Q has a negatively self-polar realization.*

PROOF. Suppose $P = -P^\circ \in \mathcal{P}_0^d$ is a pyramid over base Q. Let $a \in \mathbb{R}^d$ be the apex of the pyramid. From Lemma 4.3, we know that the dual automorphism of the face lattice that is realized by the negative transformation must pair a with Q. Furthermore, from Lemma 4.5, we know that Q is orthogonal to a.

Let H be the linear subspace of $\mathbb{R}^d$ orthogonal to a. We know $H = -H$ since this is true of any linear subspace. From Lemma 5.3 we know that $(\operatorname{proj}_H(P))^\circ = P^\circ \cap H$, where the polar is taken with respect to the space H. Then, multiplying both sides by -1, we get

$$\begin{aligned} -\left(\operatorname{proj}_H(P)\right)^\circ &= -(P^\circ \cap H) \\ &= -P^\circ \cap -H \\ &= P \cap H \end{aligned}$$

Let S be the intersection of P with H. Since the base of the pyramid, Q, is orthogonal to the apex, a, we can say that for some constant $c > 0$, $\operatorname{proj}_H(P) = cS$. Hence $-(cS)^\circ = S$. Replacing c with $\sqrt{c}\sqrt{c}$, we get $S = -(\sqrt{c}\sqrt{c})S)^\circ$ and applying Lemma 2.2, we get $S = -\frac{1}{\sqrt{c}}(\sqrt{c}S)^\circ$, which yields $\sqrt{c}S = -(\sqrt{c}S)^\circ$. Since $\sqrt{c}S$ is of the same combinatorial type as Q, we have a negatively self-polar realization of the base of the pyramid. □

6. The number of vertices of negatively self-polar polytopes

The theorems in the last section can be applied right away to the question of how many vertices are possible for negatively self-polar polytopes in dimensions higher than three. Starting in three dimensions with the constructions given in Theorem 4.8 and applying Theorem 5.1 repeatedly, we can construct negatively self-polar polytopes in higher dimensions with (almost) any number of vertices, with the only exception being $d + 2$ vertices in dimension d.

The question of whether there exist any d-dimensional negatively self-polar polytopes with $d + 2$ vertices is somewhat challenging, since there is no starting construction in $d = 3$, due to Theorem 5.5. Luckily, polytopes having this number of vertices have already received a great deal of study; see [**6**] or [**16**], for example.

THEOREM 6.1. *For $d \geq 3$, there exist negatively self-polar d-dimensional polytopes with n vertices for all values of $n \geq d + 1$ except $n = d + 2$.*

PROOF. For number of vertices $n = d + 1$ and $n \geq d + 3$, the base case of $d = 3$ has already been shown, and the higher-dimensional polytopes can be constructed inductively using the information in the proofs of Theorems 4.8 and 5.1.

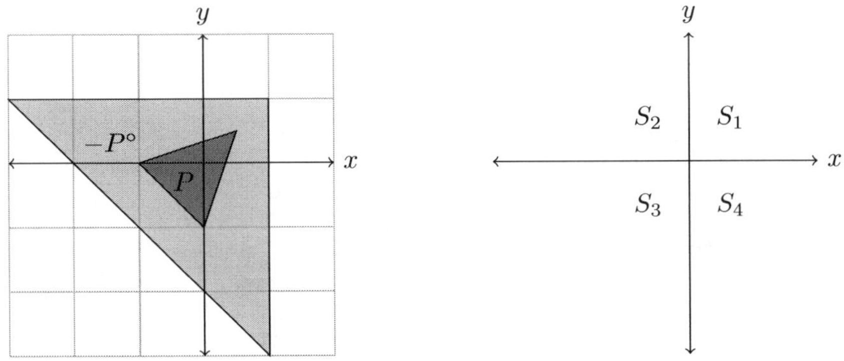

FIGURE 6. On the left, P and $-P^\circ$; on the right, the auxiliary sets S_1, S_2, S_3, S_4.

As for $n = d+2$, polytopes with this number of vertices are either simplicial, or are (multiple) pyramids over simplicial polytopes (see [**6**] or [**16**], for example). A simplicial polytope with $n = d + 2$ cannot be self-dual because it would have to be both simplicial and simple, which is only true of simplices, which have $n = d + 1$.

As for (multiple) pyramids over simplicial polytopes, we have already shown in Theorem 5.5 that a negatively self-polar pyramid must have a negatively self-polar base. Hence, for a (multiple) pyramid over a simplicial base to be negatively self-polar, the simplicial base would have to be negatively self-polar in some lower dimension. But then the base would have to be a simplex, and so the entire pyramid would have to be a simplex. □

For $d > 3$, it is an open question whether there exist self-dual d-dimensional polytopes with $d + 2$ vertices which are self-polar via some other orthogonal transformation.

7. Modifications in the same dimension

7.1. Intermediate construction. The following theorem and its proof give a method for constructing self-polar polytopes for a chosen orthogonal transformation, using a series of sets contained in the transforms of their polar sets. The theorem also establishes when a self-polar polytope exists under certain circumstances.

As the theorem statement and its proof are rather complicated, we will first look at an example in depth.

EXAMPLE 7.1. To keep things simple, we will work in $\mathbb{R}^2$. Let P be the triangle with vertices at $(\frac{1}{2}, \frac{1}{2}), (0, -1)$, and $(-1, 0)$. Then $-P^\circ$ is a triangle with vertices at $(1, 1), (1, -3)$, and $(-3, 1)$. Both are shown in the left panel of Figure 6. Our goal is to construct a polytope $Q = -Q^\circ$ that is between P and $-P^\circ$ in the sense that $P \subset Q = -Q^\circ \subset -P^\circ$.

The general strategy is to enlarge P while still ensuring that it remains a subset of $-P^\circ$. To this end, we will use four auxiliary sets, S_1, S_2, S_3, S_4. Each auxiliary set is one of the four quadrants, as shown in Figure the right panel of 6. These sets are convenient because each $S_i = -S_i^\circ$.

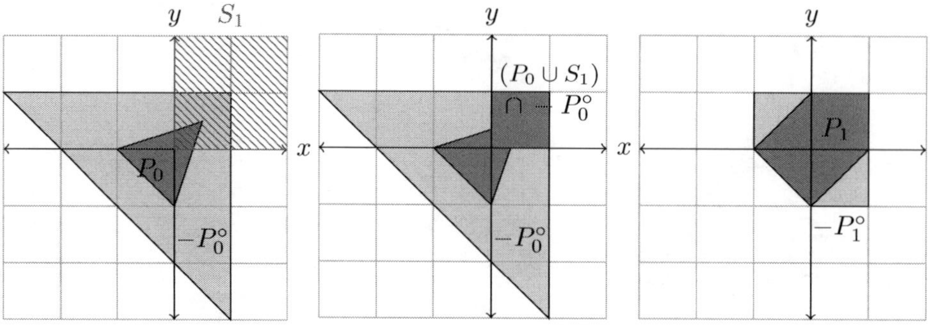

FIGURE 7. The process of constructing $P_1 = [(P_0 \cup S_1) \cap -P_0^\circ]$ and $-P_1^\circ$.

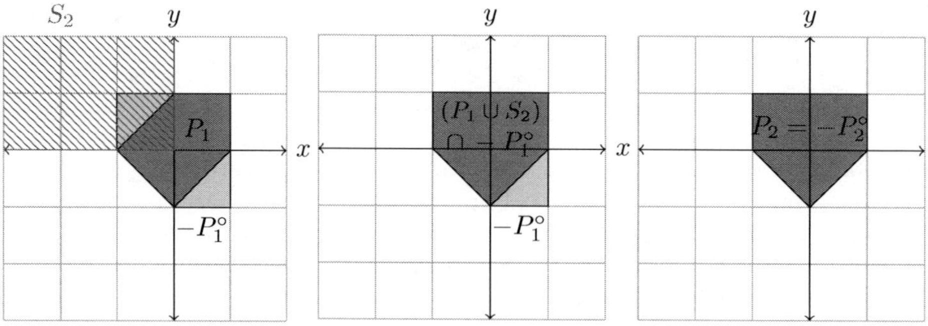

FIGURE 8. The process of constructing $P_2 = [(P_1 \cup S_2) \cap -P_1^\circ]$ and $-P_2^\circ$.

We will define $P_0 := P$ and then successively create enlarged versions of P by letting $P_1 = [(P_0 \cup S_1) \cap -P_0^\circ]$, $P_2 = [(P_1 \cup S_2) \cap -P_1^\circ]$, and so on. The results for P_1 are shown in Figure 7, and the results for P_2 are shown in Figure 8.

As it turns out, we already have our desired goal for $Q = P_2$, so S_3 and S_4 are unnecessary. Depending on which of the auxiliary sets we used, however, and in which order, we could obtain several different suitable sets to fill the role of Q.

Although this example used a polytope P containing the origin, it is not necessary to begin with a polytope $P \in \mathcal{P}_0^d$ in order to construct an intermediate self-polar set. In fact, we can begin with any set $P \subset \mathbb{R}^d$, as long as $P \subset UP^\circ$.

When reading the statement of Theorem 7.2, recall that a polyhedral set is a set which is the intersection of a finite number of closed half-spaces, and note that the theorem can be read with or without the parenthetical words.

THEOREM 7.2. *Let U be an orthogonal transformation of $\mathbb{R}^d$ and let $P \subset \mathbb{R}^d$ and $P \subset UP^\circ$. Then there exists a (polyhedral) set Q such that $P \subseteq Q = UQ^\circ \subseteq UP^\circ$ if and only if there exists a closed, convex (polyhedral) set $R \subset \mathbb{R}^d$ and a collection $S_1, S_2, ..., S_n$ of closed, convex (polyhedral) sets in $\mathbb{R}^d$ such that:*

(1) *$U^2R = R$, and*
(2) *$P \subseteq R \subseteq UR^\circ \subseteq UP^\circ$, and*
(3) *$U^2S_i = S_i$ for all $i = 1, 2, ..., n$, and*
(4) *$S_i \subseteq US_i^\circ$ for all $i = 1, 2, ..., n$, and*
(5) *$UR^\circ \subseteq S_1 \cup S_2 \cup \cdots \cup S_n$.*

PROOF. First we will proof the "if" part of the "if and only if". Assume that U is an orthogonal transformation of $\mathbb{R}^d$, that $P \subset \mathbb{R}^d$, and that $P \subset UP^\circ$. Further, assume there exists a closed, convex set $R \subset \mathbb{R}^d$ and a collection $S_1, S_2, ..., S_n$ of closed, convex sets in $\mathbb{R}^d$ which satisfy conditions (1) to (5).

Let S be one of the $S_1, S_2, ..., S_n$ and let $T = [(R \cup S) \cap UR^\circ]$. Note that if S and R are polyhedral, then T is also polyhedral. Also note that any orthogonal transformation commutes with the polar operation, with the convex hull operation, and with the closure operation.

Regarding T, we firstly find, using our assumptions and the commutativity of U, that

$$\begin{aligned} U^2T &= U^2[(R \cup S) \cap UR^\circ] \\ &= [(U^2R \cup U^2S) \cap U^3R^\circ] \\ &= [(R \cup S) \cap UR^\circ] \\ &= T \end{aligned}$$

Our next finding regarding T is that $T \subseteq UT^\circ$, as shown below. We use the theorem assumptions, the fact that [ ] is a closure operator, and Lemma 2.1 throughout, as well as basic properties of sets, and the fact that $R \subseteq UR^\circ$ is equivalent to $UR \subseteq R^\circ$ by taking the polar of both sides of the subset relation.

$$\begin{aligned} UT^\circ &= U[(R \cup S) \cap UR^\circ]^\circ \\ &= [(UR \cup US) \cap U^2R^\circ]^\circ \\ &= [(UR \cup US) \cap R^\circ]^\circ \\ &= ((UR \cup US) \cap R^\circ)^\circ \\ &= ((UR \cap R^\circ) \cup (US \cap R^\circ))^\circ \\ &= (UR \cup (US \cap R^\circ))^\circ \\ &= UR^\circ \cap (US \cap R^\circ)^\circ \\ &= UR^\circ \cap (US^{\circ\circ} \cap R^\circ)^\circ \\ &= UR^\circ \cap (US^\circ \cup R)^{\circ\circ} \\ &= UR^\circ \cap [US^\circ \cup R] \\ &\supseteq UR^\circ \cap [S \cup R] \\ &\supseteq [UR^\circ \cap (S \cup R)] \\ &= T \end{aligned}$$

Regarding T, we lastly find that

$$\begin{aligned} UT^\circ \cap S &= [R \cup S] \cap UR^\circ \cap S \\ &= ([R \cup S] \cap S) \cap UR^\circ \\ &= S \cap UR^\circ \\ &= S \cap UR^\circ \cap S \\ &\subseteq (R \cup S) \cap UR^\circ \cap S \\ &\subseteq T \cap S \end{aligned}$$

Together with $T \subseteq UT^\circ$, this implies $UT^\circ \cap S = T \cap S$.

To summarize, we have discovered T such that $R \subseteq T \subseteq UT^\circ \subseteq UR^\circ$, where T and UT° have equality on S, and T fulfills the same assumptions that we initially made about R. Hence, we proceed iteratively, using T as the new R, and constructing a new T using the same formula, but using a different S to construct T so as to extend the equality of T and UT° to a new region. Below, we describe this iterative process formally.

Let $T_0 := R$. Then for $i \in \mathbb{N}$, let $T_i := [(T_{i-1} \cup S_i) \cap UT_{i-1}^\circ]$. It then follows from our previous finding about T that $T_{i-1} \subseteq T_i \subseteq UT_i^\circ \subseteq UT_{i-1}^\circ$. Hence

$$R = T_0 \subseteq T_1 \subseteq \cdots \subseteq T_n \subseteq UT_n^\circ \subseteq \cdots \subseteq UT_1^\circ \subseteq UT_0^\circ = UR^\circ$$

From our third finding about T, we have that $T_i \cap S_i = UT_i^\circ \cap S_i$. Then by the nesting of the sets it follows that $T_n \cap S_i = UT_n^\circ \cap S_i$ for all $i = 1, 2, ..., n$. Hence $T_n \cap (S_1 \cup S_2 \cup \cdots \cup S_n) = UT_n^\circ \cap (S_1 \cup S_2 \cup \cdots \cup S_n)$. Since $T_n \subseteq UT_n^\circ \subseteq UR^\circ$ and by assumption this is a subset of $S_1 \cup S_2 \cup \cdots \cup S_n$, we have $T_n = UT_n^\circ$.

Note that so long as R and the $S_1, ..., S_n$ are polyhedral (i.e., finite intersections of closed half-spaces), then T_n is polyhedral because it is constructed in a finite number of steps from intersections, unions, convex hulls, and closures of polyhedral sets. This completes the proof of the "if" part of the "if and only if" in the theorem statement.

For the "only if" part, assume U is an orthogonal transformation of $\mathbb{R}^d$, P is a set in $\mathbb{R}^d$, and assume there exists a (polyhedral) set Q such that $P \subseteq Q = UQ^\circ \subseteq UP^\circ$. We will define $R := Q$ and $S_1 := Q$. S_1 is the only S_i we will need.

Since $Q = UQ^\circ$, we have $U^2 Q = Q$ from Theorem 3.3. This gives us conditions (1) and (3). We get conditions (2) and (4) from the assumption that $P \subseteq Q = UQ^\circ \subseteq UP^\circ$. Finally, since $UQ^\circ = Q$, we have $UR^\circ = S_1$, which gives us condition (5). □

Some of the requirements of Theorem 7.2 might seem rather difficult to fulfill, and while this is true in general, when U is the negative transformation, matters are simpler, as shown by the following corollary of Theorem 7.2. As before, the theorem statement can be read with or without the parenthetical words.

COROLLARY 7.3. *For a (polyhedral) set $P \subset \mathbb{R}^d$ such that $P \subseteq -P^\circ$, there exists a (polyhedral) set Q such that $P \subseteq Q = -Q^\circ \subseteq -P^\circ$.*

PROOF. Here, we use Theorem 7.2 and let R be P itself. Obviously $U^2 = I$, so we have fulfilled conditions (1) and (2) of Theorem 7.2. As for the auxiliary sets S_i, we can use the orthants of $\mathbb{R}^d$, which are polyhedral. Each orthant $S_i = -S_i^\circ$, and together the orthants cover all of $\mathbb{R}^d$, so they fulfill conditions (3), (4), and (5). □

7.2. Add-and-cut constructions. The next theorem gives us a way to add vertices to a set that is already self-polar while preserving that property. The key idea is to add a vertex and intersect with a half-space at the same time. See Figure 9 for an example of a successful add-and-cut operation.

Caution must taken to make sure that the order in which these operations (adding a vertex and cutting a facet) are done will make no difference to the outcome; in other words, so that $[(P \cap Uv^\circ) \cup v] = [P \cup v] \cap Uv^\circ$ for the new vertex v. See Figure 10 for an unsuccessful example where the order does make a difference.

Note that Theorem 7.4 (below) does not require P to be a polytope. It is sufficient that P is self-polar.

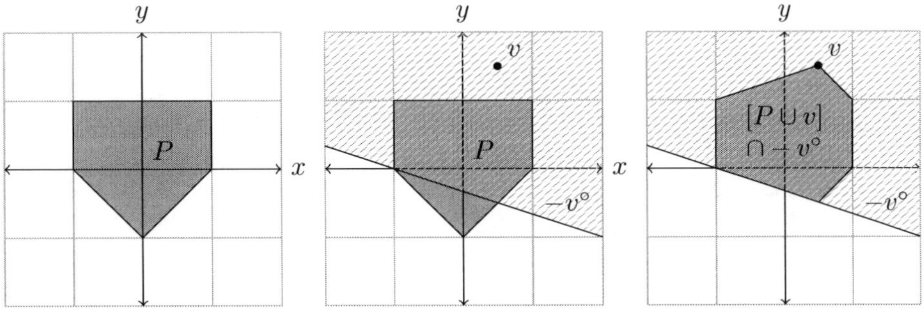

FIGURE 9. Adding a vertex and cutting a new facet while preserving self-polarity.

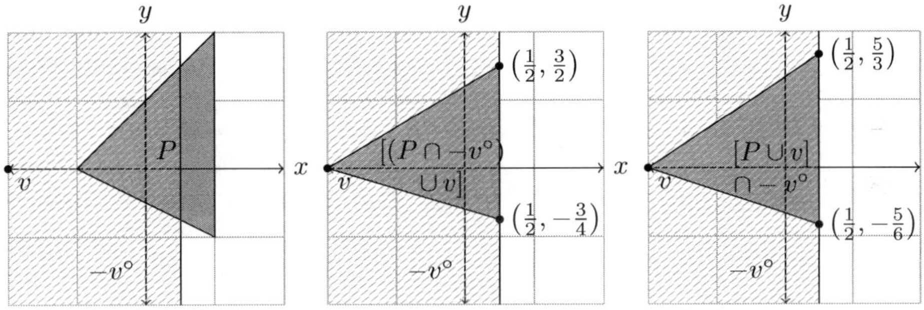

FIGURE 10. An unsuccessful choice of v that results in $[(P\cap -v^{\circ})\cup v] \neq [P\cup v]\cap -v^{\circ}$.

THEOREM 7.4. *Let $P = UP^{\circ}$ for a set $P \subset \mathbb{R}^d$ and an orthogonal transformation U. Let there be a point $x \in \mathbb{R}^d$ such that $[(P\cap Ux^{\circ})\cup x] = [P\cup x]\cap Ux^{\circ}$ and $U^2x = x$. Then $Q = UQ^{\circ}$ for the set $Q = [P\cup x]\cap Ux^{\circ}$.*

PROOF. Assume that P and x are as described. Then

$$\begin{aligned}
UQ^{\circ} &= \left([UP\cup Ux]\cap U^2x^{\circ}\right)^{\circ}\\
&= ((UP\cup Ux)^{\circ\circ}\cap x^{\circ})^{\circ}\\
&= ((UP\cup Ux)^{\circ}\cup x)^{\circ\circ}\\
&= ((UP^{\circ}\cap Ux^{\circ})\cup x)^{\circ\circ}\\
&= ((P\cap Ux^{\circ})\cup x)^{\circ\circ}\\
&= [(P\cap Ux^{\circ})\cup x]\\
&= [P\cup x]\cap Ux^{\circ}\\
&= Q
\end{aligned}$$

□

To use Theorem 7.4, an easy way to find a point x which can be added to a polytope is to choose a facet, take a point in the interior of the facet, and slightly increase the radius of that point to get x. As long as the facet chosen does not contain the vertex to which it is dual, and as long as x is still beneath the planes of all the other facets, the conditions of the theorem will be met. A pyramid will be made over the facet, with x at the apex, and the dual vertex to the facet will be cut off to form a new facet. We state this more formally as a corollary now, which will be of much use in the investigation of combinatorial types in the next section.

COROLLARY 7.5. *Let $P = UP^\circ$ for $P \in \mathcal{P}_0^d$ and an involutory orthogonal transformation U. Let vertex v not lie in its own dual facet f. Then a polytope $Q = UQ^\circ$ exists which is obtained from P by the following modifications:*

(1) *A new vertex beyond f but beneath all other facets creates a pyramid over f*
(2) *A new facet cuts off v by slicing through the interior of every face containing v*

PROOF. Using Theorem 7.4, we let x be some point which is beyond f but beneath all other facets. We have $U^2x = x$ by assumption that U is an involution. As for the other requirement, consider the following. Because x is beneath all facets except f, taking the convex hull of P with x only affects and depends on points beyond facet f. On the other hand, since x is beneath all facets except f, that means v is the only vertex that is beyond Ux°. Hence the intersection $P \cap Ux^\circ$ only affects and depends on points strictly beneath facet f. Since the two modifications are thus separated by the hyperplane through f, the result of performing the two modifications in either order is the same, and so $[(P \cap Ux^\circ) \cup x] = [P \cup x] \cap Ux^\circ$, fulfilling the conditions of the previous theorem. The description of modification 1 is a consequence of choosing x beyond f but beneath all other facets, and modification 2 is simply the dual description of modification 1. □

8. Conclusions and further questions

Combinatorial types. In general, as with all sets of polytopes, we would like to know something about which combinatorial types are possible. We have already answered the questions of the possible numbers of vertices of negatively self-polar polytopes, and we have established the possible f-vectors of negatively self-polar polytopes in two and three dimensions. But the following question remains.

QUESTION 8.1. What f-vectors are possible for negatively self-polar polytopes in $\mathbb{R}^4$ and higher dimensions?

Self-polar realizations. More generally, we would like to know about the place of self-polar polytopes within the broader class of self-dual polytopes. We established that all two and three dimensional self-dual polytopes have self-polar realizations, but we have no analogous result for higher dimensions, and the proof techniques that worked in lower dimensions will not work in higher dimensions.

QUESTION 8.2. Does every self-dual polytope have a self-polar realization?

Period of self-polar maps. By definition, every self-dual polytope has a dual automorphism on its face lattice. For negatively self-polar polytopes, this map is an involution. However, it is well-known that not all self-dual polytopes have an involutory dual automorphism.

The classic example of a self-dual polytope without an involutory dual automorphism is Jendroĺ's polytope [**9**]. Jendroĺ's polytope is self-polar, however, when realized as shown in Example 4.7, by an orthogonal transformation with period 4.

In [**7**], Grünbaum and Shephard defined the rank of a self-dual polytope as the minimum period of all such maps.

QUESTION 8.3. Are all self-polar polytopes with a rank r duality map realizable as self-polar with an orthogonal map of period r?

In two dimensions, the answer is again clearly yes, since all polygons are self-dual with rank 2, and are realizable as self-polar by reflection over a single axis.

Chromatic number of main diagonal graph. Lovász [**11**] showed that for a negatively self-polar polytope in $\mathbb{R}^d$ with vertices equidistant from the origin, the main diagonals (diagonals from a vertex to the vertices on its dual facet) all have the same length. He further showed that the graph formed by the main diagonals has chromatic number $d+1$.

QUESTION 8.4. What is the chromatic number of the graph formed by the main diagonals of a negatively self-polar polytope?

Inscribability. Lovász's study of these polytopes was motivated by questions about the chromatic number of $G(d, \alpha)$, the graph on the points of S^{d-1} formed by connecting two points iff their distance is exactly α, which is a subgraph of Borsuk's graph. Lovász showed that if there exists a negatively self-polar polytope whose vertices are all $\sqrt{2/(\alpha^2-2)}$ from the origin, then the chromatic number of $G(d, \alpha)$ is $d+1$. He left the following as an open question, however.

QUESTION 8.5. For which values of r and d does a negatively self-polar polytope exist in $\mathbb{R}^d$ with all vertices r-distant from the origin?

References

[1] Jonathan Ashley, Branko Grünbaum, G. C. Shephard, and Walter Stromquist, *Self-duality groups and ranks of self-dualities*, Applied geometry and discrete mathematics, DIMACS Ser. Discrete Math. Theoret. Comput. Sci., vol. 4, Amer. Math. Soc., Providence, RI, 1991, pp. 11–50. MR1116336

[2] Alexander Barvinok, *A course in convexity*, Graduate Studies in Mathematics, vol. 54, American Mathematical Society, Providence, RI, 2002. MR1940576

[3] Jürgen Bokowski and Bernd Sturmfels, *Polytopal and nonpolytopal spheres: an algorithmic approach*, Israel J. Math. **57** (1987), no. 3, 257–271, DOI 10.1007/BF02766213. MR889977

[4] Gabe Cunningham and Mark Mixer, *Internal and external duality in abstract polytopes*, Contrib. Discrete Math. **12** (2017), no. 2, 187–214. MR3739062

[5] Michael B. Dillencourt, *Polyhedra of small order and their Hamiltonian properties*, J. Combin. Theory Ser. B **66** (1996), no. 1, 87–122, DOI 10.1006/jctb.1996.0008. MR1368518

[6] Branko Grünbaum, *Convex polytopes*, 2nd ed., Graduate Texts in Mathematics, vol. 221, Springer-Verlag, New York, 2003. Prepared and with a preface by Volker Kaibel, Victor Klee and Günter M. Ziegler. MR1976856

[7] B. Grünbaum and G.C. Shephard, *Is selfduality involutory?*, The American Mathematical Monthly **95** (1988), no. 8, 729–733.

[8] Jacob E. Goodman, Joseph O'Rourke, and Csaba D. Tóth (eds.), *Handbook of discrete and computational geometry*, Discrete Mathematics and its Applications (Boca Raton), CRC Press, Boca Raton, FL, 2018. Third edition of [MR1730156]. MR3793131

[9] Stanislav Jendroľ, *A noninvolutory self-duality*, Discrete Math. **74** (1989), no. 3, 325–326, DOI 10.1016/0012-365X(89)90144-1. MR992743

[10] Jim Lawrence, *Valuations and polarity*, Discrete Comput. Geom. **3** (1988), no. 4, 307–324, DOI 10.1007/BF02187915. MR947219

[11] L. Lovász, *Self-dual polytopes and the chromatic number of distance graphs on the sphere*, Acta Sci. Math. (Szeged) **45** (1983), no. 1-4, 317–323. MR708798

[12] N. E. Mnëv, *The universality theorems on the classification problem of configuration varieties and convex polytopes varieties*, Topology and geometry—Rohlin Seminar, Lecture Notes in Math., vol. 1346, Springer, Berlin, 1988, pp. 527–543, DOI 10.1007/BFb0082792. MR970093

[13] Oded Schramm, *How to cage an egg*, Invent. Math. **107** (1992), no. 3, 543–560, DOI 10.1007/BF01231901. MR1150601

[14] E. Schulte, *Analogues of Steinitz's theorem about noninscribable polytopes*, Intuitive geometry (Siófok, 1985), Colloq. Math. Soc. János Bolyai, vol. 48, North-Holland, Amsterdam, 1987, pp. 503–516. MR910731

[15] Bernd Sturmfels, *On the decidability of Diophantine problems in combinatorial geometry*, Bull. Amer. Math. Soc. (N.S.) **17** (1987), no. 1, 121–124, DOI 10.1090/S0273-0979-1987-15532-7. MR888886

[16] Günter M. Ziegler, *Lectures on polytopes*, Graduate Texts in Mathematics, vol. 152, Springer-Verlag, New York, 1995. MR1311028

Department of Mathematics and Computer Science, Susquehanna University, Selinsgrove, Pennsylvania 17870

Email address: jensena@susqu.edu

Contemporary Mathematics
Volume **764**, 2021
https://doi.org/10.1090/conm/764/15358

Isomorphisms of maps on the sphere

Ken-ichi Kawarabayashi, Pavel Klavík, Bojan Mohar, Roman Nedela, and Peter Zeman

ABSTRACT. For a class of objects with a well-defined isomorphism relation the isomorphism problem asks to determine the algorithmic complexity of the decision whether two given objects are, or are not, isomorphic. Theorems by Steinitz (1916), Whitney (1933) and Mani (1971) show that the isomorphism problems for convex polyhedra, for 3-connected planar graphs, and for the spherical maps are closely related. In 1974, Hopcroft and Wong investigated the complexity of the graph isomorphism problem for polyhedral graphs. They proved that the problem can be solved in linear time. We describe a modified linear-time algorithm solving the isomorphism problem for spherical maps based on the approach by Hopcroft and Wong. The paper includes a detailed description of the algorithm including proofs. Moreover, our modified algorithm allows to determine (in linear time) the group of orientation-preserving symmetries of a spherical map.

1. Introduction

By a (topological) map we mean a 2-cell decomposition of a closed surface. A map is usually defined by a 2-cell embedding of a connected graph into a surface. A map is polyhedral if its underlying graph is combinatorially isomorphic to a 1-skeleton of a convex polyhedron. By Steinitz theorem the polyhedral maps are exactly the spherical maps with a 3-connected underlying graph. The *graph isomorphism problem* is to decide whether two given graphs are isomorphic. Similarly, *map isomorphism problem* asks whether two given maps are isomorphic. Whether or not the graph isomorphism problem can be solved in polynomial time is one of the central open problems of discrete mathematics and computer science. On the other hand, one can easily prove that one can decide the map isomorphism problem in time bounded by a quadratic function in the number of edges. The proof is based on the fact that given two maps M_1, M_2, an assignment $u_1v_1 \mapsto u_2v_2$, where

---

2010 *Mathematics Subject Classification.* Primary 05C85, 05C10; Secondary 52B10, 05C60.

The first author was supported by JST ERATO Kawarabayashi Large Graph Project JPMJER1201 and by JSPS Kakenhi JP18H05291.

The third author was supported in part by the NSERC Discovery Grant R611450 (Canada), by the Canada Research Chairs program, and by the Research Project J1-8130 of ARRS (Slovenia). On leave from IMFM, Department of Mathematics, University of Ljubljana.

The fourth author was supported by the Czech Science Foundation under grant GAČR 20-15576S and by the Slovack Research and Development Agency under grant APVV-19-0308.

The fifth author was supported by GAUK 1334217 and GAČR 20-15576S.

$u_iv_i \in M_i$, $i = 1, 2$, is an ordered pair of adjacent vertices in M_i, either extends to an (orientation preserving) map isomorphism in a unique way, or no such extension exists. For polyhedral maps the bound was first improved by Hopcroft and Tarjan (1973) [**3**] by proving that the problem can be solved in $n\log(n)$ time. In 1974 Hopcroft and Wong described an algorithm [**5**] solving the isomorphism problem for spherical maps in linear time. As a consequence of theorems by Steinitz and Whitney [**9, 10**] we get that the graph isomorphism problem restricted to 3-connected planar graphs is equivalent to the map isomorphism problem for spherical polyhedral maps. This essential corollary allows one to construct a linear-time algorithm solving the isomorphism problem for (all) planar graphs by reducing it to the 3-connected case [**1**], [**7**, Section 2].

A brief sketch of the algorithm by Hopcroft and Wong follows. The algorithm produces a series of reductions $M_0, M_1, \dots M_k$, where M_0 is the input map and M_k is an irreducible map. The irreducible maps are by definition the five platonic maps, the cycles and their duals. Each elementary reduction modifies a particular part of a map using edge- and vertex-deletions, edge-contractions and vertex-expansions. All of this is done in such a way that, when applied simultaneously to both maps in a pair of maps, the isomorphism relation is preserved. In each step the vanishing structural information is stored by labels attached to (directed) edges called darts. The two input maps are isomorphic if and only if the associated irreducible labeled maps are isomorphic. To prove the linearity one needs to solve several partial problems including the following:

(1) To introduce an efficient representation of maps and to describe the intuitive (geometric) definitions of the reductions in terms of the map representation,
(2) The number of elementary reductions is bounded by a linear function (this comes from the fact that each elementary reduction decreases the size of the map).
(3) The time spent to execute each elementary reduction is proportional to the difference between the sizes of the input and output maps, in particular, this requires an efficient storage and computation of the labels.
(4) The isomorphism problem for the irreducible maps (labeled cycles) can be solved in linear time.

The description of the algorithm in [**5**] deals with these problems only partially. In particular, all the reductions are described only informally, the management of labels is not presented and the proofs of the correctness and linearity are missing.

Other problem of the approach in [**5**] is the fact that for k-valent maps ($k \in \{3, 4, 5\}$) a different set of reductions is suggested. It transpires that it is hard to formalize them, not talking about rigorous verification that they satisfy the required properties. Moreover, by taking the medial map, for instance, we obtain a one-to-two correspondence (up to duality) between the polyhedral maps and the subfamily of the 4-valent maps with bipartite duals. Thus the isomorphism problem for the 4-valent polyhedral maps is as hard as the original problem.

The main goal of the present article is to fill in the aforementioned gaps, in particular, because the result itself is a seminal contribution (it has hundreds of citations and many applications). Perhaps the main novelty of our algorithm is that we use the same set of operations for non-k-valent maps and for k-valent maps.

Further motivation to investigate the algorithm by Hopcroft and Wong is twofold. First, we are interested whether the algorithm can be modified to compute efficiently a set of generators of the automorphism groups of polyhedra. The automorphism groups are, by Mani's theorem [**8**], the groups of isometries of the sphere, known as the *spherical groups*. Secondly, we are interested to find a modification of the Hopcroft-Wong algorithm that can be generalized to solve efficiently the map isomorphism problem for surfaces of higher genera. In both directions we succeeded; for more information the reader is referred to the final section.

2. Preliminaries

A *map* M is a 2-cell decomposition of a compact connected surface S. The i-cells of M, for $i = 0, 1, 2$ are called respectively the *vertices*, *edges* and *faces* of M. A map is usually defined by a cellular embedding of a connected graph.

An *oriented map* is a map on an orientable surface with a fixed global orientation (clockwise or counter-clockwise). Unless otherwise stated, in what follows we assume that the chosen orientation is clockwise. A given oriented map can be described as a *combinatorial map* (D, R, L), where D is a finite set and $R, L \in \mathrm{Sym}(D)$, as follows. The set D of *darts* is identified with the set of oriented edges, each edge gives rise to two darts. Let V denote the set of vertices. The permutation $R \in \mathrm{Sym}(D)$, called *rotation*, is the product $R = \Pi_{v \in V} R_v$, where each R_v cyclically permutes the darts based at $v \in V$ following the chosen global orientation. The *dart-reversing involution* $L \in \mathrm{Sym}(D)$ is a permutation of D, swapping for each edge the two associated oppositely directed darts. Formally, a *combinatorial map* is any triple $M = (D, R, L)$, where D is a finite non-empty set of *darts*, R is a permutation of darts, L is a fixed-point-free involution of D, and the group $\langle R, L\rangle \leq \mathrm{Sym}(D)$ is transitive on D. The group $\langle R, L\rangle$ is called the *monodromy group* of M. While the vertices are in correspondence with the cycles of R, the boundary walks of faces are determined by the cycles of the permutation $R^{-1}L$. By the phrase "a dart x is incident to a vertex v" we mean that $x \in R_v$. Similarly, "x is incident to a face f" means that x belongs to the boundary walk of f defined by the respective cycle of $R^{-1}L$. Note that each dart is incident to exactly one face.

The *dual* and the *mirror image* of a map $M = (D, R, L)$ are respectively the maps $M^* = (D, R^{-1}L, L)$ and $M^{-1} = (D, R^{-1}, L)$.

For convenience, we sometimes use in computations a shorthand setting $x^- = Lx$, for $x \in D$. For more information on the combinatorial maps and further references we refer the reader to [**2**, Section 7.6].

Let X be the underlying graph of M. The orientation preserving map automorphisms are the graph automorphisms (considered as incidence preserving permutations of darts) commuting with both R and L. We refer to them alternatively as the automorphisms of the oriented map. The *orientation preserving automorphism group* of M is denoted by $\mathrm{Aut}^+(M)$.

A *dart-labeling* of a map $M = (D, R, L)$ is a mapping $\ell\colon D \to \mathcal{T}$, into the set of rooted integer-valued planted trees. By a *planted tree* we mean a map (D, R, L) whose underlying graph is a tree. Alternatively, it is a spherical map with exactly one-face.

Integers are attached to the vertices of such a tree. A *labeled map* M is a 4-tuple (D, R, L, ℓ). An automorphism of a labeled map is a map automorphism preserving the labels of the darts. In what follows, we consider only labeled spherical oriented maps.

Two labeled maps $M_1 = (D_1, R_1, L_1, \ell_1)$ and $M_2 = (D_2, R_2, L_2, \ell_2)$ are *isomorphic* if there exists a bijection $\psi\colon D_1 \to D_2$ such that

$$\psi R_1 = R_2\psi, \quad \psi L_1 = L_2\psi, \quad \text{and} \quad \ell_1 = \ell_2\psi. \tag{2.1}$$

The following statement, well-known for unlabeled maps, extends easily to labeled maps.

THEOREM 2.1. *Let M_1 and M_2 be labeled maps with the sets of darts D_1 and D_2, respectively. For every $x \in D_1$ and every $y \in D_2$, there exists at most one isomorphism $M_1 \to M_2$ mapping x to y. In particular,* $\mathrm{Aut}^+(M_1)$ *is fixed-point-free on D_1.*

Let $M = (D, R, L, \ell)$ be a labeled map. The *dual map* is the map M^* defined as $M^* = (D, R^{-1}L, L, \ell)$. The *mirror image* is the map $M^{-1} = (D, R^{-1}, L, \ell^-)$, where $\ell^-(x)$ is the mirror image of $\ell(x)$ for each $x \in D$. The following statement shows how the isomorphism relations between maps and oriented maps (combinatorial maps) are related.

THEOREM 2.2. *Two topological maps described respectively by the associated oriented maps M_1 and M_2 are* isomorphic *if and only if either $M_1 \cong M_2$, or $M_1 \cong M_2^{-1}$.*

By the *degree of a face* we mean the length of the boundary walk. A face of degree d will be called a d-face. By a cyclic vector of length m we mean the orbit in the action of the cyclic group Z_m on a set of m-dimensional vectors shifting cyclically the entries of vectors. The m-dimensional vectors are endowed with the lexicographic order, therefore we can represent each cyclic vector by its minimal representative. For a vector Y we denote by $|Y|$ its length.

Let M be a spherical map and let u be a vertex of degree d. Let $v_1, \dots, v_d$ be its neighbors, listed according to the chosen orientation. The *degree type* $\mathcal{D}(u) = (\deg(v_1), \dots, \deg(v_d))$ of u is the minimal representative of the respective cyclic vector of degrees of the neighbors of u. We set $\mathcal{D}(u) \preceq \mathcal{D}(v)$ if $|\mathcal{D}(u)| < |\mathcal{D}(v)|$, or if $|\mathcal{D}(u)| = |\mathcal{D}(v)|$ and $\mathcal{D}(u)$ is lexicographically smaller than $\mathcal{D}(v)$. Following the clockwise orientation, the cyclic vector $(f_1, \dots, f_d)$ of degrees of faces incident with a vertex v is called the *local type* of v. The *refined degree* of v, denoted $\mathrm{ref}(v)$, is the minimal representative of the local type. Note that $\deg(u) = |\mathrm{ref}(u)|$.

We introduce an ordering $\preceq$ on the set of all refined degrees as follows. We put $\mathrm{ref}(u) \preceq \mathrm{ref}(v)$ if $|\mathrm{ref}(u)| < |\mathrm{ref}(v)|$, or if $|\mathrm{ref}(u)| = |\mathrm{ref}(v)|$ and $\mathrm{ref}(u)$ is lexicographically smaller than $\mathrm{ref}(v)$. The *refined degree type* $\mathcal{R}(u)$ of $u \in V(M)$ is the lexicographically minimal representative of the cyclic vector $(\mathrm{ref}(u_0), \dots, \mathrm{ref}(u_{d-1}))$, where $u_0, \dots, u_{d-1}$ are the neighbors of u listed in the order following the clockwise orientation. Similarly as above, we set $\mathcal{R}(u) \preceq \mathcal{R}(v)$ if $|\mathcal{R}(u)| < |\mathcal{R}(v)|$, or if $|\mathcal{R}(u)| = |\mathcal{R}(v)|$ and $\mathcal{R}(u)$ is lexicographically smaller than $\mathcal{R}(v)$. In what follows, the following observation will be important: both degree types and refined degree types are preserved by map isomorphisms. In particular, the respective decomposition(s) of the vertex-set of a map posses the following property: the set of vertices of the same (refined) degree type is a disjoint union of orbits in the action of $\mathrm{Aut}^+(M)$.

A map is called *face-normal*, if all its faces are of degree at least three. By the Euler formula we have

LEMMA 2.3. *A face-normal spherical map has a vertex of degree at most* 5.

We say that a vertex or a face is *light* if it has degree less or equal to 5.

A map is *uniform* if the local types (or equivalently, the refined degrees) of all vertices are the same. The face-normal uniform spherical maps are classified as follows:

(U1) Platonic maps,
(U2) the underlying maps of 13 Archimedean solids,
(U3) the underlying map of the elongated square gyrobicupola,
(U4) prisms, antiprisms, cycles, dipoles, and bouquets.

From the algorithmic point of view, the main difficulty is in the case (U4), since these are infinite families.

A map is *homogeneous* of type $\{k, \ell\}$ if every face is of degree $k > 0$ and every vertex is of degree $\ell > 0$. From Euler's fomula it follows easily that the spherical homogeneous maps are the five Platonic maps, cycles and dipoles. A *dipole* is a 2-vertex spherical map which is dual to a spherical cycle. *A bouquet* is a one-vertex map that is a dual of a *planted star* (a planted tree with at most one vertex of degree > 1).

For the convenience of the reader we close this section by the classical statements forming the framework of our paper.

THEOREM 2.4 (Steinitz [**9**]). *A finite graph is isomorphic to the* 1*-skeleton of a convex polyhedron if and only if it is a simple, planar and* 3*-connected.*

THEOREM 2.5 (Whitney [**10**]). *A finite simple* 3*-connected planar graph has a combinatorially unique embedding in the sphere.*

THEOREM 2.6 (Mani [**8**]). *If X is a finite simple* 3*-connected planar graph, then there is a convex polyhedron P in $\mathbb{R}^3$ whose edge graph is isomorphic to X such that every automorphism of X induces a symmetry of P.*

THEOREM 2.7 (Consequence of Whitney's theorem). *The automorphism group of a polyhedral graph X is equal to the automorphism group of the unique embedding of X in the sphere.*

In the remaining part of the paper all the considered maps will be oriented spherical labeled maps and we shall frequently omit the adjectives.

3. Overview of the algorithm

Our algorithm applies a set of local reductions defined in Section 4. For a given input map M it produces a sequence of labeled maps $M_0, M_1, \ldots, M_k$, where M_k is a face-normal uniform spherical map. Given input maps M and N the algorithm associates respectively two irreducible maps M' and N'. The key property of each elementary reduction is that the isomorphism relation between oriented maps is preserved, in particular, $M \cong N$ if and only if $M' \cong N'$. Hence to solve the isomorphism problem it is enough to do it for the above mentioned irreducible maps. In what follows we informally introduce particular parts of the reduction procedure.

Priorities. The elementary reductions are ordered by priority. With each reduction a list of darts, or vertices and darts is attached. The attached list determines the part of the considered map which is going to be modified by the respective reduction. In each step a reduction with highest priority with non-empty list attached is executed. When performing an elementary reduction the finite set of lists attached to the elementary reductions are reconstructed. The reduction process is multilevel. Firstly, the map is reduced to a face-normal map. Secondly, a face-normal map is reduced to a k-valent map. Thirdly, a k-valent map is reduced to a uniform map. Finally, a special set of reductions is used to transform the infinite series of uniform maps: namely, the bouquets, dipoles, prisms and antiprisms are transformed into labeled cycles. Essential are the reductions performed at the first three levels. The number of operations used there is controlled by the sum of the numbers of vertices and edges which in each step decreases. Formally the reductions are described as transformations of the combinatorial labeled maps $(D, R, L, \ell) \mapsto (D', R', L', \ell')$. This is needed to produce exact proofs that the isomorphism relation is in each step preserved. In what follows we explain them informally to help the reader.

Normalization. At the first level, a reduction called NORMALIZE is applied. The input map is forced to become face-normal. This is done by performing two types of elementary reductions: the first deletes sequences of 1-faces attached to a vertex, see Fig. 4.1. The second replaces a sequence of 2-faces by a single edge, see Fig. 4.2. Normalization terminates with a face-normal map, or with a bouquet, or with a dipole. The reader can find a detailed explanation of NORMALIZE in Subsection 4.1.

From face-normal maps to k-valent. Assume we have a face-normal map that is not k-valent. By Euler's theorem it contains a non-empty list of light vertices (vertices of degree 3, 4 or 5). Moreover, by connectivity there are edges joining a vertex of minimum degree $d \in \{3, 4, 5\}$ to a vertex of higher degree. Suppose at each vertex of minimum degree d one can canonically identify a unique edge of this sort. Then the set S of these edges forms a union of orbits in the action of the group of orientation preserving automorphisms. One can prove that S induces a disjoint union of stars, see Lemma 4.4. When these assumptions are satisfied, a reduction called APERIODIC$_{\mathcal{D}}$ is executed, where $\mathcal{D}$ is the degree type. The reduction APERIODIC$_{\mathcal{D}}$ contracts each star with pendant light vertices of degree type $\mathcal{D}$ to the central vertex. There are just finitely many classes of $\mathcal{D}$ to consider, the list of them can be found in Subsection 4.2. The algorithm executes APERIODIC$_{\mathcal{D}}$ for $\mathcal{D}$ being minimal.

It may happen that the reduction APERIODIC$_{\mathcal{D}}$ cannot be used, because we are not able to identify in a canonical way the edges that are going to be contracted. The algorithm recognises this by checking that the lists attached to APERIODIC$_{\mathcal{D}}$ are empty for all $\mathcal{D}$. Further analysis shows that this happens when all the light vertices are either joined to the vertices of higher degree (such vertices are said to be of large degree type), or they are of periodic type. Therefore we first get rid of vertices of large type by expanding each such vertex of degree d to a d-face, see Fig. 4.3. The respective reduction is called LARGE$_d$. After that the algorithm either returns to NORMALIZE, or, if the map remains face-normal, applies LARGE$_d$ again.

If NORMALIZE and LARGE$_d$ cannot be executed, the algorithm proceeds to APERIODIC$_{\mathcal{D}}$ for the smallest possible $\mathcal{D}$. Finally, we obtain a k-valent map, or the minimum degree is four and there are vertices of degree type $(4, m, 4, m)$, for some $m > 4$. In this situation the algorithm employs the reduction PERIODIC. This reduction is the most complex, see Fig. 4.5. The process of reduction of a face-normal map to a k-valent map, where $k \in \{3, 4, 5\}$ is described in detail in Subsection 4.2.

From k-valent maps to uniform. If the map is a k-valent face-normal map (for some $k \leq 5$), none of the above reductions applies. In particular, we cannot use the difference in degrees of end-vertices of edges to determine the set of edges to be contracted. The original Hopcroft-Wong algorithm at this stage introduces new kinds of reductions that are difficult to describe both formally and informally. Here our algorithm differs essentially. The main new idea consists in the observation that there is no need to introduce new reductions for face-normal k-valent maps, but the reductions APERIODIC$_{\mathcal{D}}$, LARGE$_d$ and PERIODIC can be used, where instead of degrees and degree types, we use the refined degrees (local types) and refined degree types to identify the part of map which is going to be modified. If we obtain a map which is not k-valent, then we return back to the degree type version of the reductions, or to Normalization. We repeat this process until the map becomes uniform, i.e., it has the same refined degree type at each vertex.

From uniform maps to homogeneous. The uniform maps that come as an output of the above procedure can be easily understood. There are five infinite families: the bouquets, the dipoles, cycles, prisms and antiprisms. In addition there are finitely many exceptional maps: the five Platonic maps, the 13 Archimedean maps and the elongated square gyrobicupola. The maps in the four infinite families that are not cycles are reduced to cycles applying either a special reduction (for the bouquets), or taking the dual followed by few standard reductions, see Subsection 4.4.

Irreducible maps. For the labeled maps based on the Archimedean and Platonic maps, and on the elongated square gyrobicupola, the number of darts is bounded by a constant and a brute-force algorithm is sufficient. For labeled cycles, we use the algorithm described in Section 5 (the main idea comes from [**5**]).

Labels. The elementary reductions are defined in a way that the set of darts D' of the reduced map is a subset of the set D of darts of the input map. For some of the reductions $D' = D$. Moreover, the deleted darts $D \setminus D'$ always form a union of orbits in the action of the orientation preserving automorphism group. For each reduction we show that for a given isomorphism $\psi : M_1 \to M_2$, its restriction $\psi' = \psi|_{D'_1}$ is an isomorphism between the reduced maps. To reverse the implication, labels attached to darts of the maps are introduced. It transpires that the best data-structures for the labels are planted rooted trees with nodes assigned by integers. They have several advantages. Firstly, the management of the labels is efficient. Secondly, the main concepts of the theory of oriented maps easily extend to labeled maps. Thirdly, the labels can be used for a backward reconstruction of the input map from the associated reduced map. More details on the labels can be found in Section 6.

Oriented versus general maps. The described algorithm decides the isomorphism problem for spherical maps with prescribed global orientation (clock-wise or counter-clock-wise). Since two topological maps are isomorphic if and only if the associated oriented maps $M_1 \cong M_2$, or if $M_1 \cong M_2^{-1}$ (see Theorem 2.2), to solve the isomorphism problem for general maps we need to execute the algorithm twice, first for input (M_1, M_2), secondly for input (M_1, M_2^{-1}).

4. From maps to homogeneous maps

Inspired by the article by Hopcroft and Wong [**5**], we describe a finite list of local reductions defined on labeled oriented spherical maps. The goal is to reduce an arbitrary map M to a homogeneous map, while preserving the isomorphism relation. We do this in two steps. First we reduce the map M to a uniform map and then we reduce a uniform map to a homogeneous one.

Each reduction decreases the sum $v(M)+e(M)$, where $v(M)$ denotes the number of vertices and $e(M)$ denotes the number of edges. Hence the procedure is finite. The reductions are designed in such a way that the isomorphism relation is preserved.

Among the spherical homogeneous maps, the only infinite family (up to duality) is formed by the cycles. Hence, the final step of our algorithm solves the isomorphism problem for labeled cycles. In what follows, we describe the reductions in detail.

4.1. Normalization. Even if we start with a polyhedral map, a sequence of elementary reductions may produce a map with 1-faces and/or 2-faces. In such a case, we always apply the normalization, which is the transformation of the highest priority. The reason is that the other elementary reductions apply only to face-normal maps. We denote this transformation by NORMALIZE(M). It removes faces of degree one if $v(M) > 1$ and replaces planar dipoles by edges if $v(M) > 2$. It may happen that an application of Normalization does not produce a face-normal map. In such a case we apply it repeatedly till we get either a face-normal map, or one of the following uniform maps: a bouquet or a dipole. Let $M = (D, R, L, \ell)$ be an oriented map. In short we have

Reduction:	Normalization – NORMALIZE(M)
Priority:	0
Input:	Map M which is not face-normal.
Output:	Either a face-normal map M' with $D' \subseteq D$, or a bouquet or a dipole.

For technical reasons we split the normalization into a sequence of elementary reductions of two kinds:

- $N_1(M)$ – deletion of facial loops,
- $N_2(M)$ – replacing of dipoles by edges.

In what follows, we describe the reductions N_1 and N_2 formally.

Removing facial loops. Let $M = (D, R, L, \ell)$. We are going to define $N_1(M) = M' = (D', R', L', \ell')$. First we increase the step counter s by one, i.e. we set $s := s + 1$. Let $\mathcal{L}^1$ be the list of all maximal sequences of darts of the form $\mathcal{S} = \{x_1, x_1^-, x_2, x_2^-, \ldots, x_k, x_k^-\}$, where $R(x_i) = x_i^-$, for $i = 1, \ldots, k$, $R(x_i^-) = x_{i+1}$ for $i = 1, \ldots, k-1$, and $R(x_k^-) \neq x_1$. By definition each x_i is a fixed point of $R^{-1}L$, hence it bounds a 1-face. Moreover, all the darts x_i are incident to the same vertex $v \in V(M)$. In particular, the local rotation at a vertex v has form

$$R_v = (x_0, x_1, x_1^-, x_2, x_2^-, \dots, x_k, x_k^-, x_{k+1}, \dots),$$

for some darts x_0, x_{k+1}. We shall call the darts x_0 and x_{k+1} *bounding darts*. It may be that $x_0 = x_{k+1}$, however $\{x_0, x_{k+1}\} \neq \emptyset$ by definition, otherwise the map M is a bouquet which is irreducible. The reduction applies if $\mathcal{L}^1$ is non-empty. For every element $\mathcal{S} \in \mathcal{L}^1$ we delete the darts in $\{x_1, x_1^-, x_2, x_2^-, \dots, x_k, x_k^-\}$ from D and set $R'_v = (x_0, x_{k+1}, \dots)$. Moreover, we set

$$\ell'(x_0) = \textsc{Label}(s, a_0) \quad \ell'(x_{k+1}) = \textsc{Label}(s, a_1, b_1, \dots, a_k, b_k, a_{k+1}),$$

where $a_i = \ell(x_i)$, for $i = 0, \dots, k+1$, and $b_i = \ell(x_i^-)$, for $i = 1, \dots, k$. For a dart $x \in D'$ that is not bounding we set $\ell'(x) = \ell(x)$. At the moment it is enough to require that the function Label used above is injective. Further details will be explained later. We do the reduction for every maximal sequence of 1-faces thus making the list $\mathcal{L}^1$ empty; the step counter s remains the same. We set $D' \subset D$ to be set of darts that do not appear in any sequence in $\mathcal{L}^1$. We set $L'(x) = L(x)$ for $x \in D'$, and $R'(x) = R(x)$ for a dart in D' which is not bounding. This way the map M' is completely defined, see Fig. 4.1.

Lemma 4.1. *Let $M_i = (D_i, R_i, L_i, \ell_i)$, $i = 1, 2$ and $D_1 \cap D_2 = \emptyset$, be labeled spherical maps. Then $M_1 \cong M_2$ if and only if $M'_1 = N_1(M_1) \cong N_1(M_2) = M'_2$.*

Proof. Let $\psi \colon M_1 \to M_2$ be an isomorphism. We prove that $\psi' = \psi_{\restriction D'_1}$ is an isomorphism of M'_1 and M'_2. Since ψ preserves the set of 1-faces, the mapping ψ' is a well-defined bijection. We check the commuting rules (2.1) for ψ'.

By the definition of N_1, $L'_i = L_{i\restriction D_i}$, for $i = 1, 2$. Thus, we have $\psi' L'_1 = L'_2 \psi'$. As concerns the permutations R'_1 and R'_2, it follows that we need to check the commuting rule only at the darts preceding a sequence of 1-faces (in the clockwise orientation). With the above notation, using the definition of M'_1 and M'_2, and the fact that ψ is an isomorphism, we get

$$\begin{aligned}\psi' R'_1 x_0 &= \psi' R_1 (L_1 R_1)^k x_0 = \psi R_1 (L_1 R_1)^k x_0 \\ &= R_2 (L_2 R_2)^k \psi x_0 = R_2 (L_2 R_2)^k \psi' x_0 = R'_2 \psi' x_0.\end{aligned}$$

Finally, for ℓ'_1 and ℓ'_2, we have, by the definition of N_1 and of N_2,

$$\begin{aligned}\ell'_1(x_{k+1}) &= \textsc{Label}(s, \ell_1(x_1), \ell_1(x_1^-) \dots, \ell_1(x_k), \ell_1(x_k^-), \ell_1(x_{k+1})) \\ &= \textsc{Label}(s, \ell_2(\psi x_1), \ell_2(\psi x_1^-) \dots, \ell_2(\psi x_k), \ell_2(\psi x_k^-), \ell_2(\psi x_{k+1})) = \ell'_2(\psi' x_{k+1})\end{aligned}$$

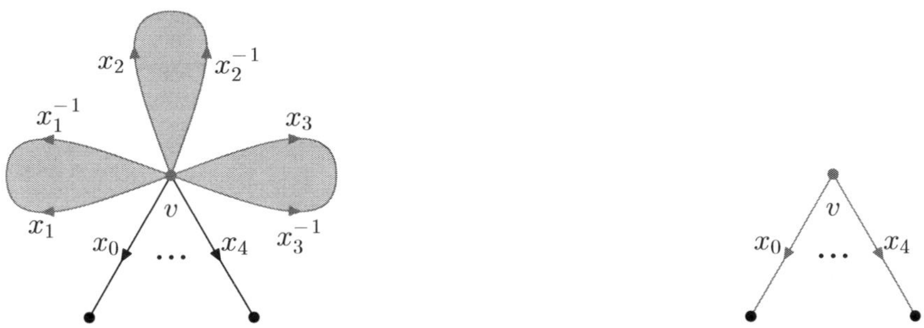

Figure 4.1. Removing loops.

if and only if

$$\ell_1(x_i) = \ell_2(\psi x_i), \text{ for } i = 0, \dots, k, k+1$$

which is satisfied since ψ is an isomorphism. Similarly, we check that $\ell_1'(x_0) = \ell_2'(\psi x_0)$.

Conversely, let $\psi'\colon M_1' \to M_2'$ be an isomorphism. With the above notation, we have

$$x_i = R_1(L_1R_1)^i x_0 \quad \text{and} \quad x_{k+1} = R_1' x_0.$$

Since LABEL is injective, it follows that there are $y_1, \dots, y_k$ in $D_2 \setminus D_2'$ such that

$$y_i = R_2(L_2R_2)^i \psi' x_0.$$

Here we employ the fact that s was changed when we applied LABEL to compute the new labels at the bounding darts. This blocks the possibility of existence of isomorphisms $\psi' : M_1' \to M_2'$ taking a bounding dart to a dart that is not bounding, i.e. ψ' takes the set of bounding darts onto the set of bounding darts.

We define an extension of ψ of ψ' by setting $\psi x_i = y_i$, for $i = 1, \dots, k$. It is straightforward to check that the extension ψ of ψ' is a well-defined isomorphism from M_1 to M_2. □

Removing dipoles. Let $M = (D, R, L, \ell)$ be a map. We are going to define $N_2(M) = M' = (D', R', L', \ell')$. First we increase s by one. Let $\mathcal{L}^2$ be the list of all maximal sequences $\mathcal{S} = \{x_1, \dots, x_k\}$ of darts, $k > 1$, satisfying $Rx_i = x_{i+1}$, $(R^{-1}L)^2 x_i = x_i$, and either $Rx_k \neq x_1$ or $Rx_1^- \neq x_k^-$. The latter condition says that the input map is not a dipole. Observe that all the darts x_i are incident to the same vertex u. Also, every $\mathcal{S} = x_1, \dots, x_k \in \mathcal{L}^2$ is paired with another sequence $\mathcal{S}^- = \{x_k^-, \dots, x_1^-\} \in \mathcal{L}^2$. Each dart in $\mathcal{S}^-$ is incident to the same vertex $v \in V(M)$, hence the union of the darts in $\mathcal{S}$ and in $\mathcal{S}^-$ induces a dipole (sub)map with the vertices u and v. In particular, the respective local rotations have form

$$R_u = (y_1, x_1, \dots, x_k, y_2, \dots) \quad \text{and} \quad R_v = (z_1, x_k^-, \dots, x_1^-, z_2),$$

for some darts $y_1, y_2, z_1, z_2 \in D$. By the assumptions at least one of the sets $\{y_1, y_2\}$, $\{z_1, z_2\}$ is non-empty. We remove the darts

$$\{x_1^-, x_2, x_2^-, \dots, x_{k-1}, x_{k-1}^-, x_k\}, \text{and set}$$

$$R_u' = (y_1, x_1, y_2, \dots), \quad R_v' = (z_1, x_k^-, z_2, \dots),$$

$$L'(x_1) = x_k^-, \quad L'(x_k^-) = x_1.$$

We shall call the darts x_1 and x_k^- the *bounding darts.* Moreover, we set

$$\ell'(x_1) = \text{LABEL}(s, a_1, \dots, a_k) \quad \text{and} \quad \ell'(x_k^-) = \text{LABEL}(s, b_k, \dots, b_1),$$

where s is the step, $a_i = \ell(x_i)$ and $b_i = \ell(x_i^-)$, for $i = 1, \dots, k$. We do the reduction for every pair $\mathcal{S}, \mathcal{S}^-$ of maximal sequences in $\mathcal{L}^2$. The step counter s remains the same during this process. We set

$$D' = (D \setminus \cup_{S \in \mathcal{L}^2} (S \cup S^-)) \cup_{S \in \mathcal{L}^2} B_S,$$

where B_S is the 2-element set of bounding darts of $S \cup S^-$. We set $L'(x) = L(x)$, $\ell'(x) = \ell(x)$ for $x \in D' \setminus \mathcal{B}$, where $\mathcal{B} = \cup_{S \in \mathcal{L}^2} B_S$. Finally, we set $R_u' = R_u$ at the vertices u that are not incident to 2-faces. This way the map M' is completely defined. See Fig. 4.2.

LEMMA 4.2. *Let $M_i = (D_i, R_i, L_i, \ell_i)$, $i = 1, 2$ and $D_1 \cap D_2 = \emptyset$, be labeled spherical maps. Then $M_1 \cong M_2$ if and only if $M_1' = N_2(M_1) \cong N_2(M_2) = M_2'$.*

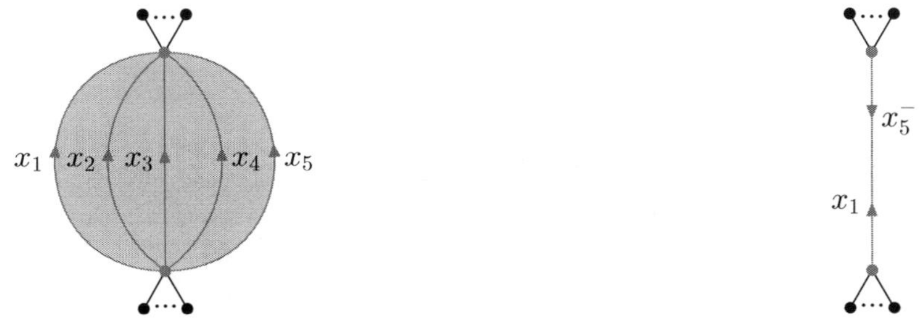

FIGURE 4.2. Removing dipoles.

PROOF. Let $\psi\colon M_1 \to M_2$ be an isomorphism. We prove that $\psi' = \psi_{\restriction D_1'}$ is an isomorphism of M_1' and M_2'. Since ψ preserves the set of 2-faces, the mapping ψ' is a well-defined bijection. We check the commuting rules (2.1) for ψ' at the set of bounding darts $\mathcal{B}$. For the other darts in D' there is nothing to prove. Let $\mathcal{S} = \{x_1, \ldots, x_k\} \in \mathcal{L}^2$. By the definition of M_1' and M_2', we have $L_1'x_1 = x_k^- = L_1R_1^{k-1}x_1$ and $L_1'x_k^- = L_1R_1^{k-1}x_k^-$. It follows that

$$\psi' L_1' x_1 = \psi' L_1 R_1^{k-1} x_1 = \psi L_1 R_1^{k-1} x_1 = L_2 R_2^{k-1} \psi x_1 = L_2' \psi' x_1,$$

and

$$\psi' L_1' x_k^- = \psi' L_1 R_1^{k-1} x_k^- = \psi L_1 R_1^{k-1} x_k^- = L_2 R_2^{k-1} \psi x_k^- = L_2' \psi' x_k^-.$$

It follows that $\psi L_1' = L_2' \psi$.

With the above notation, using the definition of M_1' and M_2', and the fact that ψ is an isomorphism, we get

$$\psi' R_1' x_1 = \psi' R_1^{k-1} x_1 = \psi R_1^{k-1} x_1 = R_2^{k-1} \psi x_1 = R_2' \psi x_1 = R_2' \psi' x_1.$$

For x_k^-, the verification of the commuting rules is similar.

Using the definition of M_1' (M_2') we obtain

$$\ell_1'(x_1) = \text{LABEL}(s, \ell_1(x_1), \ldots, \ell_1(x_k)) = \text{LABEL}(s, \ell_2(\psi x_1), \ldots, \ell_2(\psi x_k)) = \ell_2'(\psi' x_1)$$

if and only if

$$\ell_1(x_i) = \ell_2(\psi x_i), \text{ for } i = 1, \ldots, k,$$

which is satisfied since ψ is an isomorphism. Similarly, we check that $\ell'(x_k^-) = \ell'(\psi x_k^-)$.

Conversely, let $\psi'\colon M_1' \to M_2'$ be an isomorphism. With the above notation, we have

$$x_i = R_1^{i-1} x_1 \quad \text{and} \quad x_i^- = R^{i-1} x_k^-,$$

for $i = 1, \ldots, k$. Since LABEL is injective, it follows that there are $y_2, \ldots, y_k$ in $D_2 \setminus D_2'$ such that

$$y_i = R_2^{i-1} \psi' x_1$$

determining a sequence of 2-faces. We define an extension of ψ of ψ' by setting $\psi x_i = y_i$, for $i = 2, \ldots, k$. It is straightforward to check that the extension ψ of ψ' is a well-defined isomorphism from M_1 to M_2. □

When applying the elementary reduction N_1 (or N_2) new 1-faces and 2-faces may be created. Therefore after performing N_1 (or N_2) we reconstruct the lists $\mathcal{L}^1$ and $\mathcal{L}^2$ by deleting the sequences of darts which were already used and adding the new ones. Similarly, new light vertices may be created when applying N_1 and N_2.

Thus the lists $\mathcal{L}$ defined below that are associated with the following reductions of face-normal maps may be affected and have to be modified as well.

After applying the reductions N_1 and N_2 as many times as possible the output map is either face-normal, or it is a bouquet, or a dipole. The whole procedure will be called the *normalization* of a map.

4.2. Face-normal maps. Let M be a face-normal map with minimum degree d. By Lemma 2.3, $d \leq 5$. Suppose there exists a vertex of degree $> d$. Let u be a vertex of minimum degree adjacent to a vertex of degree $> d$. Let $v_0, \dots, v_{d-1}$ be the neighbors of u, and let $\mathcal{D}(u) = (m_0, \dots, m_{d-1})$ be its degree type, where $deg(v_i) = m_i$ for $i = 0, \dots, d-1$. By connectivity such a vertex u exists. We have $m_i \geq d$ for all i, $0 \leq i < d$, and for some j, $m_j > d$. We say that

- u has *large degree type* if $m_k > d$ for all k,
- u has *small degree type* if there exists i with $m_i = d$.

A small degree type is called *periodic* if it can be written in the form

$$(d, n_1 \dots, n_k, d, n_1, \dots, n_k, d, n_1, \dots, n_k)$$

where the subsequence $d, n_1 \dots, n_k$ occurs at least twice. Since $d \leq 5$, the only possible periodic degree type is $(4, m, 4, m)$, where $m > 4$. A small degree type is called *aperiodic*, if it is not periodic. Now we are ready to introduce three kinds of reductions.

Large degree type. In this reduction, the input is a labeled face-normal map $M = (D, R, L, \ell)$ and $\mathcal{L}$ is a list of all light vertices of minimum degree d with large degree type. For every vertex $v \in \mathcal{L}$ with $\mathcal{D}(v) = (m_0, \dots, m_{d-1})$ and the respective neighbors $u_0, \dots, u_{d-1}$, we delete v together with all the edges incident to it, and we add the face bounded by the cycle $(v_0, \dots, v_{d-1})$.

Reduction:	Large degree type – $\textsc{Large}_d(M)$
Priority:	$(1, d)$
Input:	Face-normal map M with minimum degree d and a non-empty list $\mathcal{L}$ of light vertices of degree d with large degree type.
Output:	Map M' with $V(M') = V(M) \setminus \mathcal{L}$ and $D' = D$.

We first increase the step counter s by one. The map $M' = \textsc{Large}_d(M) = (D', R', L', \ell')$ is defined as follows. We set $D' = D$ and $L' = L$. For $v \in \mathcal{L}$, let $u_0, u_1, \dots, u_{d-1}$ be the neighbors of v listed in the order following the chosen orientation. Denote by $x_0, x_1, \dots, x_{d-1}$ the darts based at $u_0, u_1, \dots, u_{d-1}$, joining u_j to v for $j = 0, \dots, d-1$. We have $R_{u_i} = (y_i, x_i, z_i, \dots)$, for some darts y_i, z_i, $i = 0, \dots, d-1$. We set $R'_{u_i} = (y_i, x_i, x_{i-1}^-, z_i, \dots)$. Moreover, we set $\ell'(x_i) = \textsc{Label}(s, \ell(x_i))$ and $\ell'(x_i^-) = \textsc{Label}(s, \ell(x_i^-))$, where s is the current step number. See Fig. 4.3.

Lemma 4.3. *Let $M_i = (D_i, R_i, L_i, \ell_i)$, $i = 1, 2$, be labeled spherical maps. Then $M_1 \cong M_2$ if and only if $\textsc{Large}_d(M_1) \cong \textsc{Large}_d(M_2)$.*

Proof. Let $\psi \colon M_1 \to M_2$ be an isomorphism. We prove that ψ is also an isomorphism of M_1' and M_2'. We check the commuting rules (2.1) for ψ.

We have $L_i' = L_i$, for $i = 1, 2$, so $L_1' \psi = \psi L_2'$. For R_1' and R_2', we have

$$\psi R_1' x_i = \psi R_1^{-1} L_1 x_i = R_2^{-1} L_2 \psi x_i = R_2' \psi x_i,$$

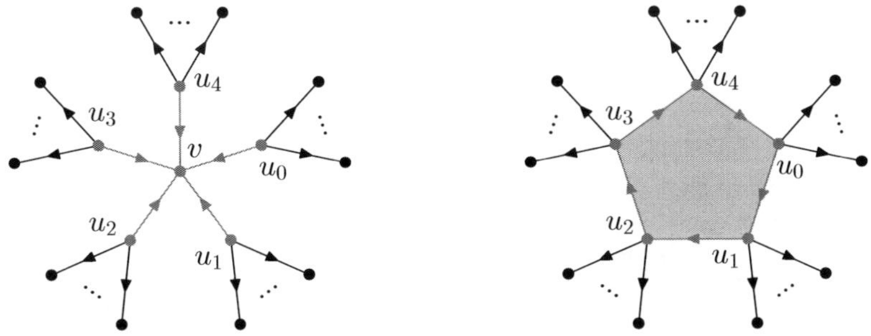

FIGURE 4.3. Removing vertices of large degree type.

$$\psi R_1' x_i^- = \psi R_1 L_1 x_i^- = R_2 L_2 \psi x_i^- = R_2' \psi x_i^-,$$

proving that $\psi R_1' = R_2' \psi$. Clearly, $\ell_1'(x_i) = \ell_2'(\psi x_i)$ if and only if $\ell_1(x_i) = \ell_2(\psi x_i)$. Similarly for x_i^-. □

Aperiodic small degree type. The following list contains all possible classes of aperiodic small degree types for spherical maps listed according to their priority. The priority is given by the ordering of the degree types. Essential is the fact that degree types of smaller lengths are of higher priority.

(1) $\mathcal{D}(u) = (2, m_1)$, $m_1 > 2$,
(2) $\mathcal{D}(u) = (3, 3, m_1)$, $m_1 > 3$,
(3) $\mathcal{D}(u) = (3, m_1, m_2)$, $m_1, m_2 > 3$,
(4) $\mathcal{D}(u) = (4, 4, 4, m_1)$, $m_1 > 4$,
(5) $\mathcal{D}(u) = (4, 4, m_1, m_2)$, $m_1, m_2 > 4$,
(6) $\mathcal{D}(u) = (4, m_1, 4, m_2)$, $4 < m_1 < m_2$,
(7) $\mathcal{D}(u) = (4, m_1, m_2, m_3)$, $m_1, m_2, m_3 > 4$,
(8) $\mathcal{D}(u) = (5, 5, 5, 5, m_1)$, $m_1 > 5$,
(9) $\mathcal{D}(u) = (5, 5, 5, m_1, m_2)$, $m_1, m_2 > 5$,
(10) $\mathcal{D}(u) = (5, 5, m_1, 5, m_2)$, $m_1, m_2 > 5$,
(11) $\mathcal{D}(u) = (5, 5, m_1, m_2, m_3)$, $m_1, m_2, m_3 > 5$,
(12) $\mathcal{D}(u) = (5, m_1, 5, m_2, m_3)$, $m_1, m_2, m_3 > 5$, $m_1 \leq m_2$,
(13) $\mathcal{D}(u) = (5, m_1, m_2, 5, m_3)$, $m_1, m_2, m_3 > 5$, $m_3 > m_1$,
(14) $\mathcal{D}(u) = (5, m_1, m_2, m_3, m_4)$, $m_1, m_2, m_3, m_4 > 5$.

Accordingly to their degree types the set of light vertices of a (face-normal) map decomposes into classes, called lists here. There are 5 lists of vertices of large degree type distinguished by the degree $d = 1, 2, 3, 4, 5$. There are 14 lists of vertices of aperiodic small degree type distinguished by the above degree types and one list of vertices of periodic type $(4, m, 4, m)$. Finally, the light vertices of homogeneous types (1), $(2, 2)$, $(3, 3, 3)$, $(4, 4, 4, 4)$ and $(5, 5, 5, 5, 5)$ are organised in lists with respect to their refined degree types. All these lists are created in the initialization, and they are effectively reconstructed after performing each elementary reduction.

In the reduction APERIODIC$_{\mathcal{D}}(M)$ we increment s and find all vertices u having $\mathcal{D}(u) = \mathcal{D}$ in the first non-empty list $\mathcal{L}$ containing the vertices of type $\mathcal{D}$. Let u be such a vertex with $\mathcal{D}(u) = (m_0, \ldots, m_{d-1})$ and let $v_0, \ldots, v_{d-1}$ be the corresponding neighbors. Since $\mathcal{D}(u)$ is aperiodic, we can canonically choose an edge $e_k = uv_k$, where k is the smallest index such that $m_k > d$. We define

$$O = \{x \in D : (x, Lx) = e_k, u \in \mathcal{L}\}$$

to be the set of darts associated to the canonically chosen edges. The transformation $\text{APERIODIC}_{\mathcal{D}}(D)$ contracts every edge in O.

Reduction:	$\text{APERIODIC}_{\mathcal{D}}(M)$
Priority:	$(2, \mathcal{D})$
Input:	Face-normal map M and a list $\mathcal{L}$ of light vertices with aperiodic degree type $\mathcal{D}$.
Output:	Map M' with $V(M') = V(M) \setminus \mathcal{L}$ and $D' \subseteq D$.

LEMMA 4.4. *The subgraph X_O of the underlying graph of M induced by O is a disjoint union of stars.*

PROOF. By definition every edge in O joins a vertex of degree d to a vertex of degree higher than d. Moreover, due to the canonical choice of edges in O, we also have that for a vertex u of type $\mathcal{D}$, there is exactly one edge incident to u. Hence u is of degree one and we are done. □

Let $M = (D, R, L, \ell)$ be a face-normal map. We define a map $M' = (D', R', L', \ell')$ $= \text{APERIODIC}_{\mathcal{D}}(M)$ as follows. We set $D' = D \setminus O$. For each $x \in D'$ we set $L'x = Lx$. If v is not in the subgraph induced by X_O, we set $R'_v = R_v$. It remains to define R' at the centers of the stars in X_O. Let $w_0, w_1, \dots, w_{k-1}$ be the vertices with aperiodic degree type $\mathcal{D}$ that have degree 1 in a connected component of X_O with the center v. Suppose that we have $R_{w_i} = (x_i, A_i)$, for some sequence of darts A_i of length $d-1$, $i = 0, \dots, k-1$, and $R_v = (Lx_0, B_0, \dots, Lx_{k-1}, B_{k-1})$, for some (possibly empty) sequences of darts B_i, $i = 0, 1, \dots, k-1$. For all such vertices v, the permutation R' is defined by setting $R'_v = (A_0, B_0, \dots, A_{k-1}, B_{k-1})$. In particular, we have $R'(R^{-1}x_i) = RLx_i$ and $R'(R^{-1}x_i^-) = Rx_i$. Moreover, we set $\ell'(Rx_i) = \text{LABEL}(s, \ell(Rx_i), \ell(x_i))$ and $\ell'(R^{-1}x_i) = \text{LABEL}(s, \ell(R^{-1}x_i), \ell(x_i^-))$. For other darts x, we set $\ell'(x) = \ell(x)$. See Fig. 4.4.

LEMMA 4.5. *Let $M_i = (D_i, R_i, L_i, \ell_i)$, $i = 1, 2$, be labeled spherical maps. Then $M_1 \cong M_2$ if and only if $\text{APERIODIC}_{\mathcal{D}}(M_1) \cong \text{APERIODIC}_{\mathcal{D}}(M_2)$.*

PROOF. Let $\psi\colon M_1 \to M_2$ be an isomorphism. We prove that $\psi' = \psi_{\restriction D'_1}$ is an isomorphism of M'_1 and M'_2. Since ψ preserves the set O, the mapping ψ' is a well-defined bijection. We check the commuting rules (2.1) for ψ'.

By the definition, $L'_i = L_{i\restriction D'_i}$, for $i = 1, 2$. Thus, we have $\psi' L'_1 = L'_2 \psi'$. For R'_1 and R'_2, it suffices to check the commuting rule at $y_i = R_1^{-1} x_i$, and at $z_i = R_1^{-1} x_i^-$ such that $z_i \notin O$. By the definition of R'_1 and R'_2, we have $R'_1 y_i = RLRy_i$ if $B_i \neq \emptyset$ and $R'_1 y_i = R_1 L R_1 L_1 R_1 y_i$ if $B_i = \emptyset$, and $R'_1 z_i = R_1 L_1 R_1 z_i$ if $B_i \neq \emptyset$. In general, for $x \in \{y_i, z_i\}$, we have

$$R'_1 x = w(R_1, L_1)x \quad \text{and} \quad R'_2 \psi x = w(R_2, L_2)\psi x,$$

where $w(R_i, L_i)$, for $i = 1, 2$, is a word in terms R_i and L_i defining an element in the respective monodromy group. We have

$$\psi' R'_1 x = \psi w(R_1, L_1)x = w(R_2, L_2)\psi x = R'_2 \psi' x.$$

For ℓ'_1 and ℓ'_2, we have

$$\ell'_1(R_1 x_i) = \text{LABEL}(s, \ell_1(R_1 x_i), \ell_1(x_i)) = \text{LABEL}(s, \ell_2(R_2 \psi x_i), \ell_2(\psi x_i)) = \ell'_2(R_2 \psi x_i)$$

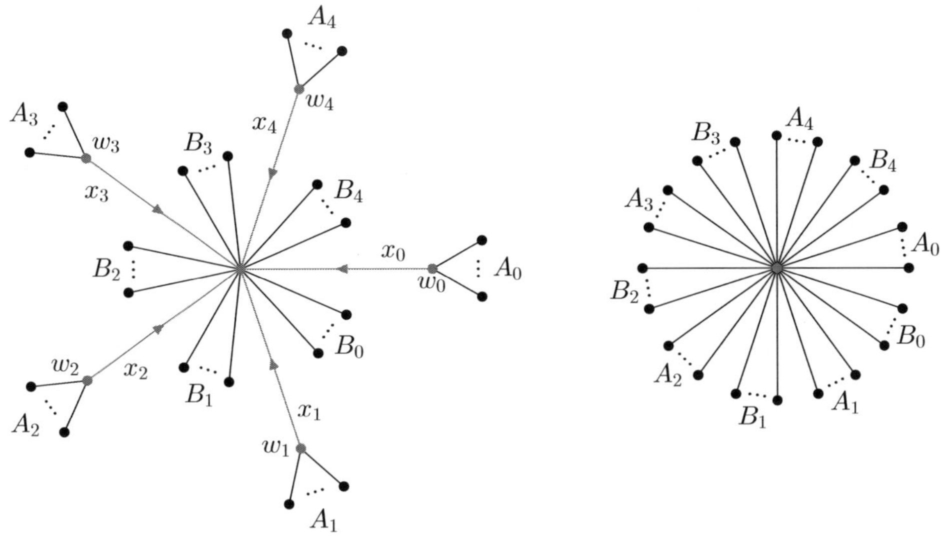

FIGURE 4.4. Removing vertices of aperiodic small degree type.

if and only if

$$\ell_1(R_1 x_i) = \ell_2(R_2 \psi x_i) \quad \text{and} \quad \ell_1(x_i) = \ell_2(\psi x_i),$$

which is true since ψ is an isomorphism.

Conversely, let $\psi' \colon M_1' \to M_2'$ be an isomorphism. By definition, we have $x_i = R_1 y_i$. Since LABEL is injective, it follows that there is $R_2\psi' y_i$ in $D_2 \setminus D_2'$. We set $\psi x_i = R_2 \psi' y_i$. It is straightforward to check that the extension ψ of ψ' is a well-defined isomorphism from M_1 to M_2. $\square$

Periodic small degree type. In the reduction PERIODIC(M) we increment s and find all vertices u in the first non-empty list $\mathcal{L}$ containing the vertices of periodic type $(4, m, 4, m)$.

Reduction:	PERIODIC(M)
Priority:	least
Input:	Face-normal map M without aperiodic vertices and a list $\mathcal{L}$ of light vertices of periodic type.
Output:	Map M' with $V(M') = V(M) \setminus \mathcal{L}$ and $D' = D$.

LEMMA 4.6. *The subgraph induced by all the periodic vertices is a disjoint union of cycles and paths.*

PROOF. Consider a component C induced by periodic vertices $v_0, \dots, v_{k-1}$. Since M has no vertices of aperiodic type and of large degree type, the component C is either a cycle, or a path whose end-vertices are of homogeneous degree type $(4, 4, 4, 4)$. $\square$

Let $C = v_0, \dots, v_{k-1}$ is either a path or a cycle induced by vertices of periodic type. Following the clockwise orientation, let the neighbors of v_i by v_{i+1}, w_i, v_{i-1}, and u_i. Informally, the reduction deletes C together with the incident edges, for each $i = 0, 1, \dots, k-1$ introduces two parallel edges joining u_i to w_i (a dipole)

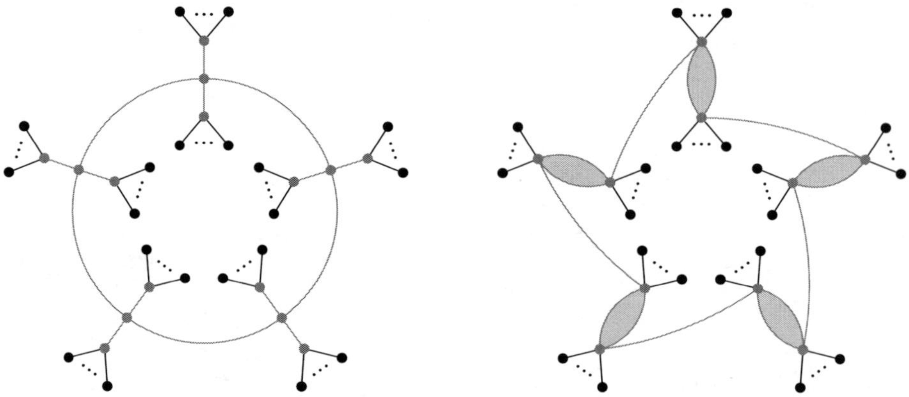

FIGURE 4.5. Global view of the periodic small degree type reduction.

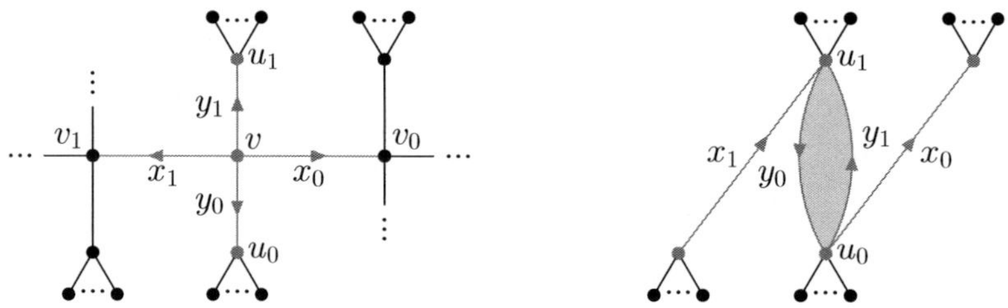

FIGURE 4.6. Local view of the periodic small degree type reduction.

together with an edge joining w_i to u_{i+1} passing through the central point of the deleted edge v_iv_{i+1}, see Fig. 4.5,4.6. A formal definition is given below.

Let $M = (D, R, L, \ell)$ be a face-normal map without vertices of large degree type and without vertices of aperiodic type. In particular, this means that each light vertex is either of type $(4, 4, 4, 4)$ or of type $(4, x, 4, x)$, where $x \geq 6$. We define the map $M' = (D', R', L', \ell') = \text{PERIODIC}(M)$ by setting $D' = D$, $L' = L$. Further we define R' in the following way. Let the cycles of R corresponding to u_i, v_i and w_i be

$$R_{u_i} = (y_i^-, \dots), \quad R_{v_i} = (x_i, z_i, x_{i-1}^-, y_i), \quad \text{and} \quad R_{w_i} = (z_i^-, \dots),$$

where the dart x_i joins v_i to v_{i+1}, the dart y_i joins v_i to u_i, and the dart z_i joins v_i to w_i. To define R', we delete R_{v_i} and set

$$R'_{u_i} = (z_i, y_i^-, x_{i-1}^-, \dots) \quad \text{and} \quad R_{w_i} = (y_i, z_i^-, x_i, \dots),$$

for every $i = 0, \dots, k-1$ and every component C, see Fig. 4.5 and Fig. 4.6.

Moreover, for every $x \in \{y_i, y_i^-, z_i, z_i^- : i = 0, \dots, k-1\}$, we set $\ell'(x) = \text{LABEL}(s, \ell(x))$, and for every $x \in \{x_i, x_i^- : i = 0, \dots, k-1\}$, we set $\ell'(x) = \text{LABEL}(s+1, \ell(x))$. For other darts the labeling remains unchanged. After finishing the PERIODIC reduction we increase the counter s by two.

LEMMA 4.7. *Let $M_i = (D_i, R_i, L_i, \ell_i)$, $i = 1, 2$, be labeled spherical maps. Then $M_1 \cong M_2$ if and only if* $\text{PERIODIC}(M_1) \cong \text{PERIODIC}(M_2)$.

PROOF. Let $\psi\colon M_1 \to M_2$ be an isomorphism. We prove that ψ is also an isomorphism of M_1' and M_2'. We check the commuting rules (2.1) for ψ'.

We have $L_i' = L_i$, for $i = 1, 2$, so $L_1'\psi = \psi L_2'$. For R_1' and R_2', it suffices check the commuting rules (2.1) at the darts x_i, x_i^-, y_i, y_i^-, z_i, and z_i^-. By the definition of R_1' and R_2', we have

- $R_1'x_i = L_1R_1L_1x_i$ and $R_2'\psi x_i = L_2R_2L_2\psi x_i$,
- $R_1'x_i^- = (R_1L_1)^2R_1^2x_i^-$ and $R_2'\psi x_i^- = (R_2L_2)^2R_2^2\psi x_i^-$,
- $R_1'y_i = R_1^2y_i$ and $R_2'\psi y_i = R_2^2\psi y_i$,
- $R_1'y_i^- = R_1^{-1}y_i$ and $R_2'\psi y_i^- = R_2^{-1}\psi y_i$,
- $R_1'z_i = R_1^2z_i$ and $R_2'\psi z_i = R_2^2\psi z_i$,
- $R_1'z_i^- = R_1^{-1}L_1z_i^-$ and $R_2'\psi z_i^- = R_2^{-1}L_2\psi z_i^-$.

In general, for $x \in \{x_i, x_i^-, y_i, y_i^-, z_i, z_i^-\}$, we have

$$R_1'x = w(R_1, L_1)x \quad \text{and} \quad R_2'\psi x = w(R_2, L_2)\psi x,$$

where $w(R_i, L_i)$, for $i = 1, 2$, is a word in the generators R_i, L_i defining an element of the respective monodromy group. We have

$$\psi R_1'x = \psi w(R_1, L_1)x = w(R_2, L_2)\psi x = R_2'\psi x.$$

Finally, we have $\ell_1'(x) = \ell_2'(\psi x)$ since $\ell_1(x) = \ell_2(\psi x)$.

To finish the proof we need to show that every isomorphism of $M_1' \to M_2'$ is an isomorphism $M_1 \to M_2$. The argument is based on the fact that the set of darts forming the disjoint union of cycles and paths induced by the periodic vertices in M_1 and M_2, respectively, is preserved by any isomorphism $M_1' \to M_2'$. This is forced by special labeling of these sets. $\square$

4.3. Refined degree type. For a light vertex u of degree d, we say the refined degree type $\mathcal{R}(u) = (\text{ref}(u_0), \dots, \text{ref}(u_{d-1}))$ is

- *large* if $\text{ref}(u) < \text{ref}(u_0)$,
- *small* if $\text{ref}(u) = \text{ref}(u_0)$ and there exists $i > 0$ such that $\text{ref}(u) < \text{ref}(u_i)$,
- *homogeneous* if $\text{ref}(u) = \text{ref}(u_i)$ for all $i = 0, \dots, d-1$.

A small refined degree type is called *periodic* if it can be written in the form

$$(r_0, r_1 \dots, r_k, r_0, r_1, \dots, r_k, r_0, r_1, \dots, r_k)$$

where the subsequence $r_0, r_1 \dots, r_k$ of refined degrees occurs at least twice. Since the refined degree of a light vertex u is of length at least five, we have that a small refined degree type $\mathcal{R}(u)$ is *periodic* if $\deg(u) = 4$ and $\text{ref}(u) = \text{ref}(u_0) = \text{ref}(u_2)$ and $\text{ref}(u_1) = \text{ref}(u_3)$, and $\mathcal{R}(u)$ is *aperiodic* otherwise.

Clearly, the application of the reductions described in Subsections 3.1 and 3.2 yields either a uniform map, or a non-uniform k-valent map with $k \in \{3, 4, 5\}$. In the second case, we will continue using the same set of reductions, but replacing degree types with refined degree types.

We say that v is a *refined light vertex* if it is incident to a light face. By Lemma 2.3 a k-valent spherical map contains a refined light vertex.

If there is a refined light vertex v with $\mathcal{R}(v)$ large, we apply reduction LARGE_d with $d = k$. If LARGE_d does no apply, then every refined light vertex v has $\mathcal{R}(v)$ small or homogeneous. If there is a refined light vertex v with $\mathcal{R}(v)$ small and aperiodic, we pick the one with the smallest $\mathcal{R}(v)$ and apply $\text{APERIODIC}_{\mathcal{D}}$ with $\mathcal{D}$ being a refined degree type. If all the refined light vertices are periodic or homogeneous, we apply PERIODIC. If none of the reductions applies, the refined degree types at all vertices are homogeneous, and consequently M is uniform. Note that application of one of these reductions to a k-valent maps typically produces a map which is not k-valent.

4.4. From uniform to homogeneous maps. We introduce a special set of reductions to reduce prisms, antiprisms, dipoles and bouquets to cycles, while preserving the isomorphism relation. Observe, that $M_1 \cong M_2$ if and only if $M_1^* \cong M_2^*$, where M^* is the dual of M.

The dual of a dipole is a cycle. The dual of an n-prism is an n-bipyramid, in particular it is a reducible map. It is easy to see that an n-bipyramid is reduced to a $3n$-dipole by applying once the periodic reduction described above. Similarly, the dual of an n-antiprism is again a reducible map, which is reduced by applying once an aperiodic reduction to $2n$-dipole. Every prism and antiprism is therefore transformed to a labeled cycle. Concerning bouquets, we transform every n-bouquet to an n-cycle based on the same set edges. Formally, let $\mathcal{B} = (D, R, L, \ell)$ be a bouquet. We set $D' = D$, $L' = L$ and $\ell'(x) = \textsc{Label}(s, \ell(x))$. By definition the rotation consists of a single cycle of the form $R = (x_0, x_0^-, x_1, x_1^-, \ldots, x_{n-1}, x_{n-1}^-)$. We set $R' = \prod_{i=0}^{n-1}(x_i^-, x_{i+1})$.

5. Homogeneous maps

In this section, we discuss how to test isomorphism of two homogeneous maps $M_1 = (D_1, R_1, L_1, \ell_1)$ and $M_2 = (D_2, R_2, L_2, \ell_2)$.

5.1. Platonic maps. For two Platonic maps M_1 and M_2, we can test for every bijection $\pi\colon D_1 \to D_2$ if it is an isomorphism. This takes only constant time since the size of D_1 and D_2 is at most 60.

5.2. Loop and dipoles. For a loop, there are only two possible bijections from D_1 to D_2. For dipoles, the dual is a cycle, for which we give an algorithm below.

5.3. Cycles. We show how to test isomorphism of two maps which have the underlying graphs isomorphic to cycles. In order to make the exposition simpler, we transform the labeling of darts $\ell\colon D \to \mathcal{U}$ of a labeled cycle M to a labeling of vertices. For every vertex v of M, we have $R_v = (x, y)$. The two-element set $\{\ell(x), \ell(y)\}$ can be considered as a new label of the vertex v. Thus, the problem reduces to testing isomorphism of two *vertex-labeled* cycles.

Testing isomorphism of cycles. Given cycles X_1 and X_2 with vertex-labellings ℓ_1 and ℓ_2, respectively, the following algorithm (suggested in [**5**]) tests if there is an isomorphism $\psi\colon V(X_1) \to V(X_2)$ such that $\ell_1(v) = \ell_2(\psi(v))$, for every $v \in V(X_1)$. For simplicity, we assume that, at the start, if X_i, for $i = 1, 2$, has k different labels, for some $k \leq |V(X_i)|$, then the labels are the integers $1, \ldots, k$ and the same coding is used in X_1 and X_2. Moreover, we fix an orientation of X_1 and X_2, so that for every vertex v its successor $\mathrm{succ}(v)$ is well defined.

STEP 1. We find an arbitrary vertex v_1 in X_1, with $\ell_1(v_1) \neq \ell_1(\mathrm{succ}(v_1))$. If no such vertex exists in X_1, then ℓ_1 is constant in which case it is easy to check if $X_1 \cong X_2$. Otherwise, we find $v_2 \in V(X_2)$ with $\ell_1(v_1) = \ell_2(v_2)$ and $\ell_1(\mathrm{succ}(v_1)) = \ell_2(\mathrm{succ}(v_2))$. If no such vertex v_2 exists, then X_1 and X_2 are not isomorphic.

STEP 2. For $i = 1, 2$, we construct the set S_i of all vertices u of X_i with $\ell_i(u) = \ell_i(v_i)$ and $\ell_i(\mathrm{succ}(u)) = \ell_i(\mathrm{succ}(v_i))$; see Fig. 5.1. The sets S_1 and S_2 form independent sets in X_1 and X_2, respectively. Every isomorphism maps S_1 bijectively to S_2. If $|S_1| \neq |S_2|$, then X_1 and X_2 are not isomorphic.

STEP 3. For every $v \in S_i$ $(i = 1, 2)$, we join v to $\mathrm{succ}(\mathrm{succ}(v))$ and remove $\mathrm{succ}(v)$. We relabel every vertex in S_i by k, where k is the smallest unused integer; see Fig. 5.1.

STEP 4. We find an arbitrary vertex $v_1 \in S_1$ with $\ell_1(v_1) \neq \ell_1(\mathrm{succ}(v_1))$. If no such vertex exists, then we have $S_1 = V(X_1)$ and ℓ_1 is constant. It is easy to check if $X_1 \cong X_2$. Otherwise, we find $v_2 \in S_2$ with $\ell_2(v_2) = \ell_1(v)$ and $\ell_2(\mathrm{succ}(v_2)) = \ell_2(\mathrm{succ}(v))$. If no such vertex exists, then X_1 and X_2 are not isomorphic, and we stop.

STEP 5. For $i = 1, 2$, we remove from S_i every u with $\ell_i(\mathrm{succ}(u)) \neq \ell_i(\mathrm{succ}(v_i))$. The sets S_1 and S_2 form independent sets in X_1 and X_2, respectively. If $|S_1| \neq |S_2|$, then X_1 and X_2 are not isomorphic and we stop. We go to Step 3.

By X_i^- we denote the labeled cycle X_i $(i = 1, 2)$ with the reverse orientation.

LEMMA 5.1. *Applying the above algorithm twice for the inputs* (X_1, X_2) *and* (X_1, X_2^-) *with* X_2 *taken with the chosen and reverse orientation, it is decided in linear time if two labeled cycles* X_1 *and* X_2 *are isomorphic as oriented maps.*

PROOF. Let X_1' and X_2' be the graphs obtained from X_1 and X_2 after applying Step 3, respectively. It suffices to show that $X_1 \cong X_2$ if and only if $X_1' \cong X_2'$.

Let T_i be the set of clockwise neighbors of S_i, for $i = 1, 2$. Formally, $T_i = \{u \in V(X_i) : u = \mathrm{succ}(v), \text{ for } v \in S_i\}$. The subgraph of X_i induced by $S_i \cup T_i$ is a matching such that all the vertices in S_i have the same label and all the vertices in T_i have the same label. Every orientation preserving isomorphism $\psi\colon X_1 \to X_2$ satisfies $\psi(S_1) = S_2$ and $\psi(T_1) = T_2$. We have $V(X_i') = V(X_i) \setminus T_i$. Therefore, the restriction of ψ to $V(X_1')$ is an isomorphism from X_1' to X_2'.

On the other hand, if $\psi'\colon X_1' \to X_2'$ is an isomorphism, then let U_i be the set of clockwise neighbors of S_i in X_i'. We have $\psi'(S_1) = S_2$. Note that we assume that S_1 and S_2 are updated before applying Step 4. Since $|S_i| = |T_i|$, we can easily extend ψ' to an isomorphism $\psi\colon X_1 \to X_2$.

We need to execute the algorithm twice to check whether X_1 is isomorphic X_2, or to a 180-degree rotation of X_2. More precisely, $ISO(X_1, X_2)$ checks map existence of map isomorphisms taking the inner face of X_1 onto the inner face X_2, while $ISO(X_1, X_2^-)$ checks map isomorphisms taking the inner face of X_1 onto the outer face of X_2. □

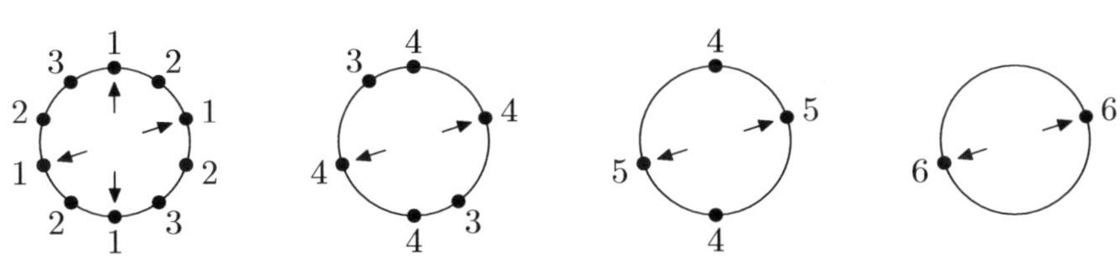

FIGURE 5.1. Illustration of the reduction procedure for cycles.

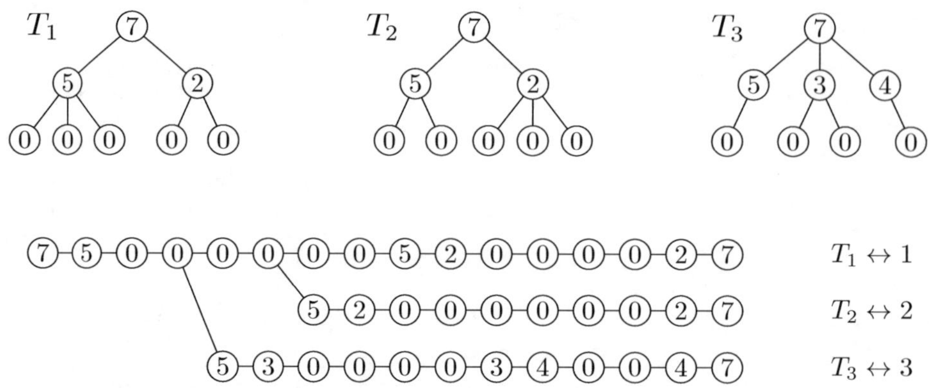

FIGURE 6.1. Labels represented as planted trees together with the associated prefix tree.

LEMMA 5.2. *The complexity of the above algorithm is $\mathcal{O}(n)$, where n is the number of vertices of X_1 and X_2.*

PROOF. Steps 1–2 take $\mathcal{O}(n)$ time. Each iteration of Steps 3–5 takes $\mathcal{O}(|S_1| + |S_2|)$ time. However, since we remove $|S_1|$ vertices from X_1 and $|S_2|$ vertices from X_2 in each iteration of Step 3, the overall complexity is $\mathcal{O}(n)$. □

6. Complexity

In this section, we investigate the algorithm for testing isomorphism of two spherical maps M_1 and M_2. We argue that it runs in time linear in the size of the input, i.e., in time $\mathcal{O}(|D_1| + |D_2|)$. We show a representation of the function ℓ such that $\ell(x)$ and $\ell(y)$ can be compared in constant time. We also describe an implementation of the function LABEL that computes the new label in time proportional to the number of its arguments.

Reductions using degree type. We analyze the complexity of the reductions from Section 5 when they use degree type. If a reduction reduces the sum $v(M) + e(M)$ by k, then the reduction must be executed in time $\mathcal{O}(k)$. This is obvious for all reductions except PERIODIC. In this case, the submap N affected by the reduction is a disjoint union of paths and cycles. The sum $v(M) + e(M)$ decreases exactly by $v(N)$. The submap N can be located using breadth-first search in time $\mathcal{O}(v(N) + e(N)) = \mathcal{O}(v(N))$.

Reductions using refined degree type. The analysis here is exactly the same as for the case when the reduction uses just the degree type. The only difficulty is with updating the refined degree type for every vertex. If there is a large face f of size $\mathcal{O}(v(M))$ incident to a vertex of small degree type, we cannot afford to update the refined degree type of every vertex incident to f, since the degree of f may decrease just by one. To overcome this obstacle we use another trick.

We define the *vertex-face incidence map* $\Gamma(M)$ of M which is a bipartite quadrangular spherical map associated to M. Its vertices are the vertices and centers of faces of M. For every vertex $v \in V(M)$ and face-center $f \in F(M)$ of a face incident to v there is an edge joining v to f. Note that f can be multiply incident to v, for each such incidence there is an edge in $\Gamma(M)$. The map $\Gamma(M)$ can be alternatively obtained as the dual of the medial map. Every reduction easily translates to a reduction in $\Gamma(M)$. We update $\Gamma(M)$ after every elementary reduction. The important property of $\Gamma(M)$ is that if v is a vertex of M, then the degree type of v in $\Gamma(M)$ is exactly the refined degree type of v in M. To update the refined degree of a light vertex after a reduction it suffices to look at its degree type in Γ. To update the refined degree type of a light vertex we look at the degree types of vertices in Γ that correspond to its neighbors in M. This allows us to update the lists $\mathcal{L}$ employed by the reductions.

Labels. In Section 2, we were using the function ℓ as the labeling of a map M and the injective function $\textsc{Label}(s, a_1, \dots, a_m)$, where $s \in \mathbb{N}$ denotes the step and every a_i is a label, for constructing new labels.

First, we describe the implementation of labels, i.e., the images of ℓ. Every label is implemented as a *rooted planted tree* with integers assigned to its nodes. A rooted planted tree is a rooted tree embedded in the plane, i.e., by permuting the children of a node we get different trees; see Fig. 6.1. Every planted tree with n nodes can be uniquely encoded by a 01-string of length $2n$. Further, we require that the children of every node N have smaller integers their nodes. This type of tree is also called a *maximum heap*. Such a tree can be uniquely encoded by a string (sequence) of integers.

Now we define $\textsc{Label}$. The integer s represents the current step of the algorithm. At the start, we have $s = 0$ and the map M has constant labeling – every dart is labeled by a one-vertex tree with 0 assigned to its only vertex. Performing a reduction increments s by 1. For labels (rooted planted trees) $a_1, \dots, a_m$, the function $\textsc{Label}(s, a_1, \dots, a_m)$ constructs a new rooted planted tree with s in the root and the root joined to the roots of $a_1, \dots, a_m$. Clearly this function is injective and can be implemented in the same running time as the corresponding reduction.

Finally, we relabel homogeneous maps by integers. This is necessary mainly for the case when the reductions terminate at cycles since in this case we need to be able to compare labels in constant time. Suppose that we have two homogeneous maps M_1 and M_2 with the corresponding sets of labels $\mathcal{T} = \{T_1, \dots, T_k\}$ and $\mathcal{T}' = \{T'_1, \dots, T'_k\}$. We construct bijections $\sigma\colon \mathcal{T} \to \{1, \dots, k\}$ and $\sigma'\colon \mathcal{T}' \to \{1, \dots, k\}$ such that after replacing T_i by $\sigma(T_i)$ and T'_i by $\sigma'(T'_i)$, we get isomorphic maps. To construct σ and σ', we replace every tree in $\mathcal{T}$ and $\mathcal{T}'$ by a string of integers. Then we find the lexicographic ordering of $\mathcal{T}$ and $\mathcal{T}'$ by constructing two *prefix trees* (sometimes in literature called *trie*); see Fig. 6.1. This lexicographic ordering gives the bijections σ and σ'. Finally, we need to check if the pre-images of every i under σ and σ' are the same planted trees, otherwise the maps are not isomorphic. This can be easily implemented in linear time.

Putting all this together, we have:

THEOREM 6.1. *Isomorphism of spherical maps can be decided in linear time.*

7. Concluding remarks

Map automorphisms. Our algorithm is designed to compare two maps. However, the reader can check that there is no obstacle to use it for a single map $M = (D, R, L, \ell)$ to derive the associated irreducible map M'. Since the irreducible maps are (labeled) Platonic maps, or labeled cycles or their duals, we get the following statement.

THEOREM 7.1. *Let M be a (labeled) spherical map and M' be the associated irreducible labeled map. Then* $\mathrm{Aut}^+(M) \cong \mathrm{Aut}^+(M')$. *In particular,* $\mathrm{Aut}^+(M)$ *is an orientation preserving spherical group.*

Mani's theorem. If we are restricted to polyhedral maps, the last statement of Theorem 7.1 follows from Mani's theorem. It is natural to ask whether it is possible to find an alternative proof of Mani's theorem employing the reduction algorithm. We think that it is possible. The main idea follows.

One can easily realize the Platonic maps, the Archimedean maps and the cycles such that the automorphisms extend to isometries of the underlying sphere. In the construction we first realize properly the irreducible map associated to the input map M. Then we apply step-by-step (in the reverse order) the reverse operations to the reductions thus getting a sequence of proper realizations terminating by a proper realization of M. Finally, we transform the proper spherical realization of M onto a polyhedron with the same automorphism group.

To verify that the above idea leads to a proof, we have to deal with problems of two sorts. First, the map automorphisms reversing the global orientation may not be preserved by the reductions, for instance by PERIODIC(M). Secondly, we need to verify that the inverse operations to the reductions can be done such that the map automorphisms of the expanded maps extend to isometries as well.

Graph isomorphism for planar graphs. Every connected graph X can be decomposed into a rooted tree T_X whose nodes are 3-connected components, cycles, and K_2 [**7**, Section 2]. This decomposition is canonical, meaning that $X \cong Y$ if and only if $T_X \cong T_Y$. Therefore, the linear-time algorithm for graph isomorphism of 3-connected planar graphs allows to solve the graph isomorphism problem of connected planar graphs. We construct both trees T_X and T_Y in linear time [**7**, Section 2] and compare them from the leaves to the roots [**7**, Section 7.32], as in the tree isomorphism algorithm [**4**]. To solve graph isomorphism of general planar graphs, we compute the canonization of each component and compare these canonizations.

Refined equivalence relations. The presented algorithm can be used to compare polyhedra with respect to "combinatorial equivalence", i.e. two polyhedra are isomorphic if the associated polyhedral graphs (maps) are isomorphic. One can introduce equivalence relations of a geometric nature, by requiring an isomorphism to preserve angles, and/or lengths of edges. A simple modification of our algorithm can be used to solve the isomorphism problem with respect to these refined equivalence relations. Information on both the angles and edge-lengths can be stored in dart-labels of the input maps. We do not need to store the real values, it is enough to require that the input labeling of the darts of the input map determines the equivalence classes on the set of angles and on the set of edges. Two angles (edges) are of the same size if and only if the corresponding labels are the same. Otherwise no essential change in the algorithm is needed.

Other surfaces. Another good question is whether one can generalize the algorithm for other closed surfaces. It was announced in [**6**] that such a generalization is possible. To do this, several subproblems have to be solved including: generalization of the concept of the light vertex and light face, generalization of the aperiodic reduction PERIODIC(M), solving the isomorphism problem for the infinite labeled toroidal maps of type $\{4,4\}$, $\{3,6\}$ and $\{6,3\}$ among others. An article containing necessary details is in preparation.

References

[1] Jiří Fiala, Pavel Klavík, Jan Kratochvíl, and Roman Nedela, *3-connected reduction for regular graph covers*, European J. Combin. **73** (2018), 170–210, DOI 10.1016/j.ejc.2018.06.002. MR3836740

[2] J. L. Gross, J. Yellen, and P. Zhang, *Handbook of graph theory*, Chapman and Hall/CRC, 2013.

[3] J. E. Hopcroft and R. E. Tarjan, *A V log V algorithm for isomorphism of triconnected planar graphs*, J. Comput. System Sci. **7** (1973), 323–331, DOI 10.1016/S0022-0000(73)80013-3. MR345442

[4] Alfred V. Aho, John E. Hopcroft, and Jeffrey D. Ullman, *The design and analysis of computer algorithms*, Addison-Wesley Publishing Co., Reading, Mass.-London-Amsterdam, 1975. Second printing; Addison-Wesley Series in Computer Science and Information Processing. MR0413592

[5] J. E. Hopcroft and J. K. Wong, *Linear time algorithm for isomorphism of planar graphs: preliminary report*, Sixth Annual ACM Symposium on Theory of Computing (Seattle, Wash., 1974), Assoc. Comput. Mach., New York, 1974, pp. 172–184. MR0433964

[6] Ken-ichi Kawarabayashi and Bojan Mohar, *Graph and map isomorphism and all polyhedral embeddings in linear time*, STOC'08, ACM, New York, 2008, pp. 471–480, DOI 10.1145/1374376.1374443. MR2582905

[7] P. Klavík, *Extension properties of graphs and structures, PhD thesis*, `http://pavel.klavik.cz/pub/phd_thesis.pdf`, 2018.

[8] P. Mani, *Automorphismen von polyedrischen Graphen* (German), Math. Ann. **192** (1971), 279–303, DOI 10.1007/BF02075357. MR296808

[9] E. Steinitz, *Bedingt konvergente Reihen und konvexe Systeme* (German), J. Reine Angew. Math. **146** (1916), 1–52, DOI 10.1515/crll.1916.146.1. MR1580921

[10] Hassler Whitney, *2-Isomorphic Graphs*, Amer. J. Math. **55** (1933), no. 1-4, 245–254, DOI 10.2307/2371127. MR1506961

NATIONAL INSTITUTE OF INFORMATICS, 2-1-2, HITOTSUBASHI, CHIYODA-KU, TOKYO, JAPAN
Email address: `k_keniti@nii.ac.jp`

NTIS, UNIVERSITY OF WEST BOHEMIA, UNIVERZITNÍ 8, 30100 PILSEN, CZECH REPUBLIC
Email address: `klavik@orgpad.com`

DEPARTMENT OF MATHEMATICS, SIMON FRASER UNIVERSITY, BURNABY, BC V5A 1S6, CANADA
Email address: `mohar@sfu.ca`

DEPARTMENT OF MATHEMATICS AND NTIS, UNIVERSITY OF WEST BOHEMIA, UNIVERZITNÍ 8, 30100 PILSEN, CZECH REPUBLIC
Email address: `nedela@savbb.sk`

DEPARTMENT OF APPLIED MATHEMATICS, FACULTY OF MATHEMATICS AND PHYSICS, CHARLES UNIVERSITY, MALOSTRANSKÉ NÁM. 25, 11800 PRAGUE, CZECH REPUBLIC
Email address: `zeman@kam.mff.cuni.cz`

Contemporary Mathematics
Volume **764**, 2021
https://doi.org/10.1090/conm/764/15334

Some enumeration relating to intervals in posets

Jim Lawrence

ABSTRACT. Given a finite ranked poset P another, $\mathcal{J}(P)$, is obtained by considering its poset of intervals. Iteration of this construction yields a sequence of ranked posets. The functions giving the number of elements of given rank in $\mathcal{J}^k(P)$ is studied by utilizing the notion of parity representation of posets. It is shown that these functions are given by polynomials in 2^k.

1. Introduction and background

It is possible to recast certain enumerative questions concerning the ranked poset of nonempty intervals of a ranked poset as questions concerning enumeration of lattice points in certain associated (not necessarily convex) polyhedra. Some previous results along this line were documented in [**2, 8, 9**]. Here the class of enumeration problems is broadened.

We consider finite posets P with partial ordering relation denoted by $\leq$. A *rank function* for P is a function $\rho : P \to \mathbb{Z}$ that maps elements of P to nonnegative integers and satisfies the single condition that, if $x < y$ in P then $\rho(x) < \rho(y)$.

Given the ranked poset P, another, denoted by $\mathcal{J}(P)$, is produced as follows. The elements of $\mathcal{J}(P)$ are the intervals $[a, b] = \{x \in P : a \leq x \leq b\}$. $\mathcal{J}(P)$ is partially ordered by inclusion. It is ranked by the function $\mathcal{J}\rho : \mathcal{J}(P) \to \mathbb{Z}$ given by $\mathcal{J}\rho([a, b]) = \rho(b) - \rho(a)$. In this paper we study the functions $F_l(k) = |\{p \in \mathcal{J}^k(P) : \mathcal{J}^k\rho(p) = l\}|$, where $\mathcal{J}^k$ denotes k-fold iteration of $\mathcal{J}$: $\mathcal{J}^0(P) = P$, and, for $k = 1, 2, \ldots$, $\mathcal{J}^k(P) = \mathcal{J}(\mathcal{J}^{k-1}(P))$; and similarly for ρ. We relate these functions to certain Ehrhart polynomials of complexes in Euclidean space and show that, as a function of k, l being fixed, $F_l(k)$ is given by a polynomial evaluated at 2^k. Additionally we show that the determination of these functions can be reduced to sums involving the special case of ranked chains.

Figure 1 shows a ranked totally ordered set (that is, a ranked chain) C and the ranked poset $\mathcal{J}(C)$.

Finite posets can be regarded as purely combinatorial objects, whereas polyhedra exude geometry. In this paper, finite ranked posets are studied by making use of a polyhedral construction, with the objective of using the seemingly rather ad hoc construction to study the posets and in particular the effects of the operator $\mathcal{J}$.

A *parity representation* of the ranked poset P is a function $z : P \to \mathbb{Z}^d$ (for some natural number d), where properties (1) - (5) below hold. Given $x = (x_1, x_2, \ldots, x_d) \in \mathbb{Z}^d$, the set of indices i for which x_i is odd will be denoted by

FIGURE 1. C and $\mathcal{J}(C)$

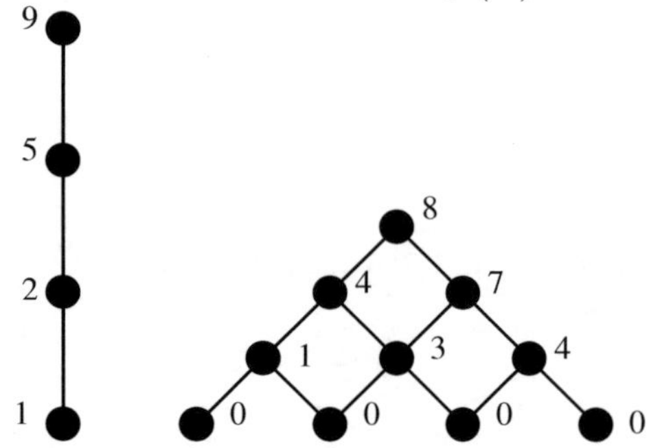

FIGURE 2. P and $\mathcal{T}(P,z)$

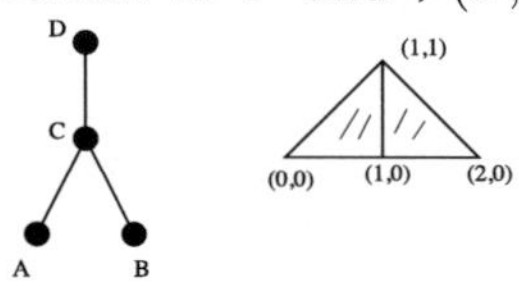

odd(x). By a *unimodular* simplex is meant a simplex in $\mathbb{R}^d$ whose vertices lie in $\mathbb{Z}^d$ and form an affine lattice basis for the intersection of its affine span with $\mathbb{Z}^d$. When the simplex is full-dimensional in $\mathbb{R}^d$ and has all vertices in $\mathbb{Z}^d$, it is unimodular provided that its Euclidean volume is $\frac{1}{d!}$.

(1) The function z is injective.
(2) Each chain $C \subseteq P$ is mapped to the vertex set $z(C)$ of a unimodular simplex. This simplex will be denoted by $s(C)$. Thus $s(C) = \operatorname{conv}(z(C))$.
(3) If C and D are chains of P then $s(C \cap D) = s(C) \cap s(D)$.
(4) If $p, q \in P$ and $p \leq q$ then $\operatorname{odd}(z(p)) \subseteq \operatorname{odd}(z(q))$.
(5) For each $p \in P$, $\rho(p) = |\operatorname{odd}(z(p))|$.

If $z : P \to \mathbb{Z}^d$ is a parity representation of P then the collection $\{s(C) : C$ is a chain of $P\}$ forms a geometric simplicial complex in $\mathbb{R}^d$. This complex will be denoted by $\mathcal{T}(P,z)$. The union of the simplexes of $\mathcal{T}(P,z)$ is the *polyhedron* of the representation and will be denoted by $\|\mathcal{T}(P,z)\|$. As shown in [**9**], every finite poset has a parity representation. In Figure 2, a poset P and the polyhedron of a parity representation are shown. The parity representation z takes the element $A \in P$ to $(0,0)$, B to $(2,0)$, C to $(1,0)$, and D to $(1,1)$. The poset is ranked by the *height function*: The height of an element x is the largest length of a chain having x as its maximum element.

Theorem 4.1 of [**9**] gives the following

THEOREM 1.1 (Theorem 4.1 of [**9**]). *If $z : P \to \mathbb{Z}^d$ is a parity representation of the ranked poset P then the function $\tilde{z} : \mathcal{J}(P) \to \mathbb{Z}^d$ given by $\tilde{z}([a,b]) = z(a) + z(b)$ is a parity representation of the ranked poset $\mathcal{J}(P)$.*

Figure 3 shows $\mathcal{J}(P)$ and the polyhedron of the parity representation obtained as in Theorem 1.1, where P is the poset of Figure 2.

If $z : P \to \mathbb{Z}^d$ is a parity representation with complex $\mathcal{T}(P,z)$ and polyhedron $||\mathcal{T}(P,z)||$ then the image $z(P)$ is $||\mathcal{T}(P,z)|| \cap \mathbb{Z}^d$. Therefore the cardinalities of P

FIGURE 3. $\mathcal{J}(P)$ and $\mathcal{T}(\mathcal{J}(P), \tilde{z})$

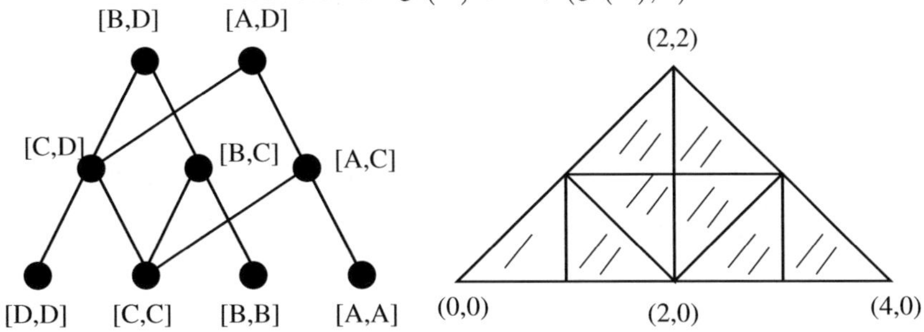

and $||\mathcal{T}(P,z)|| \cap \mathbb{Z}^d$ are equal. As a consequence of the theorem, the cardinalities of $\mathcal{J}^k(P)$ and $2^k||\mathcal{T}(P,z)|| \cap \mathbb{Z}^d$ are the same. The function taking the natural number m to $|m\,||\mathcal{T}(P,z)|| \cap \mathbb{Z}^d|$ is given by the Ehrhart polynomial of $||\mathcal{T}(P,z)|| \cap \mathbb{Z}^d$, evaluated at m. If $E(m)$ denotes this Ehrhart polynomial, then $|\mathcal{J}^k(P)| = E(2^k)$.

Fukuda and Weibel [**5**] studied counting problems related to those studied here, and that paper provided some motivation for this one. Jojiĉ [**7**] studied the effect of an operator related to $\mathcal{J}$ on the *ab*-index and *cd*-index. Lindström [**10**] posed a problem about convex polytopes that concerned an operator related to $\mathcal{J}$ and that engendered much work. Grünbaum [**6**] posed an equivalent problem phrased in terms of antiprisms. Affirmative solutions and generalizations in dimension 3 were given in Andreev [**1**], Schramm [**12**], and Schulte [**13**]. The negative conclusion in 4 and higher dimensions was found by Dobbins [**4**]. Paffenholz and Ziegler [**11**] used methods related to $\mathcal{J}$ in constructing some useful examples of convex polytopes.

For background on convex polytopes, see [**6**] and [**15**]; for posets and combinatorics more generally see [**14**]; for Ehrhart polynomials, see [**3**] and [**14**].

2. Interval posets and Ehrhart polynomials

As noted above, there is a close connection between the problems of enumeration of elements of given rank in $\mathcal{J}^k(P)$, where P is a finite ranked poset, and the topic of Ehrhart polynomials. The connection will be exploited in this section.

Let P be a finite, ranked poset having rank function ρ. We are interested in $F_l(k)$, the cardinality of the set $\{p \in \mathcal{J}^k(P) : \rho(p) = l\}$. We will relate these numbers to the Ehrhart polynomial and certain Ehrhart quasipolynomials associated to $||\mathcal{T}(P,z)||$, where $z : P \to \mathbb{Z}^d$ is a parity representation of P. First we define and review Ehrhart (quasi)polynomials. Lemmas 2.1 and 2.2 will follow from these well-known results about Ehrhart quasipolynomials. The lemmas will be used in the proof of Theorem 2.3.

As usual, a *convex polytope* in $\mathbb{R}^d$ is the convex hull of a finite subset of $\mathbb{R}^d$. The smallest set whose convex hull is the convex polytope is unique and its elements are called the *vertices* of the convex polytope.

In this paper, a *polytope* is a finite union of convex polytopes; it needn't be convex. If the polytope $\mathcal{P}$ admits a geometric triangulation into simplexes all of which have vertex sets contained in $\mathbb{Z}^d$, then $\mathcal{P}$ will be called an *integer polytope*. If $\mathcal{P}$ is a convex polytope having all vertices in $\mathbb{Z}^d$ then it admits such a triangulation, and consequently it is an integer polytope.

If $\mathcal{P}$ is an integer polytope then the function taking the positive integer k to $|\mathbb{Z}^d \cap k\mathcal{P}|$ is a polynomial whose degree is equal to the dimension of $\mathcal{P}$, which is by definition the maximum of the dimensions of the simplexes in any finite collection of simplexes whose union is $\mathcal{P}$. This polynomial is called the *Ehrhart polynomial* of $\mathcal{P}$, the result having first been obtained by Ehrhart. (Notice that this result does not hold for nonconvex polytopes that are unions of convex polytopes with vertices in $\mathbb{Z}^d$; for example, if $\mathcal{P}$ is the union of the segment joining $(0,0)$ and $(1,1)$ with the segment joining $(1,0)$ and $(0,1)$, then $|\mathbb{Z}^2 \cap k\mathcal{P}|$ is not given by a polynomial.) The Ehrhart polynomial of the integer polytope $\mathcal{P}$ is herein denoted by $E_{\mathcal{P}}$: $E_{\mathcal{P}}(k) = |\mathbb{Z}^d \cap k\mathcal{P}|$.

Ehrhart proved additionally the reciprocity result, that for a convex polytope of dimension d having vertices in $\mathbb{Z}^d$ with interior denoted by $\mathcal{P}^o$, $|\mathbb{Z}^d \cap k\mathcal{P}^o|$ is given by $(-1)^d E_{\mathcal{P}}(-k)$. We will write $E_{\mathcal{P}^o}(k) = (-1)^{\dim (\mathcal{P})} E_{\mathcal{P}}(-k)$.

In the case of a unimodular simplex S of dimension d, the Ehrhart polynomial is the same as that of the standard d-simplex. It is an elementary exercise in combinatorics to verify that this is $\binom{k+d}{d}$. It is also an elementary exercise to verify that the number of lattice points in the interior of kS, that is, $|\mathbb{Z}^d \cap kS^o|$, is given by $\binom{k-1}{d}$. This is as the second theorem of Ehrhart mentioned above dictates, since if $p(k)$ is the polynomial of degree d that coincides with $\binom{k+d}{d}$ for natural numbers k, then $(-1)^d p(-k) = \binom{k-1}{d}$.

Consequently, the Ehrhart polynomial of $\|\mathcal{T}\|$, where $\mathcal{T}$ is a geometric simplicial complex all of whose simplexes are unimodular, is easy to describe. It is:

$$E_{\|\mathcal{T}\|}(k) = \sum_{S \in \mathcal{T}, S \neq \emptyset} \binom{k-1}{\dim\ S}.$$

This holds because $\|\mathcal{T}\|$ is the disjoint union of the relative interiors of the nonempty simplexes in $\mathcal{T}$.

Recall (e.g. from [**14**], Section 4.4) that a *quasipolynomial* of *quasiperiod* m is a function on $\mathbb{Z}$ which is given on each congruence class modulo m by a polynomial.

It will be convenient to use the expression $E_{\mathcal{P}}(k)$ for $|\mathbb{Z}^d \cap k\mathcal{P}|$, with $\mathcal{P} \subseteq \mathbb{R}^d$ any polytope having rational vertices. The function is again a quasipolynomial. When the vertices of $\mathcal{P}$ lie in $\frac{1}{m}\mathbb{Z}^d$ with $m \geq 2$, the function $E_{\mathcal{P}}(k)$ may fail to be given by a polynomial, but it is indeed given by a quasipolynomial of quasiperiod m. (See, for example, [**14**], Propostion 4.4.1 and Theorem 4.6.8.)

One last well-known fact will complete this review of Ehrhart theory. When $\mathcal{P}$ is a convex polytope having rational vertices, we have seen that $E_{\mathcal{P}}(k)$ is the restriction to the natural numbers of a quasipolynomial. This quasipolynomial is unique. The Ehrhart reciprocity relation, described earlier for the case of convex polytopes having vertices in $\mathbb{Z}^d$, holds in the more general case: $|\mathbb{Z}^d \cap m\mathcal{P}^o| = (-1)^d E_{\mathcal{P}}(-m)$. (For a proof see for example [**14**], Theorem 4.6.9.)

Let $\mathcal{P} \subseteq \mathbb{R}^d$ be a convex integer polytope. Given an index set $\Lambda \subseteq \{1, 2, \ldots, d\}$, let $G_{\Lambda,\mathcal{P}}(m)$ be defined for $m = 1, 2, 3, \ldots$ as the number of points in $\mathbb{Z}^d \cap mP$ for which each coordinate with index $i \in \Lambda$ is even. The analogous function for the relative interior $\mathcal{P}^o$ of $\mathcal{P}$ will be denoted by $G_{\Lambda,\mathcal{P}^o}$.

LEMMA 2.1. *The functions $G_{\Lambda,\mathcal{P}}$ and $G_{\Lambda,\mathcal{P}^o}$ are given by quasipolynomials of quasiperiod 2. If the dimension of $\mathcal{P}$ is d, then these functions are related by $G_{\Lambda,\mathcal{P}^o}(m) = (-1)^d G_{\Lambda,\mathcal{P}}(-m)$.*

PROOF. Let $\mathcal{Q}$ be the polytope that is the image of $\mathcal{P}$ under the function taking $(x_1, \ldots, x_d) \in \mathbb{R}^d$ to $(y_1, \ldots, y_d)$, where

$$y_i = \begin{cases} \frac{1}{2}x_i & \text{if } i \in \Lambda, \\ x_i & \text{if } i \notin \Lambda. \end{cases}$$

Then $G_{\Lambda,\mathcal{P}}(m) = E_{\mathcal{Q}}(m)$. The latter is a quasipolynomial, $\mathcal{Q}$ being a polytope having rational vertices, and since the vertices of $\mathcal{Q}$ lie in $\frac{1}{2}\mathbb{Z}^d$, it is of quasiperiod 2. Also, $G_{\Lambda,\mathcal{P}^o}(m) = E_{\mathcal{Q}^o}(m)$, which is a quasipolynomial of quasiperiod 2. Since $E_{\mathcal{Q}^o}(m) = (-1)^d E_{\mathcal{Q}}(-m)$, the stated relation holds. □

Let $H_{\Lambda,\mathcal{P}}(m)$ denote the number of points in $\mathbb{Z}^d \cap m\mathcal{P}$ for which the coordinates with indices in Λ are even and the other coordinates are odd. Let $H_{\Lambda,\mathcal{P}^o}(m)$ similarly denote the number of points in $\mathbb{Z}^d \cap m\mathcal{P}^o$ for which the coordinates with indices in Λ are even and the other coordinates are odd.

LEMMA 2.2. *The functions $H_{\Lambda,\mathcal{P}}$ and $H_{\Lambda,\mathcal{P}^o}$ are given by quasipolynomials of quasiperiod 2. If the dimension of $\mathcal{P}$ is d, then these functions are related by $H_{\Lambda,\mathcal{P}^o}(m) = (-1)^d H_{\Lambda,\mathcal{P}}(-m)$.*

PROOF. By the principle of inclusion and exclusion,

$$H_{\Lambda,\mathcal{P}}(m) = \sum_{\Gamma \supseteq \Lambda} (-1)^{|\Gamma|-|\Lambda|} G_{\Gamma,\mathcal{P}}(m).$$

The analogue for $H_{\Lambda,\mathcal{P}^o}$ is also valid. Therefore the statements follow from the previous lemma. □

If P is a finite ranked poset, let $F_P(m)$ denote the number of elements of $\mathcal{J}^m(P)$ $(m = 0, 1, 2, \ldots)$. Let $F_{P,k}(m)$ denote the number of elements of rank k in $\mathcal{J}^m(P)$.

The proof of the following theorem utilizes the special case of Lemma 2.2 in which the polytope $\mathcal{P}$ is a simplex.

THEOREM 2.3. *Suppose P is a finite ranked poset, ranked by ρ. Let F_P and $F_{P,k}$ be as above. Then, for $m = 0, 1, \ldots$, the value $F_P(m)$ is given by $E_{\mathcal{P}}(2^m)$, where $E_{\mathcal{P}}$ is the Ehrhart polynomial of the polyhedron $\mathcal{P}$ of any parity representation of P. For $m = 1, 2, \ldots$, the functions $F_{P,k}$ of m are also given by polynomials in 2^m. These polynomials are of degree at most $c - 1$, where c is the maximum length of a chain in P.*

PROOF. Suppose $z : P \to \mathbb{Z}^d$ is a parity representation of P. Let $\mathcal{P} = \|\mathcal{T}(P, z)\|$ be the polyhedron of the representation. Then $|P| = |\mathbb{Z}^d \cap \mathcal{P}|$. Also it follows immediately from Theorem 1.1 that $2\mathcal{P}$ is the polyhedron of a representation of $\mathcal{J}(P)$. Iterating, $2^m\mathcal{P}$ is the polyhedron of a representation of $\mathcal{J}^m(P)$.

Therefore the number of elements $F_P(m)$ of $\mathcal{J}^m(P)$ is $|\mathbb{Z}^d \cap 2^m\mathcal{P}|$; but this is $E_{\mathcal{P}}(2^m)$.

Since $\mathcal{P}$ is the disjoint union of the nonempty simplexes of $\mathcal{T}(P, z)$,

$$F_{P,k}(m) = \sum_{S \in \mathcal{T}(P,z)} \sum_{\substack{\Lambda \subseteq \{1,\ldots,d\}, \\ |\Lambda| = d-k}} H_{\Lambda,S^o}(2^m).$$

By Lemma 2.2, the functions H_{Λ,S^o} are quasipolynomials of quasiperiod 2. For $m = 1, 2, 3, \ldots$, the number 2^m is even, so $H_{\Lambda,S^o}(2^m)$ is a polynomial function of 2^m. Since each S is of dimension at most $c - 1$, it follows that the degree of the polynomial is at most $c - 1$. □

3. Chains

From the summation in the proof of Theorem 2.3 it is clear that the evaluation of $F_{P,k}(m)$ can be reduced to a sum involving the simplexes of the geometric complex of a parity representation of the poset.

THEOREM 3.1. *Let P be a finite ranked poset. Given a chain $D \subseteq P$, let χ_D denote the Euler characteristic of the link of D in the simplicial complex of chains of P. The equation*

$$F_{P,k}(m) = \sum_{\textit{chains } D \subseteq P} (1 - \chi_D) F_{D,k}(m)$$

holds.

PROOF. As in the proof of Theorem 2.3,

$$F_{P,k}(m) = \sum_{S \in \mathcal{T}(P,z)} \sum_{\substack{\Lambda \subseteq \{1,\dots,d\},\\ |\Lambda| = d-k}} H_{\Lambda, S^o}(2^m).$$

Summing over chains of P rather than simplexes of $\mathcal{T}(P,z)$ gives

$$F_{P,k}(m) = \sum_{\text{chains } C \subseteq P} \sum_{\substack{\Lambda \subseteq \{1,\dots,d\},\\ |\Lambda| = d-k}} H_{\Lambda, s(C)^o}(2^m).$$

For each chain $C \subseteq P$,

$$H_{\Lambda, s(C)}(2^m) = \sum_{D \subseteq C} H_{\Lambda, s(D)^o}(2^m).$$

By Möbius inversion on the boolean lattice of subsets of C,

$$H_{\Lambda, s(C)^o}(2^m) = \sum_{D \subseteq C} (-1)^{|C|-|D|} H_{\Lambda, s(D)}(2^m).$$

For each $\Lambda \subseteq \{1, \dots, d\}$, reversing the order of summation yields

$$\begin{aligned}
&\sum_{\text{chains } C \subseteq P} \sum_{D \subseteq C} (-1)^{|C|-|D|} H_{\Lambda, s(D)}(2^m) \\
&= \sum_{\text{chains } D \subseteq P} \sum_{\substack{\text{chains } C: \\ P \supseteq C \supseteq D}} (-1)^{|C|-|D|} H_{\Lambda, s(D)}(2^m) \\
&= \sum_{\text{chains } D \subseteq P} \Big(\sum_{\substack{\text{chains } C: \\ P \supseteq C \supseteq D}} (-1)^{|C|-|D|} \Big) H_{\Lambda, s(D)}(2^m).
\end{aligned}$$

The Euler characteristic of the link of the chain D is

$$\sum_{\substack{\text{chains } C: \\ P \supseteq C \supset D}} (-1)^{|C|-|D|-1}$$

so

$$\sum_{\substack{\text{chains } C: \\ P \supseteq C \supseteq D}} (-1)^{|C|-|D|} = 1 - \chi_D.$$

Therefore

$$F_{P,k}(m) = \sum_{\substack{\Lambda \subseteq \{1,\dots,d\},\\ |\Lambda| = d-k}} \sum_{\text{chains } D \subseteq P} (1 - \chi_D) H_{\Lambda, s(D)}(2^m)$$

$$= \sum_{\text{chains } D \subseteq P} (1 - \chi_D) F_{D,k}(m).$$

□

References

[1] E. M. Andreev, *Convex polyhedra in Lobačevskiĭ spaces* (Russian), Mat. Sb. (N.S.) **81 (123)** (1970), 445–478. MR0259734

[2] George E. Andrews and Jim Lawrence, *Binary partitions and binary partition polytopes*, Aequationes Math. **91** (2017), no. 5, 859–869, DOI 10.1007/s00010-017-0493-8. MR3697174

[3] Matthias Beck and Sinai Robins, *Computing the continuous discretely*, 2nd ed., Undergraduate Texts in Mathematics, Springer, New York, 2015. Integer-point enumeration in polyhedra; With illustrations by David Austin. MR3410115

[4] Michael Gene Dobbins, *Antiprismlessness, or: reducing combinatorial equivalence to projective equivalence in realizability problems for polytopes*, Discrete Comput. Geom. **57** (2017), no. 4, 966–984, DOI 10.1007/s00454-017-9874-y. MR3639611

[5] Komei Fukuda and Christophe Weibel, *f-vectors of Minkowski additions of convex polytopes*, Discrete Comput. Geom. **37** (2007), no. 4, 503–516, DOI 10.1007/s00454-007-1310-2. MR2321738

[6] Branko Grünbaum, *Convex polytopes*, 2nd ed., Graduate Texts in Mathematics, vol. 221, Springer-Verlag, New York, 2003. Prepared and with a preface by Volker Kaibel, Victor Klee and Günter M. Ziegler. MR1976856

[7] Duško Jojić, *The cd-index of the poset of intervals and E_t-construction*, Rocky Mountain J. Math. **40** (2010), no. 2, 527–541, DOI 10.1216/RMJ-2010-40-2-527. MR2646456

[8] Jim Lawrence, *Dual-antiprisms and partitions of powers of 2 into powers of 2*, Discrete Comput. Geom. **61** (2019), no. 3, 465–478, DOI 10.1007/s00454-019-00070-5. MR3918544

[9] J. Lawrence, *Parity representations of posets*, (In preparation.)

[10] Bernt Lindström, *On the realization of convex polytopes, Euler's formula and Möbius functions*, Aequationes Math. **6** (1971), 235–240, DOI 10.1007/BF01819757. MR295208

[11] Andreas Paffenholz and Günter M. Ziegler, *The E_t-construction for lattices, spheres and polytopes*, Discrete Comput. Geom. **32** (2004), no. 4, 601–621, DOI 10.1007/s00454-004-1140-4. MR2096750

[12] Oded Schramm, *How to cage an egg*, Invent. Math. **107** (1992), no. 3, 543–560, DOI 10.1007/BF01231901. MR1150601

[13] E. Schulte, *Analogues of Steinitz's theorem about noninscribable polytopes*, Intuitive geometry (Siófok, 1985), Colloq. Math. Soc. János Bolyai, vol. 48, North-Holland, Amsterdam, 1987, pp. 503–516. MR910731

[14] Richard P. Stanley, *Enumerative combinatorics. Volume 1*, 2nd ed., Cambridge Studies in Advanced Mathematics, vol. 49, Cambridge University Press, Cambridge, 2012. MR2868112

[15] Günter M. Ziegler, *Convex polytopes: extremal constructions and f-vector shapes*, Geometric combinatorics, IAS/Park City Math. Ser., vol. 13, Amer. Math. Soc., Providence, RI, 2007, pp. 617–691. MR2383133

Department of Mathematical Sciences, George Mason University, 4400 University Drive, Fairfax, Virginia 22030-4444

Email address: lawrence@gmu.edu

Contemporary Mathematics
Volume **764**, 2021
https://doi.org/10.1090/conm/764/15335

String C-group representations of almost simple groups: A survey

Dimitri Leemans

ABSTRACT. This survey paper aims at giving the state of the art in the study of string C-group representations of almost simple groups. It also suggest a series of problems and conjectures to the interested reader.

CONTENTS

1. Introduction

Polytopes, and in particular regular polytopes, have been studied by mathematicians for millenia. The recent monograph of McMullen and Schulte [**45**] came as the first comprehensive up-to-date book on abstract regular polytopes after more than twenty years of rapid development and it describes a rich new theory that benefits from an interplay between several areas of mathematics including geometry, algebra, combinatorics, group theory and topology. In recent years, abstract regular polytopes whose automorphism groups are (almost) simple groups have been studied extensively. This survey paper aims to give the state of the art on the subject, and to give paths for further research in that field.

2010 *Mathematics Subject Classification.* Primary 20D06, 52B11.
Key words and phrases. String C-group representations, abstract regular polytopes, almost simple groups.

The project surveyed in this paper started in 2003 when Michael Hartley contacted us about an abstract regular polytope he had found, of type $\{5,3,5\}$, whose automorphism group is the first Janko group J_1, one of the 26 sporadic simple groups. It turned out we knew the existence of that polytope since 1999 when we published an atlas of regular thin geometries for small groups [**35**]. Unfortunately (or in fact fortunately), that polytope was not mentioned in this atlas as, at the time, we had computed all thin regular residually connected incidence geometries whose automorphism group is J_1 but there were too many of them (almost 3000) to list them in the paper. That polytope was the start of a very fruitful collaboration with Hartley and changed our research life as we then decided to focus on polytopes.

Hartley was in fact willing to determine whether the universal locally projective polytope of type $\{5,3,5\}$ was finite or infinite. That was the missing piece to finalise the classification of universal locally projective polytopes. We eventually found out that it is finite and its automorphism group is isomorphic to $J_1 \times PSL(2,19)$, a very surprising result [**31**]. Indeed, Grünbaum found in 1977 a rank four polytope which he called the 11-cell [**28**]. The automorphism group of the 11-cell is the group $PSL(2,11)$ and Grünbaum obtained it by taking the Coxeter group of type $[3,5,3]$ and adding two relations, to quotient the facets and vertex-figures and make them hemi-icosahedra and hemi-dodecahedra respectively. In the same spirit, Coxeter found the 57-cell [**19**] by adding two relations to the Coxeter group of type $[5,3,5]$ to make the facets and vertex-figures hemi-dodecahedra and hemi-icosahedra respectively. It turns out that in the case of the 11-cell, adding only one of the two relations implies the other relation, but in the case of the 57-cell, it is not the case: adding only one relation gives the group $J_1 \times PSL(2,19)$ as we found out with Hartley.

We then met in Brussels for a couple of weeks of intense work and decided at the time it would be great to build atlases of polytopes. Hartley focused on polytopes with a fixed number of flags [**29**] while we focused on polytopes whose automorphism groups are almost simple groups with Laurence Vauthier [**42**]. We started collecting a sensible amount of computer-generated data on the subject and used them to state conjectures.

The aim of this paper is to give a survey of what has been done on that subject over the last fifteen years, to also give some results that were obtained on C-groups, to state some conjectures and give open problems.

The three big questions we have tried to answer for families of groups over the years are the following ones. Given a group G (eventually belonging to a family of groups as, for instance finite simple groups, or alternating groups),

(1) determine the possible ranks of string C-group representations for G;
(2) in particular, determine the highest rank of a string C-group representation of G;
(3) enumerate all string C-group representations for G;

These questions make sense in trying to find nice geometric structures on which these groups act as automorphism groups, but they also tell us something about quotients of Coxeter groups – a string C-group representation being a smooth quotient of a Coxeter group – as well as independent generating sets of the groups G.

One can consider these same three questions for chiral polytopes. In particular, obtaining enumeration data for both types of polytopes would shed light on their relative abundance. However, we constrain this survey to questions about regular polytopes. Observe that the enumeration question is usually very hard to answer for families of groups and that, apart for shedding light on the relative abundance of regular polytopes versus chiral polytopes, its interest might be considered lower than the interest of the first two questions.

Even though almost simple groups appear scarcely among the set of all groups, trying to answer questions (1), (2) and (3) for the finite simple groups (and their automorphism groups) seems natural in order to get a better geometric understanding of some of them. Moreover, for finite simple groups, involutions have been thoroughly studied in the process of their classification. The families of (almost) simple groups not mentioned in this survey are of course also very interesting to look at but, currently, no results are known for them. So any discovery for them is most welcome.

2. Preliminaries

Let us start by defining the concepts and fixing the notation needed to understand this survey.

2.1. C-groups and string C-groups. As it is well known that abstract regular polytopes and string C-groups are in one-to-one correspondence (see for instance [**45**, Section 2E]), and since it is much easier to define string C-groups, we frame our discussion in the language of string C-groups.

Let G be a group and X a set of involutions of G. If $\langle X\rangle = G$, we say that $(G;X)$ is a *group generated by involutions* or *ggi* for short. If there is an ordering of X, say $\{\rho_0,\dots,\rho_{n-1}\}$ where $\rho_i\rho_j = \rho_j\rho_i$ for every $i,j\in\{0,\dots,n-1\}$ such that $|i-j|>1$, then $(G;\{\rho_0,\dots,\rho_{n-1}\})$ is a *string group generated by involutions* or *sggi* for short.

When $(G;\{\rho_0,\dots,\rho_{n-1}\})$ is an sggi, we assume, without loss of generality, that the involutions are ordered in such a way that $\forall i,j\in\{0,\dots,r-1\}$, if $|i-j|>1$ then $(\rho_i\rho_j)^2 = 1$ (this property is called the *commuting property*).

Let G be a group and $\{\rho_0,\dots,\rho_{n-1}\}$ be a set of involutions of G. The pair $(G;\{\rho_0,\dots,\rho_{n-1}\})$ satisfies the *intersection property* if

$$\forall I,J\subseteq\{0,\dots,n-1\},\ \langle\rho_i \mid i\in I\rangle\cap\langle\rho_j \mid j\in J\rangle = \langle\rho_k \mid k\in I\cap J\rangle.$$

If a ggi $(G;\{\rho_0,\dots,\rho_{n-1}\})$ satisfies the intersection property, it is called a *C-group representation of* G (or C-group for short). A C-group that satisfies the commuting property is called a *string C-group.*

The integer n is the *rank* of the (string) C-group $(G;\{\rho_0,\dots,\rho_{n-1}\})$.

To a string C-group $(G;\{\rho_0,\dots,\rho_{n-1}\})$, we can associate a *Schläfli type*, that is a sequence $\{p_1,\dots,p_{n-1}\}$ where $p_i = o(\rho_{i-1}\rho_i)$. Observe that if one of the p_i's is equal to 2, then the group G is a direct product of two non-trivial subgroups. If that is the case, we say that $(G;\{\rho_0,\dots,\rho_{n-1}\})$ is *reducible*; Otherwise $(G;\{\rho_0,\dots,\rho_{n-1}\})$ is called *irreducible.* Obviously, if G is simple and $(G;\{\rho_0,\dots,\rho_{n-1}\})$ is a string C-group, then $(G;\{\rho_0,\dots,\rho_{n-1}\})$ is irreducible. We take as convention that a symbol p_i appearing k times in adjacent places in a Schläfli type can be replaced by p_i^k instead of writing k times p_i. So for instance the Schläfli type of the 4-simplex can be written as $\{3,3,3\}$ or as $\{3^3\}$.

Given a group G and two sets S and T of involutions of G such that (G, S) and (G, T) are C-groups, we say that (G, S) and (G, T) are *isomorphic* if there exists an element $g \in Aut(G)$ such that g maps S onto T.

The *(string) C-rank* of a group is the largest possible rank of a (string) C-group representation for that group.

2.2. Permutation representation graphs and CPR graphs. Let G be a group of permutations acting on a set $\{1, \ldots, n\}$. Let $S := \{\rho_0, \ldots, \rho_{r-1}\}$ be a set of r involutions of G such that $G = \langle S \rangle$. We define the *permutation representation graph* $\mathcal{G}$ of G as the r-edge-labeled multigraph with vertex set $V(\mathcal{G}) := \{1, \ldots, n\}$. The edge-set of $\mathcal{G}$ is the set $\{\{a, a\rho_i\} : a \in V(\mathcal{G}), i \in \{0, \ldots, r-1\} | a \neq a\rho_i\}$ where every edge $\{a, a\rho_i\}$ has label i. When (G, S) is a string C-group representation, the permutation representation graph $\mathcal{G}$ is called a *CPR graph*, as defined in [**55**].

3. Simple groups and rank three string C-group representations

In 1980, it was asked in the Kourovka Notebook (Problem 7.30) which finite simple groups can be generated by three involutions, two of which commute. This problem was solved by Nuzhin and Mazurov in [**44, 51–54**]. The groups $PSU_4(3)$ and $PSU_5(2)$, although mentioned by Nuzhin as being generated by three involutions, two of which commute, have recently been discovered not to have such generating sets by Martin Macaj and Gareth Jones (personal communication of Jones, checked independently using MAGMA). We summarize below the accurate solution to Problem 7.30.

THEOREM 3.1 (Nuzhin - Mazurov - Macaj - Jones). *Every non-abelian finite simple group can be generated by three involutions, two of which commute, with the following exceptions:*

$$PSL_3(q),\ PSU_3(q),\ PSL_4(2^n),\ PSU_4(2^n), A_6,$$
$$A_7, PSU_4(3),\ PSU_5(2),\ M_{11},\ M_{22},\ M_{23},\ McL.$$

If G is a simple group generated by three involutions, two of which commute, G together with these three involutions form a string C-group representation by the following result due to Marston Conder and Deborah Oliveros.

THEOREM 3.2. [**12**, Corollary 4.2] *If G is a finite non-abelian simple group, or more generally any finite group with no non-trivial cyclic normal subgroup, then every smooth homomorphism from the $[k, m]$ Coxeter group onto G gives rise to a string C-group representation of rank three for G.*

It turns out that Theorem 3.1 holds when removing the hypothesis on the rank as proven by Adrien Vandenschrick in [**61**]: the exceptions in rank three do not have string C-group representations of higher rank.

4. Symmetric and alternating groups

4.1. String C-group representations of S_n. Symmetric groups gained our attention early in this research project. They were among the only ones in the data we collected that gave string C-group representations of large rank. It was known for a long time [**49**] that the group S_n has a string C-group representation of rank $n - 1$, namely $(S_n; \{(1, 2), (2, 3), \ldots, (n-1, n)\})$. A recent result of Julius Whiston [**62**], showing that the largest size of a set of independent generators of S_n is $n - 1$, implies that the string C-rank of S_n is $n - 1$.

Sjerve and Cherkassoff showed in [**56**] that S_n is a group generated by three involutions, two of which commute, provided that $n \geq 4$. Their examples satisfy the intersection property and therefore are rank three string C-group representations by Theorem 3.2.

THEOREM 4.1. [**56**, Theorem 1.2] *Every group S_n with $n \geq 4$ has at least one string C-group representation of rank three.*

Earlier work by Conder [**10, 11**] covers all but a few cases of the results of [**56**] for the symmetric groups. As Conder pointed out to us, these days, it takes a few seconds to handle the missing cases for S_n with MAGMA [**1**].

Together with Maria Elisa Fernandes, we determined the number of non-isomorphic string C-group representations of S_n of rank $n-1$ and $n-2$ [**20, 25**]. Then with Fernandes and Mark Mixer we also determined the representations of rank $n-3$ and $n-4$ [**26**].

THEOREM 4.2. [**26**, Theorem 1.1] *Let $1 \leq i \leq 4$, and $n \geq 3+2i$ when $r = n-i$. If Γ is a string C-group representation of a group G, of rank $r \geq n-i$ with a connected diagram and G is isomorphic to a transitive group of degree n then G or its dual is isomorphic to S_n and the CPR graph is one of those listed in Table 1.*

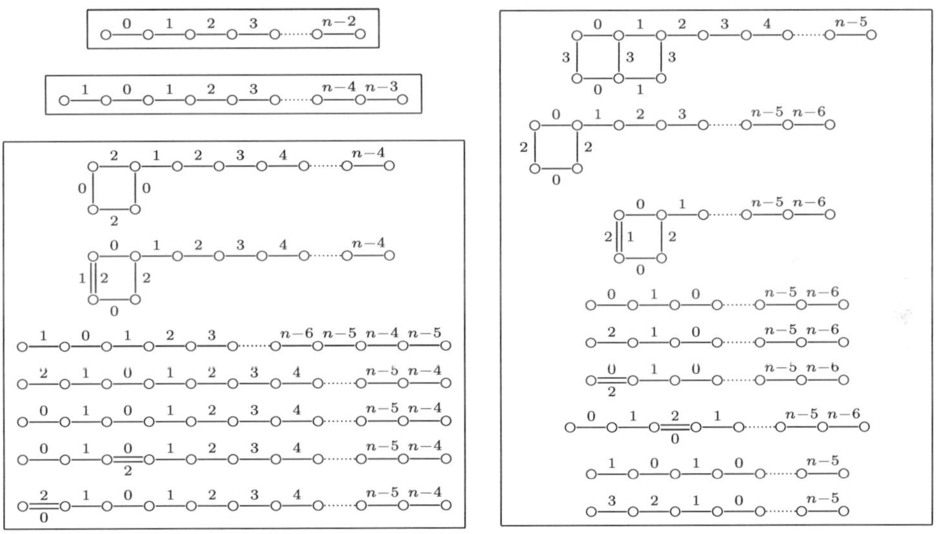

FIGURE 1. CPR graphs of string C-group representations of rank $r \geq n-4$ for S_n.

Also with Fernandes, we showed that there were no gaps in the set of possible ranks of string C-group representations of S_n.

THEOREM 4.3. [**20**, Theorem 3] *Let $n \geq 4$. For every $r \in \{3, \ldots, n-1\}$, there exists at least one string C-group representation of rank r of S_n for every $3 \leq r \leq n-1$. Its Schläfli type is $\{n-r+2, 6, 3^{r-3}\}$.*

Table 1 gives, for S_n ($5 \leq n \leq 14$), the number of pairwise nonisomorphic string C-group representations of rank r ($3 \leq r \leq n-1$). It suggests the following problem.

G	Rk 3	Rk 4	Rk 5	Rk 6	Rk 7	Rk 8	Rk 9	Rk 10	Rk 11	Rk 12	Rk 13
S_5	4	1	0	0	0	0	0	0	0	0	0
S_6	2	4	1	0	0	0	0	0	0	0	0
S_7	35	7	1	1	0	0	0	0	0	0	0
S_8	68	36	11	1	1	0	0	0	0	0	0
S_9	129	37	7	7	1	1	0	0	0	0	0
S_{10}	413	203	52	13	7	1	1	0	0	0	0
S_{11}	1221	189	43	25	9	7	1	1	0	0	0
S_{12}	3346	940	183	75	40	9	7	1	1	0	0
S_{13}	7163	863	171	123	41	35	9	7	1	1	0
S_{14}	23126	3945	978	303	163	54	35	9	7	1	1

TABLE 1. The number of pariwise nonisomorphic string C-group representations of S_n $(5 \leq n \leq 14)$.

PROBLEM 1. Enumerate the string C-group representations of rank r of S_n for every $3 \leq r \leq n-1$.

This problem is solved for the four highest values of r and n large enough as pointed out in Theorem 4.2. It suggests the following conjecture.

CONJECTURE 1. *The number of string C-group representations of rank $n-i$ for S_n with $1 \leq i \leq (n-3)/2$ is a constant independent of n.*

This conjecture suggests the existence of a new integer sequence (not appearing in the Encyclopedia of Integer Sequences) starting with 1, 1, 7, 9 and whose next number is likely 35 as the experimental data of Table 1 suggest.

A first step in trying to solve Problem 1 has been made by Kiefer and the author. We computed the number of unordered pairs of commuting involutions in S_n up to conjugacy in $Aut(S_n)$ in [**33**]. We obtained the following result.

THEOREM 4.4. [**33**, Theorem 1.1] *Let $n > 1$ be a positive integer. Define $\lambda(k)$ and $\psi(k,n)$ as follows.*

$$\lambda(k) = \left\lfloor \left(\frac{k}{2}+1\right)^2 \right\rfloor$$

$$\psi(k,n) = \begin{cases} \left[\frac{1}{2}(2k-n)\right]^2 + \frac{1}{2}(2k-n) & \text{if } n \text{ is even,} \\ \\ \left[\frac{1}{2}(2k-n-1)\right]^2 + 2k-n & \text{if } n \text{ is odd.} \end{cases}$$

There are, up to isomorphism,

$$-\frac{3n}{2} + \sum_{k=1}^{n} \lambda(k) \cdot \left(\frac{n-k}{2}+1\right) - \frac{1}{2} \cdot \sum_{k=\lfloor \frac{n}{2} \rfloor + 1}^{n} \psi(k,n)$$

unordered pairs of commuting involutions in S_{2n} and S_{2n+1} except for S_6 in which there are, up to isomorphism, five unordered pairs of commuting involutions.

A similar result was obtained for the alternating groups (see Theorem 4.10). Table 2 gives the number of pairwise nonisomorphic pairs of commuting involutions for S_n (and A_n) for some values of n. As the reader can see, the numbers appearing in this table are not very encouraging to solve Problem 1 for the cases not covered by Conjecture 1.

n	$\{\rho_0, \rho_2\}$, with $\rho_0, \rho_2 \in \mathrm{S}(n)$	$\{\rho_0, \rho_2\}$, with $\rho_0, \rho_2 \in \mathrm{A}(n)$
1,2,3	0	0
4,5	3	1
6	5	1
7	9	2
8,9	21	7
10,11	39	10
12,13	67	21
14,15	105	28
16,17	158	48
18,19	226	61
20	315	93
30	1169	315
40	3105	855
50	6774	1795

TABLE 2. Number of unordered pairs of commuting involutions in S_n and A_n, up to conjugacy in $Aut(S_n)$.

4.2. String C-group representations of A_n. Alternating groups were investigated in the same vein as symmetric groups.

Sjerve and Cherkassoff showed in [**56**] that A_n is a group generated by three involutions, two of which commute, provided that $n \geq 4$ and $n \neq 6, 7$ or 8. Their examples satisfy the intersection property and therefore are rank three string C-group representations.

THEOREM 4.5. [**56**, Theorem 1.1] *Every group A_n with $n = 5$ or $n \geq 9$ has at least one string C-group representation of rank three.*

Again, earlier work by Conder [**10, 11**] covers all but a few cases of the results of [**56**] for the alternating groups.

When we started working on string C-group representations for the alternating groups with Fernandes and Mixer, we first collected experimental data for A_n with $n \leq 12$. Table 3 gives the number of pairwise nonisomorphic string C-group representations for A_n ($5 \leq n \leq 15$). A striking observation came for A_{11}. It was the first time we found a group that had string C-group representations whose set of ranks is not an interval (in this case, the set of ranks is $\{3, 6\}$).

We proved that for each rank $r \geq 4$, there is at least one group A_n that has a string C-group representation of rank r.

THEOREM 4.6. [**23**, Theorem 1.1] *For each rank $k \geq 3$, there is a string C-group representation of rank k with group A_n for some n. In particular, for each even rank $r \geq 4$, there is a string C-group representation of Schläfli type $\{10, 3^{r-2}\}$ with group A_{2r+1}, and for each odd rank $q \geq 5$, there is a string C-group representation of Schläfli type $\{10, 3^{q-4}, 6, 4\}$ with group A_{2q+3}.*

We then managed to construct string C-group representations of rank $\lfloor \frac{n-1}{2} \rfloor$ for A_n with $n \geq 12$.

THEOREM 4.7. [**24**, Theorem 1.1] *For each $n \notin \{3, 4, 5, 6, 7, 8, 11\}$, there is a rank $\lfloor \frac{n-1}{2} \rfloor$ string C-group representation of the alternating group A_n.*

G	Rank 3	Rank 4	Rank 5	Rank 6	Rank 7	Rank 8
A_5	2	0	0	0	0	0
A_6	0	0	0	0	0	0
A_7	0	0	0	0	0	0
A_8	0	0	0	0	0	0
A_9	41	6	0	0	0	0
A_{10}	94	2	4	0	0	0
A_{11}	64	0	0	3	0	0
A_{12}	194	90	22	0	0	0
A_{13}	1558	102	25	10	0	0
A_{14}	4347	128	45	9	0	0
A_{15}	5820	158	20	42	6	0

TABLE 3. The number of pariwise nonisomorphic string C-group representations of A_n $(5 \leq n \leq 15)$.

Group	Schläfli Type	CPR Graph
A_{2r+1} (r even and ≥ 4)	$\{10, 3^{r-2}\}$	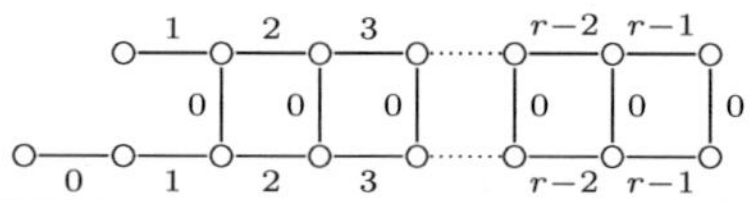
A_{2r+3} (r odd and ≥ 5)	$\{10, 3^{r-4}, 6, 4\}$	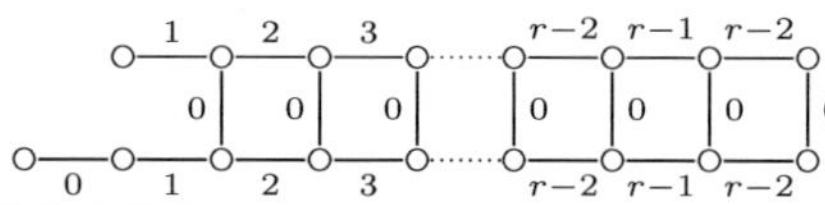

TABLE 4. String C-group representations for A_n

We were quickly convinced that this rank was the best possible but it took another five years, and Peter Cameron joining forces, to finally prove the following theorem.

THEOREM 4.8. [**9**, Theorem 1.1] *The maximum rank of a string C-group representation of A_n is 3 if $n = 5$; 4 if $n = 9$; 5 if $n = 10$; 6 if $n = 11$ and $\lfloor \frac{n-1}{2} \rfloor$ if $n \geq 12$. Moreover, if $n = 3, 4, 6, 7$ or 8, the group A_n is not a string C-group.*

Fernandes and the author then managed to construct string C-group representations of each rank $3 \leq r \leq \lfloor (n-1)/2 \rfloor$ for A_n with $n \geq 12$.

THEOREM 4.9. [**22**, Theorem 1.1] *For $n \geq 12$ and for every $3 \leq r \leq \lfloor (n-1)/2 \rfloor$, the group A_n has at least one string C-group representation of rank r.*

Table 3 together with the proof of Theorem 4.8 suggests that the enumeration results we obtained for the symmetric groups will be much harder to get for the alternating groups. Nevertheless, this is a very interesting problem as well so we list it here.

PROBLEM 2. Enumerate the string C-group representations of rank r of A_n for every $3 \leq r \leq n-1$.

As for the symmetric groups, a first step in trying to solve Problem 2 has been made by Kiefer and the author. We computed the number of unordered pairs of commuting involutions in A_n, up to conjugacy in $Aut(S_n)$ in [**33**]. We obtained the following result.

THEOREM 4.10. [**33**, Theorem 1.2] *Let $n > 1$ be a positive integer. Define $\lambda(k)$, $\phi(k,n)$ and $\mu(n)$ as follows.*

$$\lambda(k) = \left\lfloor \left(\frac{k}{2}+1\right)^2 \right\rfloor$$

$$\phi(k,n) = \begin{cases} \lambda(k) - 1 & \text{if } k \le \left\lfloor \frac{n}{2} \right\rfloor, \\ \lambda(k) - \psi(k,n) - 1 & \text{if } k > \left\lfloor \frac{n}{2} \right\rfloor, \end{cases}$$

$$\mu(n) = -2\left\lfloor \frac{n}{2} \right\rfloor + \sum_{\substack{k=1 \\ k \text{ even}}}^{n} \left[\gamma(k) \cdot \left\lceil \frac{1}{2} \cdot \left(n-k+1\right)\right\rceil + \delta(k) \cdot \left\lfloor \frac{1}{2} \cdot \left(n-k+1\right)\right\rfloor\right]$$

where

$$\gamma(k) = \frac{k^2}{8} + \frac{3k}{4} + 1,$$

$$\delta(k) = \frac{k^2}{8} + \frac{k}{4}.$$

There are, up to isomorphism,

$$\frac{1}{2}\left(\mu(n) + \sum_{\substack{k=1 \\ k \text{ even}}}^{n} \phi(k,n)\right)$$

unordered pairs of commuting involutions in A_{2n} and A_{2n+1} except for A_6 in which there is, up to isomorphism, a unique unordered pair of commuting involutions.

As mentioned in the previous section, Table 2 gives the number of pairwise nonisomorphic pairs of commuting involutions for A_n for some values of n.

5. Projective linear groups

Several projective linear groups were analyzed in [**42**]. The observation of the data collected permitted, over the years, to obtain classification results that we summarize in this section.

5.1. Groups $PSL(2,q)$. The values of q for which a $PSL(2,q)$ group has rank three polytopes were determined by Sjerve and Cherkassoff.

THEOREM 5.1. [**56**, Theorem 1.3] *The $PSL(2,q)$ group may be generated by three involutions, two of which commute, if and only if $q \neq 2,3,7$ or 9.*

For the groups $PSL(2,q)$, we quickly observed that the maximal rank of a string C-group representation for these groups is 4 as they do not possess subgroups that are direct products of two dihedral groups of order at least 6 each, obtaining the following theorem.

THEOREM 5.2. [**42**, Theorem 2] *Let $G \cong PSL(2,q)$. The rank of a string C-group representation of G is at most 4.*

More striking was the fact that, up to $q = 32$, only two groups $PSL(2,q)$ had a rank four representation, namely $PSL(2,11)$ and $PSL(2,19)$. Together with Schulte, we managed to prove the following theorem.

THEOREM 5.3. [**39**, Theorem 1] *If $PSL(2,q)$ has a string C-group representation of rank 4, then $q = 11$ or 19. Moreover, for each of these two values of q, there is, up to isomorphism, a unique string C-group representation of rank four for $PSL(2,q)$.*

This result made the 11-cell and the 57-cell even more special. They were the only two string C-group representations of rank four known for groups $PSL(2,q)$ and our result explained why.

The proof of Theorem 5.3 and subsequent theorems on groups with socle $PSL(2,q)$ (see next subsections) rely heavily on the complete knowledge of the subgroup structure of $PSL(2,q)$. As mentioned in [**6**, Section 4.1], the subgroup structure of $PSL(2,q)$ was first obtained in papers by Moore [**50**] and Wiman [**63**].

Observe also that the more recent results on $PSL(4,q)$ groups make it less surprising that the maximal rank of a string C-group representation of $PSL(2,q)$ is three in most cases. Indeed, there is not much freedom left by the commuting property to find larger sets of involutions that would give string C-group representations for these groups.

5.2. Groups $PGL(2,q)$. Again, in the rank three case, Sjerve and Cherkassoff determined for which values of q the group $PGL(2,q)$ has a string C-group representation.

THEOREM 5.4. [**56**, Theorem 1.4] *The $PGL(2,q)$ group may be generated by three involutions, two of which commute, if and only if $q \neq 2$.*

Looking at the data collected in [**42**], we conjectured with Schulte that the maximum rank of a string C-group representation of $PGL(2,q)$ should be three except when $q = 5$ and we proved the following result.

THEOREM 5.5. [**40**, Theorem 4.8] *The group $PGL(2,q)$ has string C-group representations of rank at most 4. Moreover, only $PGL(2,5)$ has a string C-group representation of rank 4.*

The only exception in rank four is the 4-simplex whose automorphism group is $S_5 \cong PGL(2,5)$.

5.3. Almost simple groups with socle $PSL(2,q)$. Later on, with Connor and De Saedeleer, we decided to study almost simple groups with socle $PSL(2,q)$. We proved the following classification theorem, in the same vein as the results obtained with Schulte, answering at the same time a conjecture we stated in [**40**].

THEOREM 5.6. [**14**, Theorem 1.1] *Let* $\mathrm{PSL}(2,q) \leq G \leq \mathrm{P\Gamma L}(2,q)$. *Suppose G has a string C-group representation. Then*

(1) *if $q = 2$ then $G \cong \mathrm{PSL}(2,2) \cong S_3$ and G has a unique rank 2 string C-group representation, namely the one coming from the triangle;*
(2) *if $q = 3$ then $G \cong \mathrm{PGL}(2,3) \cong \mathrm{S}_4$ and G only has rank three string C-group representations;*
(3) *if $q = 4$ or 5 then either $G \cong \mathrm{PSL}(2,4) \cong \mathrm{PSL}(2,5) \cong \mathrm{A}_5$ and G has rank three string C-group representations only, or $G \cong \mathrm{PGL}(2,5) \cong \mathrm{S}_5$ and G has rank three and four string C-group representations;*

(4) *if* $q = 7$ *then* $G \cong \mathrm{PGL}(2,7)$ *and* G *has rank three string C-group representations only;*
(5) *if* $q \geq 8$ *then*
   (a) *if* $q = 2^{2k+1}$, $k \geq 1$, *then* $G \cong \mathrm{PSL}(2, 2^{2k+1})$ *and* G *has rank three string C-group representations only;*
   (b) *if* $q = 9$ *then either* $G \cong \mathrm{PGL}(2,9)$, *or* $G \cong \mathrm{P\Sigma L}(2,9) \cong \mathrm{S}_6$, *or* $G \cong \mathrm{P\Gamma L}(2,9)$, *and* G *has rank three string C-group representations; moreover* $\mathrm{P\Sigma L}(2,9)$ *has string C-group representations of ranks* 3, 4 *and* 5;
   (c) *if* $q = p^{2k+1} \geq 11$, p *an odd prime and* $k \geq 0$, *then* $G \cong \mathrm{PSL}(2, p^{2k+1})$ *or* $G \cong \mathrm{PGL}(2, p^{2k+1})$; *in either case,* G *has rank three string C-group representations; if moreover* $q = 11$ *or* 19 *then* $G \cong \mathrm{PSL}(2,q)$ *has rank four string C-group representations;*
   (d) *if* $q = p^{2k} \geq 16$, p *any prime and* $k \geq 1$ *then either* $G \cong \mathrm{PSL}(2,p^{2k})$ *or* $G \cong \mathrm{PGL}(2,p^{2k})$ *or* $G \cong \mathrm{PSL}(2,p^{2k}) \rtimes \langle\beta\rangle$ *or* $G \cong \mathrm{PGL}(2,p^{2k}) \rtimes \langle\beta\rangle$, *where* β *is a Baer involution of* $\mathrm{P\Gamma L}(2,p^{2k})$; *in all four cases,* G *has rank three string C-group representations; moreover,* $\mathrm{PSL}(2,p^{2k}) \rtimes \langle\beta\rangle$ *also has string C-group representations of rank* 4.

5.4. Open problems on groups with socle $PSL(2,q)$. Enumeration of C-group representations of rank three for groups $PSL(2,q)$ and $PGL(2,q)$ can be found in [**13**]. Even though their results are for hypermaps, all the hypermaps they give satisfy the intersection property by [**12**, Corollary 4.2] as none of $PSL(2,q)$ or $PGL(2,q)$ possesses a non-trivial cyclic subgroup that is normal. String C-group representations of rank four for $PSL(2,q)$ and $PGL(2,q)$ are known thanks to Theorems 5.3 and 5.5.

PROBLEM 3. Enumerate the non-isomorphic string C-group representations of rank three and four of $\mathrm{PSL}(2,p^{2k}) \rtimes \langle\beta\rangle$ where β is a Baer involution of $\mathrm{P\Gamma L}(2,p^{2k})$.

PROBLEM 4. Enumerate the non-isomorphic string C-group representations of rank three of $\mathrm{PGL}(2,p^{2k}) \rtimes \langle\beta\rangle$ where β is a Baer involution of $\mathrm{P\Gamma L}(2,p^{2k})$.

5.5. Groups $PSL(3,q)$ and $PGL(3,q)$. In [5], Brooksbank and Vicinsky proved the following theorem.

THEOREM 5.7. [5] *If* $G \leq GL(3,q)$ *has a string C-group representation, then* q *is odd and there is a non-degenerate symmetric bilinear form* $\mathbf{f}$ *on* V *such that* $\Omega(V,\mathbf{f}) \leq G \leq I(V,\mathbf{f})$.

As a corollary of their result (see [**61**]), the groups $PSL(3,q)$ and $PGL(3,q)$ are not string C-groups.

The proof of the Brooksbank-Vicinsky theorem is very different from the ones for the groups with socle $PSL(2,q)$. It uses the fact that, for any subgroup G of $GL(3,q)$, any string C-group representation of G is of rank at most four. Then, the authors prove that if G has a string C-group representation of rank three or four, q is odd and G preserves a non-degenerate symmetric bilinear form.

5.6. Open problems on groups with socle $PSL(3,q)$. From the data collected in [**42**], it is clear that some of these groups have string C-group representations of rank three, four and five. Again, it would be nice to know what is the maximal rank for these groups and to enumerate them.

PROBLEM 5. Determine the maximal rank of a string C-group representation for G with $PSL(3,q) \leq G \leq Aut(PSL(3,q))$.

PROBLEM 6. Enumerate the non-isomorphic string C-group representations of rank three of G with $PSL(3,q) \leq G \leq Aut(PSL(3,q))$.

5.7. Groups with socle $PSL(4,q)$. In [**3**], with Brooksbank, we showed that the groups $PSL(4,q)$ have string C-group representations if and only if q is odd. Moreover, we showed that for each odd q, there is at least one string C-group representation of rank four.

THEOREM 5.8. [**3**, Theorem 1.1 and Corollary 4.5] *If $q = p^k$ for odd p, then $PSL(4,q)$ has at least one string C-group representation of rank 4. If q is even, then $PSL(4,q)$ has no string C-group representation.*

CONJECTURE 2. *The group $PSL(4,q)$, with q odd, has maximal rank 4 as a string C-group.*

PROBLEM 7. Determine the maximal rank of a string C-group representation for G with $PSL(4,q) \leq G \leq Aut(PSL(4,q))$.

6. Suzuki groups

Among the nonabelian simple groups, those with the easiest subgroup structure are the Suzuki simple groups. They are therefore the most promising groups for which the three main questions mentioned in the introduction seem solvable. This is the reason why we decided to start working on these questions with these groups. As a rule of thumb, it seems to us that if a general question on finite simple groups is not answerable for Suzuki simple groups, it is unlikely it will be for the other families.

We recall the definition of the Suzuki groups as given in [**43**]. Let $\mathcal{K}$ be a field of characteristic 2 with $\mid \mathcal{K} \mid> 2$. Let σ be an automorphism of $\mathcal{K}$ such that $x^{\sigma^2} = x^2$ for each x in $\mathcal{K}$. Let $\mathcal{B}$ be the 3-dimensional projective space over $\mathcal{K}$ and let (x_0, x_1, x_2, x_3) be the coordinates of a point of $\mathcal{B}$. Let E be the plane defined by the equation $x_0 = 0$ and put $U = (0, 1, 0, 0)\mathcal{K}$. We introduce coordinates in the affine space $\mathcal{B}_E$ by $x = \frac{x_2}{x_0}$, $y = \frac{x_3}{x_0}$, and $z = \frac{x_1}{x_0}$. Finally, let $\mathcal{D}$ be the set of points of $\mathcal{B}$ consisting of U and all those points of $\mathcal{B}_E$ whose coordinates (x, y, z) satisfy the equation

$$z = xy + x^{\sigma+2} + y^{\sigma}$$

We denote by *Sz*($\mathcal{K}$, σ) the group of all projective collineations of $\mathcal{B}$ which leave $\mathcal{D}$ invariant. When $\mathcal{K}$ is a finite field, σ is unique and in this case, $\mathcal{K}$ is isomorphic to *GF*(q) with $q = 2^{2e+1}$. The groups *Sz*(q) are the *Suzuki groups* named after Michio Suzuki who found them in 1960. The generalizations *Sz*($\mathcal{K}$, σ) are due to Rimhak Ree and Jacques Tits (see for example [**58**]). The set $\mathcal{D}$ is an *ovoid*, i.e. a non-empty point-set of a projective 3-space that satisfies the following three conditions.

(1) No three points are collinear.
(2) If $p \in \mathcal{D}$, there exists a plane E of $\mathcal{B}$ with $\mathcal{D} \cap E = \{p\}$.
(3) If $p \in \mathcal{D}$ and if E is a plane of $\mathcal{B}$ with $\mathcal{D} \cap E = \{p\}$, then all lines l through p which are not contained in E carry a point of $\mathcal{D}$ distinct from p.

Every involution of $Sz(q)$ fixes a unique point of $\mathcal{D}$. Moreover, commuting involutions have the same fixed point. Looking at the results obtained in [**42**], and thanks to the knowledge gathered on Suzuki simple groups during our PhD thesis, we quickly managed to prove the following theorem.

THEOREM 6.1. [**36**, Theorem 1] *Let $Sz(q) \leq G \leq Aut(Sz(q))$ with $q = 2^{2e+1}$ and $e > 0$ a positive integer. Then G is a C-group if and only if $G = Sz(q)$. Moreover, if $(G, \{\rho_0, \ldots, \rho_{n-1}\})$ is a string C-group, then $n = 3$.*

We give here a different proof from the original one, using centralisers of involutions.

PROOF. The group $G := Sz(q)$ has a unique conjugacy class of involutions. Suppose G is the natural permutation representation of $Sz(q)$ acting on a Suzuki-Tits ovoid. Two involutions commute in G if and only if they fix the same point on the ovoid. As G is simple, it implies that the maximal rank of a string C-group representation of G is three. Moreover, since $Aut(Sz(q)) \cong Sz(q) : C_{2e+1}$ and $2e + 1$ is odd, all involutions of $Aut(Sz(q))$ are in $Sz(q)$ and therefore, a set of involutions of $Aut(Sz(q))$ can at most generate $Sz(q)$. Finally, take two involutions ρ_0 and ρ_2 that commute in G. Pick an involution ρ_1 such that the order of $\rho_0\rho_1$ is $q - 1$. Then ρ_1 obviously does not commute with ρ_2 and $G = \langle \rho_0, \rho_1, \rho_2 \rangle$ as $\langle \rho_0, \rho_1 \rangle$ is a maximal subgroup of G. The pair $(G, \{\rho_0, \rho_1, \rho_2\})$ is a string C-group representation of G. $\square$

In 2010, with Ann Kiefer, we managed to count the number of pairwise non-isomorphic string C-group representations of rank three of a given Suzuki group $Sz(q)$.

THEOREM 6.2. [**32**, Theorem 2] *Up to isomorphism and duality, a given Suzuki group* $\mathrm{Sz}(q)$, *with $q = 2^{2e+1}$ and $e > 0$ an integer, has*

$$\frac{1}{2} \sum_{2f+1|2e+1} \mu\left(\frac{2e+1}{2f+1}\right) \sum_{\substack{n|2f+1 \\ n \neq 1}} \lambda(n)\psi(n, 2f+1)$$

string C-group representations, where $\mu(n)$ is the Moebius function,

$$\lambda(n) = \frac{1}{n} \sum_{d|n} \mu\left(\frac{n}{d}\right) \cdot 2^d \text{ and}$$

$$\psi(n, 2f+1) = \sum_{m|\frac{2f+1}{n}} \frac{\sum_{d|m} \mu(\frac{m}{d})(2^{nd} - 1)}{m}.$$

All these representations are non-degenerate, i.e. have a Schläfli type with entries ≥ 3.

This result closed the chapter of string C-group representations for Suzuki simple groups as the three main questions we gave in Section 1 were answered.

7. Small Ree groups

The small Ree groups $\mathrm{R}(q)$, defined over a finite field of order $q = 3^{2e+1}$ and $e > 0$, were discovered by Rimhak Ree [**57**] in 1960. In the literature they are also denoted by ${}^2\mathrm{G}_2(q)$. These groups have a subgroup structure quite similar to that of the Suzuki simple groups $\mathrm{Sz}(q)$, with $q = 2^{2e+1}$ and $e > 0$. Suzuki and Ree groups play a somewhat special role in the theory of finite simple groups, since they exist because of a Frobenius twist arising from a special automorphism of the field over which they are constructed. They have no counterpart in characteristic zero.

The Ree group $G := \mathrm{R}(q)$, with $q = 3^{2e+1}$ and $e \geq 0$, is a group of order $q^3(q-1)(q^3+1)$. It has a natural permutation representation on a Steiner system $\mathcal{S} := (\Omega, \mathcal{B}) = S(2, q+1, q^3+1)$ consisting of a set Ω of q^3+1 elements, the *points*, and a family of $(q+1)$-subsets $\mathcal{B}$ of Ω, the *blocks*, such that any two points of Ω lie in exactly one block. This Steiner system is also called a *Ree unital*. In particular, G acts 2-transitively on the points and transitively on the incident pairs of points and blocks of $\mathcal{S}$.

A list of the maximal subgroups of G is available, for instance, in [**60**, p. 349] and [**34**].

The group G has a unique conjugacy class of involutions (see [**57**]). Every involution ρ of G has a block B of $\mathcal{S}$ as its set of fixed points, and B is invariant under the centralizer $C_G(\rho)$ of ρ in G. Moreover, $C_G(\rho) \cong C_2 \times PSL(2,q)$, where $C_2 = \langle \rho \rangle$ and the $PSL(2,q)$-factor acts on the $q+1$ points in B as it does on the points of the projective line $PG(1,q)$. Hence the knowledge of string C-group representations of groups $PSL(2,q)$ is helpful to construct string C-group representations of large rank for G.

The automorphism group $\mathsf{Aut}(\mathrm{R}(q))$ of $\mathrm{R}(q)$ is given by

$$\mathsf{Aut}(\mathrm{R}(q)) \cong \mathrm{R}(q){:}C_{2e+1},$$

so in particular $\mathsf{Aut}(\mathrm{R}(3)) \cong \mathrm{R}(3)$.

In [**41**], using the list of maximal subgroups of G and the geometric properties of the Steiner system, we obtained, with Egon Schulte and Hendrik Van Maldeghem, the following theorem bounding the rank of a string C-group representation of a small Ree group $\mathrm{R}(q)$.

THEOREM 7.1. [**41**, Theorem 1.1] *Among the almost simple groups G with $\mathrm{R}(q) \leq G \leq \mathsf{Aut}(\mathrm{R}(q))$ and $q = 3^{2e+1} \neq 3$, only the Ree group $\mathrm{R}(q)$ itself is a C-group. In particular, $\mathrm{R}(q)$ admits a representation as a string C-group of rank 3, but not of higher rank. Moreover, the non-simple Ree group $\mathrm{R}(3)$ is not a C-group.*

In other words, the groups $\mathrm{R}(q)$ behave just like the Suzuki groups: they allow representations as string C-groups, but only of rank 3. The proof of this theorem, unlike its counterpart for the Suzuki groups, is very lengthy and uses a deep analysis of the lattice of subgroups of a Ree group.

PROBLEM 8. Enumerate the non-isomorphic string C-group representations of rank three of $\mathrm{R}(q)$.

8. Orthogonal and symplectic groups

With Brooksbank and Ferrara, we decided to investigate the orthogonal groups and to look for string C-group representations of large rank. Indeed, apart from the symmetric and alternating groups, as well as some crystallographic groups studied by Barry Monson and Egon Schulte (see [**46–48**]), no other family of groups was known to have possible large ranks that could grow linearly depending on one of its parameters, being the permutation degree or the dimension of the group or the size of the field on which the group is defined. We obtained the following theorem for orthogonal and symplectic groups.

THEOREM 8.1. [**2**, Corollary 1.3] *For each integer $k \geq 2$, positive integer m, and $\epsilon \in \{-,+\}$, the orthogonal group $\mathrm{O}^\epsilon(2m, \mathbb{F}_{2^k})$ has a string C-group representation of rank $2m$, and the symplectic group $\mathrm{Sp}(2m, \mathbb{F}_{2^k})$ has a string C-group representation of rank $2m+1$.*

Given the way we had to choose the involutions to construct the string C-group representations mentioned in Theorem 8.1, we have good reasons to believe that no higher rank can be achieved for these groups.

PROBLEM 9. Determine the maximal rank of a string C-group representation for $\mathrm{O}^\epsilon(2m, \mathbb{F}_{2^k})$ and for $\mathrm{Sp}(2m, \mathbb{F}_{2^k})$.

9. Sporadic groups

In [**42**], all string C-group representations were computed for the sporadic groups M_{11}, M_{12}, M_{22}, J_1 and J_2, as well as for their respective automorphism groups. In 2010, Hartley and Alexander Hulpke [**30**] designed more efficient algorithms that permitted them to classify all string C-group representations of the five Mathieu groups, the first three Janko groups, the Higman-Sims group, the McLaughlin group and the Held group. In 2012, with Mixer, we further improved the algorithms and managed to classify all string C-group representations of the third Conway group [**37**]. In [**18**], we proved with Connor and Mixer that the maximal rank of a string C-group representation for the O'Nan group is four. Moreover, we gave all string C-group representations of rank four for the O'Nan group. In [**17**], with Connor, we enumerated the regular maps of the O'Nan group using character theory, but we were unable to determine how many of them give a string C-group representation of rank three of the O'Nan group. Finally, in [**38**], with Jessica Mulpas, we classified all string C-group representations of the Rudvalis group and the Suzuki group.

A summary of these results is given in Table 5. No string C-group representation of rank six or higher exists for all the groups listed in that table as shown in the corresponding references given above.

PROBLEM 10. Determine the number of pairwise nonisomorphic string C-group representations of rank three for the O'Nan group.

PROBLEM 11. Try to push further the algorithms to study the remaining sporadic groups.

One of the problems of the current algorithms is that they work with a permutation representation of the group to be analyzed. This becomes a problem when such a representation is on more than, say, 150000 points. Using matrix groups instead of permutation groups might help. Another way to improve the existing algorithms would be to use parallel computing.

10. Collateral results

While working on specific families of simple groups, more general results were found. We mention the main ones in this section.

G	Order of G	Rank 3	Rank 4	Rank 5
M_{11}	7,920	0	0	0
M_{12}	95,040	23	14	0
M_{22}	443,510	0	0	0
M_{23}	10,200,960	0	0	0
M_{24}	244823040	490	155	2
J_1	175,560	148	2	0
J_2	604,800	137	17	0
J_3	50,232,960	303	2	0
HS	44,352,000	252	57	2
McL	898,128,000	0	0	0
He	4,030,387,200	1188	76	0
Ru	145,926,144,000	21594	227	0
Suz	448,345,497,600	7119	257	13
$O'N$	460,815,505,920	Unknown	16	0
Co_3	495,766,656,000	10586	873	22

TABLE 5. Number of string C-group representations for sporadic groups

10.1. Transitive groups. Transitive groups played a key role in the proof of Theorem 4.8. Indeed, to prove that theorem, the authors had to get the best possible bound for the maximal rank of a string C-group representation for a transitive group of degree n that is not S_n nor A_n. Together with Cameron, Fernandes and Mixer, we obtained the following result.

THEOREM 10.1. [**8**, Theorem 1.2] *Let Γ be a string C-group representation of rank d which is isomorphic to a transitive subgroup of S_n other than S_n or A_n. Then one of the following holds:*

(1) $d \leq n/2$;
(2) $n \equiv 2 \mod 4$, $d = n/2+1$ *and* Γ *is* $\mathrm{C}_2 \wr \mathrm{S}_{n/2}$. *The generators are*

$$\rho_0 = (1, n/2+1)(2, n/2+2)\dots(n/2, n);$$
$$\rho_1 = (2, n/2+2)\dots(n/2, n);$$
$$\rho_i = (i-1, i)(n/2+i-1, n/2+i) \text{ for } 2 \leq i \leq n/2.$$

Moreover the Schläfli type is $\{2, 3, \dots, 3, 4\}$.
(3) Γ *is transitive imprimitive and is one of the examples appearing in Table* 6.
(4) Γ *is primitive. In this case,* $n = 6$. *Moreover* Γ *is obtained from the permutation representation of degree* 6 *of* $\mathrm{S}_5 \cong \mathrm{PGL}_2(5)$ *and it is the group of the* 4*-simplex of Schläfli type* $\{3, 3, 3\}$.

This result and the experimental data obtained for the symmetric groups make us strongly believe in Conjecture 1. Indeed, it forces all the maximal parabolic subgroups of a string C-group representation of S_n of rank at least $n/2+3$ to be intransitive subgroups of S_n when n is large enough, that is $n \geq 9$.

10.2. Rank reduction. The proof of Theorem 4.9 has inspired Peter Brooksbank and the author to prove a rank reduction theorem [**4**].

THEOREM 10.2. [**4**, Theorem 1.1] *Let $(G; \{\rho_0, \dots, \rho_{n-1}\})$ be an irreducible string C-group of rank $n \geq 4$. If $\rho_0 \in \langle \rho_0\rho_2, \rho_3 \rangle$, then $(G; \{\rho_1, \rho_0\rho_2, \rho_3, \dots, \rho_{n-1}\})$ is a string C-group of rank $n-1$.*

Degree	*Number*	*Structure*	*Order*	*Schläfli type*
6	*9*	$S_3 \times S_3$	*36*	$\{3,2,3\}$
6	*11*	$2^3 : S_3$	*48*	$\{2,3,3\}$
6	*11*	$2^3 : S_3$	*48*	$\{2,3,4\}$
8	*45*	$2^4 : S_3 : S_3$	*576*	$\{3,4,4,3\}$

TABLE 6. Examples of transitive imprimitive string C-groups of degree n and rank $n/2+1$ for $n \leq 9$.

In particular, this theorem gives the following corollary.

COROLLARY 1. [**4**, Corollary 1.3] *Let $(G;\{\rho_0,\ldots,\rho_{n-1}\})$ be an irreducible string C-group representation of rank $n \geq 4$ of G. Let $\{p_1,\ldots,p_{n-1}\}$ be its Schläfli type, and put*

$$t = \max\{j \in \{0,\ldots,n-3\} \colon \forall i \in \{0,\ldots,j\},\ p_{2+i} \text{ is odd}\}.$$

Then G has a string C-group representation of rank $n-i$ for each $i \in \{0,\ldots,t\}$.

This corollary stresses the importance of trying to construct string C-group representations of large ranks as they may give representations of lower ranks for free. For instance, the corollary makes the proof of Theorem 4.3 straightforward. It can also be used to improve the proof of Theorem 4.9, and it was used by Brooksbank and the author to prove the following result.

THEOREM 10.3. [**4**, Theorem 1.4] *Let $k \geq 2$ and $m \geq 2$ be integers.*

(a) *The symplectic group $\mathrm{Sp}(2m,\mathbb{F}_{2^k})$ has a string C-group representation of rank n for each $3 \leq n \leq 2m+1$.*
(b) *The orthogonal groups $\mathrm{O}^+(2m,\mathbb{F}_{2^k})$ and $\mathrm{O}^-(2m,\mathbb{F}_{2^k})$ have string C-group representations of rank n for each $3 \leq n \leq 2m$.*

We conclude this section by recalling the following conjecture made by Brooksbank and the author in [**4**].

CONJECTURE 3. [**4**, Conjecture 5.1] *The group A_{11} is the only finite simple group whose set of ranks of string C-group representations is not an interval in the set of integers.*

11. C-groups

A natural question is to see if one could prove similar results by weakening the hypotheses. We finish this survey with what is currently known about C-group representations of almost simple groups.

Work with Thomas Connor showed that the word 'string' in the last part of the statement of the Theorem 6.1 can be removed.

THEOREM 11.1. [**16**, Classification Theorem 1.1] *Let $q = 2^{2e+1} \neq 2$ be an odd power of 2 and let $Sz(q) \leq G \leq Aut(Sz(q))$ be an almost simple group of Suzuki type. Let $(G,\{\rho_0,\ldots,\rho_{r-1}\})$ be a C-group representation of G. Then $G = Sz(q)$ and $r = 3$. Moreover there exist at least one string C-group representation and one nonstring C-group representation of G.*

C-group representations in general are also very interesting to study as they are smooth quotients of Coxeter groups, not only those with a string diagram. Moreover, they are often (but not always as in the case of string C-group representations) automorphism groups of more general geometric objects that, together with Fernandes and Weiss, we called hypertopes in [**27**]. These objects are apartments in the theory of Buildings due to Jacques Tits (see [**59**]), one more reason to study them further.

It would be interesting to have a result similar to Theorem 6.2 for C-groups.

PROBLEM 12. Determine the number of pairwise non-isomorphic C-group representations of $Sz(q)$.

Together with Connor and Sebastian Jambor, we obtained the following result for groups $PSL(2,q)$ and $PGL(2,q)$.

THEOREM 11.2. [**15**, Theorem 1.1] *Let $G \cong \mathrm{PSL}(2,q)$ for some prime power $q \geq 4$. A C-group representation of G is of rank 4 if and only if $q \in \{7, 9, 11, 19, 31\}$. Otherwise it is 3.*

Let $G \cong \mathrm{PGL}(2,q)$ for some prime power $q \geq 4$. A C-group representation of G is of rank 4 if and only if $q = 5$. Otherwise it is 3.

Our proof was in two steps: first we reduced the possible Coxeter diagrams of the C-group representations of group $PSL(2,q)$ and $PGL(2,q)$ using the subgroup structure of $PGL(2,q)$. Then we used the L_2-quotient algorithm to see which diagrams gave Coxeter groups admitting groups $PSL(2,q)$ or $PGL(2,q)$ as quotients. Unfortunately, this algorithm does not permit to recognize other almost simple groups with socle $PSL(2,q)$ so we were not able to extend our result to all almost simple groups with socle $PSL(2,q)$.

PROBLEM 13. Prove a theorem similar to Theorem 5.6 for C-groups.

The resolution of this problem would become feasible if one manages to extend the L_2-quotient algorithms to all almost simple groups with socle $PSL(2,q)$. We have no idea however on how easy or difficult the latter is.

PROBLEM 14. Enumerate C-group representations of groups with socle $PSL(2,q)$.

As pointed out earlier, small Ree groups and Suzuki groups are very similar. It would therefore be interesting to try to prove a theorem similar to Theorem 11.1 for small Ree groups. This is stated in the following problem.

PROBLEM 15. Determine the maximal rank of a C-group with group $\mathrm{R}(q)$.

As in the case of Suzuki groups and groups $PSL(2,q)$ and $PGL(2,q)$, it is possible to obtain classification theorems for the symmetric groups when removing the string condition. Theorem 1 of [**20**] was obtained as a corollary of the following result that gives the C-groups of rank $n-1$ for S_n when $n \geq 7$. It is due to Cameron and Philippe Cara but rephrased here in the framework of C-groups.

THEOREM 11.3. [**7**] *For $n \geq 7$, G a permutation group of degree n and $\{\rho_0, \ldots, \rho_{n-2}\}$ a set of involutions of G, $\Gamma := (G, \{\rho_0, \ldots, \rho_{n-2}\})$ is a C-group representation of rank $n-1$ of G if and only if the permutation representation graph of Γ is a tree $\mathcal{T}$ with n vertices. Moreover, the ρ_i's are transpositions, $G \cong S_n$ and the Coxeter diagram of Γ is the line graph of $\mathcal{T}$.*

Together with Fernandes, we managed in **[21]** to give a classification of C-groups of rank $n-2$ for S_n when $n \geq 9$. Recall that a 2-transposition is an involution that is the product of two transpositions.

THEOREM 11.4. *Let $n \geq 9$. Let $\{\rho_0, \ldots, \rho_{n-3}\}$ be a set of involutions of S_n. Then $\Gamma := (S_n, \{\rho_0, \ldots, \rho_{n-3}\})$ is a C-group representation of rank $n-2$ of S_n if and only if its permutation representation graph belongs to one of the following three families, up to a renumbering of the generators, where $\rho_2, \ldots, \rho_{n-3}$ are transpositions corresponding to the edges of a tree with $n-3$ vertices and the two remaining involutions, ρ_0 and ρ_1, are either transpositions or 2-transpositions (with at least one of them being a 2-transposition).*

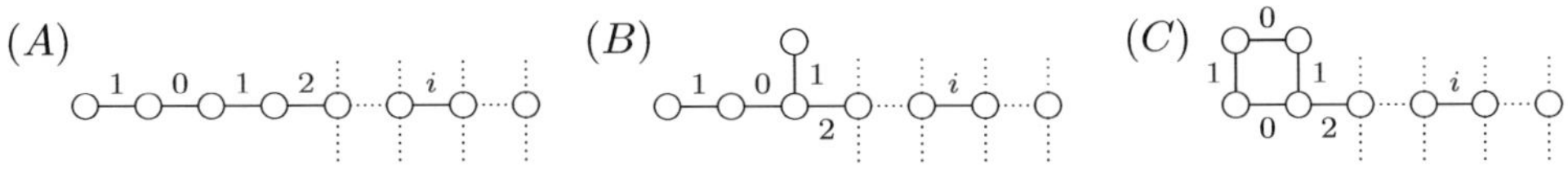

Acknowledgments

The author would like to thank two anonymous referees whose fruitful comments improved this survey.

References

[1] Wieb Bosma, John Cannon, and Catherine Playoust, *The Magma algebra system. I. The user language*, J. Symbolic Comput. **24** (1997), no. 3-4, 235–265, DOI 10.1006/jsco.1996.0125. Computational algebra and number theory (London, 1993). MR1484478

[2] P. A. Brooksbank, J. T. Ferrara, and D. Leemans, *Orthogonal groups in characteristic 2 acting on polytopes of high rank*, Discrete Comput. Geom. **63** (2020), no. 3, 656–669, DOI 10.1007/s00454-019-00083-0. MR4074338

[3] P. A. Brooksbank and D. Leemans, *Polytopes of large rank for* PSL$(4, \mathbb{F}_q)$, J. Algebra **452** (2016), 390–400, DOI 10.1016/j.jalgebra.2015.11.051. MR3461073

[4] Peter A. Brooksbank and Dimitri Leemans, *Rank reduction of string C-group representations*, Proc. Amer. Math. Soc. **147** (2019), no. 12, 5421–5426, DOI 10.1090/proc/14666. MR4021100

[5] Peter A. Brooksbank and Deborah A. Vicinsky, *Three-dimensional classical groups acting on polytopes*, Discrete Comput. Geom. **44** (2010), no. 3, 654–659, DOI 10.1007/s00454-009-9212-0. MR2679061

[6] Julie De Saedeleer and Dimitri Leemans, *On the rank two geometries of the groups* PSL$(2, q)$*: part I*, Ars Math. Contemp. **3** (2010), no. 2, 177–192, DOI 10.26493/1855-3974.107.2e3. MR2737583

[7] Peter J. Cameron and Philippe Cara, *Independent generating sets and geometries for symmetric groups*, J. Algebra **258** (2002), no. 2, 641–650, DOI 10.1016/S0021-8693(02)00550-1. MR1943939

[8] Peter J. Cameron, Maria Elisa Fernandes, Dimitri Leemans, and Mark Mixer, *String C-groups as transitive subgroups of* S_n, J. Algebra **447** (2016), 468–478, DOI 10.1016/j.jalgebra.2015.09.040. MR3427645

[9] Peter J. Cameron, Maria Elisa Fernandes, Dimitri Leemans, and Mark Mixer, *Highest rank of a polytope for* A_n, Proc. Lond. Math. Soc. (3) **115** (2017), no. 1, 135–176, DOI 10.1112/plms.12039. MR3669935

[10] Marston D. E. Conder, *Generators for alternating and symmetric groups*, J. London Math. Soc. (2) **22** (1980), no. 1, 75–86, DOI 10.1112/jlms/s2-22.1.75. MR579811

[11] Marston D. E. Conder, *More on generators for alternating and symmetric groups*, Quart. J. Math. Oxford Ser. (2) **32** (1981), no. 126, 137–163, DOI 10.1093/qmath/32.2.137. MR615190

[12] Marston Conder and Deborah Oliveros, *The intersection condition for regular polytopes*, J. Combin. Theory Ser. A **120** (2013), no. 6, 1291–1304, DOI 10.1016/j.jcta.2013.03.009. MR3044544

[13] Marston Conder, Primož Potočnik, and Jozef Širáň, *Regular hypermaps over projective linear groups*, J. Aust. Math. Soc. **85** (2008), no. 2, 155–175, DOI 10.1017/S1446788708000827. MR2470535

[14] Thomas Connor, Julie De Saedeleer, and Dimitri Leemans, *Almost simple groups with socle* $\mathrm{PSL}(2,q)$ *acting on abstract regular polytopes*, J. Algebra **423** (2015), 550–558, DOI 10.1016/j.jalgebra.2014.10.020. MR3283730

[15] Thomas Connor, Sebastian Jambor, and Dimitri Leemans, *C-groups of* $\mathrm{PSL}(2,q)$ *and* $\mathrm{PGL}(2,q)$, J. Algebra **427** (2015), 455–466, DOI 10.1016/j.jalgebra.2014.12.031. MR3312313

[16] Thomas Connor and Dimitri Leemans, *C-groups of Suzuki type*, J. Algebraic Combin. **42** (2015), no. 3, 849–860, DOI 10.1007/s10801-015-0605-2. MR3403184

[17] Thomas Connor and Dimitri Leemans, *Algorithmic enumeration of regular maps*, Ars Math. Contemp. **10** (2016), no. 2, 211–222, DOI 10.26493/1855-3974.544.1b4. MR3529287

[18] Thomas Connor, Dimitri Leemans, and Mark Mixer, *Abstract regular polytopes for the O'Nan group*, Internat. J. Algebra Comput. **24** (2014), no. 1, 59–68, DOI 10.1142/S0218196714500052. MR3189666

[19] H. S. M. Coxeter, *Ten toroids and fifty-seven hemidodecahedra*, Geom. Dedicata **13** (1982), no. 1, 87–99, DOI 10.1007/BF00149428. MR679218

[20] Maria Elisa Fernandes and Dimitri Leemans, *Polytopes of high rank for the symmetric groups*, Adv. Math. **228** (2011), no. 6, 3207–3222, DOI 10.1016/j.aim.2011.08.006. MR2844941

[21] Maria Elisa Fernandes and Dimitri Leemans, *C-groups of high rank for the symmetric groups*, J. Algebra **508** (2018), 196–218, DOI 10.1016/j.jalgebra.2018.04.031. MR3810293

[22] Maria Elisa Fernandes and Dimitri Leemans, *String C-group representations of alternating groups*, Ars Math. Contemp. **17** (2019), no. 1, 291–310, DOI 10.26493/1855-3974.1953.c53. MR4031526

[23] Maria Elisa Fernandes, Dimitri Leemans, and Mark Mixer, *Polytopes of high rank for the alternating groups*, J. Combin. Theory Ser. A **119** (2012), no. 1, 42–56, DOI 10.1016/j.jcta.2011.07.006. MR2844081

[24] Maria Elisa Fernandes, Dimitri Leemans, and Mark Mixer, *All alternating groups* A_n *with* $n \geq 12$ *have polytopes of rank* $\lfloor \frac{n-1}{2} \rfloor$, SIAM J. Discrete Math. **26** (2012), no. 2, 482–498, DOI 10.1137/110838467. MR2967478

[25] Maria Elisa Fernandes, Dimitri Leemans, and Mark Mixer, *Corrigendum to "Polytopes of high rank for the symmetric groups" [Adv. Math. 228 (2011) 3207–3222] [MR2844941]*, Adv. Math. **238** (2013), 506–508, DOI 10.1016/j.aim.2012.10.006. MR3033641

[26] Maria Elisa Fernandes, Dimitri Leemans, and Mark Mixer, *An extension of the classification of high rank regular polytopes*, Trans. Amer. Math. Soc. **370** (2018), no. 12, 8833–8857, DOI 10.1090/tran/7425. MR3864397

[27] Maria Elisa Fernandes, Dimitri Leemans, and Asia Ivić Weiss, *Highly symmetric hypertopes*, Aequationes Math. **90** (2016), no. 5, 1045–1067, DOI 10.1007/s00010-016-0431-1. MR3547707

[28] Branko Grünbaum, *Regularity of graphs, complexes and designs* (English, with French summary), Problèmes combinatoires et théorie des graphes (Colloq. Internat. CNRS, Univ. Orsay, Orsay, 1976), Colloq. Internat. CNRS, vol. 260, CNRS, Paris, 1978, pp. 191–197. MR539975

[29] Michael I. Hartley, *An atlas of small regular abstract polytopes*, Period. Math. Hungar. **53** (2006), no. 1-2, 149–156, DOI 10.1007/s10998-006-0028-x. MR2286467

[30] Michael I. Hartley and Alexander Hulpke, *Polytopes derived from sporadic simple groups*, Contrib. Discrete Math. **5** (2010), no. 2, 106–118. MR2791293

[31] Michael I. Hartley and Dimitri Leemans, *Quotients of a universal locally projective polytope of type* $\{5,3,5\}$, Math. Z. **247** (2004), no. 4, 663–674, DOI 10.1007/s00209-003-0625-9. MR2077414

[32] Ann Kiefer and Dimitri Leemans, *On the number of abstract regular polytopes whose automorphism group is a Suzuki simple group* $\mathrm{Sz}(q)$, J. Combin. Theory Ser. A **117** (2010), no. 8, 1248–1257, DOI 10.1016/j.jcta.2010.01.001. MR2677688

[33] Ann Kiefer and Dimitri Leemans, *On pairs of commuting involutions in* $\mathrm{Sym}(n)$ *and* $\mathrm{Alt}(n)$, Comm. Algebra **41** (2013), no. 12, 4408–4418, DOI 10.1080/00927872.2012.701360. MR3169527

[34] Peter B. Kleidman, *The maximal subgroups of the Chevalley groups* $G_2(q)$ *with* q *odd, the Ree groups* ${}^2G_2(q)$*, and their automorphism groups*, J. Algebra **117** (1988), no. 1, 30–71, DOI 10.1016/0021-8693(88)90239-6. MR955589

[35] Dimitri Leemans, *An atlas of regular thin geometries for small groups*, Math. Comp. **68** (1999), no. 228, 1631–1647, DOI 10.1090/S0025-5718-99-01130-8. MR1654025

[36] Dimitri Leemans, *Almost simple groups of Suzuki type acting on polytopes*, Proc. Amer. Math. Soc. **134** (2006), no. 12, 3649–3651, DOI 10.1090/S0002-9939-06-08448-6. MR2240679

[37] Dimitri Leemans and Mark Mixer, *Algorithms for classifying regular polytopes with a fixed automorphism group*, Contrib. Discrete Math. **7** (2012), no. 2, 105–118. MR3011579

[38] D. Leemans and J. Mulpas, *The string C-group representations of the Suzuki and Rudvalis sporadic groups*, Preprint, 2020.

[39] Dimitri Leemans and Egon Schulte, *Groups of type $L_2(q)$ acting on polytopes*, Adv. Geom. **7** (2007), no. 4, 529–539, DOI 10.1515/ADVGEOM.2007.031. MR2360900

[40] Dimitri Leemans and Egon Schulte, *Polytopes with groups of type* $\mathrm{PGL}_2(q)$, Ars Math. Contemp. **2** (2009), no. 2, 163–171, DOI 10.26493/1855-3974.102.290. MR2550963

[41] Dimitri Leemans, Egon Schulte, and Hendrik Van Maldeghem, *Groups of Ree type in characteristic 3 acting on polytopes*, Ars Math. Contemp. **14** (2018), no. 2, 209–226, DOI 10.26493/1855-3974.1193.0fa. MR3812018

[42] Dimitri Leemans and Laurence Vauthier, *An atlas of abstract regular polytopes for small groups*, Aequationes Math. **72** (2006), no. 3, 313–320, DOI 10.1007/s00010-006-2843-9. MR2282877

[43] Heinz Lüneburg, *Translation planes*, Springer-Verlag, Berlin-New York, 1980. MR572791

[44] V. D. Mazurov, *On the generation of sporadic simple groups by three involutions, two of which commute* (Russian, with Russian summary), Sibirsk. Mat. Zh. **44** (2003), no. 1, 193–198, DOI 10.1023/A:1022028807652; English transl., Siberian Math. J. **44** (2003), no. 1, 160–164. MR1967616

[45] Peter McMullen and Egon Schulte, *Abstract regular polytopes*, Encyclopedia of Mathematics and its Applications, vol. 92, Cambridge University Press, Cambridge, 2002. MR1965665

[46] B. Monson and Egon Schulte, *Reflection groups and polytopes over finite fields. I*, Adv. in Appl. Math. **33** (2004), no. 2, 290–317, DOI 10.1016/j.aam.2003.11.002. MR2074400

[47] B. Monson and Egon Schulte, *Reflection groups and polytopes over finite fields. II*, Adv. in Appl. Math. **38** (2007), no. 3, 327–356, DOI 10.1016/j.aam.2005.12.001. MR2301701

[48] B. Monson and Egon Schulte, *Reflection groups and polytopes over finite fields. III*, Adv. in Appl. Math. **41** (2008), no. 1, 76–94, DOI 10.1016/j.aam.2007.07.001. MR2419764

[49] Eliakim Hastings Moore, *Concerning the Abstract Groups of Order k ! and 1/2k ! Holohedrically Isomorphic with the Symmetric and the Alternating Substitution-Groups on k Letters*, Proc. Lond. Math. Soc. **28** (1896/97), 357–366, DOI 10.1112/plms/s1-28.1.357. MR1576641

[50] E H. Moore, *The subgroups of the generalized finite modular group*, Decennial Publications of the University of Chicago 9:141–190, 1904.

[51] Ya. N. Nuzhin, *Generating triples of involutions of Chevalley groups over a finite field of characteristic* 2 (Russian), Algebra i Logika **29** (1990), no. 2, 192–206, 261, DOI 10.1007/BF02001358; English transl., Algebra and Logic **29** (1990), no. 2, 134–143 (1991). MR1131150

[52] Ya. N. Nuzhin, *Generating triples of involutions of alternating groups* (Russian), Mat. Zametki **51** (1992), no. 4, 91–95, 142, DOI 10.1007/BF01250552; English transl., Math. Notes **51** (1992), no. 3-4, 389–392. MR1172472

[53] Ya. N. Nuzhin, *Generating triples of involutions of Lie-type groups over a finite field of odd characteristic. I* (Russian, with Russian summary), Algebra i Logika **36** (1997), no. 1, 77–96, 118, DOI 10.1007/BF02671953; English transl., Algebra and Logic **36** (1997), no. 1, 46–59. MR1454692

[54] Ya. N. Nuzhin, *Generating triples of involutions of Lie-type groups over a finite field of odd characteristic. II* (Russian, with Russian summary), Algebra i Logika **36** (1997), no. 4, 422–440, 479; English transl., Algebra and Logic **36** (1997), no. 4, 245–256. MR1601503

[55] Daniel Pellicer, *CPR graphs and regular polytopes*, European J. Combin. **29** (2008), no. 1, 59–71, DOI 10.1016/j.ejc.2007.01.001. MR2368614

[56] Denis Sjerve and Michael Cherkassoff, *On groups generated by three involutions, two of which commute*, The Hilton Symposium 1993 (Montreal, PQ), CRM Proc. Lecture Notes, vol. 6, Amer. Math. Soc., Providence, RI, 1994, pp. 169–185. MR1290589

[57] Rimhak Ree, *A family of simple groups associated with the simple Lie algebra of type* (G_2), Bull. Amer. Math. Soc. **66** (1960), 508–510, DOI 10.1090/S0002-9904-1960-10523-X. MR125154

[58] J. Tits, *Ovoïdes et groupes de Suzuki*, Arch. Math. **13** (1962), 187–198, DOI 10.1007/BF01650065. MR140572

[59] Jacques Tits, *Buildings of spherical type and finite BN-pairs*, Lecture Notes in Mathematics, Vol. 386, Springer-Verlag, Berlin-New York, 1974. MR0470099

[60] Hendrik van Maldeghem, *Generalized polygons*, Monographs in Mathematics, vol. 93, Birkhäuser Verlag, Basel, 1998. MR1725957

[61] A. Vandenschrick, *On string C-groups that admit a string C-group representation*, Preprint, 2019.

[62] Julius Whiston, *Maximal independent generating sets of the symmetric group*, J. Algebra **232** (2000), no. 1, 255–268, DOI 10.1006/jabr.2000.8399. MR1783924

[63] A. Wiman, *Bestimmung aller untergruppen einer doppelt unendlichen reihe von einfachen gruppen*, Bihan till K. Svenska Vet.-Akad.Handl. 25:1–147, 1899.

Université Libre de Bruxelles, Département de Mathématique, C.P.216 Algèbre et Combinatoire, Bld du Triomphe, 1050 Bruxelles, Belgium

Email address: dleemans@ulb.ac.be

Contemporary Mathematics
Volume **764**, 2021
https://doi.org/10.1090/conm/764/15360

Orientation-reversing symmetry of closed surfaces immersed in euclidean 3-space

Undine Leopold and Thomas W. Tucker

ABSTRACT. Given a finite group G of isometries of euclidean 3-space $\mathbb{E}^3$ and a closed surface S, an immersion $f : S \to \mathbb{E}^3$ is in G-general position if $f(S)$ is invariant under G, points of S have disk neighborhoods mapped homeomorphically onto their images and these images are in general position, and all double curves of f are in general position with respect to axes of rotations and reflection planes. For such an immersion, there is an induced action of the orientation-preserving subgroup G^+ on S whose Riemann–Hurwitz equation satisfies certain natural restrictions. We classify which restricted Riemann–Hurwitz equations for G^+ are realized by a G-general position immersion of S, extending results of the authors for the orientation-preserving case $G = G^+$. The analysis involves a detailed study of immersions of the quotient surface S/G^+ in the orbifold $\mathbb{E}^3/G^+ \cong \mathbb{R}^3$ having the extra 2-fold orientation-reversing symmetry required by G when G contains orientation-reversing isometries.

Motivated by our previous study [**13**] of the possible symmetries of immersed surfaces in $\mathbb{E}^3$ among the finite orientation-preserving isometry groups in $\mathbb{E}^3$, we investigate now the finite isometry groups of $\mathbb{E}^3$ containing orientation-reversing isometries. Let G be a finite subgroup of $O(3)$, and let S be a closed surface. A *G-general position immersion* of a surface S is a continuous map $f\colon S \to \mathbb{E}^3$ where every point of S possesses a disk neighborhood D such that $f|D$ is a homeomorphism onto its image, these image disks meet the way two or three coordinate planes do in $\mathbb{R}^3$, no singular points of the immersion lie on a rotation axis of G, double curves are disjoint to or transverse to any reflection planes, and $f(S)$ is invariant under the action of G. The action of G on $\mathbb{E}^3$ induces an action of G on $f(S)$, which in turn induces an action of G on S.

For other examples of symmetric embeddings of surfaces in $\mathbb{E}^3$, see [**3–6,10–12, 14,15**]. Our previous study [**13**] and the preceding note [**16**] were inspired by a question asked by B. Mohar about a sculpture envisioned by T. Pisanski [**8**].

Throughout this paper, we let G denote a finite group of isometries of $\mathbb{E}^3$ and we let G^+ denote its subgroup of orientation-preserving isometries, which has index at most two. Assume S is immersed in G-general position. The induced action of G^+ on S is pseudo-free, that is, it has isolated fixed points. We let Q denote the quotient S/G^+, which is a closed surface. The associated regular branched covering

---

2010 *Mathematics Subject Classification.* Primary 57M60, 57M12, 57M50.
The second author was supported by Grant 317689 from the Simons Foundation.

$p_+ : S \to Q$ gives rise to the *Riemann–Hurwitz equation* (abbreviated *RH*) for G^+ acting on S,

$$\chi(S) = |G^+| \left(\chi(Q) - b_r \sum \frac{r-1}{r} \right),$$

where the sum is over branch points in $Q = S/G^+$ (*not* S), r is the order of the branch point and b_r is the number of branch points of order r. The branch points come from intersections of $f(S)$ with the rotation axes of G^+. In addition to branch point information, there are three possible orientability pairings for S and Q in RH, abbreviated OO, NN, ON for S, Q both orientable, S, Q both non-orientable, S orientable and Q non-orientable, respectively. For the ON case, we note that even though G^+ is orientation-preserving on $\mathbb{E}^3$, its induced action on S may not be. For Riemann or Klein surfaces, the branching data together with the nature of S are often called the *signature* for the action of G^+.

An instance of an RH for $G = G^+$ is *realizable* if there is a G^+-general position immersion of S in $\mathbb{E}^3$ such that the action of G^+ induced on S has the given RH. There are some natural restrictions that arise for such RH coming from how the G^+-general position immersion can intersect the rotation axes for G^+. Our previous paper [**13**] determined for $G = G^+$ which such restricted RH are realizable. In this paper, we consider *orientation-reversing groups* G of isometries of $\mathbb{E}^3$, that is, groups G with G^+ as a proper subgroup, such that the induced action of G^+ on S has a given restricted RH. Briefly, if there is a G-general (*not* just G^+-general) position immersion realizing the restricted RH for G^+, then we say the RH is *realizable* for G (see Definition 2.3). We emphasize that we do not look at the Riemann–Hurwitz equation for G, which is greatly complicated by the presence of boundary for Q. Instead we continue to look at the Riemann–Hurwitz equation for G^+, to which we add further restrictions related to the extra symmetry required by G and the requirements of G-general position in the presence of any reflection planes.

The five possibilities for G^+ are a n-fold rotation, the rotation group of the right regular n-prism, or the rotation groups of the tetrahedron, octahedron/cube, icosahedron/dodecahedron. We denote these respectively $R^+(n), P^+(n), T^+, C^+, I^+$. The abstract groups are, respectively, the cyclic group C_n, the dihedral group D_n, the alternating group A_4, the symmetric group S_4, and the alternating group A_5. There are nine possibilities for G, which we will discuss in detail later.

The organization of this paper is as follows. Section 1 reviews the case $G = G^+$ from [**13**], emphasizing the role of the orbifold $\mathbb{E}^3/G^+$ and the necessary restrictions on the RH for G^+. Section 2 reviews the spherical geometry for G, relating it to the notation of various other authors. Of particular importance is the extra symmetry of the orbifold $\mathbb{E}^3/G^+ \cong \mathbb{R}^3$ required by G, either a reflection or an antipodal mapping. This symmetry adds amended restrictions on the RH for G^+. Section 3 reviews from [**13**] the standard and special models used to realize a restricted RH for G^+. Section 4 presents our main results about which amended RHs are realizable. These results must address all nine cases, but it is not hard to show that five of the nine cases are realizable only by standard models. Section 5 analyzes which amended RH are realizable for G^+ but not for G. Section 6 considers possible generalizations.

1. The orbifold and Riemann–Hurwitz equations for G^+

In this section, we review from [**13**] the Riemann–Hurwitz equations for the action of G^+ on S.

1.1. The orbifold $\mathbb{E}^3/G^+$. We visualize the action of G^+ on $\mathbb{E}^3$ in terms of the associated quotient space or *orbifold* $\mathbb{E}^3/G^+$, which is homeomorphic to 3-space $\mathbb{R}^3$, with distorted linear structure. We view the action of G^+ as a subgroup of $SO(3)$ fixing the origin. This action of G^+ on $\mathbb{E}^3$ takes the union of axes of G^+, denoted by X, to a set Y in the quotient. Y consists of a single line for $R^+(n)$ and a set of three rays from the image of the origin for $P^+(n), T^+, C^+, I^+$. We then have a second associated regular branched covering, of $\mathbb{R}^3 \cong \mathbb{E}^3/G^+$, namely $q_+ \colon (\mathbb{E}^3, X) \to (\mathbb{E}^3/G^+, Y)$. Since p_+ and q_+ are both induced by the same action of G^+, and since G^+ acts freely on singular points, there is a general position immersion $g \colon Q \to \mathbb{E}^3/G^+$, transverse to Y (no singular points on Y), completing the following commutative diagram, where $B = g^{-1}(Y)$ is the set of branch points with respect to the axes of G^+:

$$\begin{array}{ccc} (S, p_+^{-1}(B)) & \xrightarrow{\ f\ } & (\mathbb{E}^3, X) \\ \big\downarrow{\scriptstyle p_+} & & \big\downarrow{\scriptstyle q_+} \\ (Q, B) & \xrightarrow{\ g\ } & (\mathbb{E}^3/G^+, Y) \end{array}$$

When $G \neq G^+$, we could extend the above diagram to include the orbifolds with respect to the larger orientation-reversing group G. But $\mathbb{E}^3/G$ is no longer homeomorphic to $\mathbb{R}^3$, and S/G is a half-space or a projective 3-space, which require different arguments than in [**13**].

Instead, we consider the additional symmetry of the immersion of the orbifold $Q = S/G^+$ in $(\mathbb{E}^3/G^+, Y)$ forced by the additional isometries in $G - G^+$. In all cases where G contains a reflection, the extra symmetry of $\mathbb{E}^3/G^+ \cong \mathbb{R}^3$ is topologically equivalent to the reflection $r(x, y, z) = (x, y, -z)$, so we will call it a *reflection* in a *reflection plane* despite the non-euclideanness of $\mathbb{E}^3/G^+ \cong \mathbb{R}^3$. More specifically, the image in $\mathbb{E}^3/G^+ \cong \mathbb{R}^3$ of the union of all the reflection planes of G is a single surface homeomorphic to $\mathbb{R}^2$, which may be composed of half planes.

For example, suppose that G is the dihedral action generated by a rotation g of order n around a line X and a reflection in a plane containing that line. When n is even, each reflection plane of G is taken to a half plane; the reflection plane for r consists of the union of two of those half planes. When n is odd, each reflection plane of G is taken homeomorphically to a single plane in $\mathbb{R}^3$.

In the case where G contains no reflections but only an antipodal map, that map is taken to an orientation-reversing involution of $\mathbb{R}^3$ leaving only the origin fixed, which is topologically equivalent to the map $r(x, y, z) = (-x - y - z)$. We will therefore call this the *antipodal map*.

REMARK 1.1. The immersion f for S is in G-general position if and only if the immersion g for Q is in general position with respect to Y and any reflection plane for G/G^+ in $\mathbb{E}^3/G^+$.

1.2. The five possible RH for G^+. Since all branching comes from the intersection of $g(Q)$ with Y, the possible branch orders for the rays in Y are $\{n, n\}$, $\{2, 2, n\}$, $\{2, 3, 3\}$, $\{2, 3, 4\}$, $\{2, 3, 5\}$, corresponding, respectively, to $R^+(n)$, $P^+(n)$, T^+, C^+, I^+.

REMARK 1.2. We do not consider G^+ being the trivial group and we do not consider $P^+(n)$ where $n \leq 2$.

We thus get these RH, denoted in [**13**], respectively, as RRH, PRH, TRH, CRH, IRH:

$$\begin{aligned}
\mathbf{R^+(n)} &: \chi(S) = n\chi(Q) - b_n(n-1), \text{ where } G^+ \cong C_n, n > 1 \\
\mathbf{P^+(n)} &: \chi(S) = 2n\chi(Q) - nb_2 - nb_2' - 2(n-1)b_n, \text{ where } G^+ \cong D_n, n > 2 \\
\mathbf{T^+} &: \chi(S) = 12\chi(Q) - 6b_2 - 8b_3 - 8b_3', \text{ where } G^+ \cong A_4 \\
\mathbf{C^+} &: \chi(S) = 24\chi(Q) - 12b_2 - 16b_3 - 18b_4, \text{ where } G^+ \cong S_4 \\
\mathbf{I^+} &: \chi(S) = 60\chi(Q) - 30b_2 - 40b_3 - 48b_5, \text{ where } G^+ \cong A_5
\end{aligned}$$

Here, the parameter $b_i \geq 0$ refers to the number of branch points on the ray of order i for $Y = q_+(X)$. If there are two such rays of order i we call them b_i and b_i'. We say these equations have *sufficient branching* if $b_n \neq 0$ in $R^+(n)$ and if two or more of the b_i are nonzero in $P^+(n), T^+, C^+, I^+$. Otherwise, they have either *no branching* (if there are no branch points) or *insufficient branching* if exactly one of the b_i is nonzero in $P^+(n), T^+, C^+, I^+$. Generally, insufficient or no branching limits realizability of RH for $\chi(Q) \geq -2$.

1.3. RH restrictions for G^+. There are three restrictions on the parameters in the RH for G^+ [**13**]:

RH Restrictions for G^+

(1) *Order Restriction*: Each branch point must have the full order of the rotation axis.
(2) *Parity Restriction*: The numbers of branch points on each ray of Y must have the same parity. In particular, for n-fold rotational symmetry, there are an even number of branch points on the single axis Y.
(3) *Orientability Restriction*: For the ON case, $\chi(Q)$ must be even and G^+ must have an index two subgroup G^o. Furthermore, there must be insufficient branching or none at all and the rotation corresponding to any branching must be in G^o.

We call an RH for G^+ satisfying these three restrictions a *restricted RH*. The first restriction stems from the fact that intersections of $f(S)$ with rotation axes must be transverse and at regular (non-singular) points. The second comes from $g(Q)$ being a $\mathbb{Z}_2$-homology cycle in 3-space, so every curve piercing the immersed surface does so, generically, in an even number of points. The third, which applies only to the case ON, is more complicated and follows from considerations related to the representation of the fundamental group $\pi(Q - B)$ in G^+ induced by the immersion $g : Q - B \to \mathbb{E}^3/G^+ - Y$.

REMARK 1.3. (**The Epimorphism Condition**) For S to be connected, we need the homomorphism:

$$\pi(Q - B) \to \pi(\mathbb{E}^3/G^+ - Y) \to G^+$$

to be an epimorphism (see [**13**]). In particular, suppose there is no branching (B is empty) and Y is not a line. Then $\pi(\mathbb{E}^3/G^+ - Y)$ is a free group of rank 2 and there can be no such epimorphism when $\chi(Q) > -2$, since then $\pi(Q)$ is at most

2-generated and not free (see Lemma 3.4 of [**13**]). Note that when Q is the Klein bottle, there are epimorphisms to the dihedral group, but none that factor through a free group of rank 2.

In addition, when Q is non-orientable, the index two subgroup $\pi^o(Q - B)$ generated by simple closed curves with orientable annular neighborhood must be taken to G^+ in the NN case and to an index two subgroup of G^+ in the ON case (the "Index two condition" from [**13**]).

We can summarize the results of [**13**] as follows:

THEOREM 1.4. *All restricted RH are realizable with the following exceptions:*
1) $R^+(n)$, no branching: $\chi(Q) = 0$ for NN with n even;
2) $P^+(n)$, T^+, C^+, I^+, no branching: $\chi(Q) > -2$ and $\chi(Q) = -2$ for NN of $P^+(n)$ with n even;
3) $P^+(n)$, T^+, C^+, I^+, insufficient branching: $\chi(Q) \geq 0$ depending on type and branching data.

2. The nine cases for G

We have five cases for orientation-preserving G^+. There are nine cases for orientation-reversing G.

2.1. Review of spherical isometries. We recall that every spherical isometry is either a rotation around an axis through antipodal points, a reflection in a plane meeting the sphere in a great circle, or a glide rotation consisting of a rotation composed with a reflection in the plane perpendicular to the axis of rotation. A glide must have even order. A glide of order 2 is the antipodal mapping. We recall:

(1) The antipodal mapping a commutes with all other isometries.
(2) The composition of two reflections r_1, r_2 is a rotation about the axis of intersection of the two reflection planes. The direction of rotation for r_1r_2 and r_2r_1 are opposite. In particular, r_1, r_2 commute if and only if the rotation r_1r_2 has order two, that is, the reflection planes for r_1, r_2 are perpendicular.
(3) The composition ar of the antipodal mapping a and a reflection r is a rotation of order two about the line perpendicular to the reflection plane of r.

2.2. The cases T^+, C^+, I^+: The platonic solids. Three of the possibilities for G^+ are the single groups $T^+ \cong A_4, C^+ \cong S_4, I^+ \cong A_5$ for the tetrahedron, cube/octahedron, isosahedron/dodecahedron, respectively. For C^+ and I^+, we have only $C \cong S_4 \times C_2$ and $I \cong A_5 \times C_2$, respectively, with the C_2 factor represented by the antipodal mapping. We call these *fully planar* since every rotation axis is contained in a reflection plane. This means that the reflection plane in $\mathbb{E}^3/G^+$ contains the image Y of the rotation axes.

For the group T^+, the situation is different because there are two ways we can have $T^+ \cong A_4$ as a subgroup of G. For full tetrahedral symmetry, we have $G \cong S_4$ and denote this by $T1$. On the other hand, we also have $G \cong A_4 \times C_2$ as a subgroup of either $C \cong S_4 \times C_2$ or $I \cong A_5 \times C_2$. For the cube, the $A_4 \times C_2$ symmetry can be viewed with A_4 represented as the subgroup generated by the 3-fold rotations of the cube about the four diagonals between antipodal vertices. Since the antipodal mapping also leaves these diagonals invariant, we have an action of the abstract group $A_4 \times C_2$, leading to case $T2$. This is sometimes called pyritohedral symmetry [**2**, **12**].

The groups containing T^+, C^+, or I^+ as a subgroup are *irreducible* as they leave no plane invariant. The other finite groups of isometries of $\mathbb{E}^3$ are *reducible*, as they leave a plane invariant.

2.3. The cases $R^+(n), P^+(n)$: The Frieze groups. For the five remaining possibilities, we begin with the families $R^+(n)$ or $P^+(n)$ and then add extra symmetries. The resulting groups G can be visualized by frieze patterns that have been rolled up onto a cylinder. Table 1 details the five possibilities. The first five columns give notations, respectively: ours, International Union of Crystallography (IUC), Schönflies [**9**] (Sch.), Coxeter[**7**] (Cox.), and Conway[**1**, **2**] (Con.). The sixth column gives an orientation-reversing symmetry in G: reflection in a plane containing only the n-fold axis (n), reflection in a plane perpendicular to the n-fold axis (p), glide only (g). The seventh column gives the extra symmetry in the orbifold $\mathbb{E}^3/G^+$: reflection in the same plane as Y (fully planar, full), reflection in a plane not containing all of Y (not), no reflection (antipodal, ant). The eighth column gives the pair G^+, G. The last column gives a frieze pattern using letters to indicate symmetry, imagining the rotation as horizontal translation.

Our	IUC	Sch.	Cox.	Con.	$\mathbb{E}^3$	$\mathbb{E}^3/G^+$	$G^+, G \cong$	Example
R1	p1m1	C_{nv}	$[n]$	$*nn$	n	full	C_n, D_n	..AA ..
R2	p11m	C_{nh}	$[n^+, 2]$	$n*$	p	not	$C_n, C_n \times C_2$	..BB..
RG	p11g	S_{2n}	$[2^+, 2n^+]$	$n\times$	g	ant	C_n, C_{2n}	..LΓ..
P1	p2mm	D_{nh}	$[2, n]$	$*22n$	p	full	$D_n, D_n \times C_2$	..HH..
P2	p2mg	D_{nd}	$[2^+, 2n]$	$2*n$	n	not	D_n, D_{2n}	..A∀ ..
T1		T_d	$[3, 3]$	$*332$		full	A_4, S_4	
T2		T_h	$[4, 3^+]$	$3*2$		not	$A_4, A_4 \times C_2$	
C		O_h	$[4, 3]$	$*432$		full	$S_4, S_4 \times C_2$	
I		I_h	$[5, 3]$	$*532$		full	$A_5, A_5 \times C_2$	

TABLE 1. The nine cases for G when G contains orientation-reversing isometries.

2.4. Symmetry of $\mathbb{E}^3/G^+$ and amended RH. The extra orientation-reversing symmetry required for the immersion $g : T \to \mathbb{E}^3/G^+$ leads to further restrictions on the associated RH for G^+.

First, there is the issue of how the extra orientation-reversing symmetry r acts on the rays of Y. Since r is an involution, when Y has three rays, r must either fix all rays, or fix one ray while interchanging the other two.

PROPOSITION 2.1. *For $R^+(n)$, where Y consists of a line, r fixes the line for $R1$ point-wise and reverses the line for $R2, RG$. For $P^+(n), T^+, C^+, I^+$ where Y has three rays, r fixes all rays for $P1, T1, C, I$, and r fixes one ray while interchanging the two rays with the same branching order for $P2, T2$.*

Less obvious is the following

LEMMA 2.2. *$\chi(Q)$ is even.*

PROOF. In the case RG, the geometric group G is generated by a single glide reflection r with r^2 being an n-fold rotation, such that $G^+ = \langle r^2 \rangle$. Note that r does not fix any points besides the origin. Consequently, the orientation-reversing involution in in G/G^+ only fixes the origin in $\mathbb{E}^3/G^+$ and is therefore the (distorted) antipodal mapping. Any model $g(Q)$ with antipodal symmetry induces an unbranched double cover $Q \to Q'$, where Q' is a closed surface, so $\chi(Q)$ is even. If instead $g(Q)$ exhibits a reflection in a plane, then that means that there are two homeomorphic halves of the surface Q, say Q^+ and Q^-, which are immersed and glued together along boundary cycles. That is, $Q = Q^+ \cup Q^-$ with $Q^+ \cap Q^-$ a collection of cycles. We have $\chi(Q) = \chi(Q^+) + \chi(Q^-)$. As $\chi(Q^+) = \chi(Q^-)$, we have $\chi(Q)$ even. □

This leads to the following additions to our restrictions on RH:

Amended RH for G

(1) (Amended Branching) For $P2, b_2 = b_2'$, and for $T2, b_3 = b_3'$.
(2) (Amended χ) $\chi(Q)$ is even in all cases.

DEFINITION 2.3. We call an RH for G^+ satisfying both the original restrictions and the amended restrictions an *amended RH.* We say an amended RH is *realizable* for G if it there is a G-general position immersion of S having the given RH for G^+.

3. The realization problem: Standard and Klein bottle models

To realize a given amended RH, we need to give an explicit immersion for Q in $\mathbb{E}^3/G^+$ which intersects the rays of Y the right number of times and has the required extra symmetry. In addition, the representation of $\pi(Q - B)$ to G^+ induced by the immersion must be an epimorphism and, in the ON case, the representation must take $\pi^o(Q - B)$ to an index two subgroup of G^+ (satisfying the Epimorphism and Index Two Conditions).

REMARK 3.1. When there is no branching, the Epimorphism Condition requires that $\chi(Q) \le 0$ for $R^+(n)$ and $\chi(Q) \le -2$ for the other four types, see Remark 1.3. This could have been a fourth RH restriction. In particular, we will not repeat these exceptions to realizability for each of the different types.

3.1. Standard models. In most cases, as we have seen in [**13**], there is a *generic* or *standard model* which realizes a given restricted RH. These models depend on $\chi(Q)$ and on the amount of branching. The standard models for OO and NN are built by modification of the low-genus *basic* models for the OO case. The basic OO models for a restricted RH are of the following three kinds.

Basic Models for OO

(1) $\chi(Q) = 2$ (for sufficient branching): an embedded sphere intersecting the branching rays the required number of times;
(2) $\chi(Q) = 0$ (for $R^+(n)$ with no branching or $P^+(n), T^+, C^+, I^+$ with insufficient branching): an embedded torus around one ray, intersecting the branching ray the required number of times;
(3) $\chi(Q) = -2$ (for $P^+(n), T^+, C^+, I^+$ with no branching): an embedded surface of genus 2 with handles around two of the rays.

The *standard* models are obtained from these basic models by adding trivial handles for the OO case, or trivial crosscaps for the NN case to get the desired (smaller) $\chi(Q)$. By a *crosscap* we mean the Boy surface immersion of the projective plane with a disk removed. A handle or crosscap is *trivial* if it is contained in a ball in the complement of Y.

PROPOSITION 3.2. *(Use of standard models) All restricted RH for G^+ are realizable by a standard model except for:*

1) OO: insufficient branching with $\chi(Q) = 2$ (for all but $R^+(n)$);

2) NN: a) no branching with $\chi(Q) = 0$ for $R^+(n)$ and $\chi(Q) = -2$ for other types; b) insufficient branching with $\chi(Q) \geq 0$ (for all but $R^+(n)$).

Standard models cannot be used for ON since the addition of trivial crosscaps to Q forces S to be non-orientable.

PROOF. For OO, the standard models are explicitly constructed to realize a restricted RH, and there is no problem for sufficient branching. When there is no branching, the epimorphism condition implies $\chi(Q) \leq 0$ for $R^+(n)$ and $\chi(Q) \leq -2$ for the other types (see Remark 3.1), and all of these can be done with standard models as $\chi(Q)$ is even in OO. For insufficient branching (an option for all but $R^+(n)$), the standard model requires $\chi(Q) \leq 0$, so a standard model does not apply when $\chi(Q) = 2$. A special OO model (see below) can be used in this case.

For NN, standard models require adding trivial crosscaps to the basic models, which reduces $\chi(Q)$ from the values for OO. Thus for no branching, we have $\chi(Q) < 0$ for $R^+(n)$ and $\chi(Q) < -2$ for the other types, leaving $\chi(Q) = 0$ or $\chi(Q) = -2$ as exceptions. Similarly, for insufficient branching, we obtain $\chi(Q) < 0$ for standard models, leaving $\chi(Q) \geq 0$ as exceptions. □

If we want to use these models for orientation-reversing G, we want them also invariant under the reflection or antipodal map on the orbifold $\mathbb{E}^3/G^+$. The following Lemma shows this is always possible with the eight reflective cases:

LEMMA 3.3. *Each of the standard models can be arranged to have reflective symmetry. For NN, this means trivial crosscaps are added in pairs. In particular, for all nine types except RG, if the restricted RH for G^+ is realizable by a standard model, so is the amended RH for G^+.*

PROOF. For the fully planar types, this is an exercise in picturing Q (see for example [**13**]). For R2, clearly there is a torus with reflective symmetry running around the rotation axis. For P2,T2 with some branching, again this is an exercise. For P2, T2, with no branching and $\chi(Q) = -2$, the reflective symmetry forces the basic model's handles to wrap around the two axes not in the reflection plane, but the Epimorphism condition is still obeyed. □

The case RG where the orbifold $\mathbb{E}^3/G^+$ has antipodal symmetry is more complicated and is handled later.

3.2. Special models. For OO with $\chi(Q) = 2$ and insufficient branching, there is a *special spherical model.* The details are complicated by the nature of the generating rotations for G^+, see [**13**]. We will show later that for all 9 types, this model does not have the extra symmetry required of the orbifold $\mathbb{E}^3/G^+$.

For all ON cases and the NN exceptions given in Proposition 3.2, we use an immersed Klein bottle for $\chi(Q) = 0$ or two immersed Klein bottles joined by a tube for $\chi(Q) = -2$. In [**13**] the immersion is a twisted figure-eight model. It does not have, however, the reflective symmetry we need. Instead, we replace it by the usual immersion of a Klein bottle with a single double curve where the neck passes through the side of the bottle. This immersion has the needed reflective symmetry for types R2, P2, and T2, where it can be used to wrap around rays of Y not in the reflection plane. We call this a *reflective Klein bottle* model. We use it in the same situations as the figure-eight model in [**13**].

4. The theorems

We now determine which amended RH for G are realizable. By Lemma 3.3, there is no problem in any of the cases except RG if the restricted RH for G^+ is realizable by a standard model. Thus we must show for each type requiring nonstandard models that either nonstandard models cannot be used, or that they can be adapted to cover the cases where they are used in [**13**]. The case RG is treated separately.

Recall that we will not mention the cases of no branching with $\chi(Q) > 0$ for $R^+(n)$ and $\chi(Q) > -2$ for the other four types of G^+, which are never realizable.

4.1. Fully planar types. For the five fully planar types $R1, P1, T1, C, I$, we have that Y is contained in the reflection plane of $\mathbb{E}^3/G^+$.

THEOREM 4.1. *In the five fully planar types $R1$, $P1$, $Q1$, C, I, an amended RH is realizable in G-general position if and only if the corresponding restricted RH is realizable using a standard model. In particular, there is no ON case for these types.*

PROOF. We view the orbifold $\mathbb{E}^3/G^+$ as $\mathbb{R}^3$ and we let r be the reflection $r(x, y, z) = (x, y, -z)$. Let P be the xy-plane and H^+, H^- be the closed half-spaces $z \geq 0$ and $z \leq 0$, respectively. Suppose $g\colon Q \to \mathbb{R}^3$ is a general position immersion of the closed surface Q with respect to $Y \subset P$. Assume further that $r(g(Q)) = g(Q)$, that is, the immersion is invariant under the reflection r. Then the preimage $g^{-1}(g(Q) \cap P)$ cuts Q into two halves Q^+, Q^- with boundary curves $\partial Q^+ = \partial Q^-$, and the reflection r induces a reflection of Q interchanging Q^+, Q^- while fixing their boundary. The surface with boundary Q^+ can be viewed, by the classification of surfaces, as an orientable surface with possibly one or two crosscaps attached far away from the boundary, and whose images in H^+ therefore must be trivial.

Since Q^- is the reflection of Q^+, we conclude that when Q is non-orientable, we may view it as an orientable surface with 2 or 4 trivial crosscaps attached. It follows that no ON case is realizable. Moreover, the only realizable NN cases are those obtained by adding trivial crosscaps to an OO model. Lemma 3.3 says all standard models work. Thus the issue becomes which OO cases in [**13**] require a nonstandard model. By Proposition 3.2, there is only one: when $\chi(Q) = 2$ and there is insufficient branching. That case requires a special spherical model.

We shall show that no model where Q is the sphere with insufficient branching, including the special spherical model, has the symmetry required by the fully planar types. Assume that Q is a sphere and g is an immersion with $g(Q)$ meeting exactly one of the rays of Y. We view that ray as the x-axis for $x \geq 0$ and the other two rays of Y as the y-axis. We have $\chi(Q^+) = \chi(Q^-)$ and $\chi(Q) = 2 = \chi(Q^+) + \chi(Q^-)$ so for both halves $\chi(Q^+) = \chi(Q^-) = 1$. Thus Q^+ and Q^- are disks, so $g(\partial Q^+)$ is a closed curve in P not meeting the y-axis, and hence contained in the half-plane $x > 0$. Since $g(Q^+)$ is an immersed disk in H^+ not meeting the y-axis, there is an ambient isotopy of H^+ fixing the xy-plane P and moving $g(Q^+)$ to lie in the quarter space $x > 0, z \geq 0$. The same holds for Q^-. Thus we can assume that $g(Q)$ as a subset of $\mathbb{E}^3/G^+$ lies in the half-space $x > 0$. But then the Epimorphism Condition fails to hold, as no part of $g(Q)$ can wrap around the y-axis, namely the other two rays of Y. □

4.2. Types R2, P2, T2. We first treat the remaining cases that have reflective symmetry but are not fully planar: R2, P2, T2. The case R2 is straightforward.

THEOREM 4.2. *All amended RH for R2 are realizable, except NN for n even with $\chi(Q) = 0$ and no branching (ON requires n even by the Index Two Condition).*

PROOF. For OO, standard models work by Lemma 3.3. Thus standard models also work for NN if $\chi(Q) \leq -2$. For ON ($\chi(Q) \leq 0$ and even), or NN with no branching and $\chi(Q) = 0$, we use a reflective Klein bottle. For ON with $\chi(Q) \leq -2$, we attach trivial handles with the necessary reflective symmetry. The Orientability Restriction gives NN when n is odd and ON when n is even. In particular, as shown in [**13**], the restricted RH for NN with $\chi(Q) = 0$, no branching, and n even cannot be realized, as S will be orientable. □

The cases P2 and T2 are not fully planar. They both involve a reflection of $\mathbb{E}^3/G^+$ interchanging the two rays of Y with the same branching order.

THEOREM 4.3. *All amended RH for T2 are realizable with the following exceptions:*

1) OO: insufficient branching and $\chi(Q) = 2$;

2) NN: no exceptions.

There are no ON cases by the Orientability Restriction as A_4 has no index-two subgroup.

PROOF. By Proposition 3.2, the only exception for OO is insufficient branching and $\chi(Q) = 2$. In [**13**], it is shown for this case to be realizable, the branching must be 3-fold, but the Amended Branching Restriction does not allow this.

For NN, when there is sufficient branching, we use a standard model. For insufficient branching with $\chi(Q) = 0$, we use the same Klein bottle as in [**13**], but make it reflective. Note that the branching must be 2-fold by the Amended Branching Restriction. For $\chi(Q) < 0$ we just attach trivial handles with reflective symmetry. For no branching and $\chi(Q) = -2$, we again use the same kind of model as in [**13**] consisting of two Klein bottles, each running around one of the 3-fold rays, joined by a tube, and having the necessary reflective symmetry. Since A_4 has no index two subgroups, S must be non-orientable. When $\chi(Q) < -2$, we attach trivial handles with the necessary reflective symmetry. □

THEOREM 4.4. *An amended RH for P2 is realizable with the following exceptions:*

1) OO: insufficient branching and $\chi(Q) = 2$;

2) NN: insufficient branching and $\chi(Q) = 0$; no branching and $\chi(Q) = -2$;

3) ON: no exceptions.

PROOF. The analysis is very much that same as for T2 with the single 2-fold ray of Y replaced by the n-fold ray and the pair of 3-fold rays replaced by a pair of 2-fold rays. In particular, the OO exception for insufficient branching with $\chi(Q) = 2$ is still an exception. However, the change of parity from 3-fold to 2-fold on the paired rays affects NN and ON, as does the existence of an index two subgroup of the dihedral group D_n. In particular, it allows ON realizations and bans NN realizations that require Klein bottles.

We consider first the ON case. For insufficient branching with $\chi(Q) = 0$ and no branching with $\chi(Q) = -2$, we use the same Klein bottle models as we did for the NN case of T2. This time the Index Two condition is satisfied, so S is orientable. We then add trivial handles with the necessary reflective symmetry to get other values of $\chi(Q)$.

For NN, Lemma 3.3 states that standard models work in most cases. We have to show that the cases for insufficient branching with $\chi(Q) = 0$ and no branching with $\chi(Q) = -2$ cannot be realized. In both cases, we argue using the reflection plane P and half-spaces H^+, H^- as in Theorem 4.1. For $\chi(Q) = 0$, we have $\chi(Q^+) = 0$, so Q^+ has one boundary component ∂T^+, making Q^+ a projective plane with one disk removed, that is half of the reflective Klein bottle model. Although the immersion by g of Q into $\mathbb{E}^3/G^+ \cong \mathbb{R}^3$ is not the same as the reflective Klein bottle, this does not affect the algebraic topology of the representation of $\pi^o(Q - B)$ into $\pi(\mathbb{R}^3 - Y)$. Thus the Index Two condition is still satisfied forcing the ON case with S orientable.

For NN with no branching and $\chi(Q) = -2$, we have $\chi(Q^+) = -1$. Thus Q^+ is either a Klein bottle with a disk removed or a projective plane with two disks removed. For the former, just as for insufficient branching with $\chi(Q) = 0$, although the immersion by g is not the same as that for model of two Klein bottles joined by a tube, the algebraic topology of the representation is the same. Thus the Index Two condition is satisfied, forcing the ON case with S orientable. When Q^+ is a projective plane with two disks removed, we have a situation different from the model of two Klein bottles joined by a tube. In this case, both components of ∂Q^+ are taken to curves in the plane P not wrapping around Y, so both are null-homotopic in $\mathbb{E}^3/G^+ - Y$. Since these two boundary components generate $\pi^o(Q^+)$, we have that the image of $\pi(Q^+)$ in $\pi(H^+ - Y)$, which is torsion free, must be trivial. The same holds for the image of $\pi(Q^-)$. Thus the image of $\pi(Q)$ is at most infinite cyclic, so the Epimorphism Condition is not satisfied. □

4.3. Type RG. This case is somewhat different from all others since there is no reflection plane in the quotient $\mathbb{E}^3/G^+$. Instead, the extra symmetry is the antipodal map of $\mathbb{R}^3$, as explained in the proof of Lemma 2.2.

THEOREM 4.5. *Let the branching coefficient be b_n. An amended RH for RG is realizable with the following exceptions*

1) OO: $b_n + \chi(Q) \equiv 2 \mod 4$,

2) NN: $b_n \equiv 0 \mod 4$ with $\chi(Q) = 0$.

The ON case is not realizable.

PROOF. Throughout the proof, we let r be the antipodal mapping of $\mathbb{E}^3/G^+ \cong \mathbb{R}^3$ required by G; we let Q' be the quotient of Q induced by $\langle r \rangle$. For OO, we consider first the case that $b_n = 0$ and then induct on b_n. Suppose by way of contradiction that $\chi(Q) \equiv 2 \mod 4$. Then $\chi(Q') = \chi(Q)/2$ is odd. In particular $\pi(Q') = \langle a_i, b_i, c : c^2\Pi[a_i, b_i] = 1\rangle$, where c is represented by an orientation-reversing loop. Since $\pi(\mathbb{R}^3 - Y)$ is infinite cyclic, the quotient fundamental group $\pi((\mathbb{E}^3/G^+ - Y)/\langle r \rangle)$ is infinite cyclic as well. The single relation in $\pi(Q')$ thus implies that the image of c is trivial, giving a trivial crosscap on Q' which then lifts to a trivial crosscap on Q, contradicting the orientability of Q.

For $b_n \neq 0$, we just induct on b_n using the two branch points closest to the origin. That is, we remove from Q two small disks each containing one of the two opposite branch points and connect the boundaries by a tube. This cutting and pasting subtracts 2 from the Euler characteristic of Q and removes a pair of branch points, so $b_n + \chi(Q) \mod 4$ is unchanged. By induction, $b_n + \chi(Q) \equiv 0 \mod 4$.

Now, for $b_n + \chi(Q) \equiv 0 \mod 4$ we use the following models. When $b_n \equiv 2 \mod 4$ the basic model is an embedded sphere centered at the origin with antipodally symmetric "arms" reaching out to pairs of branch points on either side of the axis, and then trivial handles attached in antipodal pairs to get the standard model of higher genus. Instead, for $b_n \equiv 0 \mod 4$, we necessarily have $\chi(Q) \leq 0$. We then deviate slightly from our usual basic model to accomodate the antipodal symmetry, by starting with an embedded torus around the axis and having antipodal arms reach out to pairs of branch points on the axis.

For cases NN and ON, we first show there must be trivial crosscaps. We view the antipodal mapping r to be first reflection in a plane containing the axis of rotation, followed by a 180 rotation about a line perpendicular to this plane; if the axis of rotation is the z-axis, then the plane could be the xz plane and the perpendicular line would be the y-axis. The situation is now the same as R1, except that if Q^+ and Q^- are the two halves, we have to rotate Q^- by 180 degrees before identifying boundaries. This makes no difference, however, in the existence of a trivial crosscap on Q^+.

Since there can be no trivial crosscaps for ON, we conclude that the ON case is not realizable.

For NN with $b_n \equiv 0 \mod 4$, when there is branching, we take the "torus around axis" model for OO described above and attach antipodally symmetric trivial crosscaps to get all non-orientable Q with $\chi(Q) \leq -2$. For $b_n \equiv 2 \mod 4$, we use instead the OO sphere with antipodally symmetric trivial crosscaps for all Q with $\chi \leq 0$.

The only case left for NN is $\chi(Q) = 0$, namely the Klein bottle, with $b_n \equiv 0 \mod 4$. But we have already shown Q must have trivial crosscaps. The removal of those crosscaps gives the OO case with $\chi(Q) = 2$, which requires $b_n \equiv 2 \mod 4$. We conclude that $\chi(Q) = 0$ with $b_n \equiv 0 \mod 4$ is not realizable. □

5. Chirality

We say an amended RH for G is *chiral* if it has a realization for G^+ but none for G. For all cases but RG, this means by Lemma 3.3 that the G^+ realization in [**13**] is not achieved with a standard model.

Instead of organizing by the nine types, we organize by the cases OO, NN, ON. Each of these theorems is a summary of the theorems in Section 4 and all the results in [**13**] (since chirality first assumes the restricted RH for G^+ is realizable).

THEOREM 5.1. *An amended RH for OO is chiral if and only if the restricted RH for G^+ requires a special sphere model (i.e., $\chi(Q) = 2$ with insufficient branching for $G^+ \neq R^+(n)$) or $b_n + \chi(Q) \equiv 2 \mod 4$ for RG.*

THEOREM 5.2. *An amended RH for NN is chiral if and only if:*

1) fully planar types: $\chi(Q) = 0$ *for insufficient branching for all but* $R^+(n)$*, and* $\chi(Q) = -2$ *and no branching for all but* $R^+(n)$*;* $\chi(Q) = 0$ *with* $b_n = 0$ *for* $R^+(n)$ *(which requires n odd)*

2) R2, T2: none are chiral;

3) P2: $\chi(Q) = -2$ *with no branching (to be realizable for* G^+ *this requires n to be odd* [**13**]*);*

4) RG: $\chi(Q) = 0$ *with* $b_n \equiv 0 \mod 4$ *(this requires n odd when* $b_n = 0$*).*

THEOREM 5.3. *An amended RH for ON is chiral for all types except R2 and P2; these two types are never chiral.*

6. Questions for further study

Suppose we loosen the restriction of G-general position. In [**13**], we do that for the orientation-preserving case providing examples of various surfaces not in G^+-general position, but with singular points on a rotation axis. One could study the same issue here if one allows double curves of the immersion to lie in the reflection plane. A first attempt might be to require no singular points on rotation axes, so that S is still in G^+-general position.

A much larger project would be to look at a general orientable 3-manifold M and a finite group G acting on M. One can still define a G-general immersion of a surface S and ask which possible Riemann–Hurwitz equations for an action of G on S can be realized by a G-general immersion. An obvious place to start is embeddings of an orientable surface in $M \cong F \times I$, where F is an orientable closed surface.

Acknowledgments

The authors are very grateful for the careful reading and useful suggestions by the referees, especially the helpful requests to be more explicit about the geometry of $\mathbb{E}^3/G^+$.

References

[1] J. H. Conway, *The orbifold notation for surface groups*, Groups, combinatorics & geometry (Durham, 1990), London Math. Soc. Lecture Note Ser., vol. 165, Cambridge Univ. Press, Cambridge, 1992, pp. 438–447, DOI 10.1017/CBO9780511629259.038. MR1200280

[2] John H. Conway, Heidi Burgiel, and Chaim Goodman-Strauss, *The symmetries of things*, A K Peters, Ltd., Wellesley, MA, 2008. MR2410150

[3] William Cavendish and John H. Conway, *Symmetrically bordered surfaces*, Amer. Math. Monthly **117** (2010), no. 7, 571–580, DOI 10.4169/000298910X496705. MR2681518

[4] Antonio F. Costa, *Embeddable anticonformal automorphisms of Riemann surfaces*, Comment. Math. Helv. **72** (1997), no. 2, 203–215, DOI 10.1007/s000140050012. MR1470088

[5] Antonio F. Costa and Cam Van Quach Hongler, *Prime order automorphisms of Klein surfaces representable by rotations on the Euclidean space*, J. Knot Theory Ramifications **21** (2012), no. 4, 1250040, 9, DOI 10.1142/S0218216511009923. MR2890466

[6] Antonio Félix Costa and Ana Maria Porto, *Visualizing automorphisms of Riemann surfaces*, Atti Semin. Mat. Fis. Univ. Modena Reggio Emilia **58** (2011), 121–127 (2012). MR2987007

[7] H. S. M. Coxeter and W. O. J. Moser, *Generators and relations for discrete groups*, 4th ed., Ergebnisse der Mathematik und ihrer Grenzgebiete [Results in Mathematics and Related Areas], vol. 14, Springer-Verlag, Berlin-New York, 1980. MR562913

[8] D. Godfrey and D. Martinez (based on an idea of T. Pisanski), *Tucker's Group of Genus Two*, sculpture, Technical Museum of Slovenia, Bistra, Slovenia.

[9] R. L. Flurry, *Symmetry Groups: Theory and Chemical Applications*, Prentice–Hall, 1980.

[10] Fou Lai Lin, *Embeddable dihedral groups of Riemann surfaces*, Chinese J. Math. **7** (1979), no. 2, 133–152. MR574460

[11] Undine Leopold, *Vertex-Transitive Polyhedra in Three-Space*, ProQuest LLC, Ann Arbor, MI, 2014. Thesis (Ph.D.)–Northeastern University. MR3260050

[12] Undine Leopold, *Vertex-transitive polyhedra of higher genus, I*, Discrete Comput. Geom. **57** (2017), no. 1, 125–151, DOI 10.1007/s00454-016-9828-9. MR3589059

[13] Undine Leopold and Thomas W. Tucker, *Euclidean symmetry of closed surfaces immersed in 3-space*, Topology Appl. **202** (2016), 135–150, DOI 10.1016/j.topol.2015.12.068. MR3464156

[14] Cormac O'Sullivan and Anthony Weaver, *A Diophantine Frobenius problem related to Riemann surfaces*, Glasg. Math. J. **53** (2011), no. 3, 501–522, DOI 10.1017/S0017089511000097. MR2822795

[15] Reto A. Rüedy, *Symmetric embeddings of Riemann surfaces*, Discontinuous groups and Riemann surfaces (Proc. Conf., Univ. Maryland, College Park, Md., 1973), Princeton Univ. Press, Princeton, N.J., 1974, pp. 409–418. Ann. of Math. Studies, No. 79. MR0382630

[16] Thomas W. Tucker, *Two notes on maps and surface symmetry*, Rigidity and symmetry, Fields Inst. Commun., vol. 70, Springer, New York, 2014, pp. 345–355, DOI 10.1007/978-1-4939-0781-6_17. MR3329282

Mathematics Department, Northeastern University, Boston, Massachusetts 02115
Email address: undine.leopold@gmail.com

Mathematics Department, Colgate University (Emeritus), Hamilton, New York 13346
Email address: ttucker@colgate.edu

Contemporary Mathematics
Volume **764**, 2021
https://doi.org/10.1090/conm/764/15361

Realizations of the 120-cell

Peter McMullen

ABSTRACT. In previous papers, the realization spaces of all the classical regular polytopes have been described, except that of the 120-cell $\{5, 3, 3\}$. This omission is repaired here. It turns out that the description is long and complicated, and employs many of the techniques of the theory, including the notion of induced cosine vectors introduced here for the first time.

1. Introduction

Realizations provide geometric pictures of abstract regular polytopes, and thereby often help to elucidate their structures. Of the classical regular polytopes, which are those associated with the finite string Coxeter groups, only in the case of the 120-cell $\{5, 3, 3\}$ has the realization domain not been described so far. The others have been treated in the monograph [**10**] and papers [**5, 6**] (for cubes, the description is implicit rather than explicit); with the recent monograph [**8**] these are the general references to the background of this paper.

For several reasons, the realization domain of the 120-cell $\mathcal{H} := \{5, 3, 3\}$ is complicated. There are many different irreducible representations of the Coxeter group $[3, 3, 5]$ that give rise to pure realizations of $\mathcal{H}$; moreover, because $\mathcal{H}$ has asymmetric diagonal classes, some of the corresponding subdomains are necessarily higher-dimensional. (This is not the case with the other regular polytopes, where all diagonal classes are symmetric, and so each irreducible representation of the automorphism group yields at most one normalized pure realization up to congruence.) We shall outline the descriptions of the various subdomains that do arise, but for the most part we shall suppress many of the calculations that go into them. Fuller details will be found in [**8**, Section 7K] (some mistakes there that survived into print are corrected here).

For the convenience of the reader who may be unfamiliar with the topic, we give an outline of the salient features of realization theory in the next Section 2; proofs can be found in the previously cited material. Something that is new introduces a criterion for realizability in Section 3.

At several points we need to refer to the realization domain of the 600-cell $\mathcal{G} := \{3, 3, 5\}$, and so we briefly describe this in Section 4.

2010 *Mathematics Subject Classification.* Primary 51M20; Secondary 52B15.
Key words and phrases. Regular polytope, 120-cell, realization, cosine vector, quaternion, quotient.

In Section 5 we describe the geometric 120-cell $\{5,3,3\}$ in $\mathbb{E}^4$ and some of its symmetries; this gives us the layer vector, and tells us which layers are asymmetric.

The action of $[3,3,5]$ on various of its subgroups leads to corresponding identifications of the vertices of $\mathcal{H} := \{5,3,3\}$; these yield quotient domains $\mathcal{H}_k$ for $k = 2, 5, 8, 10, 24$. The last four of these quotients are treated in Sections 6 and 7.

The analysis then proceeds on a dimension-by-dimension basis, culminating in the listings of Tables 2 and 3. It will be seen that the techniques employed vary considerably; dimensions 25 and 36 seem to require particularly hard work, and it is here where we have skated over many details.

2. Realizations in general

The natural context for realizations is actually that of symmetric sets, namely, finite sets which admit permutation subgroups that act transitively. However, in this paper we shall confine our attention to regular polytopes, particularly since this connects more readily with the extant literature. The more general case is dealt with in [**8**, Chapter 3]. Here, we give a brief introduction to the theory, referring to the already cited literature – in particular [**6**, **8**] – for further details.

Let $\mathcal{P}$ be a (finite) regular polytope, with automorphism group $\boldsymbol{G} := \boldsymbol{G}(\mathcal{P})$, and let $\boldsymbol{H} = \boldsymbol{G}^{\boldsymbol{v}}$ be the group of its vertex-figure $\mathcal{P}^{\boldsymbol{v}}$, and thus the subgroup stabilizing an initial vertex $\boldsymbol{v}$, say. Following [**10**], we identify the vertex-set $\mathcal{V} = \mathcal{P}_0$ of $\mathcal{P}$ with the family $\boldsymbol{G}/\boldsymbol{H}$ of right cosets of $\boldsymbol{H}$, and then elements of $\mathcal{V}$ with those of $\boldsymbol{G}$. (We always think of $\mathcal{V}$ as ordered in some way, particularly since subsets of its points may fall together in a realization.) A *diagonal* is a pair $\{\boldsymbol{x},\boldsymbol{y}\}$ with $\boldsymbol{x},\boldsymbol{y} \in \mathcal{V}$; the action of $\boldsymbol{G}$ partitions the diagonals into *diagonal classes* $\mathcal{D}_0,\ldots,\mathcal{D}_r$, say, with $\mathcal{D}_0$ by convention consisting of the trivial diagonals $\{\boldsymbol{x},\boldsymbol{x}\}$. If $\boldsymbol{v}$ is fixed as before, then $\mathcal{V}$ falls into *layers* $\mathcal{L}_0,\ldots,\mathcal{L}_r$, with

$$\mathcal{L}_s := \{\boldsymbol{x} \in \mathcal{V} \mid \{\boldsymbol{v},\boldsymbol{x}\} \in \mathcal{D}_s\}$$

for $s = 0,\ldots,r$. A diagonal $\{\boldsymbol{x},\boldsymbol{y}\}$ is *symmetric* if the ordered pairs $(\boldsymbol{x},\boldsymbol{y})$ and $(\boldsymbol{y},\boldsymbol{x})$ are equivalent under $\boldsymbol{G}$, otherwise it is *asymmetric*; the corresponding layers are similarly called symmetric or asymmetric. If $\ell_s := \operatorname{card}\mathcal{L}_s$ for $s = 0,\ldots,r$, then $\Lambda = \Lambda(\mathcal{P}) := (\ell_0,\ldots,\ell_r)$ is the *layer vector* of $\mathcal{P}$; by convention, therefore, $\ell_0 = 1$. We shall always write $n = \ell_0 + \cdots + \ell_r$ for the total number of vertices of $\mathcal{P}$.

More generally, suppose that $\boldsymbol{A},\boldsymbol{B} < \boldsymbol{G}$. Then $\boldsymbol{B}$ will partition the elements of $\boldsymbol{G}/\boldsymbol{A}$ into $\boldsymbol{B}$*-cuts*, which are double cosets $\boldsymbol{A}\boldsymbol{x}\boldsymbol{B}$. We then have

THEOREM 2.1. *There is a one-to-one correspondence between the* $\boldsymbol{B}$*-cuts of* $\boldsymbol{G}/\boldsymbol{A}$ *and the* $\boldsymbol{A}$*-cuts of* $\boldsymbol{G}/\boldsymbol{B}$.

PROOF. This is obvious: the correspondence is given by $\boldsymbol{A}\boldsymbol{x}\boldsymbol{B} \leftrightarrow \boldsymbol{B}\boldsymbol{x}^{-1}\boldsymbol{A}$. $\square$

We call Theorem 2.1 the *Apollonius theorem*; it is the abstract version of an observation first made by Apollonius of Perga. (This observation, about the icosahedron and dodecahedron, is that they can be inscribed in the same sphere so that pairs of their opposite faces lie in the same planes. Alternatively, if they have the same circumsphere, then they have the same insphere. See, for instance, [**1**, p.30] for more background.) We shall say more about this later, particularly in the geometric application of Theorem 3.3.

A *realization* of $\mathcal{P}$ is obtained from an orthogonal representation $\mathbf{G}$ of $\boldsymbol{G}$ in some euclidean space E, together with a choice of an initial vertex v in its *Wythoff space*

$$W := \{x \in E \mid x\mathbf{g} = x \text{ for all } \mathbf{g} \in \mathbf{H}\}, \tag{2.2}$$

with $\mathbf{H} < \mathbf{G}$ the subgroup corresponding to $\boldsymbol{H}$. The resulting vertex-set of the realized polytope P is the image set $V := v\mathbf{G}$ of $v \in W$; we do not regard congruent realizations as different. The realization is *trivial* if $W = \{o\}$; clearly, we are only interested in non-trivial realizations. Faces of $\mathcal{P}$ are identified with their vertices as subsets of $\mathcal{V}$, and P is to be thought of as the partially ordered set of the images of these subsets in the realization. We call the realization *faithful* if P is isomorphic to $\mathcal{P}$ as a poset; observe that non-faithful realizations must be included in the classification, so that V will result from identifications of subsets in $\mathcal{V}$. We shall also use $\mathcal{P}$ to denote the family of its realizations, and so write $\mathsf{P} \in \mathcal{P}$ to mean that P is a realization of $\mathcal{P}$; hence, we can later talk about the geometry of $\mathcal{P}$.

REMARK 2.3. The realization can be thought of as a geometric picture of the action of $\boldsymbol{G}$ on the family $\boldsymbol{G}/\boldsymbol{H}$ of right cosets $\boldsymbol{H}$ in $\boldsymbol{G}$. In this sense, a non-faithful realization comes from the analogous action of $\boldsymbol{G}$ on $\boldsymbol{G}/\boldsymbol{K}$, with $\boldsymbol{H} < \boldsymbol{K} < \boldsymbol{G}$. What $\boldsymbol{K}$ does is identify certain subsets of $\mathcal{V}$, so yielding what we call a *quotient* of $\boldsymbol{G}/\boldsymbol{H}$, but while $\boldsymbol{G}/\boldsymbol{K}$ is still a poset under the induced structure of $\mathcal{P}$, it is important to note that it no longer need have the structure of a polytope itself.

If $x, y \in V$, then the inner product $\langle x, y\rangle$ depends only on the diagonal class to which x, y belong. If this is $\mathcal{D}_s$ (we use the same notation as in the abstract case), then we write $\sigma_s := \langle x, y\rangle$, and call $\Sigma = \Sigma(\mathsf{P}) := (\sigma_0, \ldots, \sigma_r)$ the *inner product vector* of the realization. Note that, if $u, v \in W$ are initial points and $\lambda, \mu \in \mathbb{R}$, then the inner product vector with $\lambda u + \mu v \in W$ as initial point is a quadratic in λ, μ. If $\mathsf{P}(u), \mathsf{P}(v)$ are the corresponding polytopes, then this defines their *linear combination* $\lambda\mathsf{P}(u) + \mu\mathsf{P}(v)$; of course,

$$(\lambda u + \mu v)\mathbf{g} = \lambda(u\mathbf{g}) + \mu(v\mathbf{g})$$

for each $\mathbf{g} \in \mathbf{G}$.

There are further ways of combining realizations. If $\mathbf{G}_j$ is a representation of $\boldsymbol{G}$ in E_j, with Wythoff space W_j and initial vertex $v_j \in W_j$ of the resulting realization P_j for $j = 1, 2$, then the *blend* $\mathsf{P}_1 \# \mathsf{P}_2$ and (*tensor*) *product* $\mathsf{P}_1 \otimes \mathsf{P}_2$ have vertices

$$\begin{aligned} &(v_1\mathbf{g}_1, v_2\mathbf{g}_2) \in E_1 \oplus E_2, \\ &v_1\mathbf{g}_1 \otimes v_2\mathbf{g}_2 \in E_1 \otimes E_2, \end{aligned} \tag{2.4}$$

respectively, with $\mathbf{g}_j \in \mathbf{G}_j$ corresponding under the representations.

The *product* ab of two vectors $a = (\alpha_1, \ldots, \alpha_k)$ and $b = (\beta_1, \ldots, \beta_k)$ is defined by $ab := (\alpha_1\beta_1, \ldots, \alpha_1\beta_1)$; that is, the vector of term-by-term products of the entries. We then have

PROPOSITION 2.5. *The inner product vectors of the blend and product are given by*

$$\begin{aligned} \Sigma(\mathsf{P}_1 \# \mathsf{P}_2) &= \Sigma(\mathsf{P}_1) + \Sigma(\mathsf{P}_2), \\ \Sigma(\mathsf{P}_1 \otimes \mathsf{P}_2) &= \Sigma(\mathsf{P}_1)\Sigma(\mathsf{P}_2). \end{aligned}$$

REMARK 2.6. Note that, up to congruence, blends and products are associative and commutative, and that the product distributes over the blend.

Observe that (up to congruence) we have *trivial* blends, of the form $\lambda\mathsf{P}\,\#\,\mu\mathsf{P} = \nu\mathsf{P}$, with $\nu^2 = \lambda^2+\mu^2$. A *pure* realization is one that cannot be expressed as a blend in a non-trivial way; the corresponding representation of $\boldsymbol{G}$ will be irreducible. The general aim is thus to classify the pure realizations of a given regular polytope.

It is more usual to work with the normalized inner product vector $\varGamma(\mathsf{P})$ or *cosine vector*, which is that obtained when the initial vertex $v \in W$ is a unit vector: $\|v\| = 1$. Thus $\varGamma(\mathsf{P}) = \sigma_0^{-1}\varSigma(\mathsf{P}) =: (\gamma_0, \dots, \gamma_r)$, say. If $\varGamma = \varGamma(\mathsf{P})$ is the cosine vector of P, then we write $d(\varGamma) := \dim \mathsf{P}$; this is always taken to be the *linear* dimension of the vertex-set.

Finally, we introduce the positive definite $\varLambda$-*inner product* by

$$\langle x, y\rangle_{\varLambda} := \frac{1}{n}\langle \varLambda, xy\rangle, \tag{2.7}$$

where $x = (\xi_0, \dots, \xi_r)$, $y = (\eta_0, \dots, \eta_r)$. Thus $\langle\cdot,\cdot\rangle_{\varLambda}$ is just the ordinary inner product with respect to the diagonal matrix $n^{-1}\operatorname{diag}(\ell_0, \dots, \ell_r)$. There is a corresponding norm, given by $\|x\|_{\varLambda}^2 := \langle x, x\rangle_{\varLambda}$. In addition, we naturally say that two vectors x, y are $\varLambda$-*orthogonal* if $\langle x, y\rangle_{\varLambda} = 0$.

REMARK 2.8. The definition of the $\varLambda$-inner product shows that, if $\varGamma_1, \varGamma_2, \varGamma_3$ are three cosine vectors, then

$$\langle \varGamma_1\varGamma_2, \varGamma_3\rangle_{\varLambda} = \langle \varGamma_1, \varGamma_2\varGamma_3\rangle_{\varLambda}.$$

We shall often use this fact without further comment.

We are now ready for the key properties of cosine vectors. First, the *simplex* realization T of $\mathcal{P}$ has as vertex-set the usual orthonormal basis $(e_1, \dots, e_n)$ of E^n; observe that $\varGamma(\mathsf{T}) = (1, 0^r)$, with (here and elsewhere) α^r standing for a string $\alpha, \dots, \alpha$ of length r. At the other extreme, we have the *henogon* realization $\{1\}$, with all vertices coinciding at the point $1 \in \mathbb{R}$; its cosine vector is $\varGamma_0 = (1^{r+1})$. We then have the *component equation* [**5**, Theorem 5.3] or [**6**, Theorem 4.10]:

THEOREM 2.9. *If the simplex realization* T *of* $\mathcal{P}$ *is expressed as a blend* $\mathsf{T} = \mathsf{P}_1 \# \cdots \# \mathsf{P}_s$ *of components* P_j *in mutually orthogonal subspaces with dimensions* $d_j := \dim \mathsf{P}_j$ *and cosine vectors* $\varGamma_j := \varGamma(\mathsf{P}_j)$ *for* $j = 1, \dots, s$, *then*

$$\sum_{i=1}^{s} d_i\varGamma_i = n\varGamma(\mathsf{T}) = n(1, 0^r).$$

If $\mathcal{P}$ is (abstractly) centrally symmetric, then T decomposes into the *staurotope* (cross-polytope) realization X and the *small simplex* realization S in which diametrically opposite vertices of $\mathcal{P}$ are identified to give $\mathcal{P}/2$. There are then corresponding analogues of Theorem 2.9 that treat these two cases.

REMARK 2.10. It often happens that $\mathcal{P}/2$ is an abstract regular polytope in its own right. In particular, this is the case for the 120-cell $\{5, 3, 3\}$.

We next note that $\langle \varGamma_0, \varGamma(\mathsf{P})\rangle_{\varLambda} \geq 0$ is the square of the distance from o to the centre of P; it comes from the scaling of the component $\{1\}$ of P with the complementary component centred. Applying this to $\mathsf{P}^2 = \mathsf{P} \otimes \mathsf{P}$ yields the *dimension inequality* [**6**, Corollary 4.9]:

LEMMA 2.11. *If* $\varGamma$ *is the cosine vector of a realization of* $\mathcal{P}$, *then*

$$d(\varGamma)\|\varGamma\|_{\varLambda}^2 \geq 1.$$

The theorem and lemma then give us *Λ-orthogonality* [**6**, Theorem 4.5]:

THEOREM 2.12. *With the assumptions of Theorem* 2.9,

$$\langle \Gamma_i, \Gamma_j \rangle_\Lambda = \frac{\delta_{ij}}{d_j}$$

for all $i, j = 1, \ldots, s$.

Here, δ_{ij} is the familiar Kronecker delta function.

REMARK 2.13. If Γ is pure with $d(\Gamma) = d$, it follows from this discussion that Γ^2 has a component $\frac{1}{d}\Gamma_0$; this can be a useful starting point. Moreover, if Γ_k is pure and Γ is an arbitrary cosine vector, then the coefficient of Γ_k in the expression of Γ as a convex combination of pure cosine vectors is $d(\Gamma_k)\langle \Gamma, \Gamma_k \rangle_\Lambda$.

We now look more closely at the Wythoff space W of an irreducible representation $\mathbf{G}$ since, if $w := \dim W > 1$, then we have added complications. Here, we shall confine our attention to the case when $\mathbf{G}$ is of real (that is, absolutely irreducible) type, which (as will emerge from our analysis) turns out to be all that we need. If $\mathbf{G}$ is of complex or quaternionic type, then the descriptions in the earlier literature need the corrections due to Ladisch [**3**]. We write $\mathcal{N} = \mathcal{N}_{\mathbf{G}}$ for the domain of normalized realizations with symmetry group $\mathbf{G}$, which we can identify with their cosine vectors. Then $\mathcal{N}$ is a compact convex subset of $\mathbb{R}^{r+1}$, lying in the hyperplane of points with first coordinate 1. Throughout, the degree of $\mathbf{G}$ will be denoted d.

If we pick an orthonormal basis $(e_1, \ldots, e_w)$ of W, then the general cosine vector of the realization $\mathsf{P}(x)$, say, with initial vertex the unit vector $x = \xi_1 e_1 + \cdots + \xi_w e_w$, is of the form

$$\Gamma(x) = \sum_{j,k} \xi_j \xi_k \Gamma_{jk}. \tag{2.14}$$

The $\Gamma_{jj} = \Gamma(e_j)$ are genuine cosine vectors; we call the $\Gamma_{jk} = \Gamma_{kj}$ for $j \neq k$ *mixed* cosine vectors. Then we have [**6**, Theorem 6.6]:

THEOREM 2.15. *The Γ_{jk} of* (2.14) *satisfy*

(a) *distinct Γ_{jk} are Λ-orthogonal;*
(b) *for each j, k,*

$$\|\Gamma_{jk}\|^2_\Lambda = \begin{cases} \frac{1}{d}, & \text{if } j = k, \\ \frac{1}{2d}, & \text{if } j \neq k. \end{cases}$$

Consequently, the Γ_{jk} comprise a Λ-orthogonal basis of the subspace spanned by the subdomain $\mathcal{N}_{\mathbf{G}}$.

Furthermore, the sum $\Gamma_{11} + \cdots + \Gamma_{ww}$ is independent of the choice of the orthonormal basis $(e_1, \ldots, e_w)$ of W.

REMARK 2.16. Calling the terms Γ_{jk} for $j \neq k$ mixed cosine vectors is a slight misnomer, since such a Γ_{jk} is not an actual cosine vector. Observe that we can attach a *notional* dimension $2d$ to Γ_{jk}. What this means is that the coefficient of Γ_{jk} in the expression of a general cosine vector Γ as a linear combination of genuine and mixed cosine vectors is $2d\langle \Gamma, \Gamma_{jk} \rangle_\Lambda$.

REMARK 2.17. We see from Theorem 2.15 that $\mathcal{N}_{\mathbf{G}}$ contributes wd to the total dimension n of the simplex realization T, and $\frac{1}{2}d(d+1)$ to that of the $r+1$ diagonals.

3. Induced cosine vectors

Inscription of one regular polytope $\mathcal{Q}$ with full symmetry in another polytope $\mathcal{P}$ (that is, $\operatorname{vert} \mathcal{Q} \subset \operatorname{vert} \mathcal{P}$) results in a general cosine vector of $\mathcal{Q}$ having entries (possibly repeated) that are a subset of those of $\mathcal{P}$. This is most relevant for cuts, as introduced in Theorem 2.1, of which special instances are vertex-figures and facets. One general result is [**6**, Theorem 8.1]:

THEOREM 3.1. *If the cut $\mathcal{Q}$ of $\mathcal{P}$ with m vertices lies in layer $\mathcal{L}_s$ of $\mathcal{P}$ from the initial vertex, then in any realization $\mathsf{P} \in \mathcal{P}$ the induced cosine vector $\Gamma(\mathsf{Q})$ of the realization $\mathsf{Q} \in \mathcal{Q}$ in P must satisfy*

$$\langle \Gamma_0(\mathsf{Q}), \Gamma(\mathsf{Q}) \rangle_{\Lambda(\mathcal{Q})} = \frac{1}{m} \langle \Lambda(\mathcal{Q}), \Gamma(\mathsf{Q}) \rangle \geq \gamma_s(P)^2.$$

Moreover, if $\mathcal{Q}$ comprises the whole of $\mathcal{L}_s$ and $w(P) = 1$, then equality holds.

For the vertex-figure, $s = 1$; we refer to the theorem in this case as the *vertex-figure criterion.* In fact, Theorem 3.1 applies equally to *strata*, the equivalents of layers under the action of the subgroup $\boldsymbol{G}(\mathcal{F}) < \boldsymbol{G} = \boldsymbol{G}(\mathcal{P})$, where $\mathcal{F}$ is the facet of $\mathcal{P}$. If $\mathcal{F}$ is realized as F, then we write

$$\eta_{\boldsymbol{f}} = \eta_{\boldsymbol{f}}(\mathsf{P}) := \frac{1}{m} \langle \Lambda(\mathcal{F}), \Gamma(\mathsf{F}) \rangle = \langle \Gamma_0(\mathsf{F}), \Gamma(\mathsf{F}) \rangle_{\Lambda(\mathcal{F})}, \tag{3.2}$$

where $\Gamma_0(\mathsf{F})$ is just Γ_0 restricted to $\mathcal{F}$. Then what we shall call the *Apollonius criterion* (which is the exact generalization of the original observation of Apollonius) is

THEOREM 3.3. *If $\mathsf{P} \in \mathcal{N}$ is a realization of $\mathcal{P}$ such that $\eta_{\boldsymbol{f}}(\mathsf{P}) > 0$, then there is a realization $\mathsf{Q} \in \mathcal{N}(\mathcal{P}^{\delta})$ of a geometric dual in the same space. Moreover, if $w(\mathsf{P}) = 1$, then Q can be chosen so that*

$$\eta_{\boldsymbol{f}}(\mathsf{Q}) = \eta_{\boldsymbol{f}}(\mathsf{P}).$$

A useful consequence of Theorem 3.3 is the *centred facet criterion*:

COROLLARY 3.4. *If P is a pure realization of $\mathcal{P}$, such that there is no pure realization of the dual $\mathcal{P}^{\delta}$ of dimension* $\dim \mathsf{P}$, *then the strata of P are centred.*

REMARK 3.5. Theorems 2.1 and 3.3 generalize: there is a one-to-one correspondence between strata of a regular polytope $\mathcal{P}$ and those of its dual $\mathcal{P}^{\delta}$, with appropriate geometric implications.

4. The 600-cell

Since we shall appeal to the result later, we first describe the realizations of the 600-cell. We repeat the result from [**5**, Section 9]; see also [**6**, Section 13] or [**8**, Theorem 7E10]. We have

PROPOSITION 4.1. *The matrix of cosine vectors of the* 600-*cell* $\mathcal{G} = \{3, 3, 5\}$ *is*

$$\begin{bmatrix} 1 & 1 & 1 & 1 & 1 & 1 & 1 & 1 & 1 \\ 1 & \frac{1}{3}\tau & 0 & -\frac{1}{3}\tau^{-1} & -\frac{1}{3} & -\frac{1}{3}\tau^{-1} & 0 & \frac{1}{3}\tau & 1 \\ 1 & -\frac{1}{3}\tau^{-1} & 0 & \frac{1}{3}\tau & -\frac{1}{3} & \frac{1}{3}\tau & 0 & -\frac{1}{3}\tau^{-1} & 1 \\ 1 & -\frac{1}{4} & \frac{1}{4} & -\frac{1}{4} & 0 & -\frac{1}{4} & \frac{1}{4} & -\frac{1}{4} & 1 \\ 1 & 0 & -\frac{1}{5} & 0 & \frac{1}{5} & 0 & -\frac{1}{5} & 0 & 1 \\ 1 & \frac{1}{2}\tau & \frac{1}{2} & \frac{1}{2}\tau^{-1} & 0 & -\frac{1}{2}\tau^{-1} & -\frac{1}{2} & -\frac{1}{2}\tau & -1 \\ 1 & -\frac{1}{2}\tau^{-1} & \frac{1}{2} & -\frac{1}{2}\tau & 0 & \frac{1}{2}\tau & -\frac{1}{2} & \frac{1}{2}\tau^{-1} & -1 \\ 1 & \frac{1}{4} & -\frac{1}{4} & -\frac{1}{4} & 0 & \frac{1}{4} & \frac{1}{4} & -\frac{1}{4} & -1 \\ 1 & -\frac{1}{6} & 0 & \frac{1}{6} & 0 & -\frac{1}{6} & 0 & \frac{1}{6} & -1 \end{bmatrix}$$

with layer and dimension vectors

$$\Lambda = (1, 12, 20, 12, 30, 12, 20, 12, 1), \qquad D = (1, 9, 9, 16, 25, 4, 4, 16, 36).$$

The *dimension vector* just lists the dimensions of the corresponding realizations. Here and throughout, $\tau = \frac{1}{2}(\sqrt{5} + 1)$ is the golden number, so that $\tau^{-1} = \frac{1}{2}(\sqrt{5} - 1)$.

Calling these cosine vectors $\Gamma_0, \dots, \Gamma_8$ in order, we note the following calculations; bear in mind here that Γ_5 is the cosine vector of the usual 600-cell $\{3, 3, 5\}$, while Γ_6 is that of the star-polytope $\{3, 3, \frac{5}{2}\}$:

$$\begin{aligned} 3\Gamma_1 &= 4{\Gamma_5}^2 - \Gamma_0, \\ 3\Gamma_2 &= 4{\Gamma_6}^2 - \Gamma_0, \\ \Gamma_3 &= \Gamma_5\Gamma_6, \\ \Gamma_8 &= \Gamma_1\Gamma_6 = \Gamma_2\Gamma_5. \end{aligned} \tag{4.2}$$

These are the key relations from which the whole realization domain can be found.

At this stage, we observe that, since $\eta_{\boldsymbol{f}} = \frac{1}{4}(1 + 3\gamma_1) > 0$ for each pure realization of $\{3, 3, 5\}$, the Apollonius criterion Theorem 3.3 implies that its dual $\mathcal{H} = \{5, 3, 3\}$ has pure realizations of dimensions $9, 9, 16, 25, 4, 4, 16, 36$, though this gives us little guidance about their construction. Of course, the two 4-dimensional realizations are just $\{5, 3, 3\}$ and $\{\frac{5}{2}, 3, 3\}$. However, to a certain extent we can parallel the constructions of the realizations of $\mathcal{G}$ to find corresponding realizations of $\mathcal{H}$ of the same dimensions. Thus it is clear that the 9-dimensional realizations are the non-trivial components of $\{5, 3, 3\} \otimes \{5, 3, 3\}$ and $\{\frac{5}{2}, 3, 3\} \otimes \{\frac{5}{2}, 3, 3\}$, while among the 16-dimensional realizations is $\{5, 3, 3\} \otimes \{\frac{5}{2}, 3, 3\}$. We can similarly identify some 36-dimensional realizations; see Section 15.

5. Coordinates and layer vector

We begin by finding the layer vector Λ of $\{5, 3, 3\}$, bearing in mind that we have the faithful realization $\{5, 3, 3\}$ in $\mathbb{E}^4$. For this, we must list the vertices, and establish which diagonal classes are asymmetric. Since $\{5, 3, 3\}$ has many layers, we compromise by writing down only the first half of Λ.

For the first result, Coxeter [**1**, Table V(v)] has done all the hard work. We write ‡ for the automorphism of the ring $\mathbb{Z}[\tau]$ that changes the sign of $\sqrt{5}$, so that $\tau^\ddagger = -\tau^{-1}$. We shall frequently employ this automorphism in what follows;

moreover, we use these shorthands (even for τ^{-1}) to save space, particularly in the final Tables 2 and 3.

PROPOSITION 5.1. *The vertex-set* H *of* $\{5,3,3\}$ *consists of all permutations with an even number of changes of sign of the coordinate vectors*

$$\begin{gathered}(4,0,0,0),\ (2,2,2,\pm 2),\ (2\tau,2,2\tau^{\ddagger},0),\\ (\sigma,\tau^{\ddagger},\tau^{\ddagger},\tau^{\ddagger}),\ (-\sigma^{\ddagger},\tau,\tau,\tau),\\ (-1,\sqrt{5},\sqrt{5},\sqrt{5}),\ (3,\sqrt{5},1,1),\\ (\varphi,-\tau,\tau^{-2},\tau^{-2}),\ (\varphi^{\ddagger},\tau^{\ddagger},\tau^{2},\tau^{2}),\end{gathered}$$

where $\varphi := \tau\sqrt{5}$ *and* $\sigma := \frac{1}{2}(3\sqrt{5}+1)$*, so that* $\varphi^{\ddagger} = \tau^{-1}\sqrt{5}$ *and* $\sigma^{\ddagger} := \frac{1}{2}(1-3\sqrt{5}) = -11\sigma^{-1}$.

We recognize the vertices in the first line as including those of an inscribed copy of the 600-cell $\{3,3,5\}$; in turn, among these are the vertices of a 24-cell $\{3,4,3\}$.

REMARK 5.2. Our coordinates are the same as Coxeter's, except that we have an *odd* number of changes of sign from his, to accord with our use of quaternions below which follows [**2**, Chapter 3]. When using quaternions, we often employ (mostly without comment) the unit vectors $\frac{1}{4}\mathsf{H}$; it will be clear from the context when we do this.

To determine the diagonals classes, and which are symmetric, we can appeal to

PROPOSITION 5.3. *The following matrices, scaled by* $\frac{1}{4}$*, are involutory symmetries of the copy of* $\{5,3,3\}$ *with the vertices of Proposition* 5.1*:*

$$\begin{bmatrix}-1 & \sqrt{5} & \sqrt{5} & \sqrt{5}\\ \sqrt{5} & 3 & -1 & -1\\ \sqrt{5} & -1 & 3 & -1\\ \sqrt{5} & -1 & -1 & 3\end{bmatrix},\quad \begin{bmatrix}\sigma & \tau^{\ddagger} & \tau^{\ddagger} & \tau^{\ddagger}\\ \tau^{\ddagger} & \varphi^{\ddagger} & -\tau^{2} & -\tau^{2}\\ \tau^{\ddagger} & -\tau^{2} & \varphi^{\ddagger} & -\tau^{2}\\ \tau^{\ddagger} & -\tau^{2} & -\tau^{2} & \varphi^{\ddagger}\end{bmatrix},\quad \begin{bmatrix}-\sigma^{\ddagger} & \tau & \tau & \tau\\ \tau & -\varphi & \tau^{-2} & \tau^{-2}\\ \tau & \tau^{-2} & -\varphi & \tau^{-2}\\ \tau & \tau^{-2} & \tau^{-2} & -\varphi\end{bmatrix}.$$

If we permute the rows of a matrix in some way, then for its inverse we must permute the columns in the same way. So, for instance, if we take $(4,0,0,0)$ to be the initial vertex of $\{5,3,3\}$, then, switching the first and second rows in the first matrix, we obtain the vertices $(\sqrt{5},3,-1,-1)$ and $(\sqrt{5},-1,\sqrt{5},\sqrt{5})$ in the same layer, but inequivalent under the symmetries of the vertex-figure (all permutations and an even number of changes of sign in coordinates $2,3,4$). In other words, this shows that the $12+12$ vertices in that layer ($\mathcal{L}_7$, as it happens) form a single asymmetric diagonal class with $(4,0,0,0)$.

The diagonal classes to $(2,2,2,\pm 2)$, $(-2,2,2,\pm 2)$ and $(0,2\tau,2\tau^{-1},\pm 2)$ are similarly asymmetric, but this is more easily shown using quaternions; see also Section 12.

We can summarize this discussion in

PROPOSITION 5.4. *The layer vector of* $\{5,3,3\}$ *is*

$$\Lambda = (1,4,12,24,12;4,24,24^{*},8^{*},24;24^{*},12,24,4,24;24^{*},24,6,48^{*},\ldots),$$

where $6,48^{*}$ *are the middle terms, the rest repeats the first half in reverse order, and* * *denotes an asymmetric diagonal class. Thus* $\{5,3,3\}$ *has* 36 *diagonal classes, including the trivial one, of which* 9 *are asymmetric.*

REMARK 5.5. So that it is easier to use Coxeter's Table V(v), we write a general cosine vector of $\mathcal{H}$ as

$$\Gamma = (\gamma_0, \ldots, \gamma_7, \gamma_{8a}, \gamma_{8b}; \gamma_9, \gamma_{10}, \gamma_{11}, \gamma_{12a}, \gamma_{12b}; \gamma_{13}, \gamma_{14}, \gamma_{15a}, \gamma_{15b}, \ldots),$$

so splitting (for example) layer 8 rather than introducing labels that are less easy to remember. Further, we have put semicolons rather than commas to split the entries into blocks of five, to help to identify which entries are which. To be specific, we have

$$\begin{aligned}
\mathcal{L}_{8a} &= \{(2,2,2,\pm 2), \ldots\}, \\
\mathcal{L}_{12a} &= \{(1,\sqrt{5},\sqrt{5},-\sqrt{5}), \ldots\}, \\
\mathcal{L}_{15a} &= \{(0,4,0,0), \ldots\}, \\
\mathcal{L}_{18a} &= \{(-1,\sqrt{5},\sqrt{5},\sqrt{5}), \ldots\}, \\
\mathcal{L}_{22a} &= \{(-2,2,2,\pm 2), \ldots\},
\end{aligned}$$

where the remaining vertices in each set are those obtained by applying the symmetries of the vertex-figure, and the layers $\mathcal{L}_{8b}$ and so on consist of the other vertices with the same first coordinate (see Proposition 5.1).

REMARK 5.6. Regarding $\mathcal{H}_2 = \{5,3,3\}/2$ as a regular polytope in its own right, its layer vector is now

$$\Lambda = (1,4,12,24,12;4,24,24^*,8^*,24;24^*,12,24,4,24;24^*,24,3,24);$$

observe that $\mathcal{L}_{15b}$ is no longer asymmetric.

Let us note here that, for $\mathcal{H}$, the centred facet criterion of Corollary 3.4 is

$$1 + 3\gamma_1 + 6\gamma_2 + 6\gamma_3 + 3\gamma_4 + \gamma_5 = 0. \tag{5.7}$$

An *allomorphism* $\boldsymbol{\lambda}$ of a regular polytope $\mathcal{P}$ is induced by an automorphism of its group $\boldsymbol{G}$ that preserves the stabilizer $\boldsymbol{H}$ of the initial vertex $\boldsymbol{v}$; we write $\mathcal{P}^{\boldsymbol{\lambda}}$ for the resulting *allomorph.* In terms of realizations, an allomorph of P will be an isomorphic copy with the same vertex-set V and symmetry group $\mathbf{G}$. For $\mathcal{H} := \{5,3,3\}$, the allomorphism $\boldsymbol{\lambda}$ corresponding to the interchange between $\{5,3,3\}$ and its starry isomorphic copy $\{\frac{5}{2},3,3\}$ is straightforward. What we observe is that the vertex-figure of $\mathcal{H}^{\boldsymbol{\lambda}}$ at the opposite vertex in $\mathcal{L}_{30}$ to the initial one $\boldsymbol{v} \in \mathcal{L}_0$ of $\mathcal{H}$ forms layer $\mathcal{L}_5$. In turn, a typical such vertex is diametrically opposite $\boldsymbol{v}$ in the initial facet $\mathcal{F} = \{5,3\}$ of $\mathcal{H}$, and so is obtained from $\boldsymbol{v}$ by the central inversion $(\boldsymbol{r}_0\boldsymbol{r}_1\boldsymbol{r}_2)^5$ of $\mathcal{F}$. Hence, if we further apply the central inversion $(\boldsymbol{r}_0\boldsymbol{r}_1\boldsymbol{r}_2\boldsymbol{r}_3)^{15}$ of $\mathcal{H}$ itself, then we arrive at

PROPOSITION 5.8. *An allomorphism of* $\{5,3,3\}$ *is given by the operation*

$$(\boldsymbol{r}_0, \ldots, \boldsymbol{r}_3) \mapsto \big((\boldsymbol{r}_0\boldsymbol{r}_1\boldsymbol{r}_2)^5(\boldsymbol{r}_0\boldsymbol{r}_1\boldsymbol{r}_2\boldsymbol{r}_3)^{15}, \boldsymbol{r}_3, \boldsymbol{r}_2, \boldsymbol{r}_1\big) =: (\boldsymbol{s}_0, \ldots, \boldsymbol{s}_3).$$

The operation is even simpler when we pass to the central quotient $\mathcal{H}/2$; this is the case that we actually need.

COROLLARY 5.9. *An allomorphism* $\boldsymbol{\tau}$ *of* $\{5,3,3\}/2$ *is given by the operation*

$$\boldsymbol{\tau}\colon (\boldsymbol{r}_0, \ldots, \boldsymbol{r}_3) \mapsto \big((\boldsymbol{r}_0\boldsymbol{r}_1\boldsymbol{r}_2)^5, \boldsymbol{r}_3, \boldsymbol{r}_2, \boldsymbol{r}_1\big) =: (\boldsymbol{s}_0, \ldots, \boldsymbol{s}_3).$$

An alternative way of looking at $\{5,3,3\}$ uses quaternions. The *binary icosahedral group* $\mathbf{I}$ consists of those quaternions $\mathbf{x} = \xi_0+\xi_1\mathbf{i}+\xi_2\mathbf{j}+\xi_3\mathbf{k}$, with $(\xi_0,\xi_1,\xi_2,\xi_3)$ obtained from

$$(\pm 1,0,0,0), \quad \tfrac{1}{2}(\pm 1,\pm 1,\pm 1,\pm 1), \quad \tfrac{1}{2}(\pm\tau,\pm 1,\pm\tau^{-1},0), \tag{5.10}$$

by even permutations and all changes of sign of the coordinates. We denote by $\widetilde{\mathbf{x}}$ the quaterion conjugate of $\mathbf{x}$, which changes the signs of all ξ_j except ξ_0. The first two sets of vectors yield the subgroup $\mathbf{A}$, the *binary octahedral group*. Then, in terms of quaterions, the vertices of Proposition 5.1 take the form $4\mathbf{I}^{\ddagger}\mathbf{I}$; recall that ‡ changes the sign of $\sqrt{5}$.

With these vertices, the symmetry group consists of the mappings

$$\mathbf{x} \mapsto \mathbf{a}^*\mathbf{x}\mathbf{b} \text{ or } \mathbf{a}^*(-\mathbf{u}\widetilde{\mathbf{x}}\mathbf{u})\mathbf{b},$$

with $\mathbf{a},\mathbf{b} \in \mathbf{I}$ and $\mathbf{u} = (1/\sqrt{2})(\mathbf{j}-\mathbf{k})$, where we write $\mathbf{g}^* := (\widetilde{\mathbf{g}})^{\ddagger} = (\mathbf{g}^{\ddagger})^{-1}$. In this context, it is well to bear in mind that $\mathbf{I} \mapsto \mathbf{I}^{\ddagger}$ under the hyperplane reflexion $\mathbf{x} \mapsto -\mathbf{u}\widetilde{\mathbf{x}}\mathbf{u}$; as an illustrative example,

$$\mathbf{g} := \tfrac{1}{2}(1-\tau^{-1}\mathbf{j}+\tau\mathbf{k}) \mapsto \tfrac{1}{2}(1+\tau\mathbf{j}-\tau^{-1}\mathbf{k}) = \mathbf{g}^{\ddagger}.$$

6. The quotient $\mathcal{H}_8$

A crucial feature of the realization domain $\mathcal{N}$ of $\mathcal{H}$ is the existence of families of quotients, which arise from the action of $[3,3,5]$ on various subgroups; see Remark 2.3. For instance, we can think of the small simplex realization S (that is, realizations of $\mathcal{H}_2$) as the action on $\mathbf{Z}_2 = \langle -I\rangle$. In general, a quotient results in identifications of subsets of vertices. In this section, the subgroups are $[3^{1,1,1}]$ and a certain subgroup $[3,4,3]^*$ of index 2 in $[3,4,3]$ (it is not the rotation subgroup $[3,4,3]^+$, but an extension of $[3^{1,1,1}]$ by a cyclic group of order 3). So, for instance, $\mathcal{H}$ has inscribed staurotopes with 8 vertices and the symmetry of $[3^{1,1,1}]$, and $\mathcal{H}_8$ results from identifying the vertices of each of these staurotopes. Similarly, $\mathcal{H}_{24}$ arises from identifying the sets of 24 vertices of certain inscribed 24-cells. As we pointed out in Remark 2.3, it may be the case (as it actually is here) that these quotients are not themselves polytopal; this is of no importance, because non-polytopal realizations must be allowed to contribute to the realization cone $\mathcal{P}$.

REMARK 6.1. Regular simplices, staurotopes and 24-cells can be inscribed in $\mathcal{H}$ in ways that do not have the full symmetry of $[3,3,3]$, $[3^{1,1,1}]$ and $[3,4,3]^*$, respectively; see [**7**] for details, particularly of compounds missed by Coxeter [**1**].

Equally naturally, we should wish to treat each quotient $\mathcal{H}_q$ in its own right, that is, with its own domain $\mathcal{N}_q$ and layer vector Λ_q, say, and corresponding cosine vectors. It is thus crucial to see how these reduced domains $\mathcal{N}_q$ embed in $\mathcal{N}$ and, here, how $\mathcal{N}_{24}$ embeds in $\mathcal{N}_8$.

From the list of vertices in Proposition 5.1, it is fairly straightforward to prove

LEMMA 6.2. *Layers in $\mathcal{H}_8$ are identified as follows:*

0	1	1	4	2	12	4	12	3	24	8a	8
15a	6	6	24	5	4	7	24	8b	24	22a	8
30	1	10	12	9	24	12a	4	11	24		
		13	24	14	24	12b	24	15b	48		
		17	24	16	24	18a	4	19	24		
		20	12	21	24	18b	24	22b	24		
		24	24	25	4	23	24	27	24		
		29	4	28	12	26	12				
	8		128		128		128		192		16

The further identifications to give $\mathcal{H}_{24}$ are $\{\mathcal{L}_0, \mathcal{L}_{8a}\}$ and $\{\mathcal{L}_1, \mathcal{L}_2, \mathcal{L}_4\}$.

REMARK 6.3. The entries in the second half of each column count the number of points in that layer. The order of the columns is chosen to emphasize the (perhaps somewhat surprising) symmetry among $\{\mathcal{L}_1, \mathcal{L}_2, \mathcal{L}_4\}$; we shall shortly see how this symmetry arises. Even though layers $12a$ and $12b$ fall together, as do layers $18a$ and $18b$, we have separated them to emphasize the correspondence with the previous two identifications. Observe as well that $\mathcal{L}_{8a}$ survives in $\mathcal{H}_8$ as an asymmetric layer; see below.

As a first step in describing $\mathcal{H}_8$, we establish

PROPOSITION 6.4. *The quotient $\mathcal{H}_{24}$ of $\mathcal{H} = \{5,3,3\}$ has two non-trivial diagonal classes. Combinatorially, $[3,3,5]$ acts on the product $\{3,3,3\} \times \{3,3,3\}$ of two 4-simplices, with edges of $\mathcal{H}$ corresponding to non-edges of the product. Consequently, the cosine matrix of $\mathcal{H}_{24}$ is*

$$\begin{bmatrix} 1 & 1 & 1 \\ 1 & -\frac{1}{4} & \frac{3}{8} \\ 1 & \frac{1}{16} & -\frac{1}{4} \end{bmatrix},$$

with layer and dimension vectors

$$\Lambda = (1, 16, 8), \;\; D = (1, 8, 16).$$

PROOF. The key to this is the observation that the identification $\mathcal{H} \searrow \mathcal{H}_{24}$ preserves inscribed 4-simplices $\{3,3,3\}$. Moreover, vert$\{5,3,3\}$ can be thought of as $\mathbf{A}$ acted on by powers of $\mathbf{p}^*$ on the left and $\mathbf{p}$ on the right, where

$$\mathbf{p} := -\tfrac{1}{2}(\tau - \mathbf{i} + \tau^{-1}\mathbf{j}); \tag{6.5}$$

this exhibits the compound of 25 copies of $\{3,4,3\}$ inscribed in $\{5,3,3\}$.

However, a more formal approach is the following (for reasons that will become clear, we cannot appeal to the vertex-figure criterion of Theorem 3.1 here). Since $\mathcal{H}_{24}$ has $600/24 = 25$ vertices, the dimensions of its two non-trivial pure components sum to 24, which is not a sum of dimensions of realizations of $\{3,3,5\}/2$. Thus Corollary 3.4 implies that the facet of at least one component must be centred, leading us to solve

$$1 + 16\alpha + 8\beta = 0,$$
$$1 + 13\alpha + 6\beta = 0$$

for the cosine vector $(1, \alpha, \beta)$, where we have applied (5.7) to the collected terms in the second equation using the table of Lemma 6.2. This yields $\Gamma_1 := (1, -\frac{1}{4}, \frac{3}{8})$

(in contracted form), with dimension 8 from the Λ-orthogonality Theorem 2.12, and then complementarity using the component equation of Theorem 2.9 gives the other cosine vector $(1, \frac{1}{16}, -\frac{1}{4})$ with dimension 16. □

The indices (such as that of Γ_1) refer to the final tables. These begin with Γ_0 in Table 2, while Table 3 carries on from it, with its first row giving Γ_{19}.

Observe that, for the 16-dimensional realization, we have $\eta_{\boldsymbol{f}} = \frac{1}{64}$, whereas we have $\eta_{\boldsymbol{f}} = \frac{1}{16}$ for the 16-dimensional realization of $\{3,3,5\}/2$. This already tells us that $w > 1$ in this case, and so explains why we cannot appeal to Theorem 3.1. (Here, $\eta_{\boldsymbol{f}}$ is as in (3.2), namely, the squared distance to the centre of the facet.)

We now move on to the rest of $\mathcal{H}_8$; we shall prove the following.

PROPOSITION 6.6. *The pure cosine vectors in $\mathcal{H}_8$ that are Λ-orthogonal to $\mathcal{N}_{24}$ are of the form $(1, \alpha, \beta, \gamma, 0, -\frac{1}{2})$, with*

$$\alpha + \beta + \gamma = 0, \qquad \alpha^2 + \beta^2 + \gamma^2 = \tfrac{3}{32}.$$

The common dimension is 25.

PROOF. First note that the layer vector of $\mathcal{H}_8$ is $(1, 16, 16, 16, 24, 2^*)$; as we said before, the asymmetric diagonal class $\mathcal{D}_{8a}$ survives the identification, because no automorphism of $\mathcal{H}$ fixes one copy of $\{3,3,4\}$ inscribed in $\{3,4,3\}$ while interchanging the other two. The injection of cosine vectors of $\mathcal{H}_{24}$ into $\mathcal{H}_8$ is given by

$$(1, \alpha, \beta) \hookrightarrow (1, \alpha, \alpha, \alpha, \beta, 1).$$

Applying the Λ-orthogonality to what we know from Proposition 6.4 shows that the remaining cosine vectors of $\mathcal{H}_8$ satisfy

$$\begin{aligned} 1 + 16\gamma_1 + 16\gamma_2 + 24\gamma_3 + 16\gamma_4 + 2\gamma_5 &= 0, \\ 1 - \gamma_1 - \gamma_2 - \gamma_4 + 2\gamma_5 &= 0, \\ 1 - 3\gamma_3 + 2\gamma_5 &= 0. \end{aligned}$$

From this, we easily see that they are of the form $(1, \alpha, \beta, \gamma, 0, -\frac{1}{2})$, with $\alpha+\beta+\gamma = 0$. The asymmetric diagonal class suggests that we should expect a Wythoff space with $w > 1$, and this is reinforced by the symmetry among α, β and γ.

In describing the groups of the regular star-polytopes in [**10**, Theorem 7D16], vertex-figure replacements were used to lead from one polytope to another. In our context, we have Proposition 5.8 which replaces $\mathcal{H} = \{5,3,3\}$ by its allomorph, and Corollary 5.9 which does the same for its central quotient $\mathcal{H}_2$. In particular, the latter holds for the quotient $\mathcal{H}_8$, where it swaps layers $\mathcal{L}_1$ and $\mathcal{L}_2$.

Now a typical automorphism of $\mathcal{H}_8$ of period 3 that permutes the points of $\mathcal{L}_0$ and $\mathcal{L}_5$ cyclically corresponds on the vertex-set H of Proposition 5.1 to the (double) rotation of $\mathbb{E}^4$ given by right multiplication by $\mathbf{u} := -\frac{1}{2}(1+\mathbf{i}+\mathbf{j}+\mathbf{k})$. If we multiply $-1 + \sqrt{5}(\mathbf{i}+\mathbf{j}+\mathbf{k})$ by $\mathbf{u}$ and $\mathbf{u}^2 = \widetilde{\mathbf{u}}$, then we obtain in turn

$$\sigma - \tau^{-1}(\mathbf{i}+\mathbf{j}+\mathbf{k}), \quad \sigma^{\ddagger} - \tau(\mathbf{i}+\mathbf{j}+\mathbf{k}).$$

In other words, the multiplication permutes representatives of layers $\mathcal{L}_1$, $\mathcal{L}_2$ and $\mathcal{L}_3$, and so induces a corresponding automorphism $\boldsymbol{\varrho}$, say, of $\mathcal{H}_8$.

It follows that, if we conjugate the operation $\boldsymbol{\tau}$ of Corollary 5.9 by $\boldsymbol{\varrho}$, then we are enabled to swap $\mathcal{L}_1$ or $\mathcal{L}_2$ with $\mathcal{L}_3$ as well, thus making the corresponding diagonal class $\mathcal{D}_3$ the edge-class of realizations. For instance, $\boldsymbol{\varrho}^{-1}\boldsymbol{\tau}\boldsymbol{\varrho}$ fixes the initial vertex and interchanges $\mathcal{L}_2$ and $\mathcal{L}_3$.

This symmetry makes it clear that the Wythoff space W of any realization of $\mathcal{H}_8$ must contain $\mathcal{L}_5$ also. In the present case, this implies that $w = \dim W = 2$, and the relation of the theorem results from applying Theorem 2.15 to calculate the dimension 25, namely,

$$\tfrac{1}{75}\big(1 + 16(\alpha^2 + \beta^2 + \gamma^2) + 2(-\tfrac{1}{2})^2\big) = \tfrac{1}{25}.$$

This completes the proof. □

REMARK 6.7. We can now see from Proposition 6.6 that

$$|\beta| \le \tfrac{1}{4} \implies 0 \le \eta_f \le \tfrac{1}{10},$$

with η_f as defined in (3.2). This is still short of the value $\eta_f = \frac{1}{4}$ for the 25-dimensional realization of $\{3,3,5\}/2$, and so we know that we must have yet more 25-dimensional realizations of $\{5,3,3\}$.

There is much freedom to choose a basis in standard form. With Section 12 in mind, we pick

$$\begin{aligned} \Gamma_{13} &= (1, -\tfrac{1}{8}, -\tfrac{1}{8}, \tfrac{1}{4}, 0, -\tfrac{1}{2}), \\ \Gamma_{14} &= (0, -\tfrac{\sqrt{3}}{8}, \tfrac{\sqrt{3}}{8}, 0, 0, 0), \\ \Gamma_{15} &= (1, \tfrac{1}{8}, \tfrac{1}{8}, -\tfrac{1}{4}, 0, -\tfrac{1}{2}), \end{aligned} \tag{6.8}$$

7. The quotient $\mathcal{H}_5$

We now move on to the action on the subgroups $[3,3,3]$ and $[3,3,3] \times \mathbf{Z}_2$, which yield $\mathcal{H}_5$ and $\mathcal{H}_{10}$. We first have

LEMMA 7.1. *Layers in $\mathcal{H}_5$ are identified as follows:*

0	1	1	4	3	24	2	12
18a	4	4	12	12b	24	7	24
		9	24	16	24	10	12
		17	24	19	24	15ab	54
		22ab	32	24	24	20	12
		25	4			23	24
						28	12
	5		100		120		150
12a	4	5	4	6	24		
30	1	8ab	32	11	24		
		13	24	14	24		
		21	24	18b	24		
		26	12	27	24		
		29	4				
	5		100		120		

The identifications to give $\mathcal{H}_{10}$ are in the same columns, which are arranged to suggest the central symmetry of $\mathcal{H}_5$.

Note that corresponding layers of $\{5,3,3\}$ and $\{\frac{5}{2},3,3\}$ (that is, under the change of sign of $\sqrt{5}$) fall into the same layers of $\mathcal{H}_5$.

The technique here is quite different. Recall that the basic geometric 4-simplex $\{3,3,3\}$ inscribed with full symmetry in $\{5,3,3\}$ has vertex-set $\mathbf{P} = \{\mathbf{p}_k^*\mathbf{p}_k \mid k = 0,\ldots,4\}$, where $\mathbf{p}_k := \mathbf{p}^k$ with $\mathbf{p}$ as in (6.5) (again, see [**2**] for the background here

– note the use of unit quaternions). We see at once that the 600 points $\mathbf{Pg}$ with $\mathbf{g} \in \mathbf{I}$ are distinct, because $\arccos(-\frac{1}{4})$ is not among the angles subtended by pairs of vertices of $\{3,3,5\}$, and from this we conclude that each copy of $\mathbf{P}$ inscribed in the 120-cell is uniquely determined by the point $\mathbf{g} = 1\mathbf{g}$. Moreover, from this we easily obtain the identifications.

Exactly the same argument applies to left multiplication by elements of $\mathbf{I}^{\ddagger}$. However, if $\mathbf{h}^{\ddagger}\mathbf{P}$ and $\mathbf{Pg}$ have any point in common, then they coincide. Since we can write

$$\mathbf{g} = \mathbf{g}^{\ddagger} \cdot \mathbf{g}^{*}\mathbf{g} = \mathbf{g}^{\ddagger} \cdot \mathbf{p}_k^{*}\mathbf{p}_k$$

for some k, it follows that $\mathbf{g}$ and $\mathbf{g}^{\ddagger}$ determine the same copy of the simplex. As a consequence, we have

LEMMA 7.2. *Any realization of $\{3,3,5\}$ that is invariant under the change of sign of $\sqrt{5}$ induces a realization of $\{5,3,3\}/5$ of the same dimension.*

Confining our attention to the pure realizations of $\{3,3,5\}$, Lemma 7.2 permits $\Gamma_3, \Gamma_4, \Gamma_7, \Gamma_8$ (for the moment, the notation refers to the realizations of $\{3,3,5\}$). However, we must replace two pairs by half their sums:

$$\Gamma_{1,2} := \tfrac{1}{2}(\Gamma_1 + \Gamma_2), \quad \Gamma_{5,6} := \tfrac{1}{2}(\Gamma_5 + \Gamma_6).$$

We can justify this in another way by an appeal to the component equation Theorem 2.9, since $\Gamma_{1,2}$ and $\Gamma_{5,6}$ are the complements of the remaining components in the small simplex S and staurotope X, respectively. Effectively, then, the cosine vectors are of the form $(1, \gamma_6, \gamma_{1,5}, \gamma_4, \gamma_{3,7}, \gamma_2, \gamma_8)$, with the double indices for entries that are forced to be equal; the re-ordering is deliberate. To see how these are injected into $\mathcal{H}_5$, we just look for the layers of $\{5,3,3\}$ containing $\mathbf{I}$; these are $\mathcal{L}_0, \mathcal{L}_3 = \mathcal{L}_{20}, \mathcal{L}_{8ab}, \mathcal{L}_{11} = \mathcal{L}_{27}, \mathcal{L}_{15ab}, \mathcal{L}_{22ab}, \mathcal{L}_{30}$. Taking the layer vector of $\mathcal{H}_5$ to be as implied by Lemma 7.1, we see that the induced cosine vector is exactly of the form we chose. Our analysis has led us to

PROPOSITION 7.3. *The cosine matrix of $\{5,3,3\}/5$ is*

$$\begin{bmatrix} 1 & 1 & 1 & 1 & 1 & 1 & 1 \\ 1 & 0 & \frac{1}{6} & -\frac{1}{3} & \frac{1}{6} & 0 & 1 \\ 1 & \frac{1}{4} & -\frac{1}{4} & 0 & -\frac{1}{4} & \frac{1}{4} & 1 \\ 1 & -\frac{1}{5} & 0 & \frac{1}{5} & 0 & -\frac{1}{5} & 1 \\ 1 & -\frac{1}{2} & \frac{1}{4} & 0 & -\frac{1}{4} & \frac{1}{2} & -1 \\ 1 & \frac{1}{4} & \frac{1}{4} & 0 & -\frac{1}{4} & -\frac{1}{4} & -1 \\ 1 & 0 & -\frac{1}{6} & 0 & \frac{1}{6} & 0 & -1 \end{bmatrix},$$

with layer and dimension vectors

$$\Lambda = (1, 20, 24, 30, 24, 20, 1), \quad D = (1, 18, 16, 25, 8, 16, 36).$$

REMARK 7.4. We have already seen in Remark 6.7 that the Wythoff space of the 25-dimensional realizations must have dimension at least 3; Proposition 7.3 adds confirmation to this. The cosine vectors for dimensions 18 and 8 are those of Γ_7 and Γ_{21} in the final tables, appropriately contracted.

8. Small dimensions

We now begin working through dimension by dimension. Apart from $\Gamma_0 = (1^{36})$, we have the 4-dimensional starting points $\Gamma_{19} = \Gamma(\{5,3,3\})$ itself, and $\Gamma_{20} = {\Gamma_{19}}^{\ddagger} = \Gamma(\{\frac{5}{2},3,3\})$.

We have already seen the 8-dimensional cases $\Gamma_1 \in \mathcal{N}_{24} \subset \mathcal{N}_8$ and $\Gamma_{21} \in \mathcal{N}_5$. Though quite unrelated, they play surprisingly parallel rôles in the analysis.

Next, we have the 9-dimensional Γ_2 and $\Gamma_3 = {\Gamma_2}^{\ddagger}$, given by

$$4{\Gamma_{19}}^2 = \Gamma_0 + 3\Gamma_2,$$
$$4{\Gamma_{20}}^2 = \Gamma_0 + 3\Gamma_3.$$

Finally, there is the 18-dimensional $\Gamma_7 \in \mathcal{N}_{10}$, again already noted.

9. Dimension 16 – small simplex

So far, we know three 16-dimensional realizations. First, we have $\mathrm{Q}_1 := \{5,3,3\} \otimes \{\frac{5}{2},3,3\}$. Second, we have the 16-dimensional component Q_2 in $\mathcal{H}_{24}$. Third, there is the 16-dimensional component Q_3 in $\mathcal{H}_{10}$. Their respective cosine vectors are given by

$$\begin{aligned} 16\Delta_1 &= (16,-11,5,-4,9;-11,1,-5,4,4;-1,5,-4,1,1;-1,1,0,0), \\ 16\Delta_2 &= (16,1,1,-4,1;1,1,1,16,-4;1,1,-4,1,1;1,1,16,-4), \\ 4\Delta_3 &= (4,1,0,-1,1;1,-1,0,1,1;1,0,-1,4,-1;1,-1,0,0). \end{aligned} \tag{9.1}$$

The notation is temporary, but will be widely referred to later; we shall extensively use inner product rather than cosine vectors to save space and make for easier reading. The corresponding values of $\eta_{\boldsymbol{f}}$ are $\frac{1}{64}$, $\frac{1}{64}$, $\frac{1}{16}$.

REMARK 9.2. The last tells us that $\mathrm{Q}_3 \in \mathcal{H}_{10}$ is the geometric dual of the 16-dimensional realization of $\{3,3,5\}/2$; it is interesting that, in this dual, the vertices are identified in tens. It is also worth noting that, even though Q_2 and Q_3 lie in disjoint quotients of $\mathcal{H}$ (except for $\{1\}$), their cosine vectors Δ_2 and Δ_3 are not Λ-orthogonal; in fact, $16\langle \Delta_j, \Delta_k\rangle_\Lambda = \frac{1}{4}$ whenever $j \neq k$.

Knowing that $w = 2$, and expecting that there should be a 16-dimensional realization with $\eta_{\boldsymbol{f}} = 0$, it is straightforward to see that we have

PROPOSITION 9.3. *With the notation of* (9.1), *the general cosine vector* Δ *of a pure* 16*-dimensional realization of* $\{5,3,3\}/2$ *is*

$$\Delta = \lambda_1\Delta_1 + \lambda_2\Delta_2 + \lambda_3\Delta_3,$$

with

$$\lambda_1 + \lambda_2 + \lambda_3 = 1, \quad \lambda_1^2 + \lambda_2^2 + \lambda_3^2 = 1.$$

Indeed, the cosine vector of the realization Q_4 with $\eta_{\boldsymbol{f}} = 0$ is $\Delta_4 = \frac{2}{3}\Delta_1 + \frac{2}{3}\Delta_2 - \frac{1}{3}\Delta_3$. Moreover, if we set $\Gamma_4 := \Delta_3$, $\Gamma_6 := \Delta_4$ and define Γ_5 by $\sqrt{3}\Gamma_5 := \Delta_2 - \Delta_1$, then we have the Λ-orthogonal basis vectors $\Gamma_4, \Gamma_5, \Gamma_6$ in standard form; the indices refer to Table 2.

REMARK 9.4. The entries in Δ_4 are certainly nicer than those in Δ_1, and arguably nicer than the ones in Δ_2. It is curious, therefore, that Q_4 seems not to arise in a natural way; rewriting the definition of Δ_4 as $3\Delta_4 + \Delta_3 = 2\Delta_1 + 2\Delta_2$ does not shed any more light on the matter.

REMARK 9.5. Observe that Δ_2, Δ_3 occur in

$$8{\Gamma_1}^2 = \Gamma_0 + 3\Gamma_1 + 4\Delta_2,$$
$$8{\Gamma_{21}}^2 = \Gamma_0 + 4\Delta_3 + 3\Gamma_7;$$

moreover,

$$16{\Delta_2}^2 = \Gamma_0 + 6\Gamma_1 + 9\Delta_2,$$
$$16{\Delta_3}^2 = \Gamma_0 + 4\Delta_3 + 6\Gamma_7 + 5\Psi,$$

where Ψ is the 25-dimensional component of $\mathcal{H}_{10}$, as we can see using Propositions 6.4 and 7.3.

10. Dimension 16 – staurotope

We already have the cosine vector Γ_{22} of the 16-dimensional component of $\mathcal{H}_5$ in X, namely,

$$\Phi_1 := \Gamma_{22} = \tfrac{1}{4}(4, 1, 0, 1, 1, -1, -1, 0, -1, -1, 1, 0, -1, -4, -1, -1, -1, 0, 0)$$

(the notation Φ_1 is temporary). If $x \in \mathbb{E}^4$, then $x \otimes x \otimes x$ has 4 components of type ξ_i^3, 12 of type $\xi_i^2\xi_j$ and 4 of type $\xi_i\xi_j\xi_k$, giving total dimension $4 + 12 + 4 = 20$. The crucial observation is then that $d({\Gamma_{19}}^3) \leq 20$ and, since

$$\begin{aligned} 4\langle {\Gamma_{19}}^3, \Gamma_{19}\rangle_\Lambda &= 4\langle {\Gamma_{19}}^2, {\Gamma_{19}}^2\rangle_\Lambda \\ &= \tfrac{1}{4}\langle \Gamma_0 + 3\Gamma_2, \Gamma_0 + 3\Gamma_2\rangle_\Lambda = \tfrac{1}{4}(1+1) = \tfrac{1}{2}, \end{aligned}$$

we can write

$${\Gamma_{19}}^3 = \tfrac{1}{2}\Gamma_{19} + \tfrac{1}{2}\Phi_2,$$

where $d(\Phi_2) \leq 16$. In exactly the same way, we have

$${\Gamma_{20}}^3 = \tfrac{1}{2}\Gamma_{20} + \tfrac{1}{2}{\Phi_2}^\ddagger.$$

Suppressing the messy calculations, we find that

$$16\langle \Phi_1, \Phi_2\rangle_\Lambda = \tfrac{3}{8} = 16\langle \Phi_1, {\Phi_2}^\ddagger\rangle_\Lambda;$$

this shows, incidentally, that $\Phi_2, {\Phi_2}^\ddagger$ have the same group as Φ_1. Because

$$16\langle \Phi_2, {\Phi_2}^\ddagger\rangle_\Lambda = \tfrac{1}{16},$$

again from calculations that we do not give, we can complete the picture in the following way. Since

$$32\|\Phi_2 - {\Phi_2}^\ddagger\|_\Lambda^2 = 2 - \tfrac{1}{4} + 2 = \tfrac{15}{4},$$

we see that the mixed cosine vector Γ_{23} is given by

$$\sqrt{15}\Gamma_{23} = 2(\Phi_2 - {\Phi_2}^\ddagger),$$

and the sub-basis is completed by

$$\Gamma_{24} = \tfrac{1}{5}(4\Phi_2 + 4{\Phi_2}^\ddagger - 3\Phi_1);$$

it is easily verified that $32\|\Gamma_{23}\|_\Lambda^2 = 1 = 16\|\Gamma_{24}\|_\Lambda^2$, as required.

REMARK 10.1. The 16-dimensional component of $\{3,3,5\}$ in X has $\eta_{\boldsymbol{f}} = \frac{7}{16}$. We easily find the corresponding values in $\mathcal{H}$: $\eta_{\boldsymbol{f}} = (\frac{3}{16}, \frac{\sqrt{3}}{8}, \frac{1}{4})$ for $(\Gamma_{23}, \Gamma_{24}, \Gamma_{25})$. The realization achieving $\eta_{\boldsymbol{f}} = \frac{7}{16}$ has cosine vector $\alpha^2\Gamma_{23} + 2\alpha\beta\Gamma_{24} + \beta^2\Gamma_{25}$ when $(\alpha, \beta) = \frac{1}{\sqrt{7}}(\sqrt{3}, 2)$, which leads to cosine vector beginning

$$\tfrac{1}{28}(28, 23, 16, 7, 3; 1, \ldots)$$

(we have given enough to check that the correct value of $\eta_{\boldsymbol{f}}$ is attained).

11. Dimension 24

We obtain the realizations in $\mathbb{E}^{24}$ by methods which are curiously parallel, even though the 8-dimensional realizations employed – one in $\mathcal{H}_5$ and the other in $\mathcal{H}_{24}$ – seem completely unrelated. Thus we obtain $\Gamma_8, \Gamma_9, \Gamma_{25}, \Gamma_{26}$ from

$$\begin{aligned} 4\Gamma_{20}\Gamma_{21} &= \Gamma_1 + 3\Gamma_8 \\ 4\Gamma_{19}\Gamma_{21} &= \Gamma_1 + 3\Gamma_9 \\ 4\Gamma_1\Gamma_{20} &= \Gamma_{21} + 3\Gamma_{25} \\ 4\Gamma_1\Gamma_{19} &= \Gamma_{21} + 3\Gamma_{26} \end{aligned}$$

Observe that we have exact counts of dimensions: $4 \cdot 8 = 32 = 8 + 24$. More specifically, if the dimension is d, then $d + 8 \leq 32$ giving $d \leq 24$. In the other direction, we have (for instance)

$$\|\Gamma_{19}\Gamma_{21}\|_\Lambda^2 = \langle \Gamma_{19}{}^2, \Gamma_{21}{}^2 \rangle_\Lambda = \langle \tfrac{1}{4}\Gamma_0 + \tfrac{3}{4}\Gamma_2, \tfrac{1}{8}\Gamma_0 + \tfrac{3}{8}\Gamma_7 + \tfrac{1}{2}\Delta_3 \rangle_\Lambda = \tfrac{1}{4} \cdot \tfrac{1}{8} = \tfrac{1}{32},$$

where we have used Remark 9.5 (the Λ-orthogonality of the other terms is due to their different dimensions). But then

$$\tfrac{1}{32} = \|\Gamma_{19}\Gamma_{21}\|_\Lambda^2 = \|\tfrac{1}{4}\Gamma_1 + \tfrac{3}{4}\Gamma_9\|_\Lambda^2 = \tfrac{1}{16} \cdot \tfrac{1}{8} + \tfrac{9}{16} \cdot \|\Gamma_9\|_\Lambda^2$$

by Λ-orthogonality of the construction gives $\|\Gamma_9\|_\Lambda^2 = \frac{1}{24}$, which leads to $d \geq 1/\|\Gamma_9\|_\Lambda^2 = 24$ by Lemma 2.11.

The purity of these new realizations can be checked by their Λ-orthogonality to all the lower-dimensional realizations.

12. Dimension 25

In this case, we mimic the way that $[3,3,5]$ acts by quaternions. We work here with 6×6 matrices Z having row and column sums 0; in other words, we are in $\mathbb{L}^5 \otimes \mathbb{L}^5$, with $\mathbb{L}^5 = \{(\xi_0, \ldots, \xi_5) \in \mathbb{E}^6 \mid \xi_0 + \cdots + \xi_5 = 0\}$. The elements of the group are mappings $Z \mapsto AZB$ or $AZ^{\mathsf{T}}B$, with A, B matrices of permutations and Z^{T} the transpose of Z; observe that $Z \mapsto AZ^{\mathsf{T}}A$ is involutory. The generatrix $(R_0, \ldots, R_3)$ of the symmetry group is given by

$$\begin{aligned} &R_0 \colon \sigma_0 = (1\ 2\ 3\ 4\ 5), \\ &R_1 \colon \sigma_1 = \iota \text{ (the identity)}, \\ &R_2 \colon \sigma_2 = (0\ 1\ 2)(3\ 5\ 4), \\ &R_3 \colon \sigma_3 = (1\ 5)(2\ 4); \end{aligned}$$

it may be checked that these generators indeed satisfy the relations of $[3,3,5]$ (and, moreover, are consistent with Table 1 below).

REMARK 12.1. If the matrix R corresponds to the permutation σ, then $Z \mapsto RZR$ will result in σ acting on the right permuting the columns of Z, while its inverse σ^{-1} acts on the left permuting the rows of Z.

A little work shows that the Wythoff space W consists of those matrices

$$\begin{bmatrix} \alpha & \beta & \gamma & \delta & \gamma & \beta \\ \beta & \gamma & \alpha & \beta & \delta & \gamma \\ \gamma & \alpha & \beta & \gamma & \beta & \delta \\ \delta & \beta & \gamma & \alpha & \gamma & \beta \\ \gamma & \delta & \beta & \gamma & \beta & \alpha \\ \beta & \gamma & \delta & \beta & \alpha & \gamma \end{bmatrix},$$

with $\alpha+2\beta+2\gamma+\delta = 0$. In particular, we see that $w = \dim W = 3$, thus confirming the observation of Remark 7.4.

We adopt the temporary notation A for the matrix with $\alpha = 1$ and $\beta = \gamma = \delta = 0$, with B, C, D defined analogously in the obvious way; thus A, B, C, D are mutually orthogonal, with 6×6 matrices regarded as vectors in $\mathbb{E}^{36} = \mathbb{E}^6 \otimes \mathbb{E}^6$. The condition for W leads to an initial orthogonal (but not orthonormal) basis consisting of

$$A - D,\ B - C,\ \tfrac{1}{3}(2A - B - C + 2D) = A + D - \tfrac{1}{3}J,$$

where J is the 6×6 matrix all of whose entries are 1, with square norms 12, 24 and 8, respectively.

We start with the third initial vertex. A possible image $A' + D'$ of $A + D$ (we can ignore $-\frac{1}{3}J$ for the moment) is determined by three disjoint pairs of row indices, and three such pairs of column indices, giving at most

$$15 \cdot 15 = 225 < 300.$$

Hence we must be in a further quotient of $\mathcal{H}_2$, and since 75 is the only plausible divisor of 225, we must actually be in $\mathcal{H}_8$. Indeed, the corresponding entry of the cosine vector will be

$$\tfrac{1}{8}\langle A + D - \tfrac{1}{3}J, A' + D' - \tfrac{1}{3}J\rangle = \tfrac{1}{8}\langle A + D, A' + D'\rangle - \tfrac{1}{2},$$

and working out the first few entries, or using calculations like those following, yields Γ_{13} of (6.8).

The situation for $B - C$ is similar, if not quite so clear. There are 225 choices for B, and each choice of B seems to allow 2 choices for C (the other, in the initial matrix, being $A + D$), thus giving $2 \cdot 225 = 450$ possibilities. Since 300 is not a divisor of 450, again we must be in the quotient $\mathcal{H}_8$. This is confirmed by direct calculation, leading to the cosine vector Γ_{15} of (6.8), exactly as to be expected from Λ-orthogonality.

Moreover, the matrix representation also leads to the corresponding mixed cosine vector Γ_{14} of (6.8) as well.

There remain the cosine vectors derived from $A - D$; note that, with this as initial vertex, we do indeed obtain all 300. We approach this case in three steps. First, for each layer $\mathcal{L}_s$, we choose $\mathbf{a} \in \mathbf{I}^\dagger$ and $\mathbf{b} \in \mathbf{I}$ such that (regarded as a quaternion) $4\widetilde{\mathbf{a}}\mathbf{b} \in \mathcal{L}_s$. In fact, as we shall see, we can choose either $\mathbf{a} = 1$ or $\mathbf{a}$ as a fixed pure imaginary quaternion; this simplifies the later calculations. Second, we

$\mathcal{L}_0$:	1	ι,	$\mathcal{L}_{8a}$:	$\frac{1}{2}(1+\mathbf{i}+\mathbf{j}+\mathbf{k})$	(0 2 5)(1 3 4),
$\mathcal{L}_3$:	$\frac{1}{2}(\tau+\mathbf{i}+\tau^{-1}\mathbf{j})$	(0 5 2 1 3),		$\frac{1}{2}(1-\mathbf{i}-\mathbf{j}-\mathbf{k})$	(0 5 2)(1 4 3),
$\mathcal{L}_{8b}$:	$\frac{1}{2}(1+\tau^{-1}\mathbf{i}+\tau\mathbf{j})$	(0 4 3)(1 2 5),	$\mathcal{L}_{15b}$:	$\frac{1}{2}(\tau\mathbf{i}+\tau^{-1}\mathbf{j}+\mathbf{k})$	(1 4)(2 3),
$\mathcal{L}_{11}$:	$\frac{1}{2}(\tau^{-1}+\tau\mathbf{i}+\mathbf{j})$	(0 2 3 5 1),		$-\frac{1}{2}(\tau\mathbf{i}+\tau^{-1}\mathbf{j}+\mathbf{k})$	(1 4)(2 3),
$\mathcal{L}_{15a}$:	$\mathbf{i}$	(1 5)(2 4),			
$\mathcal{L}_1$:	$\frac{1}{2}(\tau\mathbf{i}+\tau^{-1}\mathbf{j}+\mathbf{k})$	(1 4)(2 3),	$\mathcal{L}_7$:	$\frac{1}{2}(\tau+\mathbf{i}+\tau^{-1}\mathbf{j})$	(0 5 2 1 3),
$\mathcal{L}_2$:	$\frac{1}{2}(\mathbf{i}+\tau\mathbf{j}+\tau^{-1}\mathbf{k})$	(0 1)(3 4),		$\frac{1}{2}(\tau-\mathbf{i}-\tau^{-1}\mathbf{j})$	(0 3 1 2 5),
$\mathcal{L}_4$:	$\frac{1}{2}(\tau^{-1}\mathbf{i}+\mathbf{j}+\tau\mathbf{k})$	(1 3)(4 5),	$\mathcal{L}_9$:	$\frac{1}{2}(\tau^{-1}+\tau\mathbf{i}-\mathbf{j})$	(0 4 3 1 5),
$\mathcal{L}_5$:	$\frac{1}{2}(\mathbf{i}+\tau\mathbf{j}-\tau^{-1}\mathbf{k})$	(0 4)(3 5),		$\frac{1}{2}(\tau^{-1}-\tau\mathbf{i}+\mathbf{j})$	(0 5 1 3 4),
$\mathcal{L}_6$:	$\frac{1}{2}(\tau\mathbf{i}+\tau^{-1}\mathbf{j}-\mathbf{k})$	(0 2)(4 5),	$\mathcal{L}_{13}$:	$\frac{1}{2}(\tau+\mathbf{j}-\tau^{-1}\mathbf{k})$	(1 5 4 3 2),
$\mathcal{L}_{10}$:	$\frac{1}{2}(\tau\mathbf{i}-\tau^{-1}\mathbf{j}-\mathbf{k})$	(2 5)(3 4),		$\frac{1}{2}(\tau-\mathbf{j}+\tau^{-1}\mathbf{k})$	(1 2 3 4 5),
$\mathcal{L}_{12a}$:	$\frac{1}{2}(\tau^{-1}\mathbf{i}-\mathbf{j}+\tau\mathbf{k})$	(0 1)(2 5),			
$\mathcal{L}_{12b}$:	$\frac{1}{2}(\tau^{-1}\mathbf{i}+\mathbf{j}-\tau\mathbf{k})$	(0 5)(1 4),			
$\mathcal{L}_{14}$:	$\frac{1}{2}(\mathbf{i}-\tau\mathbf{j}+\tau^{-1}\mathbf{k})$	(0 2)(1 3).			

TABLE 1. Layers, quaternions and permutations.

relate these quaternions **a**, **b** to permutations in $\boldsymbol{A}_5$ acting on antipodal vertices of the icosahedron $\{3,5\}$, with the identifications

$$\begin{aligned} 0 &:= \pm(\tau,-1,0), & 3 &:= \pm(\tau,1,0),\\ 1 &:= \pm(0,\tau,1), & 4 &:= \pm(1,0,\tau),\\ 2 &:= \pm(-1,0,\tau), & 5 &:= \pm(0,\tau,-1). \end{aligned}$$

In this context, recall that the quaternion $\mathbf{q} = \cos\vartheta + \sin\vartheta\mathbf{u}$, with $\mathbf{u}$ a pure imaginary unit quaternion identified with a unit vector in $\mathbb{E}^3$, induces a rotation of -2ϑ about the axis $\mathbf{u}$ by $\mathbf{x} \mapsto \widetilde{\mathbf{q}}\mathbf{x}\mathbf{q}$. Third, we lift these permutations to actions on our 6×6 matrices or, rather, their basis elements. As a check on the workings, we can use Λ-orthogonality with respect to $\mathcal{H}_8$ (we cannot use $\mathcal{H}_{10}$, because we still have a component there).

With $\mathbf{a} = 1$, we are essentially dealing with an inscribed copy of $\{3,3,5\}$. In Table 1, asymmetric diagonals are listed on the right; for the first half, $\mathbf{a} = 1$, while for the second half, $\mathbf{a} = \frac{1}{2}(\tau\mathbf{i}+\mathbf{j}+\tau^{-1}\mathbf{k}) \in \mathbf{I}^\dagger$ is a fixed involution, with corresponding permutation (1 4)(2 3). The other entries in the table are the corresponding **b** with induced permutations (and their inverses, which are needed for the mixed cosine vectors). Of course, since we are in $\mathcal{H}_2$ here, the layer $\mathcal{L}_{15b}$ has become symmetric.

We shall not give the details of calculating the entries of the cosine vectors. However, it is worth illustrating why we may need to find two inner products when asymmetric diagonal classes are involved. For instance, the unsymmetrized mixed vectors corresponding to Γ_{14} are actually (before scaling)

$$(0,-1,1,0,0;1,-1,0,\pm 4,0;1,-1,0,0,0;-1,1,0,0),$$

illustrating the asymmetry of $\mathcal{L}_{8a}$. The cosine vectors (including mixed ones) resulting from $A - D$ are then $\Gamma_{10}, \Gamma_{11}, \Gamma_{12}$ of Table 2.

13. Dimensions 30 and 40

We treat these cases together, because their calculations are exactly parallel. At this stage, considering components of the small simplex S, we have three diagonal classes to account for, and a dimension deficit of 100. The remaining cosine vectors $\Gamma_{16}, \Gamma_{17}, \Gamma_{18}$ are given by

$$\begin{aligned} 12\Gamma_{19}\Gamma_{26} &= 3\Gamma_1 + 3\Gamma_7 + \Gamma_9 + 5\Gamma_{16}, \\ 12\Gamma_{20}\Gamma_{25} &= 3\Gamma_1 + 3\Gamma_7 + \Gamma_8 + 5\Gamma_{17}, \\ 12\Gamma_{19}\Gamma_{25} &= 4\Delta_6 + 3\Gamma_8 + 5\Gamma_{18}, \\ 12\Gamma_{20}\Gamma_{26} &= 4\Delta_6 + 3\Gamma_9 + 5\Gamma_{18}, \end{aligned}$$

where

$$\Delta_6 := \tfrac{3}{4}\Gamma_4 + \tfrac{\sqrt{3}}{2}\Gamma_5 + \tfrac{1}{4}\Gamma_6$$

is pure of dimension 16. (We give both constructions for Γ_{18}, to complete the pattern.)

Observe that dimensions do not match up here. For the first two, $8+18+24+30 = 80 < 96 = 4\cdot 24$, while for the others we have $16+24+40 = 80 < 96$. Thus these do not help to pin down the dimensions. However, we have $\|\Gamma_{16}\|_\Lambda^2 = \|\Gamma_{17}\|_\Lambda^2 = \frac{1}{30}$, giving $d_{16}, d_{17} \geq 30$, while $\|\Gamma_{18}\|_\Lambda^2 = \frac{1}{40}$ gives $d_{18} \geq 40$, all by Lemma 2.11. But now $d_{16} + d_{17} + d_{18} = 100$ imposes equality.

14. Dimension 48

We treat dimension 48 before dimension 36, because we need the results of the former case to feed into the latter. However, it is worth referring to the initial remarks of Section 15, particularly Remark 15.1 showing that the Wythoff space in that case has dimension at least 3. The dimension deficit in the staurotope realization X to bear in mind is thus at most 96.

We define $K_1, \ldots, K_4$ by

$$\begin{aligned} 3\Gamma_7\Gamma_{19} &= \Gamma_{26} + 2K_1, \\ 3\Gamma_7\Gamma_{20} &= \Gamma_{25} + 2K_2, \\ 3\Gamma_2\Gamma_{21} &= \Gamma_{26} + 2K_3, \\ 3\Gamma_3\Gamma_{21} &= \Gamma_{25} + 2K_4. \end{aligned}$$

In each case, we have an exact dimension count: $24 + 48 = 72 = 18 \cdot 4 = 9 \cdot 8$. Indeed, $d(K_j) = 48$ for each j, since we have $\|K_j\|_\Lambda^2 = \frac{1}{48}$, yielding $48 \leq d_j \leq 48$ as for dimension 24. Alternatively, for instance,

$$\|\Gamma_7\Gamma_{19}\|_\Lambda^2 = \langle {\Gamma_7}^2, {\Gamma_{19}}^2 \rangle_\Lambda = \tfrac{1}{18} \cdot \tfrac{1}{4} = \tfrac{1}{72},$$

since

$${\Gamma_7}^2 = \tfrac{1}{18}\Gamma_0 + \tfrac{1}{6}\Gamma_7 + \tfrac{2}{9}\Delta_2 + \tfrac{5}{9}H, \quad {\Gamma_{19}}^2 = \tfrac{1}{4}\Gamma_0 + \tfrac{3}{4}\Gamma_2,$$

with Δ_2 pure of dimension 16 as before and H pure of dimension 25, and the other inner products vanish because the dimensions are different. On the other hand,

$$\tfrac{1}{72} = \|\Gamma_7\Gamma_{19}\|_\Lambda^2 = \|\tfrac{1}{3}\Gamma_7 + \tfrac{2}{3}K_1\|_\Lambda^2 = \tfrac{1}{9}\|\Gamma_7\|_\Lambda^2 + \tfrac{4}{9}\|K_1\|_\Lambda^2 = \tfrac{1}{9} \cdot \tfrac{1}{24} = \tfrac{4}{9} \cdot \tfrac{1}{d},$$

yielding $d = 48$, as claimed.

In a similar way, the various inner products of the K_j can be found:

$$\langle \Gamma_7\Gamma_{19}, \Gamma_7\Gamma_{20}\rangle_\Lambda = \langle {\Gamma_7}^2, \Gamma_{19}\Gamma_{20}\rangle_\Lambda = \tfrac{2}{9}\langle \Delta_2, \Delta_1\rangle_\Lambda = \tfrac{2}{9}\cdot\tfrac{1}{64} = \tfrac{1}{288},$$

from which it follows that $48\langle K_1, K_2\rangle_\Lambda = \frac{3}{8}$. Appealing to similar calculations, or directly, we find that the matrix of inner products $48\langle K_i, K_j\rangle_\Lambda$ is

$$\begin{bmatrix} 1 & \frac{3}{8} & \frac{1}{16} & \frac{3}{8} \\ \frac{3}{8} & 1 & \frac{3}{8} & \frac{1}{16} \\ \frac{1}{16} & \frac{3}{8} & 1 & \frac{27}{32} \\ \frac{3}{8} & \frac{1}{16} & \frac{27}{32} & 1 \end{bmatrix}.$$

The fact that these inner products are positive confirms that the K_j belong to the same family.

If we define

$$K_0 = \tfrac{3}{10}(K_1 + K_2) + \tfrac{1}{5}(K_3 + K_4),$$

then $\|K_0\|_\Lambda^2 = \frac{1}{96}$, so that K_0 gives the centre of the subdomain; compare Theorem 2.15. The general pure cosine vector Γ in the family must be an affine combination of the K_j which satisfies $\|\Gamma - K_0\|_\Lambda^2 = \frac{1}{96}$. A little work then shows that

$$\Gamma = (1 - \lambda - \mu)K_0 + \lambda K_1 + \mu K_2,$$

for some λ, μ such that $\lambda^2 - \frac{1}{2}\lambda\mu + \mu^2 = 1$.

The best choice for a basis seems to be to take $\lambda = -\mu = \pm\sqrt{2/5}$ to obtain Γ_{33}, Γ_{35}, while $K_1 + K_2 - 2K_0$, appropriately normalized, yields Γ_{34} (even so the entries are not very nice). We shall not write out these vectors here, but instead refer to Table 3.

15. Dimension 36

We begin by listing what we already know. First, Proposition 7.3 gives the 36-dimensional component $\Theta_1 := \Gamma_{27}$ in $\mathcal{H}_5$, namely,

$$6\Theta_1 = (6, 0, 0, -1, 0; 0, 1, 0, 0, 0; 0, 0, 1, -6, -1; 0, 1, 0, 0).$$

The notation is, of course, temporary.

Next, we know that $\Theta_2 := \Gamma_2\Gamma_{20}$ and ${\Theta_2}^\ddagger = \Gamma_3\Gamma_{19}$ are pure 36-dimensional, in analogy to the corresponding construction for $\{3, 3, 5\}$; see (4.2), but note here that $\Theta_2 \neq {\Theta_2}^\ddagger$. Suppressing the calculations, we find that

$$36\langle \Theta_1, \Theta_2\rangle_\Lambda = 36\langle \Theta_1, {\Theta_2}^\ddagger\rangle_\Lambda = \tfrac{1}{4}, \quad 36\langle \Theta_2, {\Theta_2}^\ddagger\rangle_\Lambda = \tfrac{1}{16}.$$

REMARK 15.1. It is not too hard to see that the initial vertices of the three corresponding polytopes are not coplanar in the Wythoff space, which is thus at least 3-dimensional.

The strongest tool is provided by the products $\Delta\Gamma_{19}$ (and $\Delta\Gamma_{20}$), with Δ a pure 16-dimensional realization of $\mathcal{H}_2$; these must all have the same group. We first have

$$4\langle \Delta\Gamma_{19}, \Gamma_{20}\rangle_\Lambda = 4\langle \Delta, \Gamma_{19}\Gamma_{20}\rangle_\Lambda = 4\langle \Delta, \Delta_1\rangle_\Lambda,$$

which takes the values $\frac{1}{4}, \frac{1}{16}, \frac{1}{16}$ for $\Delta = \Delta_1, \Delta_2, \Delta_3$, respectively.

Next, recall that $\Gamma_1, \Delta_2 \in \mathcal{H}_{24}$, with $\Gamma_1\Delta_2 = \frac{1}{4}\Gamma_1 + \frac{3}{4}\Delta_2$. From this, with $\Delta = \Delta_2$ and using $3\Gamma_{25} = 4\Gamma_1\Gamma_{20} - \Gamma_{21}$ from Section 11, we find after some messy calculations that

$$24\langle \Delta_2\Gamma_{19}, \Gamma_{25}\rangle_\Lambda = \tfrac{3}{8}.$$

Taking $\Delta = \Delta_1 = \Gamma_{19}\Gamma_{20}$, we then have

$$\Delta_1\Gamma_{19} = {\Gamma_{19}}^2\Gamma_{20} = \tfrac{3}{4}\Gamma_2\Gamma_{20} + \tfrac{1}{4}\Gamma_{20}.$$

We know that $\Gamma_2\Gamma_{20}$ is pure of dimension 36 from the beginning of the section, and the coefficient of Γ_{20} is that given above. We have arrived at

LEMMA 15.2. *If Δ is a pure 16-dimensional component of $\mathcal{H}_2$, then $\Delta\Gamma_{19}$ can only have components Γ_{20}, Γ_{25} and a pure component of dimension* 36.

PROOF. We have seen that, among the products $\Delta\Gamma_{19}$, we have found such components. Since $d(\Delta\Gamma_{19}) \leq 16 \cdot 4 = 64 = 4 + 24 + 36$, giving an exact count, it follows that there can be no others. □

We now consider $\Delta_3 \in \mathcal{H}_5$. Since $\Theta_1 \in \mathcal{H}_5$ also, we see that

$$\langle \Delta_3\Gamma_{19}, \Theta_1\rangle_\Lambda = \langle \Gamma_{19}, \Delta_3\Theta_1\rangle_\Lambda = 0,$$

since no component of $\mathcal{H}_5$ has dimension 4. In other words,

LEMMA 15.3. *The 36-dimensional components of $\Delta_3\Gamma_{19}, \Delta_3\Gamma_{20}$ are Λ-orthogonal to Θ_1.*

We calculate directly that

$$24\langle \Delta_3\Gamma_{19}, \Gamma_{25}\rangle_\Lambda = \tfrac{3}{8},$$

and since we already know that $4\langle \Delta_3\Gamma_{19}, \Gamma_{20}\rangle_\Lambda = \frac{1}{16}$, it follows that the pure 36-dimensional component Θ_3, say, is given by

$$16\Delta_3\Gamma_{19} = \Gamma_{20} + 6\Gamma_{25} + 9\Theta_3; \tag{15.4}$$

similarly,

$$16\Delta_3\Gamma_{20} = \Gamma_{19} + 6\Gamma_{26} + 9{\Theta_3}^\ddagger. \tag{15.5}$$

Now,

$$\begin{aligned} 256\langle \Delta_3\Gamma_{19}, \Delta_3\Gamma_{20}\rangle_\Lambda &= 256\langle \Gamma_{19}\Gamma_{20}, {\Delta_3}^2\rangle_\Lambda \\ &= 16\langle \Delta_1, \Gamma_0 + 4\Delta_3 + 6\Gamma_7 + 5\Psi\rangle_\Lambda \\ &= 4 \cdot \tfrac{1}{4} = 1, \end{aligned}$$

where we have employed Remarks 9.2 and 9.5. If we substitute $\Delta_3\Gamma_{19}$ and $\Delta_3\Gamma_{20}$ using (15.4) and (15.5), then we obtain

$$1 = \langle \Gamma_{20} + 6\Gamma_{25} + 9\Theta_3, \Gamma_{19} + 6\Gamma_{26} + 9{\Theta_3}^\ddagger\rangle_\Lambda = 81\langle \Theta_3, {\Theta_3}^\ddagger\rangle_\Lambda,$$

since the remaining terms in the Λ-inner product vanish. In other words, we have shown that

$$36\langle \Theta_3, {\Theta_3}^\ddagger\rangle_\Lambda = \tfrac{36}{81} = \tfrac{4}{9}. \tag{15.6}$$

We now appeal to what we have done previously. The complement $108\Theta_0$ of the remaining terms in the component equation Theorem 2.9 for the staurotope realization X gives the centre Θ_0 of the realization domain $\mathcal{N}$, or (then multiplied by 3) the sum of any 3 mutually Λ-orthogonal pure cosine vectors in $\mathcal{N}$; see Theorem 2.15. Here, the entry is

$$36\Theta_0 = (36, 3, 3, -6, 2; -3, -1, 0, 0, 0; -1, 3, 6, -6, 0; 1, -1, 0, 0).$$

It follows that the centre of the Λ-circle in $\mathcal{N}$ that contains $\Delta_3\Gamma_{19}, \Delta_3\Gamma_{20}$ and is Λ-orthogonal to Θ_1 is $\Theta_4 := \frac{1}{2}(3\Theta_0 - \Theta_1)$, whose square Λ-norm is $\frac{1}{72}$, as should be expected.

Since $\langle\Theta_3, \Theta_4\rangle_\Lambda = \langle\Theta_3{}^\ddagger, \Theta_4\rangle_\Lambda = \frac{1}{72}$ also, the Λ-orthogonal sub-basis is formed by $\lambda(\Theta_3 + \Theta_3{}^\ddagger) + (1 - 2\lambda)\Theta_4$ and $\mu(\Theta_3 - \Theta_3{}^\ddagger)$ for suitable λ and μ. For λ, we must solve

$$\begin{aligned} 1 &= 36\|\lambda(\Theta_3 + \Theta_3{}^\ddagger) + (1 - 2\lambda)\Theta_4\|_\Lambda^2 \\ &= \lambda^2(1 + \tfrac{8}{9} + 1) + 2\lambda(1 - 2\lambda)(\tfrac{1}{2} + \tfrac{1}{2}) + (1 - 2\lambda)\tfrac{1}{2} = \tfrac{8}{9}\lambda^2 + \tfrac{1}{2}, \end{aligned}$$

whence $\lambda = \pm\frac{3}{4}$. Similarly, for μ, we need

$$1 = 72\|\mu(\Theta_3 - \Theta_3{}^\ddagger)\|_\Lambda^2 = \mu^2(2 - \tfrac{16}{9} + 2) = \tfrac{20}{9}\mu^2,$$

giving $\mu = \frac{3}{2\sqrt{5}}$ (or its negative). In other words, setting

$$\begin{aligned} 4\Gamma_{30} &:= 10\Theta_4 - 3(\Theta_3 + \Theta_3{}^\ddagger), \\ 2\sqrt{5}\Gamma_{31} &:= 3(\Theta_3 - \Theta_3{}^\ddagger), \\ 4\Gamma_{32} &:= 3(\Theta_3 + \Theta_3{}^\ddagger) - 2\Theta_4, \end{aligned}$$

gives us a Λ-orthogonal basis of the subdomain of $\mathcal{N}$ that is Λ-orthogonal to Γ_{27}.

We next bring Δ_2 into play. Exactly parallel to what we had for Δ_3, we find that

$$16\Delta_2\Gamma_{19} = \Gamma_{20} + 6\Gamma_{26} + 9\Theta_5,$$

where Θ_5 is pure 36-dimensional. Then

$$36\Theta_5 = (36, -3, 5, -6, 2; 3, 1, 0, 18, -12; -3, 5, 6, -6, 4; 3, 1, 0, 0).$$

Since this is rational, we get the same if we substitute Γ_{20} for Γ_{19}. We calculate that

$$36\langle\Gamma_{27}, \Theta_5\rangle_\Lambda = \tfrac{1}{3},\ 36\langle\Gamma_{30}, \Theta_5\rangle_\Lambda = \tfrac{2}{3},\ \langle\Gamma_{31}, \Theta_5\rangle_\Lambda = \langle\Gamma_{32}, \Theta_5\rangle_\Lambda = 0.$$

The appropriate mixed cosine vector Γ_{28} is thus given by

$$3\Theta_5 = \Gamma_{27} + \sqrt{8}\Gamma_{28} + 2\Gamma_{30}.$$

For the last step, we return to the pure $\Theta_2 := \Gamma_2\Gamma_{20}$ and $\Gamma_3\Gamma_{19}$. Again after suppressed calculations, we have

$$\begin{aligned} 36\langle\Theta_2, \Gamma_{30}\rangle_\Lambda &= \tfrac{1}{8}, \quad 36\langle\Theta_2, \Gamma_{32}\rangle_\Lambda = \tfrac{5}{8}, \\ 72\langle\Theta_2, \Gamma_{28}\rangle_\Lambda &= \tfrac{1}{\sqrt{8}}, \quad 72\langle\Theta_2, \Gamma_{30}\rangle_\Lambda = -\tfrac{\sqrt{5}}{4}. \end{aligned}$$

Since Θ_2 is pure, we know that the remaining mixed cosine vector Γ_{29} must satisfy (up to sign)

$$8\Theta_2 = 2\Gamma_{27} + \sqrt{8}\Gamma_{28} + 2\sqrt{10}\Gamma_{29} + \Gamma_{30} - 2\sqrt{5}\Gamma_{31} + 5\Gamma_{32},$$

and so we have completed the basis. We shall not write out these basis vectors, but instead refer to Table 3.

	0	1	2	3	4	5	6	7	8*a*	8*b*	9	10	11	12*a*	12*b*	13	14	15*a*	15*b*
d	1	4	12	24	12	4	24	24*	8*	24	24*	12	24	4	24	24*	24	3	24
1	1	1	1	1	1	1	1	1	1	1	1	1	1	1	1	1	1	1	1
8	8	−2	−2	3	−2	−2	−2	−2	8	3	−2	−2	3	−2	−2	−2	−2	8	3
9	12	3φ	ψ	4τ	5	$3\varphi^\ddagger$	$-\sigma^\ddagger$	1	0	0	$-\varphi^\ddagger$	$\psi^\ddagger$	$4\tau^\ddagger$	−3	−3	$-\varphi$	$-\sigma$	−4	−4
9	12	$3\varphi^\ddagger$	$\psi^\ddagger$	$4\tau^\ddagger$	5	3φ	$-\sigma$	1	0	0	$-\varphi$	ψ	4τ	−3	−3	$-\varphi^\ddagger$	$-\sigma^\ddagger$	−4	−4
16	4	1	0	−1	1	1	−1	0	1	1	1	0	−1	4	−1	1	−1	0	0
$32^\dagger$	$(8\sqrt{3})$	6	−2	0	−4	6	0	3	6	−4	1	−2	0	0	0	1	0	8	−2
16	12	−6	3	−3	4	−6	2	−2	9	−1	−1	3	−3	−3	2	−1	2	8	−2
18	6	0	−2	1	0	0	1	−2	0	0	0	−2	1	6	1	0	1	−2	−2
24	24	$6\sqrt{5}$	2	$-\rho$	−10	$-6\sqrt{5}$	$-2\tau^\ddagger$	2	0	5	$2\sqrt{5}$	2	$-\rho^\ddagger$	−6	4	$-2\sqrt{5}$	-2τ	−8	−3
24	24	$-6\sqrt{5}$	2	$-\rho^\ddagger$	−10	$6\sqrt{5}$	-2τ	2	0	5	$-2\sqrt{5}$	2	$-\rho$	−6	4	$2\sqrt{5}$	$-2\tau^\ddagger$	−8	−3
25	12	3	3	0	−2	3	−1	−2	0	−4	−1	3	0	6	2	−1	−1	−4	4
$50^\dagger$	$(4\sqrt{6})$	3	−1	0	2	3	−1	−2	0	0	1	−1	0	−6	2	1	−1	0	0
$50^\dagger$	$(4\sqrt{2})$	3	1	0	0	−3	−1	0	0	0	−1	−1	0	0	0	1	1	0	0
25	8	−1	−1	0	2	−1	−1	2	−4	0	−1	−1	0	2	2	−1	−1	8	0
$50^\dagger$	$(\frac{8}{\sqrt{3}})$	−1	1	0	0	1	−1	0	0	0	1	−1	0	0	0	−1	1	0	0
25	8	1	1	0	−2	1	1	−2	−4	0	1	1	0	−2	−2	1	1	8	0
30	12	$-3\tau^{-2}$	$-\sigma$	2τ	1	$-3\tau^2$	$-\tau$	1	0	−2	τ^2	$-\sigma^\ddagger$	$2\tau^\ddagger$	−3	−1	τ^{-2}	$-\tau^\ddagger$	−4	0
30	12	$-3\tau^2$	$-\sigma^\ddagger$	$2\tau^\ddagger$	1	$-3\tau^{-2}$	$-\tau^\ddagger$	1	0	−2	τ^{-2}	$-\sigma$	2τ	−3	−1	τ^2	$-\tau$	−4	0
40	24	6	−6	−3	2	6	4	2	0	1	−2	−6	−3	−6	−2	−2	4	−8	5

TABLE 2. Components of the small simplex S.

	0	1	2	3	4	5	6	7	8a	8b	9	10	11	12a	12b	13	14	15a	15b
d	1	4	12	24	12	4	24	24*	8*	24	24*	12	24	4	24	24*	24	6	48*
4	4	σ	φ	2τ	3	$-\sigma^\ddagger$	τ^2	$\sqrt{5}$	2	2	τ	$\varphi^\ddagger$	$-2\tau^\ddagger$	1	1	$-\tau^\ddagger$	τ^{-2}	0	0
4	4	$\sigma^\ddagger$	$\varphi^\ddagger$	$2\tau^\ddagger$	3	$-\sigma$	τ^{-2}	$-\sqrt{5}$	2	2	$\tau^\ddagger$	φ	-2τ	1	1	$-\tau$	τ^2	0	0
8	4	−2	0	1	−2	2	−1	0	2	2	−2	0	−1	−4	1	2	−1	0	0
16	4	1	0	1	1	−1	−1	0	−1	−1	1	0	−1	−4	−1	−1	−1	0	0
$32^\dagger$	$(\frac{8}{\sqrt{3}})$	2	2	0	0	2	0	−1	0	0	−1	−2	0	0	0	−1	0	0	0
16	4	2	1	1	0	−2	0	0	−1	−1	−1	1	−1	1	−2	1	0	0	0
24	12	3τ	$-\varphi^\ddagger$	$\sigma^\ddagger$	−1	$-3\tau^\ddagger$	$-\tau^\ddagger$	$\sqrt{5}$	6	1	τ^2	$-\varphi$	$-\sigma$	3	−2	$-\tau^{-2}$	$-\tau$	0	0
24	12	$3\tau^\ddagger$	$-\varphi$	σ	−1	-3τ	$-\tau$	$-\sqrt{5}$	6	1	τ^{-2}	$-\varphi^\ddagger$	$-\sigma^\ddagger$	3	−2	$-\tau^2$	$-\tau^\ddagger$	0	0
36	6	0	0	−1	0	0	1	0	0	0	0	0	1	−6	−1	0	1	0	0
$72^\dagger$	$(12\sqrt{2})$	−6	0	0	4	6	0	0	6	−4	−1	0	0	0	0	1	0	0	0
$72^\dagger$	$(12\sqrt{2})$	−6	4	0	0	−6	0	−2	0	0	3	−4	0	0	0	3	0	0	0
36	24	9	5	−4	−6	−9	−1	0	6	−4	−1	5	4	6	6	1	−1	0	0
$72^\dagger$	(24)	9	−1	0	0	9	−5	−2	0	0	3	1	0	0	0	3	−5	0	0
36	24	−3	1	−4	10	3	−5	0	−6	4	−1	1	4	6	−2	1	−5	0	0
48	48	$-\alpha$	β	2	−4	$\alpha^\ddagger$	$\gamma^\ddagger$	$8\sqrt{2}$	−12	−2	$-2\sqrt{2}$	$\beta^\ddagger$	−2	−12	4	$-2\sqrt{2}$	γ	0	0
$96^\dagger$	$(8\sqrt{6})$	6	4	0	−4	6	−1	0	0	0	2	4	0	0	−2	−2	−1	0	0
48	48	$-\alpha^\ddagger$	$\beta^\ddagger$	2	−4	α	γ	$-8\sqrt{2}$	−12	−2	$2\sqrt{2}$	β	−2	−12	4	$2\sqrt{2}$	$\gamma^\ddagger$	0	0

TABLE 3. Components of the staurotope X.

16. The classification

Tables 2 and 3 list the realizations of $\mathcal{H}$. Instead of cosine vectors, to save space we list inner product vectors, with notional scaling of mixed vectors indicated by brackets replacing what should be 0; this has been done to fit each table into a single page. (Alternatively, they can be viewed as providing a merely Λ-orthogonal basis.) Therefore, to obtain the genuine (mixed) cosine vector, divide by the scaling factor in the second column. The first column gives the dimension d of the realization; for the mixed vectors, recall that the notional dimension (indicated by †) is $2d$ (see Remark 2.16).

Where $w > 1$, the cosine vectors (including the mixed ones) fall into a block of $\frac{1}{2}w(w+1)$. Such a block is easily identified in the tables; it begins and ends with genuine cosine vectors labelled with some dimension d, and in between are genuine and mixed cosine vectors with attached dimensions d or $2d$, the latter being notional. So, for instance, rows 10–15 of Table 2 correspond to the 25-dimensional realizations.

In a similar way, the first two rows of Table 3 are $4\Gamma_{19}$ and $4\Gamma_{20}$, giving the cosine vectors Γ_{19} of the original 120-cell $\{5,3,3\}$ and Γ_{20} of its starry allomorph $\{\frac{5}{2},3,3\}$ in $\mathbb{E}^4$.

We have adopted the following abbreviations, again for reasons of space:

$$\varphi := \tau\sqrt{5}, \quad \rho := 1+2\sqrt{5}, \quad \psi := \tfrac{1}{2}(7+5\sqrt{5}).$$

As usual, ‡ changes the sign of $\sqrt{5}$. Note that $\psi = -\tau^2\rho^\ddagger$; we continue to employ $\tau := \frac{1}{2}(1+\sqrt{5})$ (but write $\tau^\ddagger$ rather than $-\tau^{-1}$) and $\sigma := \frac{1}{2}(1+3\sqrt{5}) = \tau + \sqrt{5}$.

For dimension 48, we also write

$$\alpha := 12+6\sqrt{2}, \quad \beta := -4+4\sqrt{2}, \quad \gamma := 4+5\sqrt{2},$$

denoting changing the sign of $\sqrt{2}$ by ‡ as well; this should cause no confusion.

Acknowledgments

We wish to express our gratitude for the contribution of Frieder Ladisch [**4**], who provided us with some invaluable calculations found using representation theory; see also [**3**], which (as we said earlier) corrected claims made in [**5, 6, 9**]. While, in the end, we have not needed to use his calculations, nevertheless they provided a guide without which we could not have carried through our analysis.

We also wish to thank the two referees, whose useful suggestions have led to substantial improvements in the presentation of the material of the paper.

References

[1] H. S. M. Coxeter, *Regular polytopes*, 3rd ed., Dover Publications, Inc., New York, 1973. MR0370327

[2] P. Du Val, *Homographies, quaternions and rotations*, Oxford Mathematical Monographs, Clarendon Press, Oxford, 1964. MR0169108

[3] F. Ladisch, *Realizations of abstract regular polytopes from a representation theoretic view*, Aequationes Math. **90** (2016), no. 6, 1169–1193, DOI 10.1007/s00010-016-0434-y. MR3575585

[4] F. Ladisch, *Cosine vectors of the* 120*-cell*, (private communication).

[5] P. McMullen, *Realizations of regular polytopes, III*, Aequationes Math. **82** (2011), no. 1-2, 35–63, DOI 10.1007/s00010-010-0063-9. MR2807032

[6] P. McMullen, *Realizations of regular polytopes, IV*, Aequationes Math. **87** (2014), no. 1-2, 1–30, DOI 10.1007/s00010-013-0187-9. MR3175095

[7] P. McMullen, *New regular compounds of 4-polytopes*, New trends in intuitive geometry, Bolyai Soc. Math. Stud., vol. 27, János Bolyai Math. Soc., Budapest, 2018, pp. 307–320. MR3889265

[8] P. McMullen, *Geometric regular polytopes*, (Encyclopedia of Mathematics and Its Applications **172**). Cambridge University Press (Cambridge, 2020).

[9] P. McMullen and B. Monson, *Realizations of regular polytopes. II*, Aequationes Math. **65** (2003), no. 1-2, 102–112, DOI 10.1007/s000100300007. MR2012404

[10] P. McMullen and E. Schulte, *Abstract regular polytopes*, Encyclopedia of Mathematics and its Applications, vol. 92, Cambridge University Press, Cambridge, 2002. MR1965665

University College London, Gower Street, London WC1E 6BT, England
Email address: p.mcmullen@ucl.ac.uk

Contemporary Mathematics
Volume **764**, 2021
https://doi.org/10.1090/conm/764/15359

Prescribing symmetries and automorphisms for polytopes

Egon Schulte, Pablo Soberón, and Gordon Ian Williams

ABSTRACT. We study finite groups that occur as combinatorial automorphism groups or geometric symmetry groups of convex polytopes. When Γ is a subgroup of the combinatorial automorphism group of a convex d-polytope, $d \geq 3$, then there exists a convex d-polytope related to the original polytope with combinatorial automorphism group *exactly* Γ. When Γ is a subgroup of the geometric symmetry group of a convex d-polytope, $d \geq 3$, then there exists a convex d-polytope related to the original polytope with both geometric symmetry group and combinatorial automorphism group *exactly* Γ. These symmetry-breaking results then are applied to show that for every abelian group Γ of even order and every involution σ of Γ, there is a centrally symmetric convex polytope with geometric symmetry group Γ such that σ corresponds to the central symmetry.

1. Introduction

The study of convex polytopes is largely motivated by their symmetries. With every convex polytope P are associated two finite groups: the (geometric) symmetry group $G(P)$ consisting of the Euclidean isometries of the ambient space that preserve P, and the (combinatorial) automorphism group $\Gamma(P)$ consisting of the combinatorial symmetries of the face lattice of P. It is natural to ask about whether or not the converse is true: is every finite group the symmetry group or automorphism group of a convex polytope?

For automorphism groups this question was answered positively by Schulte and Williams [**SW15**], and later a simpler proof was found by Doignon [**Doi18**]. In this paper we are studying variations of this question with additional restrictions imposed on the polytopes in question. We are particularly interested in *centrally symmetric* convex polytopes in Euclidean d-space $\mathbb{E}^d$. By definition these admit the reflection in the origin, $x \mapsto -x$, as a geometric symmetry and thus have an automorphism group (as well as symmetry group) that contains an involution. The main motivation for this paper was to characterize the pairs (Γ, σ), consisting of a finite group Γ and an involution σ in Γ, with the property that Γ is the automorphism group of a *centrally symmetric* convex polytope such that σ corresponds to the central symmetry. In Theorem 4.1 we show that every abelian group Γ of

---

2010 *Mathematics Subject Classification.* Primary 52B15; Secondary 52B11, 51M20.

The research of the first author was partially supported by Simons Foundation award no. 420718.

The research of the second author was partially supported by NSF Grant DMS 1851420.

even order has the desired property: for every involution σ in Γ there is a centrally symmetric polytope with automorphism group Γ such that σ acts like the central symmetry.

Along the way we generalize the methods of [**SW15**] to establish the following two symmetry-breaking results for arbitrary convex polytopes, which are applicable in a wider context and are of independent interest. When Γ is a subgroup of the automorphism group of some convex d-polytope Q, $d \geq 3$, then there exists a convex d-polytope P related to Q with automorphism group *exactly* Γ. When Γ is a subgroup of the geometric symmetry group of some convex d-polytope Q, $d \geq 3$, then there exists a convex d-polytope P related to Q with both symmetry group and automorphism group *exactly* Γ. Our symmetry-breaking constructions are described in Section 3 and generalize to some extent to abstract polytopes (see [**MS02**]). In Section 4 we investigate centrally symmetric polytopes. Finally, in Section 5 we discuss some open problems and point to recent solutions.

The question of finding polytopes with prescribed automorphism group has also been asked as motivated by representation theory, see [**Lad16, BL18, FL18**]. These articles study orbit polytopes, that is, convex hulls of single point orbits under finite groups acting affinely on a real vector space. In this context it is natural to additionally consider the "affine symmetry group" (sometimes also called the "affine automorphism group") of a convex polytope, consisting of all non-singular affine transformations of the ambient space that preserve the polytope. As not every finite group is the affine automorphism group of an orbit polytope, it seems that symmetry-breaking processes as described here cannot be completely avoided to settle the above problem.

The question whether or not a given group is the automorphism group or symmetry group of a geometric, combinatorial, algebraic, or topological structure of a specified kind has been studied quite extensively. For a recent article describing the common characteristics of the approaches see the recent article [**Jon18**] by Jones.

2. Basic notions

We begin by recalling some basic definitions from the theory of convex and abstract polytopes (see [**Grü03, MS02, Zie95**]).

An *abstract polytope* of rank d is a ranked poset $\mathcal{P}$ with the following properties. The elements of $\mathcal{P}$ are called *faces*, and the possible face ranks are $-1, 0, \ldots, d$. A face is a *j-face* if its rank is j. Faces of ranks 0, 1 or $d-1$ are also called *vertices*, *edges* or *facets* of $\mathcal{P}$, respectively. The poset $\mathcal{P}$ has a smallest face (of rank -1) denoted F_{-1} and a largest face (of rank d) denoted F_d. Each *flag* (maximal totally ordered subset) Φ of $\mathcal{P}$ contains exactly $d+2$ faces, one for each rank j. Two flags are said to be *adjacent* if they differ in just one face; they are *j-adjacent* if this face has rank j. The poset $\mathcal{P}$ is *strongly flag-connected*, meaning that any two flags Φ and Ψ can be joined by a sequence of flags $\Phi = \Phi_0, \Phi_1, ..., \Phi_k = \Psi$, all containing $\Phi \cap \Psi$, such that any two successive flags Φ_{i-1} and Φ_i are adjacent. Finally, $\mathcal{P}$ satisfies the *diamond condition*: whenever $F \leq G$, with $\text{rank}(F) = j-1$ and $\text{rank}(G) = j+1$, there are exactly two faces H of rank j such that $F \leq H \leq G$. Thus, for $j = 0, \ldots, d-1$, a flag of $\mathcal{P}$ has exactly one j-adjacent flag.

If F and G are faces with $F \leq G$, then $G/F := \{H \mid F \leq H \leq G\}$ is called a *section* of $\mathcal{P}$. This is a polytope in its own right. For a face F, we also call F_d/F the *co-face* of $\mathcal{P}$ at F, or the *vertex-figure* of $\mathcal{P}$ at F if F is a vertex.

The face lattice of a convex polytope is an example of an abstract polytope. Recall that a *convex polytope* P is the convex hull of finitely many points in Euclidean d-space $\mathbb{E}^d$. A (proper) *face* of a convex d-polytope P is the intersection of P with a supporting hyperplane of P; the latter is a hyperplane H in $\mathbb{E}^d$ such that P lies entirely in one of the two closed half-spaces bounded by H and has points in common with H. The empty set $\emptyset$, and P itself, are also called (improper) faces of P. The set of all (proper and improper) faces of a convex polytope P, ordered by inclusion, forms a lattice called the *face lattice* of P. This is an abstract polytope, of rank d if P has dimension d. The *boundary complex* of a convex d-polytope P, denoted $\mathrm{bd}(P)$, is the set of all faces of P of rank less than d, partially ordered by inclusion (see [**Grü03**, p. 40]); this complex tessellates the boundary ∂P of P and is topologically a $(d-1)$-sphere.

Recall that a convex d-polytope is called *simple* if all its vertices have valency d, and *simplicial* if all its facets are $(d-1)$-simplices.

Let P be a convex d-polytope. The (standard) *barycentric subdivision* of P is the geometric simplicial complex of dimension d, whose d-simplices are precisely the convex hulls of the centroids of the non-empty faces in a flag of P (see [**Bay88**], [**TGOR17**, p. 642] or [**MS02**, Sect. 2C]). We use the term "barycentric subdivision" more broadly and allow the centroid of a face to be replaced by a relative interior point of that face. Thus, a barycentric subdivision of P is a d-dimensional geometric simplicial complex with one vertex in the relative interior of each non-empty face of P, and with one d-dimensional simplex per flag of P, such that the vertices of a d-simplex are precisely the relative interior points chosen in the faces of the corresponding flag. Each barycentric subdivision of P is isomorphic (as an abstract simplicial complex) to the order complex of the face lattice of P (with the empty face removed); in particular, any two barycentric subdivisions are isomorphic.

There is a similar notion of barycentric subdivision for the boundary complex of a convex polytope. By $\mathcal{C}(P)$ we denote the barycentric subdivision of the boundary complex $\mathrm{bd}(P)$ of P. This is a $(d-1)$-dimensional simplicial complex.

The order complex of an abstract polytope $\mathcal{P}$ similarly can be viewed as a "combinatorial barycentric subdivision" of $\mathcal{P}$ (see [**MS02**, Sect. 2C]).

The *k-skeleton* $\mathrm{skel}_k(\mathcal{P})$ of an abstract polytope $\mathcal{P}$ is the poset consisting of all proper faces of $\mathcal{P}$ of rank at most k (together with the induced partial order).

3. Preassigning symmetry groups

We begin this section with the following theorem about symmetry-breaking in convex polytopes.

THEOREM 3.1. *Let $d \geq 3$, let Q be a convex d-polytope with (combinatorial) automorphism group $\Gamma(Q)$, and let Γ be a subgroup of $\Gamma(Q)$. Then there exists a finite abstract d-polytope $\mathcal{P}$ with the following properties:*

(a) $\Gamma(\mathcal{P}) = \Gamma$.

(b) $\mathcal{P}$ is isomorphic to a face-to-face tessellation $\mathcal{T}$ of the $(d-1)$-sphere $\mathbb{S}^{d-1}$ by spherical convex $(d-1)$-polytopes.

(c) $\mathrm{skel}_{d-2}(\mathcal{C}(Q))$ *is a subcomplex of* $\mathrm{skel}_{d-2}(\mathcal{P})$.

(d) If Γ is a subgroup of the (geometric) symmetry group $G(Q)$ of Q, then the tessellation $\mathcal{T}$ on $\mathbb{S}^{d-1}$ in part (b) can be chosen in such a way that $G(\mathcal{T}) = \Gamma = \Gamma(\mathcal{T})$.

First notice that the conclusion of the theorem above may fail for $d = 2$, since the combinatorial automorphism group of a finite abstract 2-polytope (polygon) is necessarily dihedral and so in particular cannot be cyclic. Hence we must require $d \geq 3$.

PROOF. We begin with the second and third parts of the theorem, then settle the first part, and later refine our arguments to settle the fourth part. Our strategy is to refine the structure of the given convex polytope Q in such a way that all automorphisms in $\Gamma(Q)$ outside Γ are destroyed. The result will be a spherical abstract polytope whose automorphism group is given by Γ. This symmetry-breaking process is interesting in its own right.

Parts (b,c). Consider the (standard or any other) barycentric subdivision $\mathcal{C}(Q)$ of the boundary complex $\mathrm{bd}(Q)$ of Q in d-space $\mathbb{E}^d$. This is a simplicial $(d-1)$-complex that refines $\mathrm{bd}(Q)$ and is a realization of the order complex of $\mathrm{bd}(Q)$ (see [**MS02**, Sect. 2C]). Its simplices correspond to the chains (totally ordered subsets) in the poset $\mathrm{bd}(Q)$, with the chambers (maximal simplices) corresponding to the flags of $\mathrm{bd}(Q)$; here, inclusion of simplex faces in $\mathcal{C}(Q)$ corresponds to inclusion of chains in $\mathrm{bd}(Q)$. In particular, $\mathcal{C}(Q)$ has the structure of a *labelled* simplicial complex, in which every simplex is labelled by the set of ranks of the faces in the chain of $\mathrm{bd}(Q)$ represented by the simplex. Thus the vertices of $\mathcal{C}(Q)$ can be labelled by $0, \ldots, d-1$. The vertices of Q are exactly the vertices of $\mathcal{C}(Q)$ with label 0. The vertices of each chamber are labelled $0, \ldots, d-1$ such that no two vertices have the same label. Note that $\Gamma(Q)$ and hence Γ act on $\mathcal{C}(Q)$ as groups of automorphisms of a labelled simplicial complex (labels of simplices are preserved), and that the action on the chambers is free.

As in the proof of [**SW15**, Theorem 1], a key step in the construction consists of chamber replacement by complexes made up of Schlegel diagrams of convex polytopes. These complexes are inserted into the chambers of $\mathcal{C}(Q)$ in such a way that the $(d-2)$-skeleton $\mathrm{skel}_{d-2}(\mathcal{C}(Q))$ of $\mathcal{C}(Q)$ stays intact, unrefined. In fact, our proof basically consists of adapting the proof of [**SW15**, Theorem 1] to the more general situation at hand. (In the proof of that theorem, the corresponding subgroup Γ acted simply vertex-transitively on a special convex d-polytope Q constructed from a suitable permutation representation of Γ. In the present context, Q can be an arbitrary d-polytope and Γ need not act vertex-transitively.)

The complexes inserted into the chambers are constructed in exactly the same manner as in [**SW15**]. We will review the properties of these complexes below. Each complex is built from a Schlegel diagram $\mathcal{D}$ of a d-crosspolytope supported on a $(d-1)$-dimensional simplex D with vertices $u_0, \ldots, u_{d-1}$, by inserting affine images of Schlegel diagrams of certain convex d-polytopes (the polytopes R_i and L described below) into the $(d-1)$-simplices of $\mathcal{D}$ that correspond to certain facets of the d-crosspolytope. The resulting $(d-1)$-dimensional complex, which as in [**SW15**] is denoted $\mathcal{R}^L$, is also supported on D and has the boundary complex of D as a subcomplex. The particular choice of the polytopes R_i and L is quite delicate and is taken in such a way that the vertices $u_0, \ldots, u_{d-1}$ of the outer simplex D acquire very high valencies in $\mathcal{R}^L$ compared with the vertices in the interior of D, and that the valencies of these vertices in $\mathcal{R}^L$ are integers "very far apart" from each other. These conditions on the insertion process later prevent the existence of unwanted automorphisms. In particular, $\mathcal{R}^L$ itself will have no automorphism other than the trivial automorphism.

Before moving on to the actual chamber insertion process we briefly review the construction and properties of the complexes $\mathcal{R}^L$. First recall that the Schlegel diagram $\mathcal{D}$ of a d-crosspolytope consists of an *outer* $(d-1)$-simplex D, tiled in a face-to-face manner by $(d-1)$-simplices, the *simplex tiles* of $\mathcal{D}$. Among these simplex tiles is a *central* $(d-1)$-simplex Z, corresponding to the facet of the crosspolytope opposite to the facet defining D. The simplices D and Z have no vertices in common. The simplex tiles of $\mathcal{D}$ adjacent to Z (i.e., intersecting Z in a common facet) share precisely one vertex with D; conversely, every vertex u of D is a vertex of precisely one simplex tile, F_u (say), that is adjacent to Z. In the course of the construction we often require affine images of Schlegel diagrams supported on $(d-1)$-simplices. Clearly, any affine transformation that carries the supporting $(d-1)$-simplex of a Schlegel-diagram to another $(d-1)$-simplex, also carries the Schlegel diagram on the first simplex to a "diagram" on the second simplex (this also is a Schlegel diagram of some polytope). This is true no matter how the vertices of the first $(d-1)$-simplex are assigned by the affine transformation to the vertices of the second.

The next step is to modify $\mathcal{D}$ in such a way that the vertices in the outer simplex D acquire very high valencies compared with those in the interior, and that the valencies of the vertices of D are very far apart from each other. To this end, consider the simplex tiles $F_{u_0}, \ldots, F_{u_{d-1}}$ of $\mathcal{D}$ determined by the vertices $u_0, \ldots, u_{d-1}$ of D, and replace every simplex tile F_{u_i} by an affine image of the Schlegel diagram of a suitable convex d-polytope R_i. All vertices of this polytope R_i, save one, have small valencies but the exceptional vertex (which is mapped to u_i) has valency given by a large integer m_i to be determined. For example, for R_i we could take the pyramid over a *simple* convex $(d-1)$-polytope with m_i vertices and with at least one facet which is a simplex; then R_i itself has a simplex facet, with the apex of R_i as a vertex of valency m_i in R_i. Suppose R_i is a pyramid of this kind. Then R_i admits a Schlegel diagram $\mathcal{R}_i$ whose outer $(d-1)$-simplex corresponds to a simplex facet of R_i containing the apex of R_i. In this diagram, the outer vertex representing the apex has valency m_i while all other vertices have (small) valency d. Now take an affine transformation that maps the outer simplex of $\mathcal{R}_i$ to the simplex tile F_{u_i} of $\mathcal{D}$ such that the vertex corresponding to the apex is mapped to u_i, and then insert the corresponding affine image of the Schlegel diagram $\mathcal{R}_i$ into the simplex F_{u_i} such that F_{u_i} becomes the outer simplex. If this procedure is performed for each $i = 0, \ldots, d-1$, the result is a $(d-1)$-dimensional complex $\mathcal{R}$ supported on D, in which each vertex u_i of D has large valency, namely $m_i + d - 1$, while all vertices of $\mathcal{R}$ that are not vertices of D have small valencies.

We require one additional type of modification to complete the construction of $\mathcal{R}^L$, now targeting the $(d-1)$-simplex of $\mathcal{R}$ that was the central simplex of $\mathcal{D}$. Suppose L is any *simplicial* convex d-polytope. Then we let $\mathcal{R}^L$ denote the $(d-1)$-dimensional complex supported on D, in which the central simplex has been replaced by a suitable affine copy of a Schlegel diagram of L.

At this point of the construction we still have the choice of the parameters $m_0, \ldots, m_{d-1}$ and the polytopes L at our disposal. These will be chosen as we move along and will depend on the given polytope Q.

The chamber insertion process for the polytope Q employs the action of Γ as a group of label preserving automorphisms on the barycentric subdivision $\mathcal{C}(Q)$. For a chamber C of $\mathcal{C}(Q)$, we let $o(C)$ denote the orbit of C under Γ in its action on $\mathcal{C}(Q)$.

The chamber replacement now proceeds as follows. We first settle the choice of the polytopes L. For each chamber orbit $o(C)$ choose a simplicial convex d-polytope $L_{o(C)}$ in such a way that no two such polytopes have the same number of vertices. Then, for any fixed choice of parameters $m_0, \ldots, m_{d-1}$ (and associated complex $\mathcal{R}$), no two of the corresponding $(d-1)$-dimensional complexes $\mathcal{R}^{L_{o(C)}}$ have the same number of vertices, and thus no two complexes are combinatorially isomorphic.

In the final step of the chamber insertion process we first replace, for each chamber orbit $o(C)$, one of its chambers, C (say), by an affine copy of the corresponding complex $\mathcal{R}^{L_{o(C)}}$ such that, for each $i = 0, \ldots, d-1$, the vertex u_i of D is mapped onto the vertex of C labelled i in $\mathcal{C}(Q)$. We then exploit Γ to carry this new structure to all the other chambers in an orbit, and therefore to all chambers of $\mathcal{C}(Q)$. Recall that $\Gamma(Q)$, and hence Γ, acts freely and in a label preserving manner on the chambers of $\mathcal{C}(Q)$ (flags of Q). More explicitly, if C' is a chamber in the same orbit as C, that is, $o(C') = o(C)$, we replace C' by an affine copy of the complex $\mathcal{R}^{L_{o(C)}}$ that we used for C, such that, for each $i = 0, \ldots, d-1$, the vertex u_i of D is mapped onto the vertex of C' labelled i in $\mathcal{C}(Q)$. In short, with respect to insertion of diagrams we treat C and C' in the same manner, and we can do so without destroying the action of Γ because of the existence of label preserving transfer maps from Γ between chambers in the same orbit under Γ. The resulting $(d-1)$-dimensional complex $\mathcal{C}'$ is a refinement of $\mathcal{C}(Q)$ and has the full $(d-2)$-skeleton of $\mathcal{C}(Q)$ as a subcomplex, unrefined. In particular, $\mathcal{C}'$ tiles the boundary ∂Q of Q and hence is topologically a $(d-1)$-sphere. By construction, Γ acts on $\mathcal{C}'$ as a group of automorphisms.

Clearly we may project the complex $\mathcal{C}'$ radially onto any sphere about the centroid of Q, and rescale the sphere (if need be) to obtain an isomorphic complex $\mathcal{T}$ which tiles the unit sphere $\mathbb{S}^{d-1}$ in a face-to-face manner by spherical convex polytopes.

Finally, by adjoining suitable improper faces (of ranks -1 and d) to $\mathcal{C}'$ we arrive at a spherical abstract d-polytope, denoted $\mathcal{P}$. Then the properties of $\mathcal{P}$ described in parts (b) and (c) of the theorem are clear by construction. It remains to establish parts (a) and (d).

Part (a). For the proof of part (a), a more subtle choice of the parameters $m_0, \ldots, m_{d-1}$ is needed to guarantee that the polytope $\mathcal{P}$ has the property that $\Gamma(\mathcal{P}) = \Gamma$. Suppose Q and $\mathcal{C}(Q)$ are as before. For a vertex u of $\mathcal{C}(Q)$, we let s_u denote the number of chambers containing u, and note that this is just the number of flags of Q containing the face of Q represented by u. If x is a vertex of any complex $\mathcal{S}$, we also write $\text{val}_{\mathcal{S}}(x)$ for the valency of x in the edge graph (1-skeleton) of $\mathcal{S}$.

It is straightforward to compute the valencies of the vertices of $\mathcal{P}$ (or $\mathcal{C}'$). The valencies of the vertices of $\mathcal{P}$ in $\mathcal{C}(Q)$ depend on $m_0, \ldots, m_{d-1}$, while those of the vertices of $\mathcal{P}$ outside of $\mathcal{C}(Q)$ do not depend on $m_0, \ldots, m_{d-1}$ but are bounded by a constant depending on d and the polytopes $L_{o(C)}$. The details are as follows. For each $i = 0, \ldots, d-1$, each vertex x of $\mathcal{C}(Q)$ labelled i is the vertex labelled i in every chamber that contains it, and therefore

$$\mathrm{val}_{\mathcal{P}}(x) = \mathrm{val}_{\mathcal{C}(Q)}(x) + s_x m_i. \tag{3.1}$$

If x is a vertex of the central simplex in the complex $\mathcal{R}^{L_{o(C)}}$ inserted into a chamber C, then

$$\mathrm{val}_{\mathcal{P}}(x) = 2(d-1) + (\mathrm{val}_{L_{o(C)}}(x) - (d-1)) = \mathrm{val}_{L_{o(C)}}(x) + d - 1.$$

If x is a vertex of the copy of a polytope R_i inside a chamber C that is not a vertex of C or of the central simplex inside C, then $\mathrm{val}_{\mathcal{P}}(x) = d$. Finally, if x is a vertex of the copy of $L_{o(C)}$ in a chamber C that is not a vertex of the central simplex in C, then $\mathrm{val}_{\mathcal{P}}(x) = \mathrm{val}_{L_{o(C}}(x)$, In particular, there exists a constant m (depending on d and our choice of polytopes $L_{o(C)}$) such that

$$\mathrm{val}_{\mathcal{P}}(x) \leq m \tag{3.2}$$

for all vertices x of $\mathcal{P}$ outside of $\mathcal{C}(Q)$.

The parameters m_i are chosen inductively for $i = d-1, d-2, \ldots, 0$, beginning with $m_{d-1} := m$ where m is a fixed constant as in (3.2). Suppose for a moment that a specific parameter value m_i has been chosen and then substituted on the left side of equation (3.1) to give certain integers, $\mathrm{val}_{\mathcal{C}(Q)}(x)+s_x m_i$, representing vertex valencies in $\mathcal{P}$. In this situation we write a_i and b_i for the minimum or maximum of these integers $\mathrm{val}_{\mathcal{C}(Q)}(x) + s_x m_i$, respectively, taken over all vertices x in $\mathcal{C}(Q)$ labelled i, as given in (3.1). Thus $a_i \leq \mathrm{val}_{\mathcal{C}(Q)}(x) + s_x m_i \leq b_i$ for each vertex x of $\mathcal{C}(Q)$ labelled i. In particular, we trivially have $m < a_{d-1} \leq b_{d-1}$.

Proceeding inductively, we next choose m_{d-2} in such a way that $b_{d-1} < a_{d-2}$. More generally, if $j \leq d-1$ and m_j has already been chosen, we pick m_{j-1} in such a way that $b_j < a_{j-1}$. At the final step when $j = 1$, we are choosing m_0. Our choice of $m_0, \ldots, m_{d-1}$ then guarantees that

$$m < a_{d-1} \leq b_{d-1} < a_{d-2} \leq b_{d-2} < \ldots\ldots < a_1 \leq b_1 < a_0 \leq b_0. \tag{3.3}$$

Now set $M_i := [a_i, b_i]$ for each i, and observe that $M_0, \ldots, M_{d-1}$ are mutually disjoint intervals.

We now are ready to prove part (a) of the theorem. We show that if the parameters $m_0, \ldots, m_{d-1}$ are chosen in such a way that (3.3) is satisfied, then $\Gamma(\mathcal{P}) = \Gamma$. Suppose $m_0, \ldots, m_{d-1}$ are chosen such that (3.3) holds.

For the proof of part (a) we can mostly proceed as in [**SW15**], specifically Lemma 2. By construction, Γ is a subgroup of $\Gamma(\mathcal{P})$, so we only need to prove the opposite inclusion. The initial steps of the proof are the same (almost word for word) as those in [**SW15**, pp. 451-452]. To make the present paper reasonably self-contained we reproduce here some of the arguments.

The first step is to show that every automorphism of $\mathcal{P}$ is induced by an automorphism of Q. Suppose γ is an automorphism of $\mathcal{P}$. We want to show that γ lies in Γ. The vertices of $\mathcal{P}$ corresponding to vertices of $\mathcal{C}(Q)$ have higher valency than other vertices of $\mathcal{P}$ and hence must be permuted among each other by γ. Thus γ maps vertices of $\mathcal{C}(Q)$ to vertices of $\mathcal{C}(Q)$. Moreover, by our choice of $m_0, \ldots, m_{d-1}$, the valency of each vertex of $\mathcal{C}(Q)$ labelled i lies in M_i for each i, and the sets $M_0, \ldots, M_{d-1}$ are mutually disjoint. Hence γ must map vertices of $\mathcal{C}(Q)$ labelled i to vertices of $\mathcal{C}(Q)$ labelled i, for each i. In particular, since the vertices of $\mathcal{C}(Q)$ labelled 0 are precisely the vertices of Q, the vertices of Q must be permuted by γ. Since the full $(d-2)$-skeleton of $\mathcal{C}(Q)$ is an (unrefined) subcomplex of the

$(d-1)$-dimensional complex $\mathcal{C}'$ and already contains all the information about $\mathcal{C}(Q)$ (only the chambers need to be added to the $(d-2)$-skeleton to obtain $\mathcal{C}(Q)$), it then follows that γ induces a label preserving automorphism of $\mathcal{C}(Q)$ mapping vertices of Q to vertices of Q. Here it helps to bear in mind that $\mathcal{C}(Q)$ lies on a sphere.

We show that γ induces an automorphism γ_Q (say) of Q itself, and that γ_Q determines γ uniquely. Since every face of the polytope Q is uniquely determined by the set of flags of Q containing this face, it is clear that every vertex of $\mathcal{C}(Q)$ is uniquely determined by the chambers of $\mathcal{C}(Q)$ containing this vertex. Now if F is an i-face of Q and w_F is the corresponding vertex labelled i in $\mathcal{C}(Q)$, then $\gamma(w_F)$ is also a vertex labelled i in $\mathcal{C}(Q)$ and hence must corresponds to an i-face of Q. This i-face is simply $\gamma(F)$. Note here that γ induces an isomorphism between the vertex-stars of w_F and $\gamma(w_F)$ in $\mathcal{C}(Q)$; in particular, chambers of $\mathcal{C}(Q)$ containing w_F are mapped in a one-to-one and label preserving manner to chambers containing $\gamma(w_F)$.

It remains to show that γ_Q determines γ uniquely. To this end suppose γ_Q is the identity map on Q. Then the automorphism induced by γ on $\mathcal{C}(Q)$, $\gamma_{\mathcal{C}(Q)}$ (say), is also the identity map on $\mathcal{C}(Q)$, since the simplices in $\mathcal{C}(Q)$ just represent the chains of the boundary complex of Q, such that vertices of $\mathcal{C}(Q)$ labelled i correspond to faces of Q of rank i. With regards to chamber replacement in $\mathcal{C}(Q)$ by complexes like $\mathcal{R}^{L_{o(C)}}$, note that γ maps a complex like $\mathcal{R}^{L_{o(C)}}$ placed into a chamber, to a similar such complex placed into the image chamber under γ. But since γ fixes every face of a chamber of $\mathcal{C}(Q)$, which in a complex like $\mathcal{R}^{L_{o(C)}}$ becomes the outer simplex, γ then must also fix the entire complex inserted into the chamber. This follows from a simple connectedness argument. The outer simplex of a complex $\mathcal{R}^{L_{o(C)}}$ can be joined to every tile in $\mathcal{R}^{L_{o(C)}}$ by a finite sequence of successively adjacent tiles (successive tiles meet in a facet). Beginning with the outer simplex on which γ is the identity map, we then can move along the sequence to show that γ is also the identity map on every tile in the sequence. Hence γ is the identity map on the entire complex $\mathcal{C}'$ and therefore also on $\mathcal{P}$.

Thus $\Gamma(\mathcal{P})$ can be viewed as a subgroup of $\Gamma(Q)$ containing Γ. The final step consists of showing that $\Gamma(\mathcal{P}) = \Gamma$. Here the arguments of [**SW15**, pp. 452-453] need to be modified as follows.

Suppose that Γ is a proper subgroup of $\Gamma(\mathcal{P})$. Then since $\Gamma(Q)$ acts freely on the chambers of $\mathcal{C}(Q)$, and $\Gamma(\mathcal{P})$ is a subgroup of $\Gamma(Q)$, the orbits of chambers C of $\mathcal{C}(Q)$ under $\Gamma(\mathcal{P})$ are strictly larger than those under Γ. In particular, there are two different orbits $o(C_1)$ and $o(C_2)$ of chambers C_1 and C_2 under Γ, which lie in the same orbit under $\Gamma(\mathcal{P})$. Any automorphism γ of $\Gamma(\mathcal{P})$ which maps a chamber C_1' in $o(C_1)$ to a chamber C_2' in $o(C_2)$ induces an isomorphism between the corresponding complexes $\mathcal{R}^{L_{o(C_1)}}$ and $\mathcal{R}^{L_{o(C_2)}}$ inserted into C_1' and C_2', respectively. However, this is impossible, since the complexes $\mathcal{R}^{L_{o(C)}}$ are mutually non-isomorphic, by our choice of the polytopes $L_{o(C)}$. Thus $\Gamma(\mathcal{P}) = \Gamma$. This completes the proof of part (a) of the theorem.

Part (d). For the proof of part (d) we must further refine our arguments. So let Γ be a subgroup of the geometric symmetry group $G(Q)$ of Q. In this case we choose the standard barycentric subdivision for $\mathcal{C}(Q)$ (with the vertices of $\mathcal{C}(Q)$ at the centroids of the faces of Q). Then $\mathcal{C}(Q)$ is invariant under Γ, since geometric symmetries of convex polytopes map face centroids to face centroids. Next we proceed as before and replace, for each chamber orbit $o(C)$ under Γ, one of its chambers, C (say), by an affine copy of the corresponding complex $\mathcal{R}^{L_{o(C)}}$ such that, for each $i = 0, \ldots, d-1$, the vertex u_i of D is mapped onto the vertex of C labelled i. For the chamber replacement of the remaining chambers of $\mathcal{C}(Q)$ we use as transfer maps the elements of Γ, which now are geometric symmetries of $\mathcal{C}(Q)$. More explicitly, if C' is a chamber of $\mathcal{C}(Q)$ with $o(C') = o(C)$, and γ is the (unique, labeling preserving) symmetry that maps C to C', we replace C' by the image of $\mathcal{R}^{L_{o(C)}}$ under γ. Then the overall structure is also invariant under Γ, and the same holds for its (scaled) projected image $\mathcal{T}$ on $\mathbb{S}^{d-1}$. Note that $\mathcal{T}$ cannot acquire geometric symmetries which do not belong to Γ, since these would also give combinatorial symmetries, which is impossible by part (a). This completes the proof of (d). □

Our next theorem is based on Theorem 3.1 and deals with geometric symmetry breaking results for convex polytopes.

THEOREM 3.2. *Let $d \geq 3$, let Q be a convex d-polytope, and let Γ be a subgroup of $\Gamma(Q)$. The abstract polytope $\mathcal{P}$ of Theorem 3.1 may be realized by a convex d-polytope P. Moreover, if Γ is a subgroup of $G(Q)$, then P can be chosen in such a way that $G(P) = \Gamma = \Gamma(P)$.*

PROOF. The proof of the first statement is the same as the proof of [**SW15**, Theorem 4.2]: first the complex $\mathcal{C}(Q)$ is realized by a convex d-polytope R, and then all subsequent modifications to the boundary of R required for the construction of $\mathcal{P}$ are achieved by gluing projective copies of convex polytopes to the facets of R that are sufficiently thin in the direction of the outward facing normal to the facet. The result is a convex d-polytope P.

The proof of the second statement is similar. First observe that R can be chosen in such a way that Γ lies in $G(R)$. In fact, the construction of R described in the proof of [**SW15**, Lemma 3] respects symmetries and leads to a convex d-polytope R whose symmetry group contains Γ as a subgroup. The chamber replacement can again be realized by gluing thin projective copies of convex polytopes to facets of R. Now this is done in two steps. First, we only glue copies to the facets of R which correspond to chambers in a system of representatives for the chamber orbits $o(C)$ on $\mathcal{C}(Q)$ under Γ. Second, we use the symmetries in Γ to attach copies to the remaining facets of R, such that facets of R equivalent under Γ receive projective copies which are also equivalent under Γ. Bear in mind that the boundary complex of R has the structure of a labeled simplicial complex on which Γ acts freely in a label preserving manner. If the projective copies used in the first step are sufficiently thin, then the resulting structure is a convex d-polytope. By construction this polytope is invariant under Γ. □

Parts of Theorem 3.1 hold more generally for finite abstract polytopes. With a very similar proof we can establish the following theorem.

THEOREM 3.3. *Let $d \geq 3$, let $\mathcal{Q}$ be a finite abstract d-polytope, and let Γ be a subgroup of $\Gamma(\mathcal{Q})$. Then there exists a finite abstract d-polytope $\mathcal{P}$ with the following properties:*

(a) $\Gamma(\mathcal{P}) = \Gamma$.

(b) $\mathcal{P}$ is isomorphic to a face-to-face tessellation on the topological space $|\mathcal{C}(\mathcal{Q})|$ of the order complex $\mathcal{C}(\mathcal{Q})$ of $\mathcal{Q}$ by topological copies of convex polytopes.

(c) $\mathrm{skel}_{d-2}(\mathcal{C}(\mathcal{Q}))$ is a subcomplex of $\mathrm{skel}_{d-2}(\mathcal{P})$.

4. Prescribing involutions as central symmetries

As an application of Theorem 3.2 we consider the following problem. Given a finite group Γ and a subgroup Λ of Γ, can we find a convex polytope P such that

- $\Gamma(P) = \Gamma$ and
- Λ acts on P in a predetermined way?

We are particularly interested in the case where $\Lambda = C_2$ and Λ is generated by a central involution σ of Γ. We wish to find a polytope P such that σ acts on P as a central symmetry; that is, abusing notation, $\sigma(x) = -x$ for all $x \in P$. Thus P would be centrally symmetric under the central symmetry σ. A positive answer would give a centrally symmetric version of the results of [**SW15**]. Here we show that the answer is always positive for finite abelian groups containing an involution, that is, for abelian groups of even order.

THEOREM 4.1. *Let Γ be a finite abelian group of even order, and let σ be an involution of Γ. Then there is a positive integer d and a centrally symmetric convex d-polytope P in $\mathbb{E}^d$, such that $G(P) = \Gamma(P) = \Gamma$ and σ is realized as the central symmetry of P, that is, $\sigma(x) = -x$ for all $x \in \mathbb{E}^d$.*

PROOF. Let us begin with the case where Γ is a cyclic group of even order with generator γ. Thus $\Gamma = C_{2m}$ for some $m \geq 1$, and $\sigma = \gamma^m$. We show that there exists a polytope of the desired kind in dimension $d = 4$. Consider the action of Γ as a group of isometries on $\mathbb{E}^4$, here viewed as complex 2-space $\mathbb{C}^2$ (with $x \in \mathbb{E}^4$ corresponding to $(u, v) \in \mathbb{C}^2$), defined by letting γ act as the mapping

$$(u, v) \mapsto (e^{\pi i/m}\, u, e^{\pi i/m}\, v).$$

Notice that for all $x \in \mathbb{E}^4$, $||\gamma(x)|| = ||x||$. If we take a large enough finite set of points S in $\mathbb{S}^3$ (a five-element subset S in general position suffices if $m \geq 3$, although one can do with less), then the convex hull of the orbit set

$$\Gamma{\cdot}S := \{\varphi(x) \mid \varphi \in \Gamma,\ x \in S\}$$

is a convex 4-polytope Q such that $\Gamma \leq G(Q)$ and $\sigma(x) = -x$ for all $x \in Q$. We then apply the construction process underlying Theorem 3.2 to construct the desired convex 4-polytope P. In other words, we get rid of all excess combinatorial symmetries outside of Γ while preserving each element of Γ as a geometric symmetry for P, including in particular the involution σ as the central symmetry for P. Thus $G(P) = \Gamma = \Gamma(P)$. This settles the case when Γ is cyclic. (Note that we cannot work with $\mathbb{E}^2$ in place of $\mathbb{E}^4$ since the corresponding statement of Theorem 3.2 fails to be true for $n = 2$.)

If Γ is abelian but not cyclic, then, by the fundamental theorem of abelian groups, we can write Γ as a direct product of $k+1$ abelian groups $\Gamma = \Gamma_1 \times \ldots \times \Gamma_k \times \Gamma_{k+1}$ for some $k \geq 1$, so that

- $\Gamma_1, \ldots, \Gamma_k$ are cyclic and of even order, and
- $\sigma = (\sigma_1, \ldots, \sigma_k, 1)$, where σ_i is an involution in Γ_i for all $1 \leq i \leq k$.

The idea is to manufacture a suitable polytope for each direct factor of Γ and then combine these polytopes into a single polytope for Γ itself.

We know from the above that for each direct factor Γ_i, with $1 \leq i \leq k$, there is a centrally symmetric 4-polytope P_i in $\mathbb{E}^4$ such that $G(P_i) = \Gamma_i = \Gamma(P_i)$ and σ_i is the central symmetry for P_i. Consider the cartesian product polytope $P' := P_1 \times \ldots \times P_k$ in $\mathbb{E}^{4k}$, whose vertex set is the cartesian product of the vertex sets of the component polytopes. Clearly, P' is a centrally symmetric $4k$-polytope, and the direct product $\Gamma' := \Gamma_1 \times \ldots \times \Gamma_k$ acts on P' as a group of symmetries such that each factor Γ_i acts on the ambient 4-dimensional subspace of the component polytope P_i. Under this action, $(\sigma_1, \ldots, \sigma_k)$ is the central symmetry for P'.

For the last direct factor, Γ_{k+1}, we embed $\Gamma'' := \Gamma_{k+1}$ into a symmetric group, S_{l+1} for some l, and then take a regular l-simplex P'' in $\mathbb{E}^l$ centered at the origin. Then Γ'' is a (generally proper) subgroup of $G(P'') = \Gamma(P'') = S_{l+1}$. Any excess symmetries that P'' might have, will be trimmed at a later stage.

We now combine these polytopes. Set $d := 4kl$. Let V denote the set of points in $\mathbb{E}^d = \mathbb{E}^{4k} \otimes \mathbb{E}^l$ of the form $u \otimes v$, where u and v are vertices of P' and P'', respectively, and $\otimes$ denotes the standard tensor product (given by $u \cdot v^T$ if u and v are viewed as column vectors). Then V is a centrally symmetric point set, since the vertex set of P' is centrally symmetric and $(-u) \otimes v = -u \otimes v$. Hence the convex hull of V in $\mathbb{E}^d$ is a centrally symmetric convex d-polytope P. Note that for the central symmetry of P it is not required that P'' is centrally symmetric.

By construction, the actions of Γ' on P' and Γ'' on P'' induce an action of $\Gamma = \Gamma' \times \Gamma''$ on P as a group of geometric symmetries. Thus Γ is a subgroup of $G(P)$. If we write the given involution σ of Γ in the form $\sigma = (\sigma', \sigma'')$ with $\sigma' = (\sigma_1, \ldots, \sigma_k) \in \Gamma'$ and $\sigma'' := 1 \in \Gamma''$, then under this action, σ maps each vertex $u \otimes v$ of P to $(-u) \otimes v = -u \otimes v$ and thus acts on P as central symmetry $-\mathrm{id}$, as desired.

In the final step, if P has any extra symmetries outside of Γ (as will usually be the case), we can trim them down using Theorem 3.2. This finally produces the desired centrally symmetric polytope. □

For cyclic groups (of even order), the construction underlying Theorem 4.1 produced convex polytopes in dimension 4. The reader might wonder if a suitable geometric representation of these groups in 3-space $\mathbb{E}^3$ can not also give a convex 3-polytope. As the following theorem shows, the answer is negative for many abelian groups. Dimension 4 is optimal in many cases.

THEOREM 4.2. *If $\Gamma = C_{4m} = \langle \gamma \rangle$ for some $m \geq 1$, and $\sigma := \gamma^{2m}$, then there is no centrally symmetric 3-polytope P in $\mathbb{E}^3$ such that $\Gamma(P) = \Gamma$ and σ is realized as the central symmetry of P.*

PROOF. Suppose to the contrary that such a 3-polytope P exists. Then, since ∂P is homeomorphic to $\mathbb{S}^2$, we can view Γ as a group of homeomorphisms of $\mathbb{S}^2$. In particular, $\lambda := \gamma^m$ is a homeomorphism of $\mathbb{S}^2$ with $\lambda^2 = \sigma = -\mathrm{id}$ and thus its topological degree must be positive. On the other hand, the topological degree of the homeomorphism $-\mathrm{id}$ of the k-sphere $\mathbb{S}^k$ is $(-1)^{k+1}$, which is -1 when $k = 2$. This leaves no possibility for the topological degree of λ. Thus P cannot exist (and dimension 4 is optimal if $\Gamma = C_{4m}$ and $\sigma := \gamma^{2m}$). □

Note that if in Theorem 4.2 we had insisted on achieving $G(P) = \Gamma$ (rather than $\Gamma(P) = \Gamma$), we could have argued similarly by using the determinant of linear mappings (rather than the topological degree of homeomorphisms) to rule out the existence of P. In fact, the determinant of λ^2 would have to be positive, but the central inversion $-\mathrm{id}$ has determinant -1 in dimension 3.

On the other hand, for cyclic groups of the form $\Gamma = C_{2m} = \langle \gamma \rangle$ with m odd, and $\sigma := \gamma^m$, we can indeed find a convex 3-polytope in $\mathbb{E}^3$ such that $\Gamma(P) = \Gamma$ and σ is realized as the central symmetry of P. This can be obtained as follows.

Consider a bipyramid P' in $\mathbb{E}^3$ over a regular $2m$-gon in the xy-plane centered at the origin o, where the two apices lie symmetrically on the z-axis on different sides of the xy-plane. Clearly, P' is invariant under the rotatory reflection γ of order $2m$ which is the product of the rotation by π/m about the z-axis and the reflection in the xy-plane. Thus $C_{2m} = \langle\gamma\rangle \leq G(P')$ and $\gamma^m = -\,\mathrm{id}$. Note that $G(P')$ is strictly larger than C_{2m}, since it also contains the reflection in the xy-plane but C_{2m} does not. (In fact, $G(P') \cong D_{2m} \times C_2$.) Thus P' itself does not have the required properties. However, a simple application of Theorem 3.2 allows us to find a polytope P by getting rid of the additional symmetries while preserving the action of C_{2m}. Alternatively, we can construct a polytope P directly from P' by attaching sufficiently thin pyramids to the facets of P' in one facet orbit of P' under C_{2m}.

5. Some open problems

Our previous discussion invites a number of open problems concerning the dimension of polytopes with preassigned symmetry groups or automorphism groups. Usually, given the group Γ the interest is in finding polytopes of small dimension realizing Γ. After the first version of this manuscript was uploaded to public repositories, independently of our work the three open questions below have been answered affirmatively [**CLS19**]. We present the questions here since they may lead to more directions of research.

For a finite group Γ, we define the (combinatorial) *convex polytope dimension* of Γ, denoted $\mathrm{cpd}(\Gamma)$, as the smallest dimension d for which there exists a convex d-polytope P whose combinatorial automorphism group is Γ, that is, $\Gamma(P) = \Gamma$. Note that the results of [**SW15, Doi18**] are saying that for every finite group Γ, we have $\mathrm{cpd}(\Gamma) < \infty$.

Similarly, the *geometric convex polytope dimension* of Γ, denoted $\mathrm{gcpd}(\Gamma)$, is defined to be the smallest dimension d for which there is a convex d-polytope P whose geometric symmetry group is Γ, that is, $G(P) = \Gamma$. The results of [**Doi18**] also imply $\mathrm{gcpd}(\Gamma) < \infty$.

OPEN QUESTION 1. For each n, is there a finite group Γ_n such that $\mathrm{cpd}(\Gamma_n) \geq n$?

OPEN QUESTION 2. For each n, is there a finite group Γ_n such that $\mathrm{gcpd}(\Gamma_n) \geq n$?

OPEN QUESTION 3. Does Theorem 4.1 hold for non-abelian groups Γ and central involutions σ of Γ? In other words, given a finite group Γ of even order and a central involution σ of Γ, is there a centrally symmetric convex polytope P with $G(P) = \Gamma(P) = \Gamma$ such that σ is realized as the central symmetry $-\,\mathrm{id}$ of P?

Note that the proof of Theorem 4.1 carries over to finite groups of the form $\Gamma = \Gamma_1 \times \Gamma_2$ where Γ_1 is abelian, and central involutions of Γ of the form $\sigma = (\sigma_1, 1)$ where σ_1 is a central involution of Γ_1.

Acknowledgments

The authors would like to thank the anonymous referees for their valuable comments and suggestions that have improved the paper.

References

[BL18] Barbara Baumeister and Frieder Ladisch, *A property of the Birkhoff polytope*, Algebr. Comb. **1** (2018), no. 2, 275–281, DOI 10.1007/s42001-018-0018-9. MR3856525

[Bay88] Margaret M. Bayer, *Barycentric subdivisions*, Pacific J. Math. **135** (1988), no. 1, 1–16. MR965681

[CLS19] A. Chirvasitu, F. Ladisch, and P. Soberón, *Non-commutative groups as prescribed polytopal symmetries*, arXiv preprints `arXiv:1907.10022` [math.MG]

[Cox73] H. S. M. Coxeter, *Regular polytopes*, 3rd ed., Dover Publications, Inc., New York, 1973. MR0370327

[Doi16] Jean-Paul Doignon, *Any finite group is the group of some binary, convex polytope*, Discrete Comput. Geom. **59** (2018), no. 2, 451–460, DOI 10.1007/s00454-017-9945-0. MR3755730

[Doi18] Jean-Paul Doignon, *Any finite group is the group of some binary, convex polytope*, Discrete Comput. Geom. **59** (2018), no. 2, 451–460, DOI 10.1007/s00454-017-9945-0. MR3755730

[FL18] Erik Friese and Frieder Ladisch, *Classification of affine symmetry groups of orbit polytopes*, J. Algebraic Combin. **48** (2018), no. 3, 481–509, DOI 10.1007/s10801-017-0804-0. MR3864737

[Fr38] R. Frucht, *Herstellung von Graphen mit vorgegebener abstrakter Gruppe* (German), Compositio Math. **6** (1939), 239–250. MR1557026

[Grü03] Branko Grünbaum, *Convex polytopes*, 2nd ed., Graduate Texts in Mathematics, vol. 221, Springer-Verlag, New York, 2003. Prepared and with a preface by Volker Kaibel, Victor Klee and Günter M. Ziegler. MR1976856

[Jon18] G. A. Jones, *Realisation of groups as automorphism groups in categories*, arXiv Preprint `arXiv:1807.00547` [math.GR]

[Lad16] Frieder Ladisch, *Realizations of abstract regular polytopes from a representation theoretic view*, Aequationes Math. **90** (2016), no. 6, 1169–1193, DOI 10.1007/s00010-016-0434-y. MR3575585

[MS02] Peter McMullen and Egon Schulte, *Abstract regular polytopes*, Encyclopedia of Mathematics and its Applications, vol. 92, Cambridge University Press, Cambridge, 2002. MR1965665

[SW15] Egon Schulte and Gordon Ian Williams, *Polytopes with preassigned automorphism groups*, Discrete Comput. Geom. **54** (2015), no. 2, 444–458, DOI 10.1007/s00454-015-9710-1. MR3372119

[TGOR17] Jacob E. Goodman, Joseph O'Rourke, and Csaba D. Tóth (eds.), *Handbook of discrete and computational geometry*, Discrete Mathematics and its Applications (Boca Raton), CRC Press, Boca Raton, FL, 2018. Third edition of [MR1730156]. MR3793131

[Zie95] Günter M. Ziegler, *Lectures on polytopes*, Graduate Texts in Mathematics, vol. 152, Springer-Verlag, New York, 1995. MR1311028

Northeastern University, Department of Mathematics, Boston, Massachusetts 02115

Email address: `e.schulte@northeastern.edu`

Baruch College, City University of New York, Department of Mathematics, New York, New York 10010

Email address: `pablo.soberon-bravo@baruch.cuny.edu`

University of Alaska Fairbanks, Department of Mathematics and Statistics, Fairbanks, Alaska 99709

Email address: `giwilliams@alaska.edu`

Contemporary Mathematics
Volume **764**, 2021
https://doi.org/10.1090/conm/764/15337

The rhombic triacontahedron and crystallography

Marjorie Senechal and Jean E. Taylor

ABSTRACT. We show that subsets of the rhombic triacontahedron (hereinafter RT) tile R^3 and correspond to the combinatorial types of lattice Voronoi cells. In the process, we relax the hypothesis of convexity in the classification of parallelohedra. Finally, we show that this provides a uniform description of periodic approximants to a large class of quasicrystals.

1. The overlap puzzle

Quasicrystals, first discovered in nature in 1982, are crystals whose atomic arrangements exhibit long-range aperiodic order, as shown (for example) in sharp diffraction patterns with icosahedral or decagonal symmetry. This discovery overturned the long-standing lattice paradigm in crystallography, which held that atomic groupings repeated, tile-like, periodically in all directions.

Where are the atoms in aperiodic crystals? What are the local arrangements that give rise to global patterns that we call aperiodic yet orderly? These questions are difficult and far from resolved, both theoretically and for actual crystals. Indeed, though many different quasicrystalline materials were quickly found, the question "where are the atoms?" was not settled for any of them until 2007 [**1**], 25 years after the first sighting, and four years before the discoverer was awarded a Nobel prize for it [**2**].

That first "solved" structure (see §4) dashed the expectation that the atoms in aperiodic crystals could always be meaningfully grouped into tiles: it is better described as a covering by interlocking nested polyhedral clusters. In many such cluster models, the atomic pattern as a whole is described by rhombic triacontahedra (RTs) overlapping in golden obtuse rhombohedra.

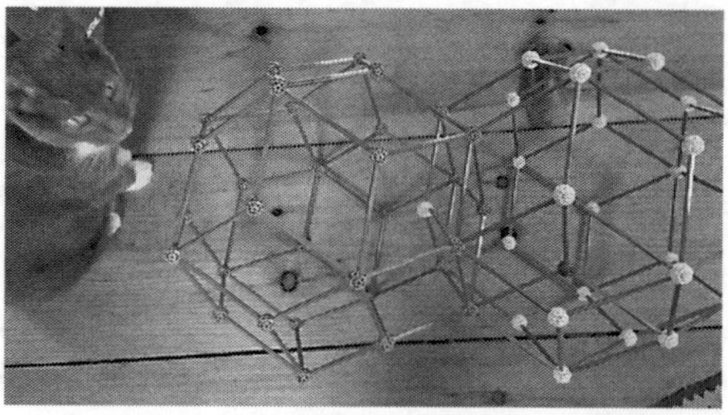

FIGURE 1. Father Frost contemplating overlapping RTs.

But, as appealing as the description of overlapping RTs is, there are problems with it. Obviously it is unrealistic for any growth model; one cannot force an overlap of two RTs without removing atoms and replacing bonds. Yet overlapping RTs are ubiquitous in periodic intermetallic crystals too. We begin by re-examining the classical (mathematical) crystal structure model, tiling space by translation.

2. Polyhedra and paralellohedra

The following definitions are tailored for our purposes, and thus are somewhat less general than usual.

DEFINITION 1. A finite polygon, or n-gon, is the figure formed by $n > 2$ distinct points (vertices) $V_1, \cdots, V_n$ in R^3 together with the line segments (edges) (V_i, V_{i+1}) for $i = 1, 2, \ldots, n-1$, and (V_n, V_1), with the requirements that adjacent edges are not colinear and the figure, *as a graph*, is planar.

DEFINITION 2. A polyhedron P is any finite family of polygons (called its faces) such that

(1) each edge of each polygon is the edge of just one other polygon, and the spans (defined below) of the two polygons are not coplanar along that edge, and
(2) the union of the spans of the polygons of P forms the boundary of a solid which is topologically a ball (we also refer to that solid as the polyhedron P).

We will use the phrase "vertex of P" to mean a vertex of a polygon of P and "edge of P" to mean an edge of a polygon of P.

A (translation) lattice $L \in R^d$ is a set of points spanned by the integral linear combination of d linearly independent vectors. Each point of L is a center of symmetry for L, that is, $L = -L$. The midpoints of the vectors in L are also centers of symmetry for it. In keeping with our goal of modeling real materials, we assume in this paper that $d = 3$.

The following proposition is an easy exercise (see, for example [**3**]).

PROPOSITION 1. *The order of a rotation symmetry about a lattice point can be only 2, 3, 4 or 6.*

Before the discovery of quasicrystals, Proposition 1 was thought to explain the apparent absence of icosahedral symmetry in crystals. Indeed it was (and in some quarters still is) known as "The Crystallographic Restriction."

Crystal growth was modeled by stacking congruent polyhedra in rows and layers, parallel and face-to-face.

DEFINITION 3. A parallelohedron is a polyhedron that tiles R^3 face-to-face by translations.

Which polyhedra are parallelohedra? M. C. Escher's periodic drawings show how unexpectedly complex parallelohedra can be even in R^2, if no constraints are placed on them. That may be why, in the mathematics literature, convexity is usually part of their definition. But, as we will show, nonconvex parallelohedra can bridge local "non-crystallographic" order, global periodic order, and aperiodic order. Still, some constraints are necessary. We do not require convexity but consider instead the (possibly) special case of polyhedra all of whose faces are centrosymmetric.[1]

Thus we will consider only centro-symmetric polygons in the remainder of this paper, and only ones which are the boundary of a centro-symmetric surface that is topologically an embedded disk. (Except in the case of intersection of parts of triangles of opposite orientation, such surfaces are the union of the n (two-dimensional) triangles from its centroid to its edges.) We call such a surface the *span* of the polygon.

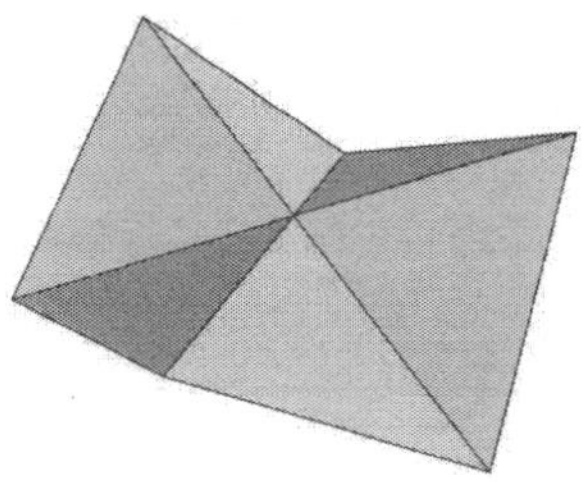

FIGURE 2. A skew hexagon, spanned.

It is important that we do not consider the centroid of a span of a polygon of P to be a vertex of P, nor do we consider the intersections of the triangles to be edges of P. The purpose of spanning is to provide the boundary of the solid object which can (or cannot) be used to tile space.

DEFINITION 4. A cs-polyhedron P is any polyhedron such that

(1) each face is centro-symmetric[2] and can be spanned by a centro-symmetric topological disk; we call this centrosymmetric disk the span of that polygon;
(2) P is itself centro-symmetric (about its centroid).

Thus, immediately, the number edges of every face of a cs-polyhedron is even. Each edge e of a cs-polyhedron P defines a *belt*:

DEFINITION 5. A *belt* of a cs-polyhedron P given by an edge e of P is the set of edges of P obtained from e by repeatedly inverting through the centroids of faces

[1] That is, the polygons are symmetric about their centroids.

[2] For convex polyhedra, the centro-symmetry of the faces implies that of the polyhedron itself ([**4**]).

to obtain opposite edges, moving from one face to the next via that opposite edge, together with the faces (possibly skew) to which these edges are incident.[3] The cardinality of a belt is the number of edges (or, equivalently, faces) it contains.

Thus, for example, a cube has three belts, each consisting of quadrilaterals and all of cardinality four, while the hexagonal prism has one belt of cardinality six (all quadrilateral) and three belts of cardinality four (alternating quadrilaterals and hexagons). Each of the RT's six belts is of cardinality ten (Figure 3.) The cs-polyhedron of Figure 4 has one belt of cardinality 8 and four of cardinality 4; observe that non-opposite but parallel edges of the octagonal face are in different belts.

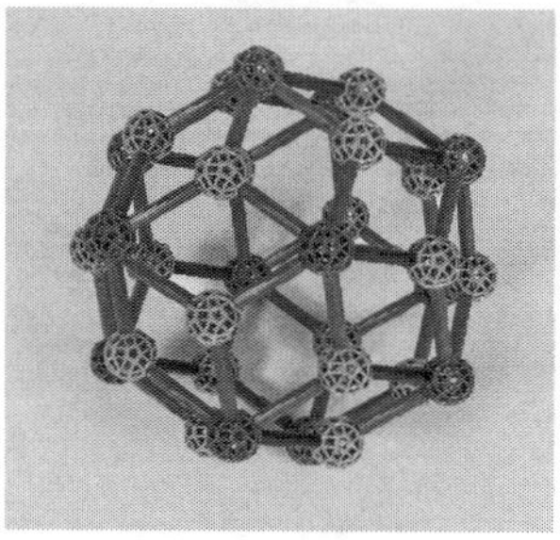

FIGURE 3. A ZomeTool model of the RT. The gray vertices lie on 3-fold rotation axes, the red on 5-fold. The ten black edges constitute a belt.

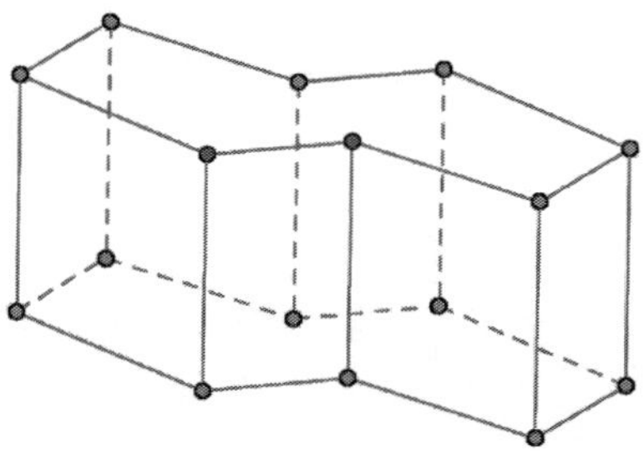

FIGURE 4. This cs-polyhedron has one belt of cardinality 8 and four of cardinality 4.

Convex parallelohedra (in any dimension d) are characterized by the following theorem, due to McMullen [**5**] (special cases had been studied by Minkowski and many others).

THEOREM 1. *A convex polyhedron in R^d is a parallelohedron if and only if*

(1) *the polyhedron is centro-symmetric;*
(2) *its faces are centro-symmetric; and*
(3) *its belts have cardinality four or six.*

[3]The term *zone* is also used for belt in the literature.

The combinatorial types of parallelohedra in R^3 were first enumerated in 1885 by the crystallographer E. S. Fedorov, who discovered one of them (the elongated dodecahedron) in the process. His list is a corollary to Theorem 1 (see, e.g., [**6**]).

COROLLARY 1. *There are exactly five combinatorial types of convex parallelohedra: the rhombic hexahedron RH (also known as the cube), the rhombic dodecahedron (RD), the hexagonal prism (HP), the elongated dodecahedron (ED), and the truncated octahedron (TO) (shown in Figure* 5*).*

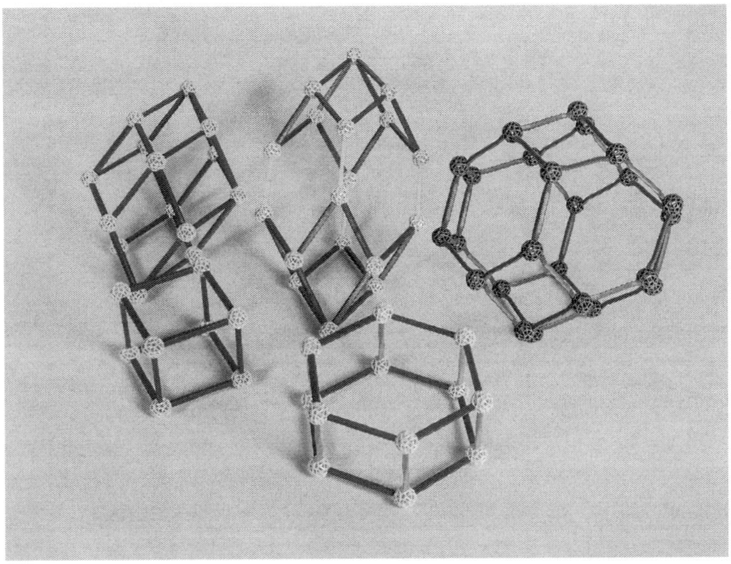

FIGURE 5. The five convex parallelohedra. Clockwise from upper left: the rhombic dodecahedron RD, the elongated dodecahedron ED (the elongating belt of four edges is shown in white), the truncated octahedron TO, the hexagonal prism HP, and the rhombic hexahedron RH (cube).

Because of its crucial role in mathematical crystallography, B. N. Delone ranked Fedorov's list in importance with the five regular solids.

Voronoi cells of lattices are parallelohedra. Recall that the Voronoi cell of a given lattice point is the closure of the set of points closer to it than to any other lattice point. By construction, Voronoi cells are convex and centro-symmetric, and their faces are centro-symmetric about their centroids. Since the lattice is an orbit of a translation group, the Voronoi cells of its points are congruent. The symmetry group of the Voronoi cell is the stabilizer (in $O(3)$) of a lattice point.

Voronoi conjectured that every convex parallelohedron is an affine image of a lattice Voronoi cell. This has been proved for R^3, R^4, and other important special cases but in its full generality the conjecture remains open.

The distribution of the five Voronoi cell types among the seven symmetry classes of translation lattices is not one-to-one, as the cell types vary with lattice parameters. We note here (see the upper left entry in the table in Figure 6) that the TO is the Voronoi cell for the body-centered cubic (bcc) lattice, in which lattice points are situated at the vertices and centers of congruent cubes.

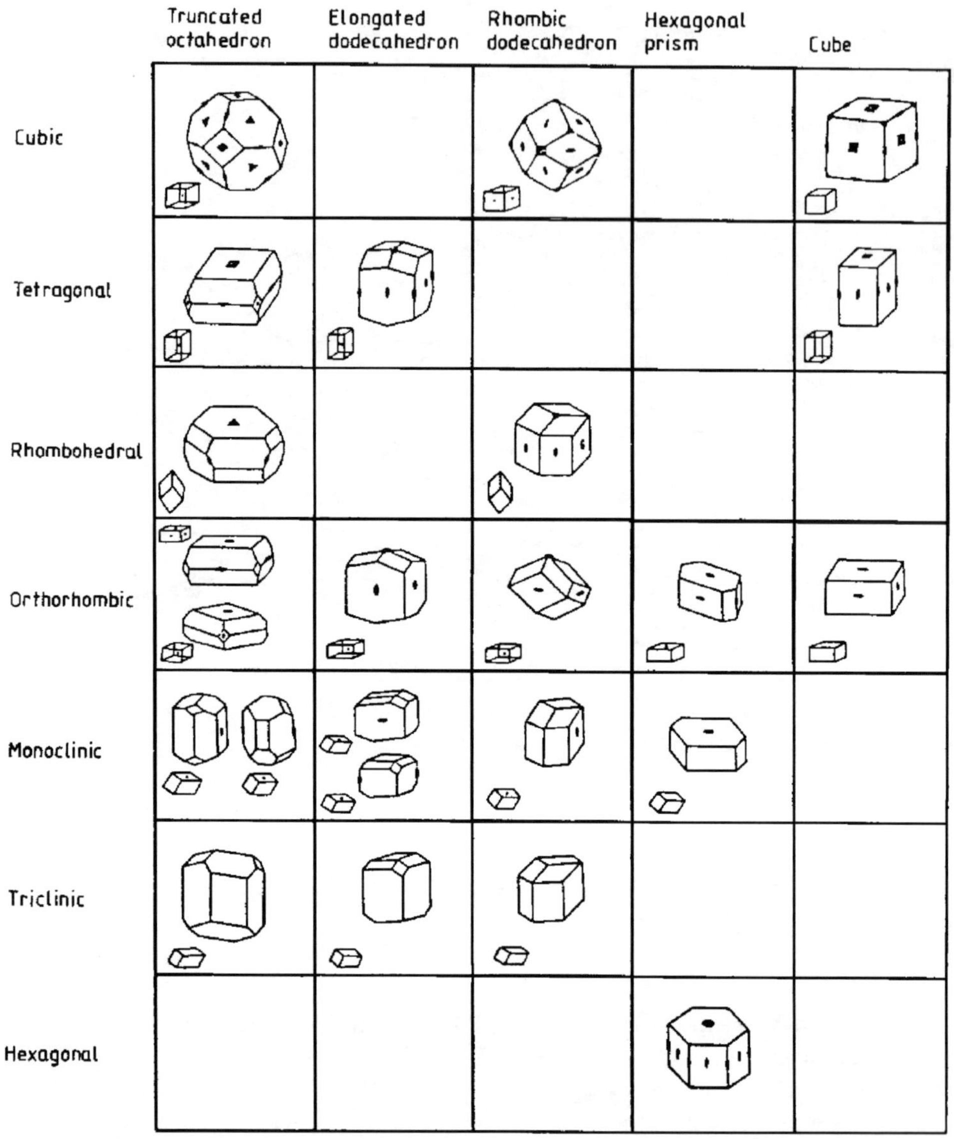

FIGURE 6. The distribution of Voronoi cell types among the three-dimensional lattices. From [1], adapted from [4].

3. The RT and non-convex parallelohedra

Uniformly lengthening or shrinking the edges of a belt to any finite (nonzero) length changes the symmetry group of the polyhedron but not its status *vis a vis* Theorem 1. Shrinking the length of the edges of a belt to zero, however – an operation known as *belt reduction*– changes its number and kinds of faces. For example, reducing the edges of any belt of a cube collapses it to a square. Reducing the edges of the hexagonal prism's belt of six rectangles collapses it to a hexagon, but reducing any of its belts of cardinality four collapses the two rectangles to edges and the two hexagons to quadrilaterals; thus the hexagonal prism reduces to an RH. Reducing the belt of four hexagons of the ED, we get the RD.

Now consider the RT, which is perhaps the quintessential cs-polyhedron that is *not* a parallelohedron (its belts have cardinality ten). The RT, as its name declares, has 30 rhombic faces, and so 60 edges. Twenty of its 32 vertices (gray in 3) are 3-valent, and the remaining 12 are 5-valent. When its faces are congruent golden rhombs, the RT has icosahedral symmetry: the 3-valent vertices are those of a regular dodecahedron, and the 5-valent vertices those of a regular icosahedron.

The RT is a polyhedral peacock: it's Catalan, zonohedral, multi-inscribable, and multi-stellatable (in 358,833,072 ways!) Moreover, it is a measure polyhedron (see [6]); that is, it is a projection of an d-cube into 3-space. In this case, $d = 6$. When $d = 5$, the measure polytope is a rhombic icosahedron (RI):

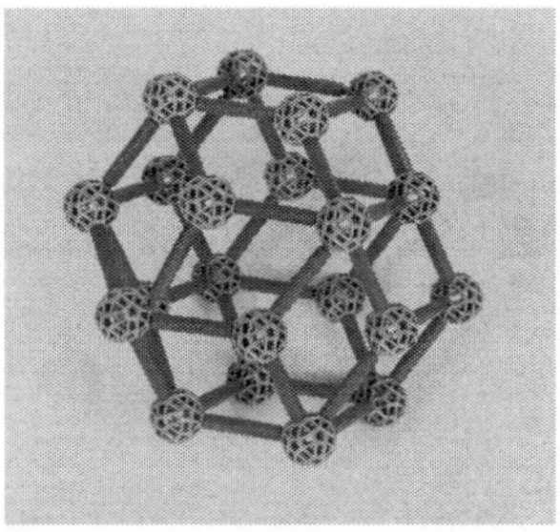

FIGURE 7. A ZomeTool model of the RI, obtained from the RT by reducing the belt of black edges in Figure 3.

When $d = 4$, the measure polyhedron is a rhombic dodecahedron (RD).

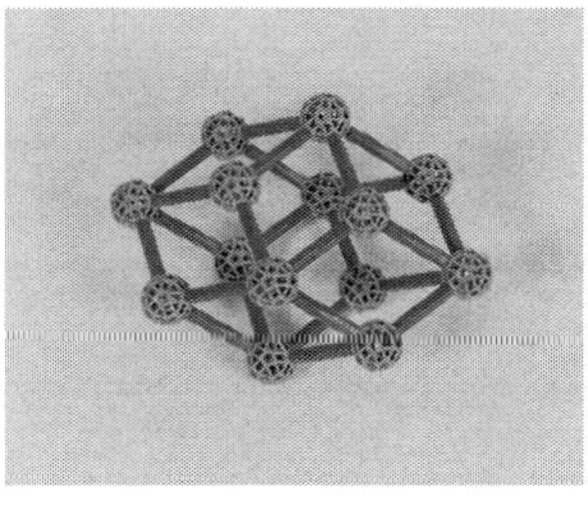

FIGURE 8. A ZomeTool model of the RD, obtained from the RI by deleting a belt in Figure 7.

When $d = 3$, the measure polyhedron is the cube (RH).

The RT, the RI, the RD, and the RH are related by belt-reduction, in that order. They inherit this property from the d- cubes (6, 5, 4,3) from which they are projected. Belt-reduction commutes with that projection, since the d-cubes themselves are related that way.

We now introduce another pair of operations on a cs-polyhedron which, as far as we know, have not been discussed elsewhere.

Let P be a cs-polyhedron that has a 3-valent vertex V shared by three quadrilaterals. We remove V, its three incident edges, and its three incident quadrilaterals, leaving behind a "bare" skew hexagon (composed of the remaining two edges of each of the three quadrilaterals), which we span. We call this operation *uncapping* P at V. Note that uncapping both a vertex and its opposite vertex results in another cs-polyhedron. Conversely, if a cs-polyhedron P has a skew hexagonal face, we can *cap* it by removing that face and adding three edges, a vertex V which is outside of P, and three parallelogram faces. By the centro-symmetry of P, there will be another skew hexagonal face opposite to it, which we also cap.

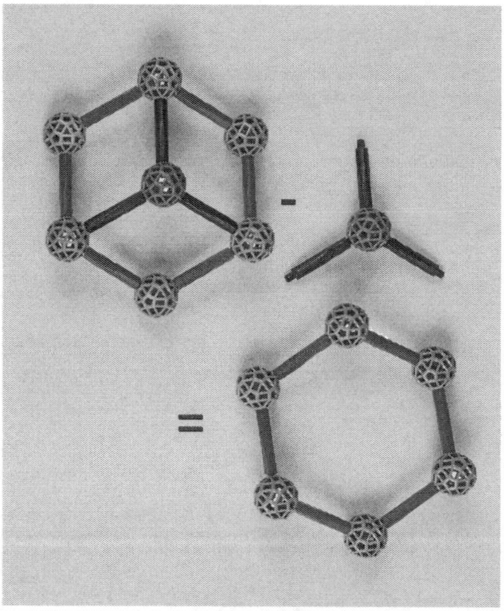

FIGURE 9. Uncapping: removing a 3-valent vertex and its incident edges, we have a skew hexagon.

Note that capping and uncapping opposite faces increase and decrease, respectively, the cardinality of a belt, but not the number of belts.

Now we will show that though they are not themselves parallelohedra, the RT and the RI contain them. First, we consider the RT. Imagine a cube inscribed in the dodecahedron formed by its 12 3-valent vertices, matching the 8 cube vertices to 8 of the dodecahedron's (there are five ways to do this; any of them will serve our purpose here.)

FIGURE 10. A cube inscribed in an RT.

Uncappping the RT at those 3-valent vertices, we are left with 8 skew centrosymmetric hexagons sharing their edges with the RT's 6 remaining rhombic faces; each of these rhombs is joined to two others by edges meeting their acute-angled vertices. This non-convex cs-polyhedron is, combinatorially, a truncated octahedron; we will call it a skew truncated octahedron, or STO.

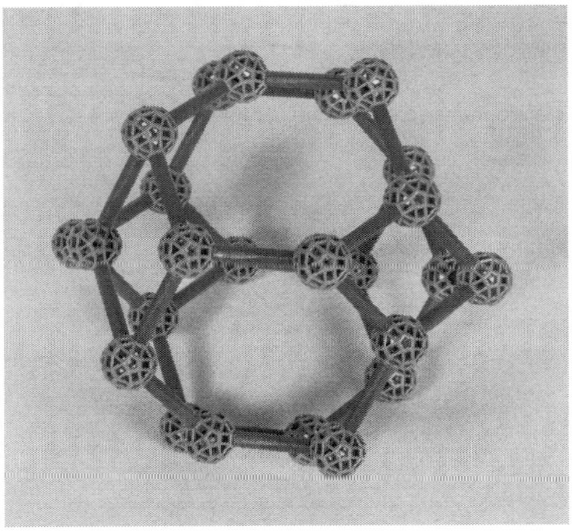

FIGURE 11. Uncapping eight of the RT's vertices, we have a skew truncated octahedron, or STO.

Obviously, we can transform the STO back into an RT by capping these hexagons.

Like the TO, the STO tiles R^3 by translation. To see this, consider first a single TO. If you pinch its six square faces so that they become rhombs, with squares pinched in the same direction if (and only if) they are antipodal, and similarly tilt all the edges in the belts of the edges of those rhombs so that the result is a cs-polyhedron, then the hexagonal faces start to crumple and become non-convex and non-planar.

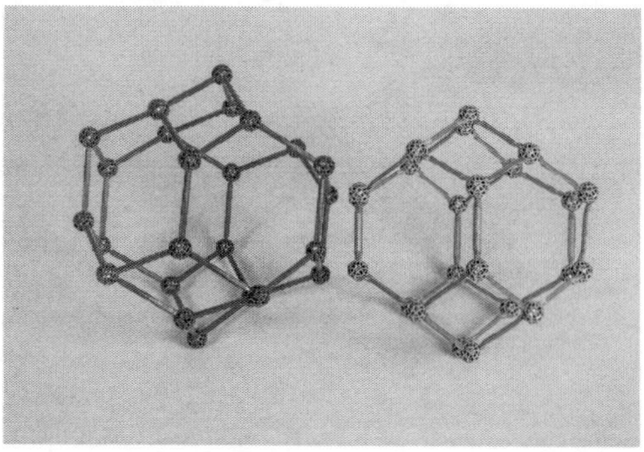

FIGURE 12. An STO (left) results when the square faces of a TO (right) are pinched.

Next perform this same operation simultaneously on each of the TO's in the tiling of R^3 just described. If you pinch all similarly oriented square faces in the same way, then the hexagonal faces of each tile crumple as before and the TOs transform continuously through cs-polyhedra into STOs.[4]

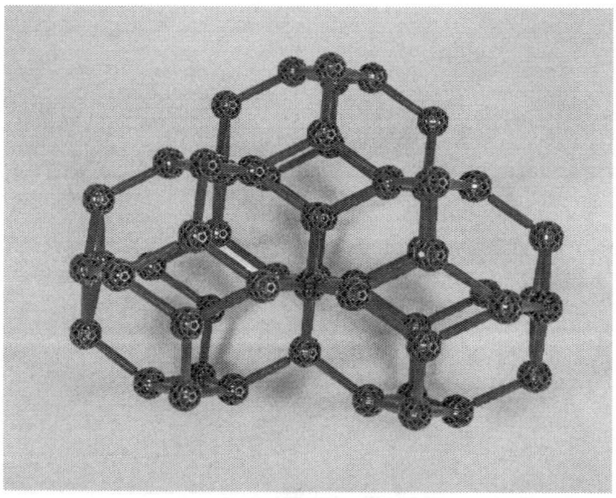

FIGURE 13. Translates of STO tile R^3.

Are there any other skew parallelohedra? Yes, there are two right before our eyes. Three of the five combinatorial types of Voronoi cells (Figure 5), the TO, the ED, and the HP, have hexagonal as well as quadrilateral faces. Replacing their hexagons with our skew hexagons, and pinching their square faces into rhombs as above, we obtain skew variants of all of them: the STO as above, the SHP (skew hexagonal prism) and the SED (skew elongated dodecahedron). In fact, all of the cs-polyhedra creating during this pinching are also skew parallelohedra.

[4]We thank Egon Schulte for this observation (email to Marjorie Senechal, September 23, 2017).

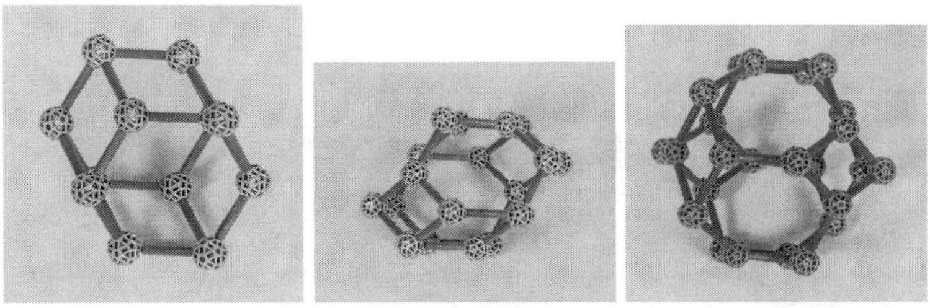

FIGURE 14. Skewed versions of the hexagonal prism, the elongated dodecahedron, and the truncated octahedron).

Conversely, the STO, the SHP and SED can be capped to form polyhedra with rhombic faces only. We find that

- the SHP becomes a rhombic dodecahedron (RD);
- the SED becomes a rhombic icosahedron (RI); and as we have seen,
- the STO becomes an RT.

Convexity aside, the SHP and SED as well as the STO satisfy the conditions of Theorem 1. Also obviously, uncapping (and capping) commute with belt-reduction.

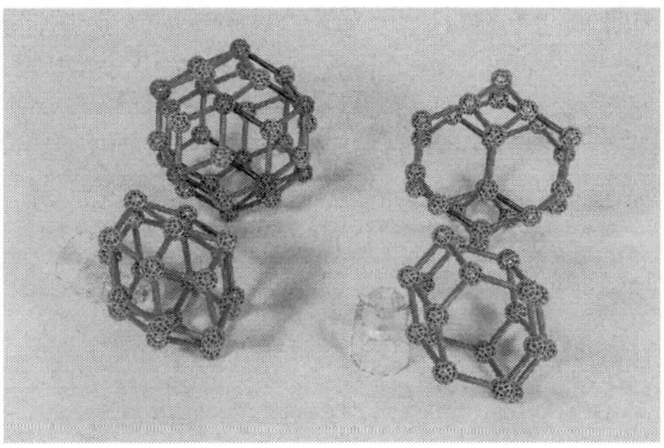

FIGURE 15. Two paths from RT (upper left) to SED (lower right): one via belt reduction first and then uncapping, the other by uncapping first and then belt reduction.

4. The RT and complex crystals

The relation we've found between non-convex parallelohedra and the related convex capped non-parallelohedra may play a role in understanding many complex crystals. In particular, clusters with icosahedral symmetry fit naturally inside an RT with its icosahedral symmetry, yet appropriately uncapping the RT produces a STO which, like the TO, tiles space by translates (taking, in this case, its center to the vertices of a body-centered cubic (bcc) lattice). The STO thus provides a bridge between local icosahedral symmetry and global bcc periodicity. This bridge is a key to understanding quasicrystals because, much as irrational numbers can be approximated by rationals, quasicrystals have periodic approximants with the same local clusters.

Previous descriptions have required overlaps of RTs or distortions of positions of atoms in the outer shells of the clusters; our description shows these are unnecessary.

There are three common types of clusters in icosahedral quasicrystals, called Bergman, Mackay, and Tsai clusters. These clusters also appear in bcc crystals, whether as approximants to known quasicrystals or not. In yet other stoichiometries, they can appear in even larger unit cells of periodic crystals. How such clusters are glued together in all these configurations have all had their own descriptions in the literature; we unify them here through uncapped RTs.

The first completely solved quasicrystal structure, an Ytterbium-Cadmium (YbCd) quasicrystal, was described [**1**] using a nesting of five polyhedra: a tetrahedron inside a dodecahedron inside an icosahedron inside an icosidodecahedron inside an RT with extra atoms at the midpoints of the RT edges. The atomic pattern as a whole was described by RTs overlapping in obtuse golden rhombohedra, together with additional acute and obtuse golden rhombohedra. This nested cluster as far out as its fourth shell, but omitting the RT (and its edge midpoints) is now called a Tsai cluster.

FIGURE 16. Jean Taylor with her ZomeTool model of the Tsai cluster (including the RT with edge midpoints) of a YbCd quasicrystal.

But, in addition to being an unrealistic growth model, this RT cluster model leads to contradictions. In the overlap description, atoms that are in the icosidodecahedral shell inside one RT become midpoints of RT edges in the other RT, and that is geometrically impossible without distorting the positions of those atoms.

Uncapping resolves this dilemma: if one uncaps eight vertices of each RT to form STOs, they pack just like truncated octahedra. The problem of midpoint versus icosidodecahedral shell is eliminated; in each STO, the inner cluster has only the latter type.

A 2-shell Mackay cluster consists of a regular icosahedron inside a larger regular icosahedron, together with atoms at the midpoints of the edges of that larger icosahedron.[5] For a 2-shell Mackay cluster, the 30 midpoints of the 30 edges in the

[5] In theory, there could be a third regular icosahedron shell outside the second one, with atoms at its vertices and two atoms added along its edges and one added at the center of each face. This begins to look like a face centered cubic (fcc) lattice developing on each of the inner icosahedron's faces, but in fact the interatomic distances cannot be made to be the same; either one builds out fcc lattices, which have gaps rather than shared atoms at the edges of icosahedronshells (Mackay's

outer icosahedron are the vertices of an icosidodecahedron just like the outer shell of a Tsai cluster, and thus this 2-shell Mackay cluster fits inside an RT exactly as a Tsai cluster does.

A Bergman cluster consists of an icosahedron inside a dodecahedron inside a larger icosahedron inside a truncated dodecahedron. To explain why so many Bergman-cluster crystals have bcc lattices, Bergman and subsequent authors distorted the positions of the atoms in the outer Bergman shell to lie on the surface of a TO, the bcc lattice's Voronoi cell. But no distortion of those atomic positions is necessary if we replace the TO with the STO.

There are periodic crystals with slightly different stoichiometry which have unit cells that contain more than one of these icosahedral clusters; these are further approximants to quasicrystals. The 2:1 approximant in the YbCd systems has a lattice that is body-centered hexagonal. Here the Tsai clusters are in RTs that have had six vertices uncapped, in such a way that there are three pairwise adjacent skew hexagonal faces at two opposite poles. In the crystal, the uncapped RTs share these hexagonal faces; they also share six of the remaining rhombic faces. The long diagonals of these shared rhombi form a skew hexagon around the equator of each uncapped RT. The result does not fill space but rather leaves acute golden rhombohedral vacancies, with one per pair of clusters. In the periodic crystal with 5.8 Cd atoms per Yb atom, these acute golden rhombohedra are occupied by two Yb atoms along the main diagonal. The structure is more naturally described by SHP than by the standard hexagonal prism, because of those parallel skew hexagonal networks.

Any quasicrystal descriptions involving overlapping RTs can be converted to non-overlapping descriptions by uncapping both RTs in the overlap. Whether this will lead to any insights remains to be seen.

5. Next steps

One next step will be to study atomic clusters in decagonal quasicrystals and their approximants, to see whether the SED plays a similar role in crystal structures which might otherwise be described by overlapping RIs.

idea), or the atoms do not have the same distances and thus they are not in fcc lattices. Only 2-shell Mackay clusters have been found experimentally in periodic and quasiperiodic crystals.

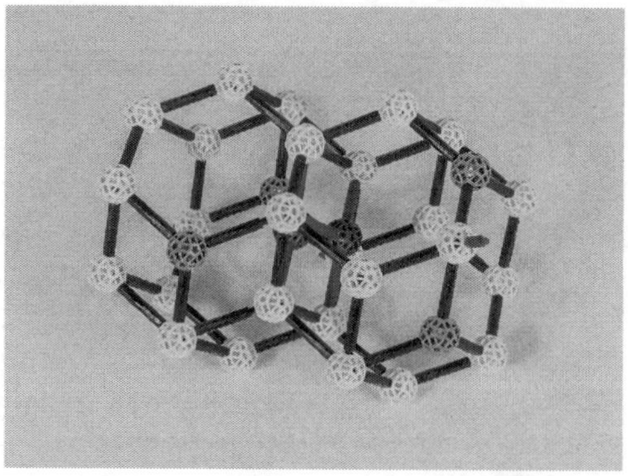

FIGURE 17. Two RI's overlap in an obtuse golden rhombohedron(interior red struts). The small red struts emerging from the surfaces aligned with the 5-fold axes of the infinite pattern.

We also hope to understand better the class of nonconvex parallelohedra. We have the following partial result; the notion of essential cardinality arises naturally within the proof and is defined there.

THEOREM 2. *If a cs-polyhedron in R^3 is a parallelohedron, then its belts have essential cardinality four or six.*

PROOF. Let P be the cs-polyhedron in question and consider the tiling of R^3 by parallel copies of P. Since the tiling is face-to-face and thus edge-to-edge (and vertex to vertex), each belt of P defines a layer of the tiling, and the tiling is a countable stack of congruent layers. The midpoints of the consecutive edges of a belt of P form the vertices of an n-gon that loops around P (passing through the centers of the faces of that belt), where n is the cardinality of the belt. Constructing this n-gon for each tile in the layer, we have a periodic network of congruent n-gons in R^3. Projecting this network onto a plane orthogonal to the chosen edge, the midpoints become vertices and we get an edge-to-edge translation tiling of the plane by congruent centro-symmetric n-gons. However, a vertex of a tile *in a tiling* is usually defined as a point at which three or more tiles meet. In an arbitrary tiling, not all vertices of a polygon *per se* are necessarily vertices of the tiling in this sense. For example, the belt corresponding to a vertical edge of the cs-polyhedron of Figure 4 leads to a tiling of the plane by octagons, but *as a tiling* it is regarded as having only four vertices, and the two sets of three edges between these vertices are regarded as two edges of the tiling. Using this definition of vertex, only quadrilaterals and hexagons can tile the plane by translation ([**7**]). Returning to the belts, we see that there are distinguished edges in the belt whose midpoints do not project to vertices of the tiling. The *essential cardinality* of the belt neglects them and counts only those edges whose midpoints do project to vertices of the tiling. Hence the essential cardinality of the belt is four or six. □

Acknowledgments

We would like to thank AIM (the American Institute for Mathematics) for hosting both the 2016 workshop on "Soft Packings, Nested Clusters, and Condensed Matter" and the SQuaRe research project on "Geometric Frustration in the Growth and Form of Complex Crystals" that grew out of it. We would like to thank our fellow SQuaRE members, in particular Erin Teich, for useful discussions, Stan Sherer for photographing the models, and an anonymous referee for astute suggestions.

References

[1] H. Takakura, C. Pay Gomez, A, Yamamoto, M. De Boissieu, A. P. Tsai, A.P. *Atomic structure of the binary icosahedral Yb?Cd quasicrystal*, Nature Materials, 2007, 6, 58-63.

[2] See `https://www.nobelprize.org/prizes/chemistry/2011/press-release/`.

[3] Marjorie Senechal, *Crystalline symmetries*, Adam Hilger, Ltd., Bristol, 1990. An informal mathematical introduction. MR1100479

[4] A. D. Aleksandrov, *Vypuklye mnogogranniki* (Russian), Gosudarstv. Izdat. Tehn.-Teor. Lit., Moscow-Leningrad, 1950. MR0040677

[5] P. McMullen, *Convex bodies which tile space by translation*, Mathematika **27** (1980), no. 1, 113–121, DOI 10.1112/S0025579300010007. MR582003

[6] This derivation appears in the appendix to A. Bravais, *Collected Scientific Works*, edited by B. N. Delone and I. I Shafronovski, Science Publishers, Leningrad, 1974 (in Russian). See also [3].

[7] Branko Grünbaum and G. C. Shephard, *Tilings and patterns*, W. H. Freeman and Company, New York, 1987. MR857454

Contemporary Mathematics
Volume **764**, 2021
https://doi.org/10.1090/conm/764/15338

Tilings with congruent edge coronae

Mark D. Tomenes and Ma. Louise Antonette N. De Las Peñas

Abstract. In this paper, we discuss properties of a normal tiling of the Euclidean plane ($\mathbb{E}^2$) with congruent edge coronae. We prove that the congruence of the first edge coronae is enough to say that the tiling is isotoxal.

1. Introduction

The work of D. Schattschneider and N. Dolbilin [**1**] establishes local conditions for a monohedral tiling of d-dimensional space to be isohedral. In particular, they have shown in [**4**] that for polygonal tilings of the Euclidean plane ($\mathbb{E}^2$), pairwise congruence of centered first coronae of the tiles implies the tiling is isohedral. In this paper, we consider an extension of these studies and prove that the congruence of centered first edge coronae in a normal tiling implies the tiling is isotoxal.

2. Preliminaries

A *tiling* $\mathcal{T}$ of $\mathbb{E}^2$ is a countable collection of closed topological disks called *tiles* $\mathcal{T} = \{T_i : i \in \mathbb{N}\}$ that is a covering ($\bigcup_i T_i = \mathbb{E}^2$) as well as a packing ($Int(T_i) \cap Int(T_j) = \emptyset$ if $i \neq j$, $Int(T)$ denotes the interior of tile T). The intersection of any two distinct tiles can be a set of isolated arcs and points. The points are called *vertices* of $\mathcal{T}$ and the arcs are called *edges* of $\mathcal{T}$.

A vertex with w number of edges incident to it is said to have *valence* w. A *prototile set* for $\mathcal{T}$ is a minimal subset of tiles of $\mathcal{T}$ such that each tile of $\mathcal{T}$ is a congruent copy of one of those in the prototile set. The tiles in this set are called the *prototiles* of $\mathcal{T}$.

The *symmetry group* $S(\mathcal{T})$ of a tiling $\mathcal{T}$ is the group of isometries of $\mathbb{E}^2$ that leave $\mathcal{T}$ invariant; elements of $S(\mathcal{T})$ are called *symmetries* of $\mathcal{T}$. $\mathcal{T}$ is called *isotoxal* or *edge-transitive* if its symmetry group acts transitively on the set of all edges of $\mathcal{T}$.

The *edge corona* $C(E)$ of an edge E of $\mathcal{T}$ is the set of all tiles of $\mathcal{T}$ that have nonempty intersection with E. It is possible for two edges E_1, E_2 to have the same edge corona, that is, $C(E) = C(E')$ but $E \neq E'$. To avoid ambiguity, we make the convention that by an edge corona of an edge E of $\mathcal{T}$ we mean a *"centered*

2010 *Mathematics Subject Classification.* Primary 52C20, 05B45; Secondary 58D19.
Key words and phrases. Edge coronae, Isotoxal tilings.
The first author would like to acknowledge the Department of Science and Technology (DOST) through the Accelerated Science and Technology Human Resource Development Program (ASTHRDP) for the scholarship grant during his graduate studies.

edge corona" which refers to the pair consisting of the edge corona $C(E)$ and its fixed center E, that is, $C(E) := (E, C(E))$. Two edge coronae $C(E)$ and $C(E')$ are congruent if there is an isometry σ of $\mathbb{E}^2$ such that $\sigma(C(E)) = C(E')$ and $\sigma(E) = E'$. $\mathcal{T}$ is called a *tiling with congruent edge coronae* if every pair of edge coronae of $\mathcal{T}$ are congruent. $\mathcal{T}$ is necessarily *monotoxal*, that is $\mathcal{T}$ has congruent edges.

The kth *edge figure* of an edge E is denoted by $F_k(E)$ and is defined as follows: $F_0(E) := \{E\}$ and $F_k(E)$, $k \geq 1$, is the set of all edges with non-empty intersection with an edge in $F_{k-1}(E)$. We write $F_1(E)$ as $F(E)$ which we refer to as the *edge figure* of E. By a kth edge figure $F_k(E)$ we mean a "centered k^{th} edge figure" which is a pair $(E, F_k(E))$ consisting of the edge figure $F_k(E)$ and its fixed center E, that is, $F_k(E) := (E, F_k(E))$. Two k^{th} edge figures $F_k(E)$ and $F_k(E')$ are congruent if there exists an isometry σ of $\mathbb{E}^2$ such that $\sigma(F_k(E)) = F_k(E')$ and $\sigma(E) = E'$. Note that if E and E' are edges of a tiling such that $C(E)$ is congruent to $C(E')$, then $F(E)$ is congruent to $F(E')$.

The *symmetry group* $S_k(E)$ of a kth edge figure $F_k(E)$ of E is defined as $S_k(E) = \{\sigma \in S_0(E) | \sigma(F_k(E)) = F_k(E)\}$ where $S_0(E)$ is the symmetry group of the edge E, the group of isometries of $\mathbb{E}^2$ that leave E invariant.

The group $S_k(E)$ is a subgroup (not necessarily proper) of $S_{k-1}(E)$ for all $k \geq 1$ because if σ leaves E invariant and sends $F_k(E)$ onto itself, then σ must send every finite path of edges in $F_k(E)$ that begins with E to another path of edges in $F_k(E)$ that has the same length and begins with E. Thus, σ must leave $F_i(E)$ invariant for $1 \leq i \leq k$. If E is an edge of a tiling, then the group $S_0(E)$ is finite, so the chain $S_0(E) \geq S_1(E) \geq \cdots \geq S_k(E) \geq \cdots$ can contain only a finite number of distinct subgroups. It is easy to see that $|S_0(E)|$ must be a divisor of 4, and thus must be 4, 2 or 1.

In this study, we will assume that all tilings under consideration are normal tilings of $\mathbb{E}^2$ with congruent edge coronae. We define a *normal* tiling as a tiling whose tiles are uniformly bounded in diameter and have inradii uniformly bounded away from zero and for which the intersection of any two tiles is either empty, or an edge, or a vertex of each [**2**]. We thus exclude tilings with digons or tilings with vertices of valence two.

A normal tiling $\mathcal{T}$ is locally finite (see 3.2.1 of [**3**]). By *locally finite* we mean for every $x \in \mathbb{E}^2$, there exists a circular disk centered at x which intersects only finitely many tiles of $\mathcal{T}$. Then $\mathcal{T}$ satisfies the *finite path property*, that is, for every pair of edges E and E' of $\mathcal{T}$, we can always find a path of edges of $\mathcal{T}$, $E = E_0, E_1, E_2, \cdots, E_n = E'$ that begins with E and ends with E'.

3. Properties of a normal tiling with congruent edge coronae

LEMMA 3.1. *If a normal tiling has congruent edge coronae then it has at most two prototiles. If the tiling has exactly two prototiles then each of its vertices has even valence.*

PROOF. Consider tiles T_1, T_2 of a normal tiling $\mathcal{T}$ with congruent edge coronae. Suppose T_1, T_2 share an edge E. If T is any tile of $\mathcal{T}$ and E' any edge of T, then any isometry taking E' to E must map T to T_1 or T_2. Hence there are at most two prototiles.

Suppose $\mathcal{T}$ has exactly two prototiles T_1 and T_2. Then congruent copies of T_1 and T_2 must meet at every edge of the tiling since the tiling has congruent edge coronae. Thus, congruent copies of T_1 and T_2 must alternate in a vertex. Hence, the valence of every vertex is even. □

REMARK 3.2. From the above result, a normal tiling $\mathcal{T}$ with congruent edge coronae is a tiling consisting of m-gons or n-gons (where $m, n > 2$). In the case when $\mathcal{T}$ has two prototiles, an edge is shared by an m-gon and an n-gon. Moreover, the vertices in an edge are of valence p or q $(p, q > 2)$. We write $\mathcal{T}$ as $\langle m, n; p, q\rangle$, its *configuration*. It is possible that $p = q$, but if $p \neq q$ then the vertices with valence p and q must alternate around a tile since every edge must have endpoints of valence p and q. This means that the tiles of $\mathcal{T}$ must have an even number of edges.

Before we proceed with the proof of the next theorem we first define the following. A *patch* $\mathcal{A}(r, P)$ *of* $\mathcal{T}$ *generated by a disc* $D(r, P)$ with radius r centered at P is derived from the union of all tiles of $\mathcal{T}$ which meet $D(r, P)$ by adjoining just enough tiles to fill up the "holes" of this union and turn it into a topological disc. By *circumparameter* U of a tiling $\mathcal{T}$ we mean every tile of $\mathcal{T}$ is contained in some circular disk of radius U.

THEOREM 3.3. *If a normal tiling has congruent edge coronae, then it has one of the following configurations:* $\langle 6^2; 3^2\rangle$, $\langle 3, 6; 4^2\rangle$, $\langle 3^2; 6^2\rangle$, $\langle 4^2; 4^2\rangle$ *or* $\langle 4^2; 3, 6\rangle$.

PROOF. Let $\mathcal{T}$ be a normal tiling with congruent edge coronae having configuration $\langle m, n; p, q\rangle$. Every edge of $\mathcal{T}$ has endpoints of valences p and q, $(p, q > 2)$ and is shared by an m-gon and an n-gon.

Consider a patch $\mathcal{A}(r, P)$ of $\mathcal{T}$. Let $e(r, P)$ and $v_p(r, P)$ be the number of edges and vertices of valence p in $\mathcal{A}(r, P)$, respectively. The product $pv_p(r, P)$ is the number of edges with an endpoint of valence p in $\mathcal{A}(r, P)$.

Suppose $p \neq q$. Take $pv_p(r, P) - e(r, P)$ (e.g. dotted edges in Figure 1). Since $pv_p(r, P)$ may possibly include an edge not in $\mathcal{A}(r, P)$, we have $pv_p(r, P) - e(r, P) \geq 0$. Let $E(r, P)$ be the number of edges of $\mathcal{T}$ with one endpoint (of valence p or q) in $\mathcal{A}(r, P)$ (e.g. dotted and dashed edges). We have

$$0 \leq pv_p(r, P) - e(r, P) \leq E(r, P). \tag{3.1}$$

Now an edge counted in $E(r, P)$ belongs to a tile in $\mathcal{A}(r + 2U, P)$, U is the circumparameter of $\mathcal{T}$. Then,

$$0 \leq E(r, P) \leq k(t(r + 2U, P) - t(r, P)) \tag{3.2}$$

where $k = \max(m, n)$ and $t(r, P)$ is the number of tiles in $\mathcal{A}(r, P)$.

Since $\mathcal{T}$ is a normal plane tiling, by the Normality Lemma [**3**], we have

$$\lim_{r\to\infty} \frac{t(r + 2U, P) - t(r, P)}{t(r, P)} = 0.$$

From (3.2) and by the squeeze theorem we have

$$\lim_{r\to\infty} \frac{E(r, P)}{t(r, P)} = 0.$$

Moreover, from (3.1) and also the squeeze theorem we have

$$\lim_{r\to\infty} \frac{pv_p(r, P) - e(r, P)}{t(r, P)} = 0.$$

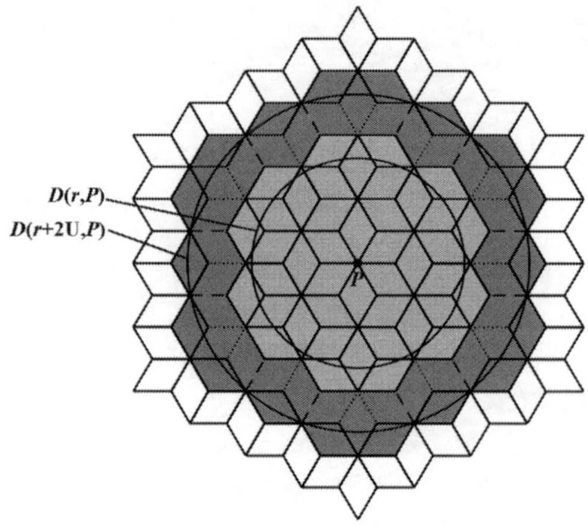

FIGURE 1. An example of a patch $\mathcal{A}(r,P)$ generated by $D(r,P)$ (light shaded tiles) and a patch $\mathcal{A}(r+2U,P)$ generated by $D(r+2U,P)$ (light and dark shaded tiles).

By normality of $\mathcal{T}$, $e(\mathcal{T}) := \lim_{r\to\infty} \dfrac{e(r,P)}{t(r,P)}$ exists. This implies that $v_p(\mathcal{T}) := \lim_{r\to\infty} \dfrac{v_p(r,P)}{t(r,P)}$ also exists. Hence, $pv_p(\mathcal{T}) = e(\mathcal{T})$. By a similar argument, we also have $qv_q(\mathcal{T}) = e(\mathcal{T})$ where $v_q(\mathcal{T}) := \lim_{r\to\infty} \dfrac{v_q(r,P)}{t(r,P)}$ and $v_q(r,P)$ is the number of vertices of valence q in $\mathcal{A}(r,P)$. Moreover, if we consider $t_m(r,P)$ and $t_n(r,P)$, the number of m-gonal and n-gonal tiles in $\mathcal{A}(r,P)$; it can be shown that if $m \neq n$, then $mt_m(\mathcal{T}) = e(\mathcal{T})$ and $nt_n(\mathcal{T}) = e(\mathcal{T})$ where $t_m(\mathcal{T}) := \lim_{r\to\infty} \dfrac{t_m(r,P)}{t(r,P)}$ and $t_n(\mathcal{T}) := \lim_{r\to\infty} \dfrac{t_n(r,P)}{t(r,P)}$. Note that $t_m(\mathcal{T})$ and $t_n(\mathcal{T})$ exists since $e(\mathcal{T})$ exists.

Now, consider $v(r,P)$ consisting of vertices in $\mathcal{A}(r,P)$ and define $v(\mathcal{T}) := \lim_{r\to\infty} \frac{v(r,P)}{t(r,P)}$. Then we have

$$v(r,P) = v_p(r,P) + v_q(r,P).$$

This implies

$$v(\mathcal{T}) = \lim_{r\to\infty} \frac{v(r,P)}{t(r,P)} = \lim_{r\to\infty} \frac{v_p(r,P)}{t(r,P)} + \lim_{r\to\infty} \frac{v_q(r,P)}{t(r,P)}.$$

Or equivalently,

$$v(\mathcal{T}) = v_p(\mathcal{T}) + v_q(\mathcal{T}) = \frac{e(\mathcal{T})}{p} + \frac{e(\mathcal{T})}{q}. \tag{3.3}$$

We also have,

$$t(r,P) = t_m(r,P) + t_n(r,P).$$

This implies

$$1 = \lim_{r\to\infty} \frac{t(r,P)}{t(r,P)} = \lim_{r\to\infty} \frac{t_m(r,P)}{t(r,P)} + \lim_{r\to\infty} \frac{t_n(r,P)}{t(r,P)}.$$

Or equivalently,

$$1 = t_m(\mathcal{T}) + t_n(\mathcal{T}) = \frac{e(\mathcal{T})}{m} + \frac{e(\mathcal{T})}{n}. \tag{3.4}$$

Now suppose $p = q$, then corresponding to (3.1), we have

$$0 \le pv_p(r, P) - 2e(r, P) \le E(r, P). \tag{3.5}$$

This is because in this case, every edge has endpoints of valence p and hence every edge is counted twice in $pv_p(r, P)$. This will imply that $pv_p(\mathcal{T}) = 2e(\mathcal{T})$. Moreover, if $m = n$, it can be shown that $mt_m(\mathcal{T}) = 2e(\mathcal{T})$. Then corresponding to (3.3) and (3.4), we have

$$v(\mathcal{T}) = v_p(\mathcal{T}) = \frac{2e(\mathcal{T})}{p} \tag{3.6}$$

and

$$1 = t_m(\mathcal{T}) = \frac{2e(\mathcal{T})}{m} \tag{3.7}$$

respectively.

Euler's Theorem for tilings [**3**] states that $e(\mathcal{T}) - v(\mathcal{T}) = 1$. Suppose $p = q$, then substituting (3.6) to the Euler's Formula, we have $e(\mathcal{T}) - \dfrac{2e(\mathcal{T})}{p} = 1$. If $m = n$, then we get $\dfrac{2}{m} + \dfrac{2}{p} = 1$ by (3.7). Solving this, we obtain the configurations $\langle 6^2; 3^2\rangle$, $\langle 3^2; 6^2\rangle$ and $\langle 4^2; 4^2\rangle$. On the other hand, if $m \neq n$, we get $\dfrac{1}{m} + \dfrac{1}{n} + \dfrac{2}{p} = 1$ by (3.4). By Lemma 3.1, p is even. Then solving the equation results to the configuration $\langle 3, 6; 4^2\rangle$.

Suppose $p \neq q$, then substituting (3.3) to the Euler's Formula, we have $e(\mathcal{T}) - \dfrac{e(\mathcal{T})}{p} + \dfrac{e(\mathcal{T})}{q} = 1$. By Remark 3.2, m and n are even. If $m = n$, then we get $\dfrac{2}{m} + \dfrac{1}{p} + \dfrac{1}{q} = 1$ by (3.7). Solving this, we obtain the configuration $\langle 4^2; 3, 6\rangle$. On the other hand, if $m \neq n$, we get $\dfrac{1}{m} + \dfrac{1}{n} + \dfrac{1}{p} + \dfrac{1}{q} = 1$ by (3.4) which yields no integral solutions such that m, n, p, q are even.

Thus, we obtain the following five configurations: $\langle 6^2; 3^2\rangle$, $\langle 3, 6; 4^2\rangle$, $\langle 3^2; 6^2\rangle$, $\langle 4^2; 4^2\rangle$ or $\langle 4^2; 3, 6\rangle$. □

4. The main result

Essential to the proof of the isotoxality of a normal tiling with congruent edge coronae is Theorem 4.2 which is an analogous result to the "if direction" of The Local Theorem for Tilings [**1**]. We reformulate the conditions so they apply to edges of the tiling, rather than to faces or tiles. We proceed by first proving Lemma 4.1 which will be used in the proof of Theorem 4.2.

LEMMA 4.1. *Let $\mathcal{T}$ be a tiling of $\mathbb{E}^2$ such that for some integer $k > 0$, $\mathcal{T}$ satisfies the following two conditions:*

(i) *All k^{th} edge figures in $\mathcal{T}$ are pairwise congruent; and*
(ii) *$S_{k-1}(E) = S_k(E)$, for any edge E of $\mathcal{T}$.*

If E and E' are edges of $\mathcal{T}$ and σ is an isometry of $\mathbb{E}^2$ that satisfies $\sigma(E) = E'$ and $\sigma(F_{k-1}(E)) = F_{k-1}(E')$, then σ must also satisfy $\sigma(F_k(E)) = F_k(E')$.

PROOF. Let E and E' be edges of $\mathcal{T}$. Suppose $\sigma(E) = E'$ and $\sigma(F_{k-1}(E)) = F_{k-1}(E')$, where σ is an isometry of $\mathbb{E}^2$. By condition (i), there is an isometry σ' of $\mathbb{E}^2$ such that $\sigma'(F_k(E)) = F_k(E')$ and $\sigma'(E) = E'$. Thus $\sigma'(E) = \sigma(E)$ which implies that $\sigma^{-1}\sigma'(E) = E$.

Now, $\sigma'(F_k(E)) = F_k(E')$ implies $\sigma'(F_{k-1}(E)) = F_{k-1}(E')$. Moreover, $\sigma(F_{k-1}(E)) = F_{k-1}(E')$. Thus, $\sigma^{-1}\sigma'(F_{k-1}(E)) = F_{k-1}(E)$. Observe that the equations $\sigma^{-1}\sigma'(E) = E$ and $\sigma^{-1}\sigma'(F_{k-1}(E)) = F_{k-1}(E)$ imply that $\sigma^{-1}\sigma' \in S_{k-1}(E)$. Let $\rho = \sigma^{-1}\sigma'$. Then by condition (ii), $\sigma^{-1}\sigma' = \rho \in S_k(E)$. Since $\sigma = \sigma'\rho^{-1}$ and $\rho^{-1} \in S_k(E)$, it follows that $\sigma(F_k(E)) = \sigma'\rho^{-1}(F_k(E)) = \sigma'(F_k(E)) = F_k(E')$. □

THEOREM 4.2. *A locally finite tiling $\mathcal{T}$ with symmetry group G is isotoxal if there is an integer $k > 0$ such that*

(i) *all k^{th} edge figures in $\mathcal{T}$ are pairwise congruent; and*
(ii) *$S_{k-1}(E) = S_k(E)$, for an edge E of $\mathcal{T}$.*

PROOF. Let E, E' be edges of $\mathcal{T}$. We must show that there is a symmetry of $\mathcal{T}$ that maps E to E'. By assumption, there is an isometry σ of $\mathbb{E}^2$ such that $\sigma(E) = E'$ and $\sigma(F_k(E)) = F_k(E')$. We will show that $\sigma \in G$, that is, σ maps the edges of $\mathcal{T}$ to other edges of $\mathcal{T}$.

Let $\overline{E}$ be any edge of $\mathcal{T}$ and assume $\overline{E}' = \sigma(\overline{E})$. We show that $\overline{E}'$ is an edge of $\mathcal{T}$. Since $\mathcal{T}$ is locally finite, it satisfies the finite path property. Thus, there is a path of edges of $\mathcal{T}$, $E = E_0, E_1, E_2, \dots, E_m = \overline{E}$ such that $E_{i-1} \cap E_i \neq \emptyset, 1 \leq i \leq m$, that begins with E and ends in $\overline{E}$. We will show that for each E_i in this path, $i = 1, \dots m, \sigma(E_i) = E_i'$ is an edge of $\mathcal{T}$.

Now, $E_1 \in F(E) \subseteq F_k(E)$ so $\sigma(E_1) = E_1' \in \sigma(F_k(E)) = F_k(E')$. Hence E_1' is an edge of $\mathcal{T}$. Every edge of $F_{k-1}(E_1)$ is also an edge of $F_k(E)$ since E_1 and E are adjacent edges of $\mathcal{T}$. Since $\sigma(F_k(E)) = F_k(E')$, σ sends an edge of $F_{k-1}(E_1)$ to an edge of $F_k(E')$. Now every edge of $F_{k-1}(E_1')$ is also an edge of $F_k(E')$ since $\sigma(E) = E'$, $\sigma(E_1) = E_1'$, and E_1 and E are adjacent. This implies $\sigma(F_{k-1}(E_1)) = F_{k-1}(E_1')$. By Lemma 4.1, we have $\sigma(F_k(E_1)) = F_k(E_1')$.

Since $E_2 \in F_k(E_1)$, $\sigma(E_2) = E_2' \in \sigma(F_k(E_1)) = F_k(E_1')$, so E_2' is an edge in $\mathcal{T}$. Continuing in this manner, we see that $\sigma(E_i) = E_i' \in \sigma(F_k(E_{i-1})) = F_k(E_{i-1}')$; hence, E_i' is an edge of $\mathcal{T}$. Thus, $\sigma(\overline{E}) = \overline{E}'$ is an edge of $\mathcal{T}$. □

We now prove our main result as follows.

THEOREM 4.3. *If a normal tiling has congruent edge coronae, then it is isotoxal.*

PROOF. Consider a normal tiling $\mathcal{T}$ with congruent edge coronae. Suppose we take an edge E such that $S_0(E) \cong$ (isomorphic) C_1, D_1, C_2 or D_2 (Figure 2) where C_i is the cyclic group of order i and D_i is the dihedral group of order $2i$ ($i = 1, 2$). Note that $\mathcal{T}$ is monotoxal so that the edges of $\mathcal{T}$ have isomorphic symmetry group.

If $S_0(E) \cong C_1$, then $S_1(E) = S_0(E) \cong C_1$ since $S_1(E) \leq S_0(E)$. Since $\mathcal{T}$ has congruent edge coronae, then it has congruent edge figures so both conditions of Theorem 4.2 are true for $k = 1$ and $\mathcal{T}$ is isotoxal. For the remainder of the proof, we consider $S_0(E) \cong D_1$, C_2 or D_2.

FIGURE 2. Edges with their corresponding symmetry groups.

We divide the proof assuming one of the five configurations in Theorem 3.3. We always assume an edge E of $\mathcal{T}$ with adjacent angles $\theta_1\theta_2$ and $\rho_1\rho_2$ that meet in (vertices) endpoints of E. If E' is another edge of $\mathcal{T}$, then $C(E)$ is congruent to $C(E')$, so the pair of adjacent angles that meet in endpoints of E' will also be $\theta_1\theta_2$ and $\rho_1\rho_2$; this will also be true for another edge E'' of $\mathcal{T}$, and so on. That is, every pair of edges of $\mathcal{T}$ has the same two pairs of adjacent angles. This assumption will be employed in completing an edge figure of a given edge.

The computer program *Geometer's Sketchpad* was an important tool in the case-by-case analysis as it helped with an accurate construction of the edge figure: taking into consideration the type of edge and its orientation, and the adjacent angles that may occur; or even show that the construction of an edge figure is impossible.

CASE 1. $\mathcal{T}$ has configuration $\langle 6^2; 3^2\rangle$.

(a) Consider an edge $E = PQ$ of $\mathcal{T}$ such that $S_0(E) \cong D_1$ with adjacent angles $\theta_1\theta_2$ and $\rho_1\rho_2$ (Figure 3a). Let $E' = PR$ be another edge. Since $C(E')$ is congruent to $C(E)$, E' has adjacent angles $\theta_1\theta_2$ and $\rho_1\rho_2$ (Figure 3b). Vertex P has valence three so that θ_1, θ_2 satisfy $\theta_1 + 2\theta_2 = 2\pi$. In this case, the only possibility is $\theta_1 = \theta_2$ (Figure 3c). Now, $C(E'')$ is congruent to $C(E)$ where $E'' = QS$ (Figure 3d). Following the same argument, we have $\rho_1 = \rho_2$. Since $C(E')$ is congruent to $C(E)$ and $C(E'')$ is congruent to $C(E)$, we have $\theta_1 = \rho_1$. We get the edge figure in Figure 3e. There is only one type of edge figure obtained. Given that congruent angles meet at a vertex, there is only one way to orient the edges around a vertex. We have $S_0(E) = S_1(E) \cong D_1$ so that $\mathcal{T}$ is isotoxal.

(b)-(c) For the cases where we have $S_0(E) \cong C_2$ or $S_0(E) \cong D_2$ we follow the arguments given in 1(a) and we obtain the edge figures in Figures 4a and 4b, respectively. In the case of Figure 4a, we have $S_0(E) = S_1(E) \cong C_2$ and in the case of Figure 4b, we have $S_0(E) = S_1(E) \cong D_2$. Thus, in either case $\mathcal{T}$ is isotoxal.

CASE 2. $\mathcal{T}$ has configuration $\langle 3, 6; 4^2\rangle$.

(a) Suppose $\mathcal{T}$ has an edge $E = PQ$ such that $S_0(E) \cong D_1$ with adjacent angles $\theta_1\theta_2$ and $\rho_1\rho_2$. Let $E' = PR$ be another edge with adjacent angles $\theta_1\theta_2$ and $\rho_1\rho_2$ (Figure 5a). With this construction, θ_1 and θ_2 must alternate around P. Now, consider the edge E'' where $E'' = QR$ (Figure 5b); PQR is a 3-gon. Since $C(E')$ is congruent to $C(E)$ and $C(E'')$ is congruent to $C(E)$, then it is not possible for E' and E'' to have adjacent angles $\rho_1\rho_2$ unless $\rho_1 = \theta_1$ and $\rho_2 = \theta_2$. We obtain the edge figure in Figure 5c. If we consider another orientation for edge E' and a different angle measure for θ_1; (this case E' an edge of a 6-gon) we obtain the edge figure in Figure 5d. Both edge figures satisfy $S_0(E) = S_1(E) \cong D_1$ and $\mathcal{T}$ is isotoxal.

(b) We now consider an edge E of $\mathcal{T}$ such that $S_0(E) \cong C_2$. Following the same arguments in 2(a) used to obtain Figures 5c and 5d, we get the edge figures in Figures 6 a and 6b, respectively. Unlike in 2(a) where only two edge figures can

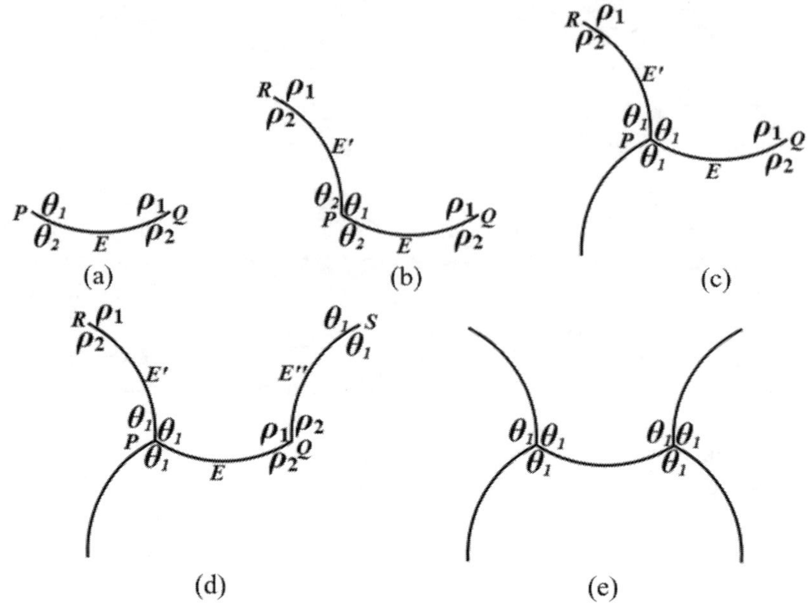

FIGURE 3. Construction of edge figure when $\mathcal{T}$ has configuration $\langle 6^2; 3^2 \rangle$ and $S_0(E) \cong D_1$.

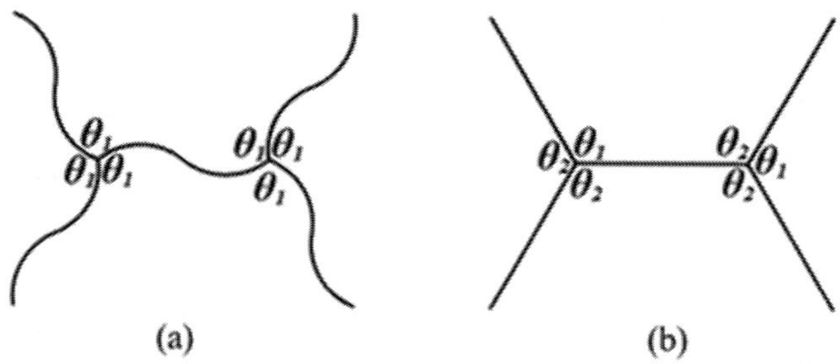

FIGURE 4. Edge figures when $\mathcal{T}$ has configuration $\langle 6^2; 3^2 \rangle$ and (a) $S_0(E) \cong C_2$; (b) $S_0(E) \cong D_2$.

be constructed, two additional possibilities here are when there are three distinct angles meeting in a vertex (e.g. letting $\rho_1 = \theta_1, \rho_2 = \theta_3$; or $\rho_1 = \theta_3, \rho_2 = \theta_2$). We obtain the edge figures shown in Figures 6c and 6d.

Observe that in either of the edge figures, $S_1(E) \cong C_1$ so that $S_0(E) \neq S_1(E)$. However, $S_1(E) = S_2(E) \cong C_1$ since $S_1(E) \geq S_2(E)$. It is now left to show that the second edge figures are congruent. We are going to build the second edge figure for the edge figure in Figure 6a. A similar approach can be used to show that the corresponding second edge figures for the cases in Figures 6b, 6c and 6d are congruent. With this, we see that $\mathcal{T}$ is isotoxal.

From $F(E)$ shown in Figure 7a, we see that the isometry that sends E to E' is the 3-fold rotation with center at the black triangle. This rotation must carry the other edges of $F(E)$ to edges of $F(E')$ and maintain the pairs of adjacent angles

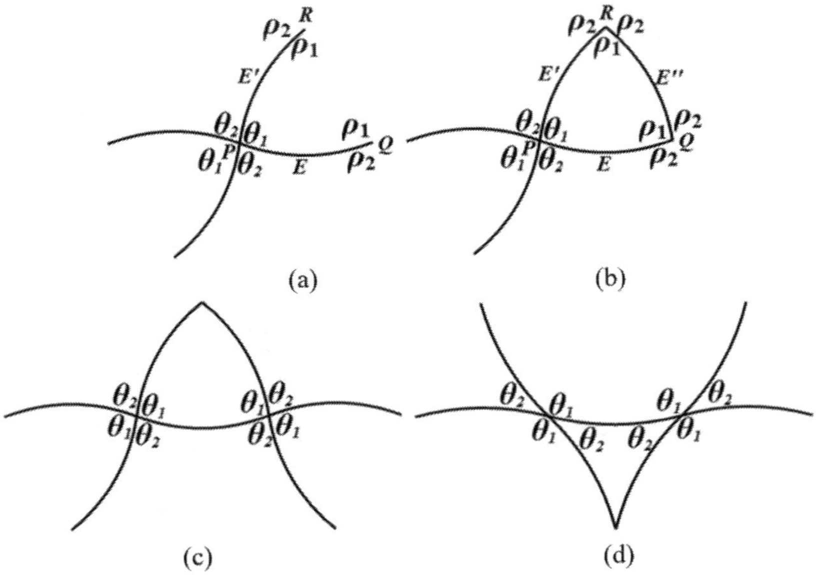

FIGURE 5. Construction of the edge figures when $\mathcal{T}$ has configuration $\langle 3, 6; 4^2 \rangle$ and $S_0(E) \cong D_1$.

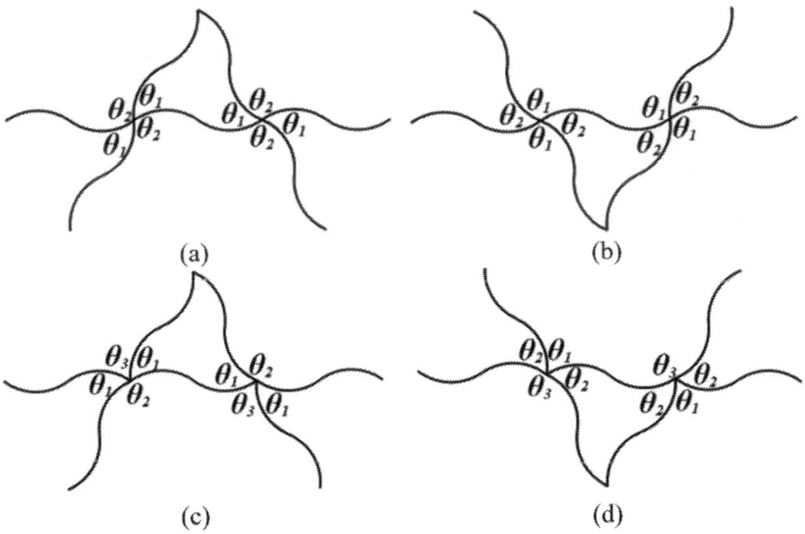

FIGURE 6. Edge figures when $\mathcal{T}$ has configuration $\langle 3, 6; 4^2 \rangle$ and $S_0(E) \cong C_2$.

(Figure 7b). On the other hand, the isometry that sends E to E_1 is the 2-fold rotation centered on the black circle (Figure 7c) and this carries the edges of $F(E)$ to the edges of $F(E_1)$. Moreover, the isometry that sends E to either E_2 or E_3 is the 6-fold rotation centered at the black hexagon and this carries the edges of $F(E)$ to the edges of $F(E_2)$ (Figure 7d) or $F(E_3)$ (Figure 7e) respectively. Finally,

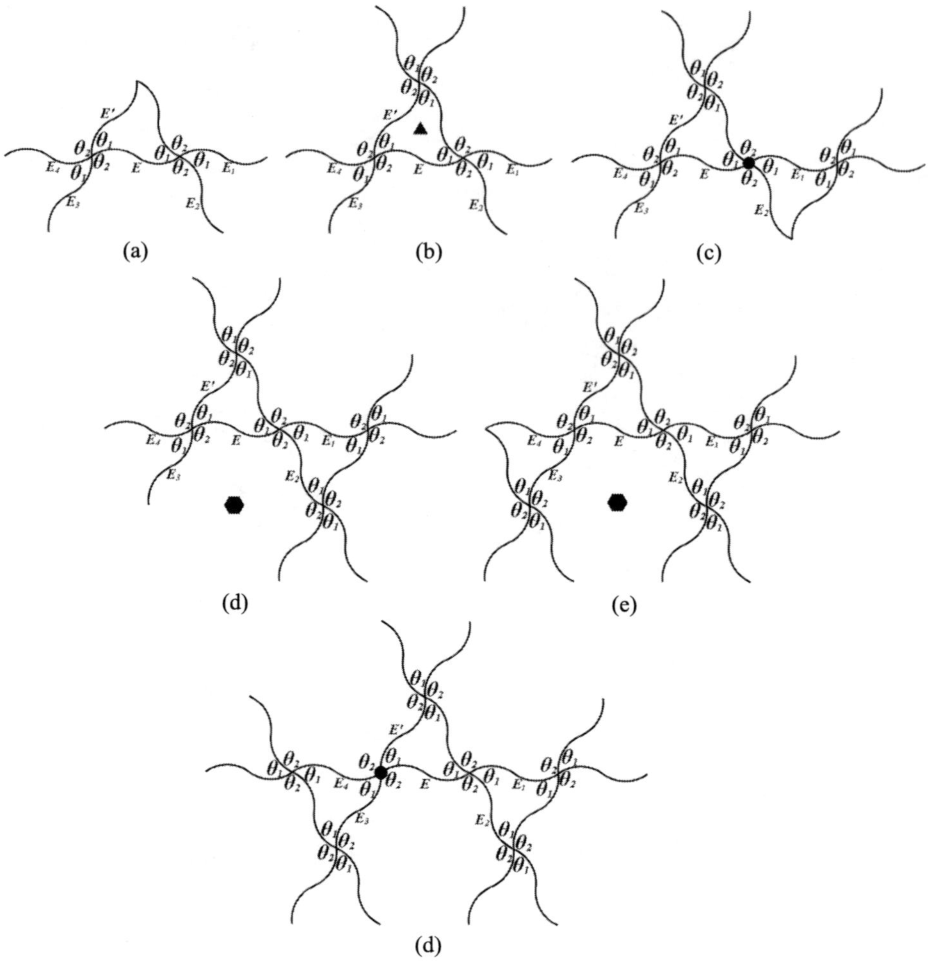

FIGURE 7. Construction of second edge figure when $\mathcal{T}$ has configuration $\langle 3,6;4^2\rangle$ and $S_0(E)\cong C_2$: (a) $F(E)$; (b) $F(E)\cup F(E_1)$; (c) $F(E)\cup F(E_1)\cup F(E_2)$; (d) $F(E)\cup\bigcup_{i=1}^{3}F(E_i)$; (e) $F(E)\cup\bigcup_{i=1}^{4}F(E_i)$; (f) $F(E)\cup\bigcup_{i=1}^{5}F(E_i)$.

the isometry sending E to E_4 is the 2-fold rotation centered at the black circle and will send edges of $F(E)$ to the edges of $F(E_4)$ (Figure 7f). The matching of pairs of edges and pairs of adjacent angles in $F(E)$ by isometries forces the creation of $F_2(E)$ shown in Figure 7f. Thus, $F_2(E)$ is uniquely determined by $F(E)$ and the condition that all edge figures are congruent. It will then follow that all second edge figures of $\mathcal{T}$ are congruent.

(c) Suppose $\mathcal{T}$ has an edge E such that $S_0(E)\cong D_2$. Following the construction in 2(b) we get four edge figures (Figure 8). In either case $S_0(E)\neq S_1(E)$ but $S_1(E)=S_2(E)$ and forming the second edge figures using the same approach in 2(b), it can be shown that the second edge figures are congruent. Thus, $\mathcal{T}$ is isotoxal.

CASE 3. $\mathcal{T}$ has configuration $\langle 3^2;6^2\rangle$.

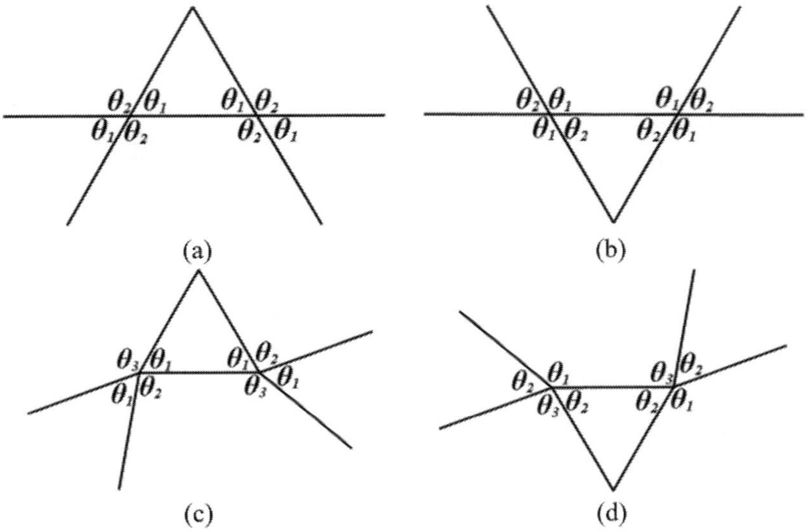

FIGURE 8. Edge figures when $\mathcal{T}$ has configuration $\langle 3,6;4^2\rangle$ and $S_0(E)\cong D_2$.

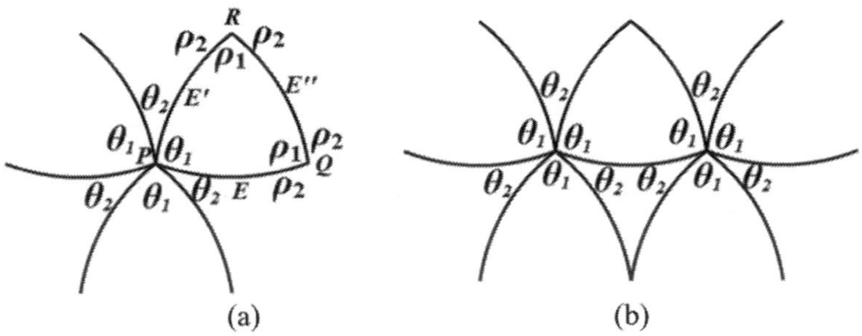

FIGURE 9. Construction of edge figure when $\mathcal{T}$ has configuration $\langle 3^2;6^2\rangle$ and $S_0(E)\cong D_1$.

(a) We consider an edge $E = PQ$ of $\mathcal{T}$ such that $S_0(E)\cong D_1$ with adjacent angles $\theta_1\theta_2$ and $\rho_1\rho_2$. Following a similar construction in 2(a), θ_1 and θ_2 must alternate around P (Figure 9a). Again, since $C(E')$ is congruent to $C(E)$ and $C(E'')$ is congruent to $C(E)$ where $E'' = QS$ then we have $\rho_1 = \theta_1$ and $\rho_2 = \theta_2$. We obtain the edge figure shown in Figure 9b. We obtain $S_0(E) = S_1(E)\cong D_1$ and $\mathcal{T}$ is isotoxal.

(b) Consider an edge $E = PQ$ of $\mathcal{T}$ such that $S_0(E)\cong C_2$. If θ_1 and θ_2 alternate around P and Q, a problem occurs, since an angle θ arises in the 3-gon not congruent to either θ_1 or θ_2, a contradiction to congruence of edge coronae (Figure 10a). Hence, we must have $\theta_1 = \theta_2$ and we get the edge figure in Figure 10b. In this case, $S_0(E) = S_1(E)\cong C_2$ so that $\mathcal{T}$ is isotoxal.

(c) Suppose $\mathcal{T}$ has an edge E such that $S_0(E)\cong D_2$. Since $\mathcal{T}$ is monotoxal and the edges are straight line segments; all 3-gons must be equilateral and hence

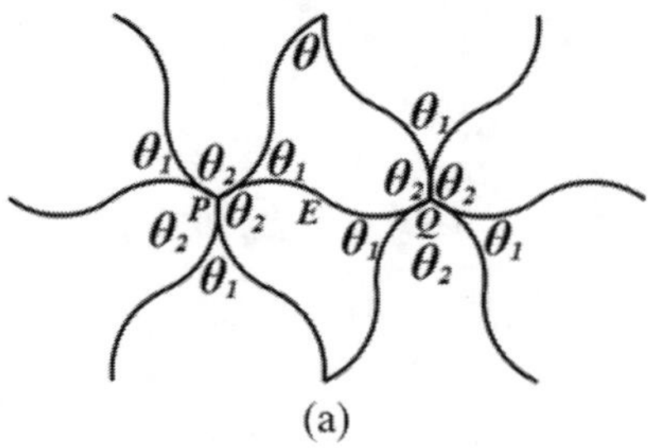

FIGURE 10. Construction of edge figure when $\mathcal{T}$ has configuration $\langle 3^2; 6^2 \rangle$ and $S_0(E) \cong C_2$.

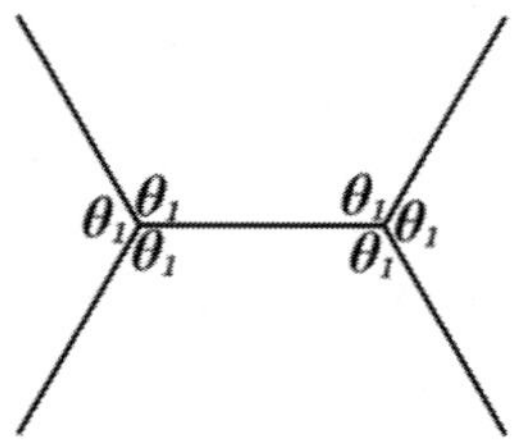

FIGURE 11. Edge figure when $\mathcal{T}$ has configuration $\langle 3^2; 6^2 \rangle$ and $S_0(E) \cong D_2$.

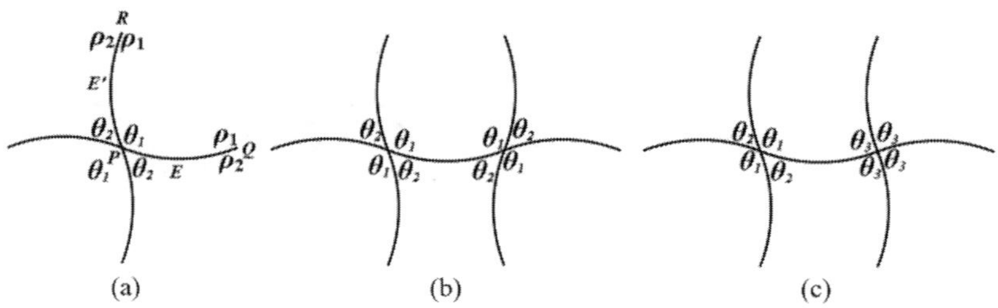

FIGURE 12. Construction of edge figures when $\mathcal{T}$ has configuration $\langle 4^2; 4^2 \rangle$ and $S_0(E) \cong D_1$.

must have congruent angles, that is, $\theta_1 = \frac{\pi}{3}$. We obtain the edge figure in Figure 11. Here, $S_0(E) = S_1(E) \cong D_2$ so that $\mathcal{T}$ is isotoxal.

CASE 4. $\mathcal{T}$ has configuration $\langle 4^2; 4^2 \rangle$.

(a) Consider an edge $E = PQ$ of $\mathcal{T}$ such that $S_0(E) \cong D_1$ with adjacent angles $\theta_1\theta_2$ and $\rho_1\rho_2$. Let $E' = PR$ be another edge. One possibility is θ_1 and θ_2 alternate around vertex P (Figure 12a). We take $\rho_1 = \theta_1, \rho_2 = \theta_2$ and obtain the edge figure in Figure 12b. If $\rho_1 = \rho_2 = \theta_3$ we obtain the edge figure in Figure 12c.

Another possibility is when there are three distinct angles meeting in a vertex, in particular we take $\rho_1 = \theta_3, \rho_2 = \theta_1$ and obtain the edge figure in Figure 13a. The last case is when $\rho_1 = \rho_2 = \theta_2 = \theta_1$. This results in the edge figure in Figure 13b.

Now, for the edge figures in Figures 12b and 13b, $S_0(E) = S_1(E) \cong D_1$ and so $\mathcal{T}$ is isotoxal. However for the edge figures in Figures 12c and 13a, $S_0(E) \neq S_1(E)$

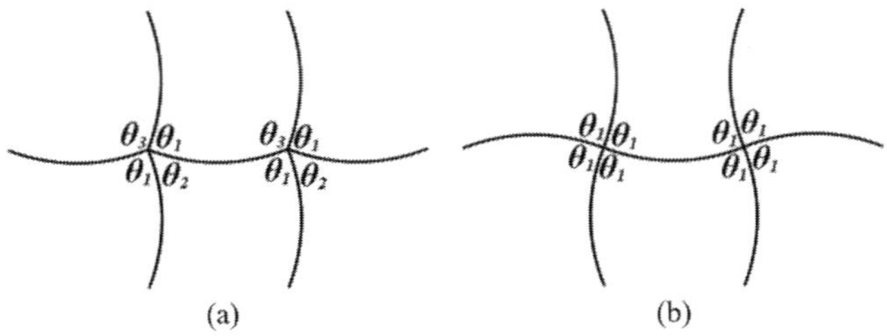

FIGURE 13. Edge figures when $\mathcal{T}$ has configuration $\langle 4^2; 4^2\rangle$ and $S_0(E) \cong D_1$.

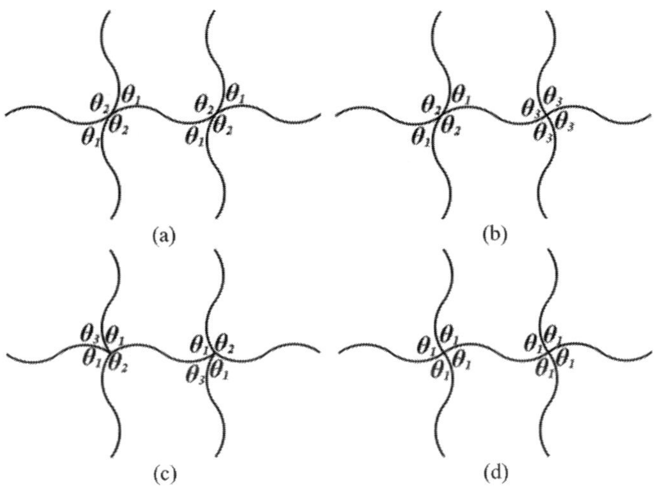

FIGURE 14. Edge figures when $\mathcal{T}$ has configuration $\langle 4^2; 4^2\rangle$ and $S_0(E) \cong C_2$.

and forming the second edge figures using the same approach as in 2(b), we can verify that the second edge figures are congruent. Thus, $\mathcal{T}$ is isotoxal.

(b) We now consider an edge E of $\mathcal{T}$ such that $S_0(E) \cong C_2$. Following 4(a), we also obtain four possible edge figures, shown in Figure 14. For the edge figures in Figures 14a and 14d, we see that $S_0(E) = S_1(E) \cong C_2$ so that $\mathcal{T}$ is isotoxal. For the edge figures in Figures 14b and 14c, $S_0(E) \neq S_1(E)$ and forming the second edge figures is necessary.

(c) We now consider an edge E of $\mathcal{T}$ such that $S_0(E) \cong D_2$. We get two edge figures when θ_1 and θ_2 alternate around a vertex (Figures 15a and 15b); two additional edge figures when there are three distinct angles meeting in a vertex (Figures 15c and 15d) and one edge figure when congruent angles meet a vertex (Figure 15e). For the edge figure in Figure 15e, we see that $S_0(E) = S_1(E) \cong D_2$. For the edge figures in Figures 15a, 15b, 15c and 15d, $S_0(E) \neq S_1(E)$. Thus, forming the second edge figures is necessary. In all cases, we have $\mathcal{T}$ is isotoxal.

CASE 5. $\mathcal{T}$ has configuration $\langle 4^2; 3, 6\rangle$.

(a) Suppose $\mathcal{T}$ has an edge $E = PQ$ such that $S_0(E) \cong D_1$ with adjacent angles $\theta_1\theta_2$ and $\rho_1\rho_2$. We assume that P has valence three and Q has valence six.

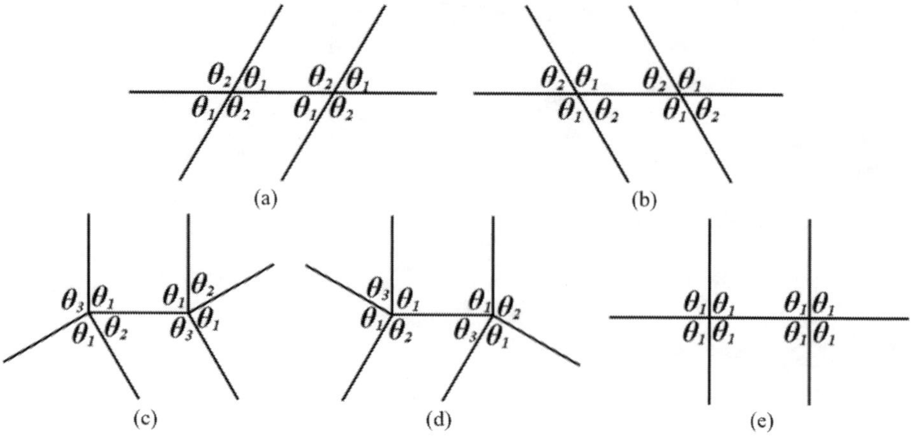

FIGURE 15. Edge figures when $\mathcal{T}$ has configuration $\langle 4^2; 4^2 \rangle$ and $S_0(E) \cong D_2$.

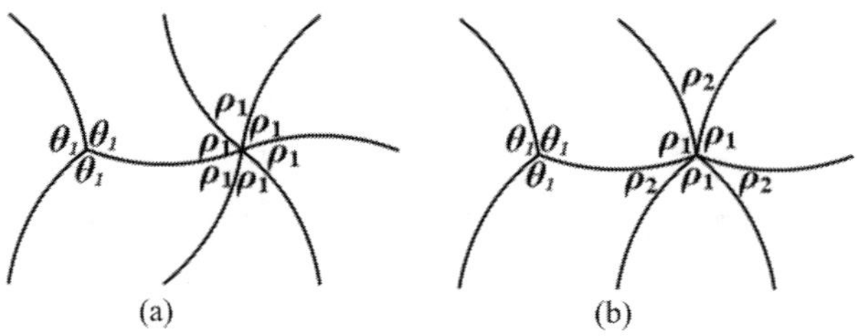

FIGURE 16. Edge figures when $\mathcal{T}$ has configuration $\langle 4^2; 3, 6 \rangle$ and $S_0(E) \cong D_1$.

As in the previous cases when a vertex has valence three, the angles are congruent around the vertex so we have $\theta_1 = \theta_2$. For the vertex with valence six, following the constructions in 3(a)-(b), we have either $\rho_1 = \rho_2$ which will result to the edge figure in Figure 16a or $\rho_1 \neq \rho_2$ which will result to the edge figure in Figure 16b. In either case $S_0(E) \neq S_1(E)$ but $S_1(E) = S_2(E)$ and forming the second edge figures using the same approach as in 2(b), we can verify that the second edge figures are congruent. Thus, $\mathcal{T}$ is isotoxal.

(b) Suppose $\mathcal{T}$ has an edge E such that $S_0(E) \cong C_2$. As in 5(a), there are two edge figures obtained (Figure 17), and the same approach as in 2(b) can be used to prove isotoxality of the resulting tilings.

(c) Suppose $\mathcal{T}$ has an edge E such that $S_0(E) \cong D_2$. Following the construction in 1(c) and 3(c), we must have $\theta_1 = \frac{2\pi}{3}$ and $\rho_1 = \frac{\pi}{3}$ and we get the edge figure in Figure 18. Here, $S_0(E) \neq S_1(E)$ but $S_1(E) = S_2(E)$ and we construct the second edge figure as in 2(b) to show isotoxality of $\mathcal{T}$. □

REMARK 4.4. Note that in every case considered in the proof of Theorem 4.3, we obtained either $S_0(E) = S_1(E)$ or $S_1(E) = S_2(E)$. That is, the conditions of Theorem 4.2 are satisfied for $k = 1$ or $k = 2$. This means that in the proof, we need not to go beyond the second edge figure to check for isotoxality of the tiling.

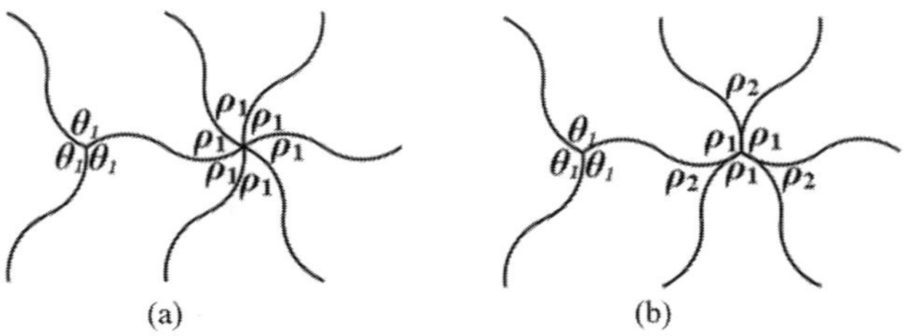

FIGURE 17. Edge figures when $\mathcal{T}$ has configuration $\langle 4^2; 3, 6\rangle$ and $S_0(E) \cong C_2$.

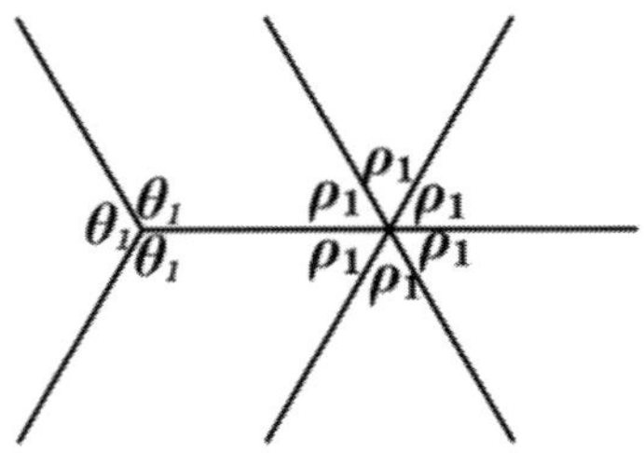

FIGURE 18. Edge figure when $\mathcal{T}$ has configuration $\langle 4^2; 4^2\rangle$ and $S_0(E) \cong D_2$.

From the discussion in the proof of Theorem 4.3, we list in Table 1 the isotoxal tilings with congruent edge coronae. For a given subcase in each configuration, if the edge figures that arise are congruent, then only one edge figure appears in the list.

5. Conclusion and recommendations

In this work we have shown that a normal tiling of $\mathbb{E}^2$ with congruent edge coronae is isotoxal. This has been proven by verifying that certain local conditions involving edge figures of the tiling are satisfied. Conversely, if a tiling is isotoxal then any symmetry of a tiling that maps one edge to another edge automatically maps the tiles containing the first edge to the tiles containing the second edge, and hence provides a symmetry between the edge coronae. Thus, for a normal tiling of $\mathbb{E}^2$, isotoxality also means congruence of its edge coronae.

A question posed by E. Schulte [**5**] was whether or not congruent edge coronae with a mirror of symmetry in the perpendicular bisector of the edges implies {0,1}-transitivity. The term *{0,1}-transitivity* means that the symmetry group of the tiling acts transitively on the incident vertex-edge pairs. We give an affirmative reply to this question in the case of normal tilings. Since a tiling is isotoxal if and only if it has congruent edge coronae, we just need to consider the isotoxal tilings. From the list of isotoxal tilings [**2**], the tilings that have a mirror symmetry in the perpendicular bisector of the edge are IT4, IT5, IT8, IT23, IT24, IT25, IT29 and IT30. Observe that each of these tilings are {0,1}-transitive. For an edge E with endpoints u and v, the incident vertex-edge pair $u - E$ can be sent to $v - E$ via the reflection symmetry of the tiling passing through the perpendicular bisector

FIGURE 19. A non-normal tiling with congruent edge coronae but is not isotoxal.

of the edge. Since the tiling is isotoxal, the symmetry group acts transitively on the incident vertex-edge pairs.

Throughout this paper, we restrict our study to normal tilings. For non-normal tilings, it is not the case that congruence of edge coronae implies isotoxality. For instance, the tiling shown in Figure 19 is a non-normal tiling with congruent edge coronae but is not isotoxal.

It would be interesting as a next step in the study to find the conditions for which non-normal tilings with congruent edge coronae are isotoxal. It is also seems fruitful to study tilings with congruent edge coronae in other spaces like the spherical and hyperbolic spaces.

TABLE 1. Isotoxal tiling with congruent edge coronae (Column 4) corresponding to a given configuration (Column 1), and edge E such that $S_0(E) \cong C_2, D_1$ or D_2. The first edge figure (Column 2) or second edge figure (Column 3) of each tiling is shown.

Configuration	First edge figure	Second edge figure	Tiling
1. $\langle 6^2; 3^2 \rangle$	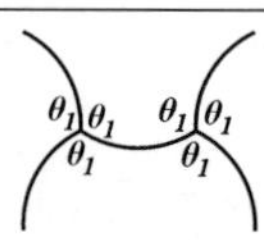 $S_0(E) \cong D_1$ $S_1(E) \cong D_1$		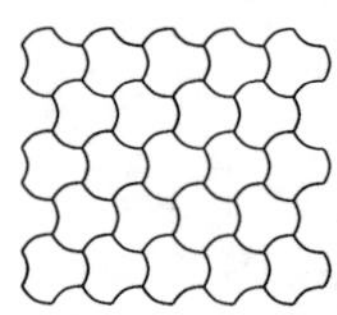
	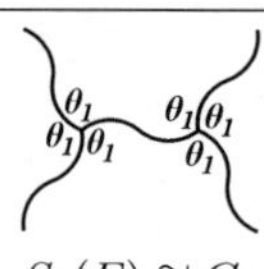 $S_0(E) \cong C_2$ $S_0(E) \cong C_2$		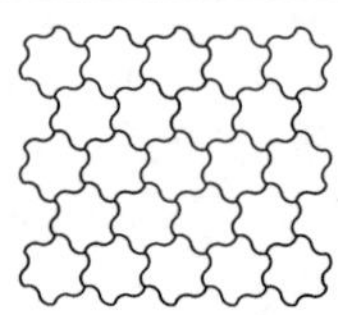

Table 1 (continued)

Configuration	First edge figure	Second edge figure	Tiling
	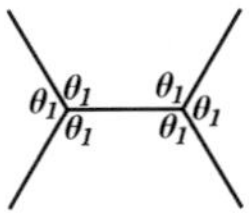 $S_0(E) \cong D_2$ $S_1(E) \cong D_2$		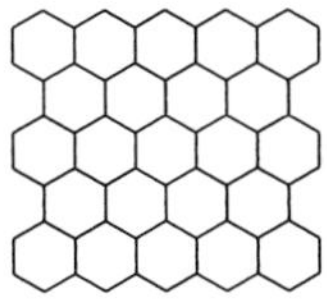
2. $\langle 3, 6; 4^2 \rangle$	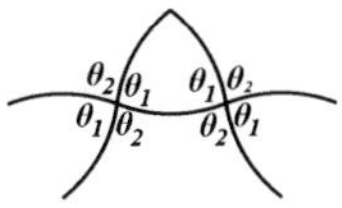 $S_0(E) \cong D_1$ $S_1(E) \cong D_1$		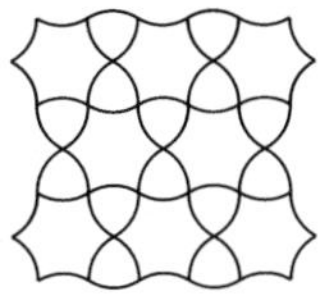
	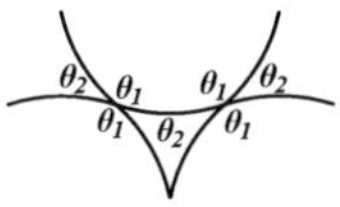 $S_0(E) \cong D_1$ $S_1(E) \cong D_1$		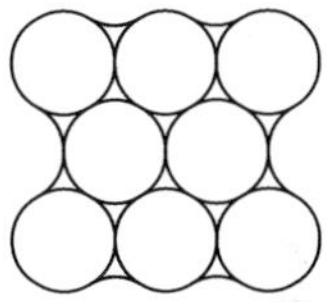
	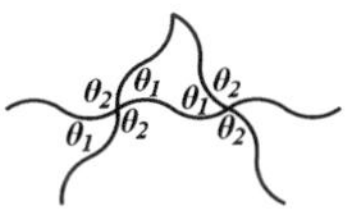 $S_0(E) \cong C_2$ $S_1(E) \cong C_1$	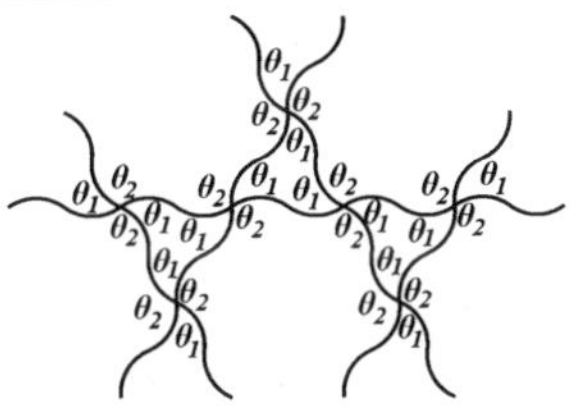 $S_2(E) \cong C_1$	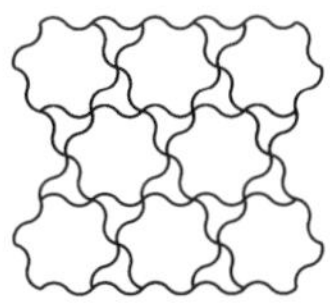
	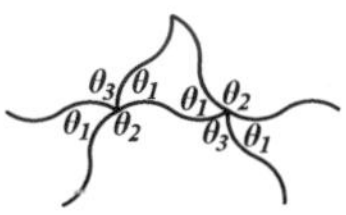 $S_0(E) \cong C_2$ $S_1(E) \cong C_1$	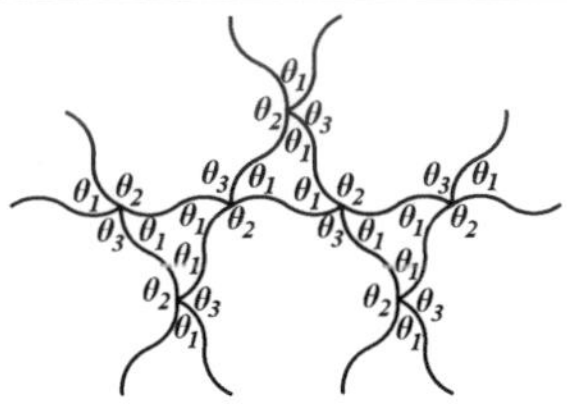 $S_2(E) \cong C_1$	
	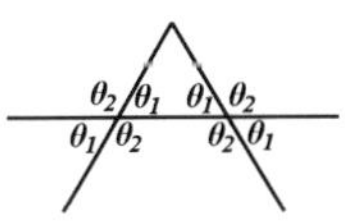 $S_0(E) \cong D_2$ $S_1(E) \cong D_1$	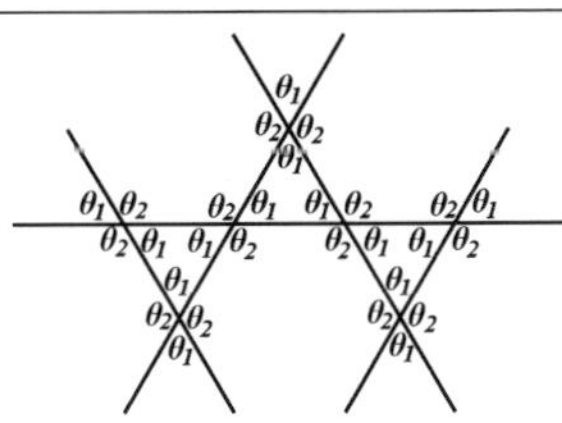 $S_2(E) \cong D_1$	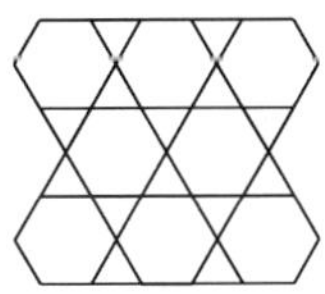

Table 1 (continued)

Configuration	First edge figure	Second edge figure	Tiling
	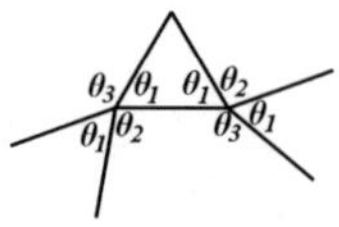 $S_0(E) \cong D_2$ $S_1(E) \cong C_1$	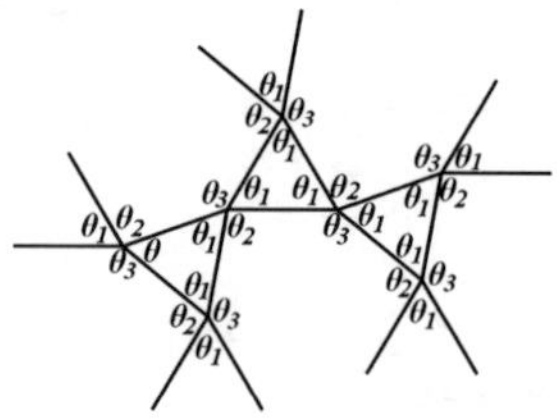 $S_2(E) \cong C_1$	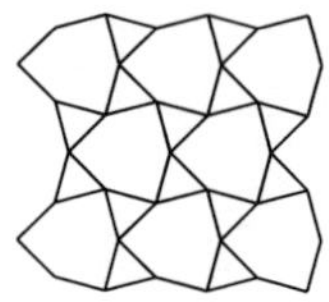
3. $\langle 3^2; 6^2 \rangle$	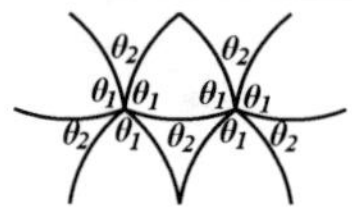 $S_0(E) \cong D_1$ $S_1(E) \cong D_1$		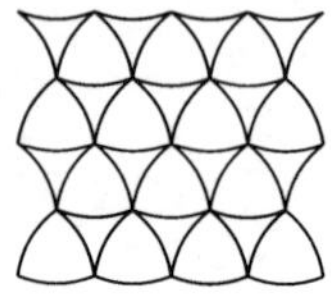
	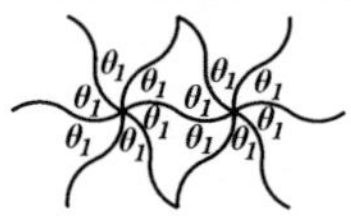 $S_0(E) \cong C_2$ $S_1(E) \cong C_2$		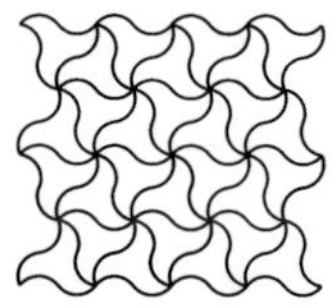
	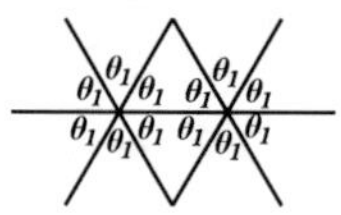 $S_0(E) \cong D_2$ $S_0(E) \cong D_2$		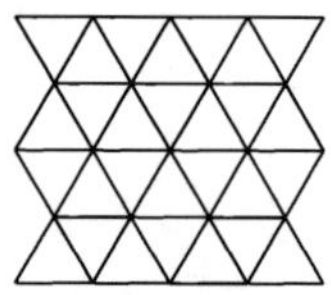
4. $\langle 4^2; 4^2 \rangle$	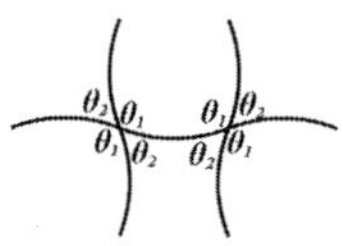 $S_0(E) \cong D_1$ $S_1(E) \cong D_1$		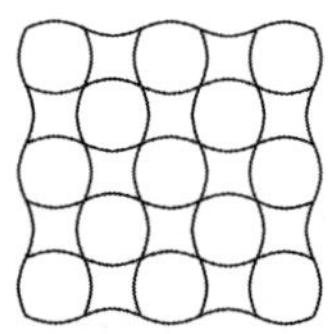
	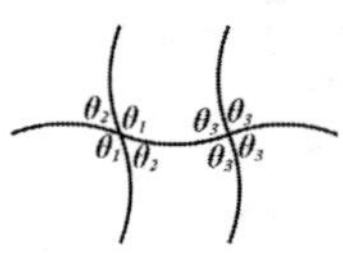 $S_0(E) \cong D_1$ $S_1(E) \cong C_1$	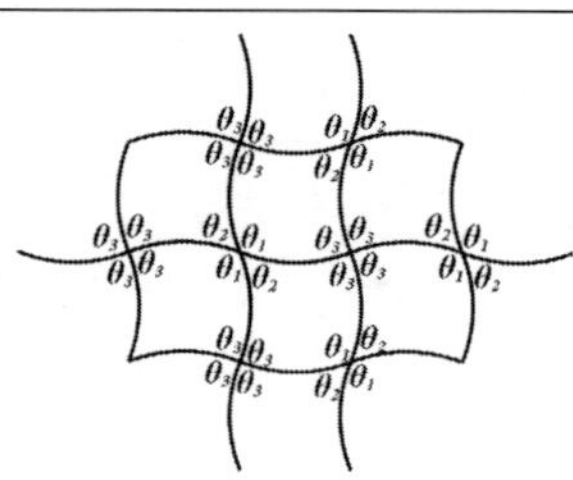 $S_2(E) \cong C_1$	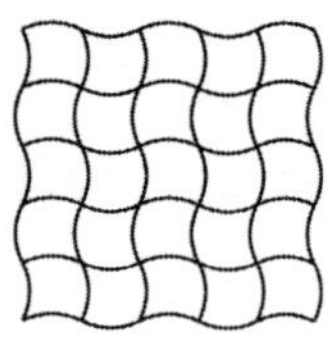

Table 1 (continued)

Configuration	First edge figure	Second edge figure	Tiling
	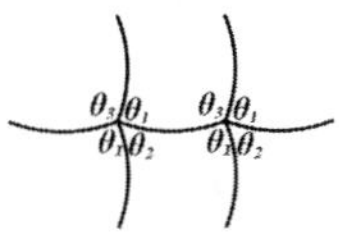 $S_0(E) \cong D_1$ $S_1(E) \cong C_1$	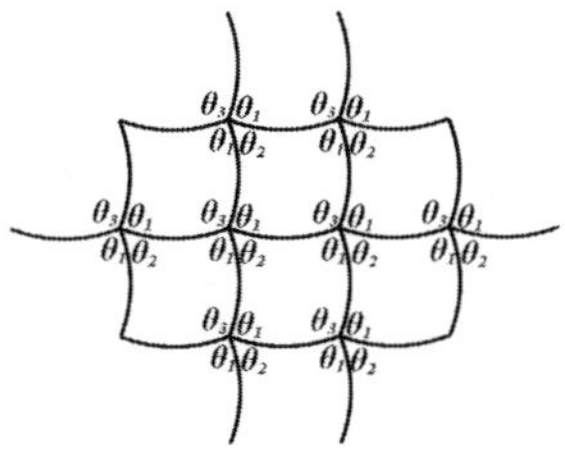 $S_2(E) \cong C_1$	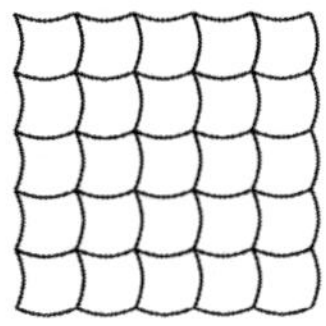
	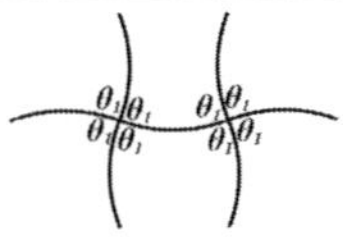 $S_0(E) \cong D_1$ $S_1(E) \cong D_1$		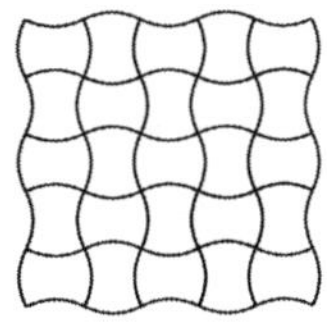
	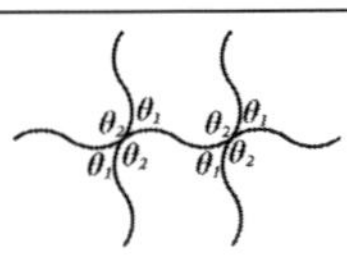 $S_0(E) \cong C_2$ $S_1(E) \cong C_2$		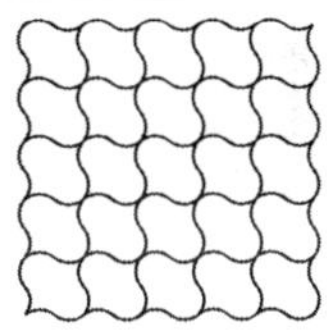
	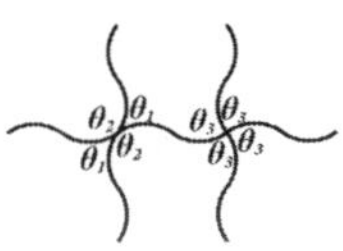 $S_0(E) \cong C_2$ $S_1(E) \cong C_1$	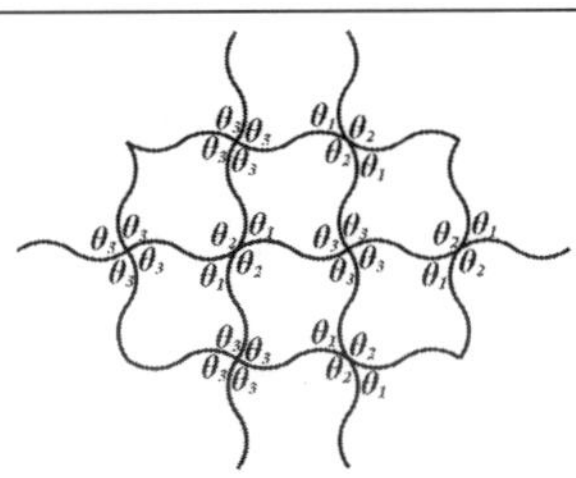 $S_2(E) \cong C_1$	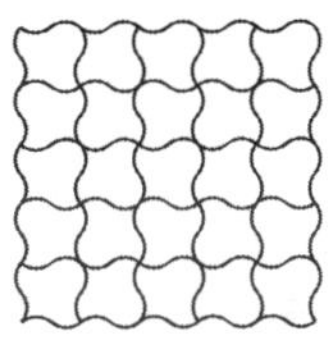
	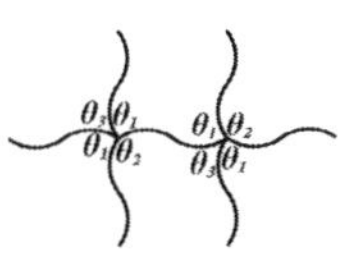 $S_0(E) \cong C_2$ $S_1(E) \cong C_1$	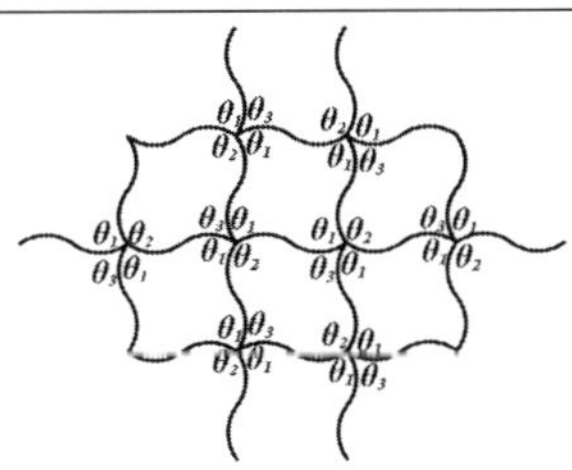 $S_2(E) \cong C_1$	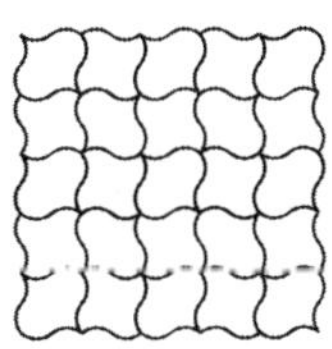
	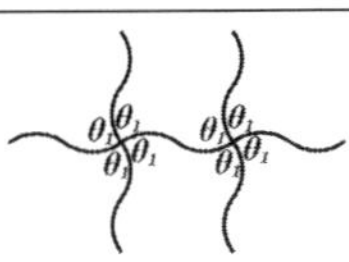 $S_0(E) \cong C_2$ $S_1(E) \cong C_2$		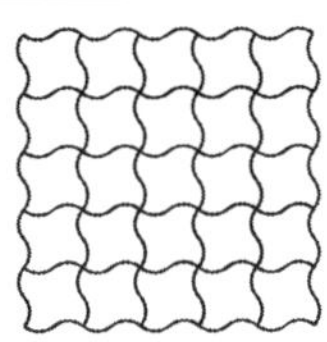

Table 1 (continued)

Configuration	First edge figure	Second edge figure	Tiling
	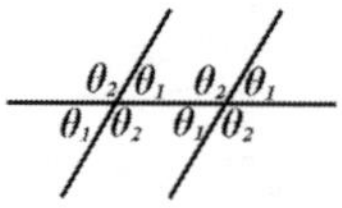 $S_0(E) \cong D_2$ $S_1(E) \cong C_2$	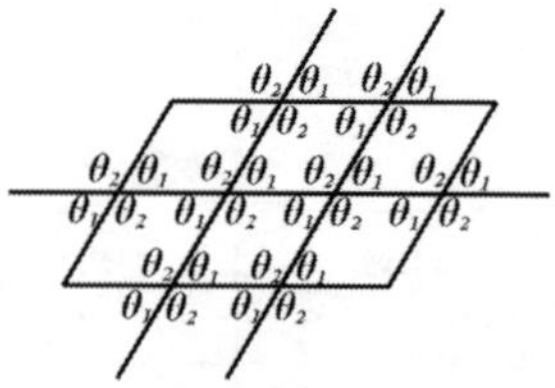 $S_2(E) \cong C_2$	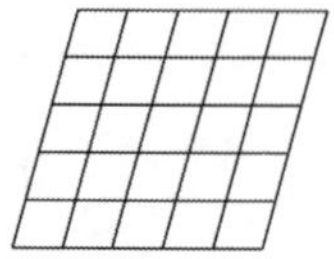
	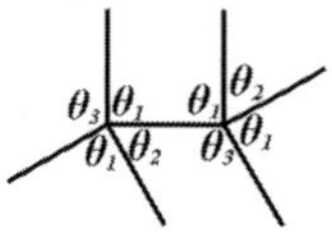 $S_0(E) \cong D_2$ $S_1(E) \cong C_1$	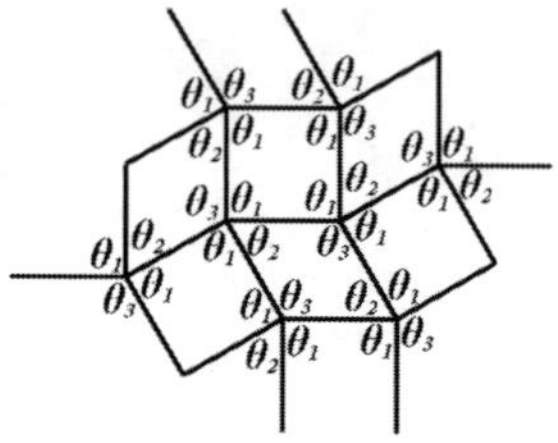 $S_1(E) \cong C_1$	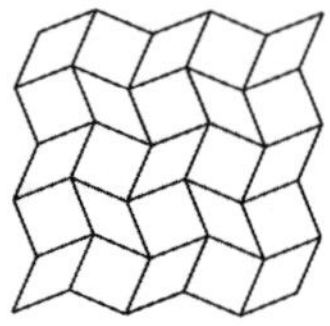
	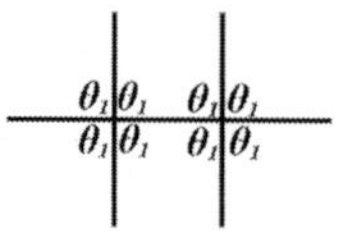 $S_0(E) \cong D_2$ $S_1(E) \cong D_2$		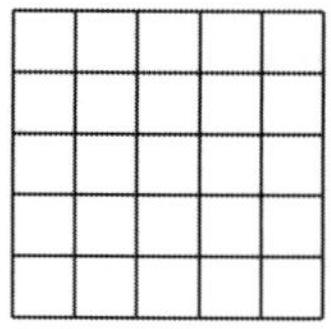
5. $\langle 4^2; 3, 6\rangle$	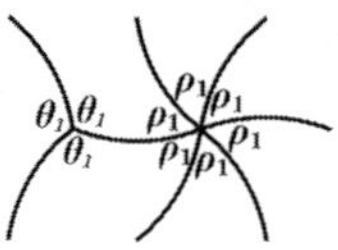 $S_0(E) \cong D_1$ $S_1(E) \cong C_1$	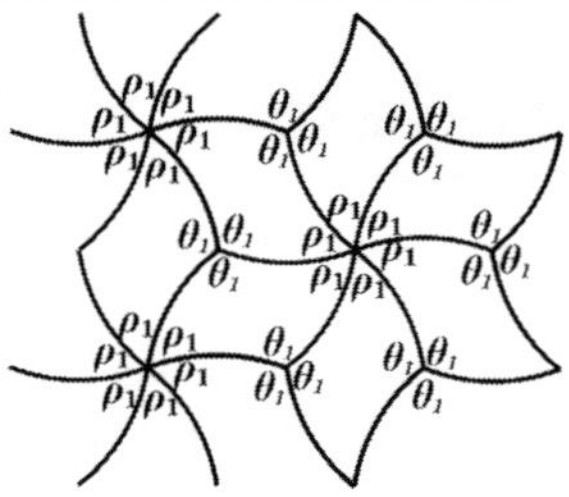 $S_2(E) \cong C_1$	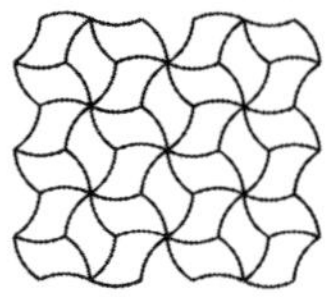
	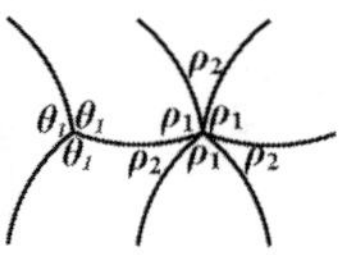 $S_0(E) \cong D_1$ $S_1(E) \cong C_1$	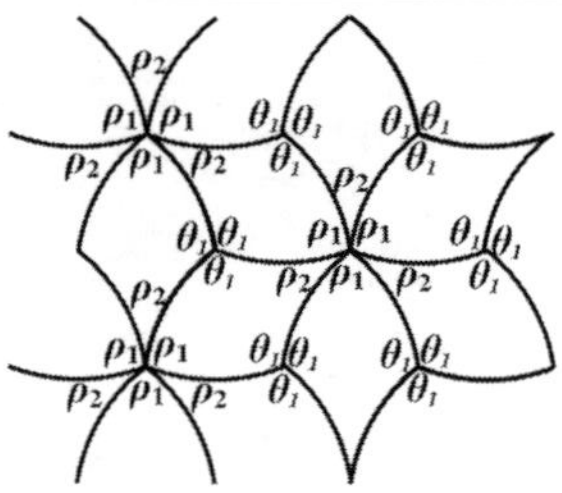 $S_2(E) \cong C_1$	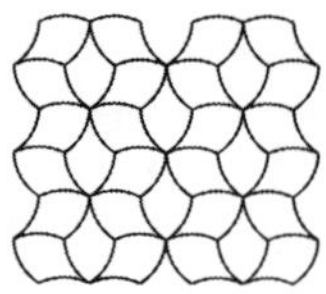

Table 1 (continued)

Configuration	First edge figure	Second edge figure	Tiling
	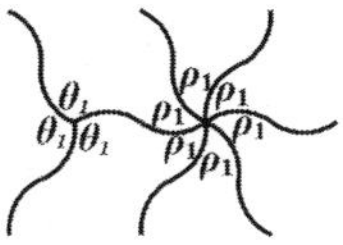 $S_0(E) \cong C_2$ $S_1(E) \cong C_1$	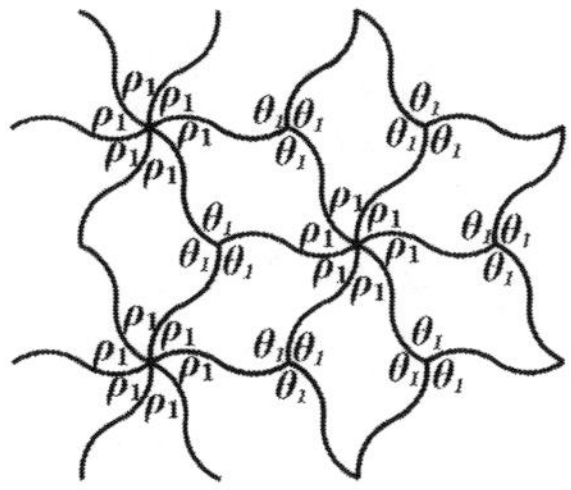 $S_2(E) \cong C_1$	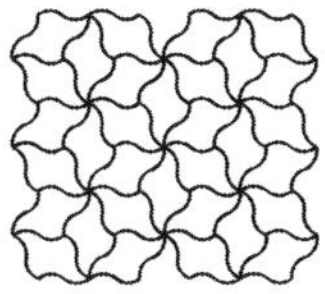
	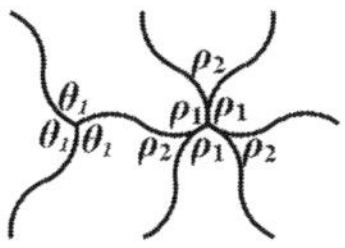 $S_0(E) \cong C_2$ $S_1(E) \cong C_1$	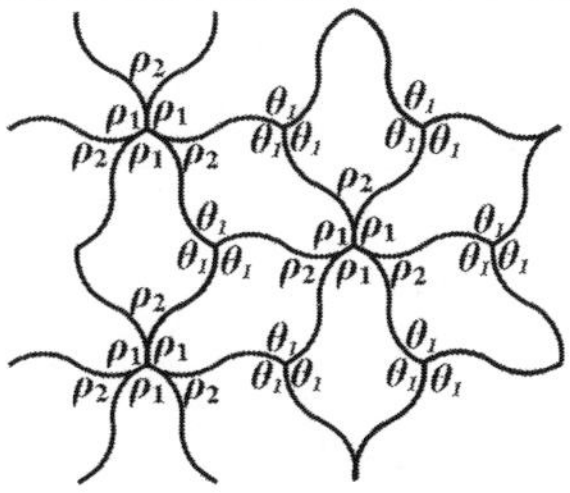 $S_2(E) \cong C_1$	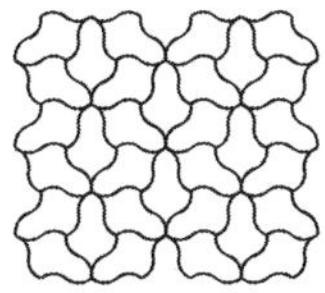
	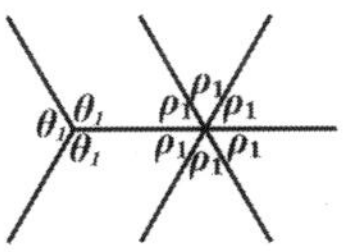 $S_0(E) \cong D_2$ $S_1(E) \cong D_1$	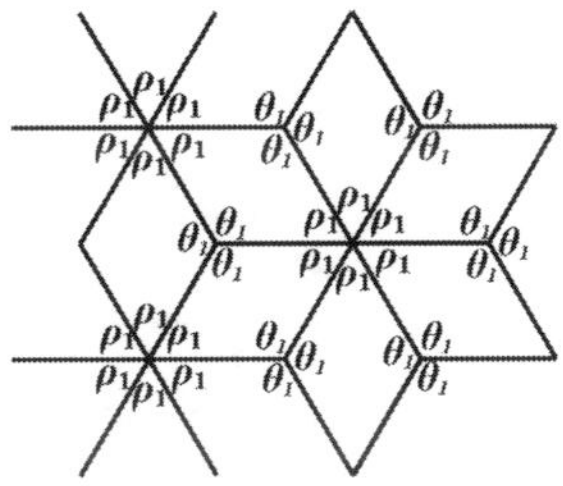 $S_2(E) \cong D_1$	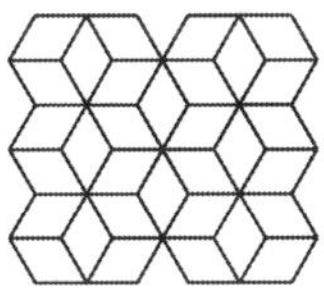

Acknowledgments

The authors wish to thank Eduard Taganap for his insights on tilings with congruent edge coronae. The authors would also like to thank the reviewers for their helpful comments and suggestions which greatly improved this work.

References

[1] Nikolai Dolbilin and Doris Schattschneider, *The local theorem for tilings*, Quasicrystals and discrete geometry (Toronto, ON, 1995), Fields Inst. Monogr., vol. 10, Amer. Math. Soc., Providence, RI, 1998, pp. 193–199. MR1636779

[2] Branko Grünbaum and G. C. Shephard, *Isotoxal tilings*, Pacific J. Math. **76** (1978), no. 2, 407–430. MR506143

[3] Branko Grünbaum and G. C. Shephard, *Tilings and patterns*, W. H. Freeman and Company, New York, 1987. MR857454

[4] Doris Schattschneider and Nikolai Dolbilin, *One corona is enough for the Euclidean plane*, Quasicrystals and discrete geometry (Toronto, ON, 1995), Fields Inst. Monogr., vol. 10, Amer. Math. Soc., Providence, RI, 1998, pp. 207–246. MR1636781

[5] E. Schulte, private comunications, (2017).

Ateneo de Manila University
Email address: mark.tomenes@obf.ateneo.edu

Ateneo de Manila University
Email address: mdelaspenas@ateneo.edu

Selected Published Titles in This Series

764 **Gabriel Cunningham, Mark Mixer, and Egon Schulte, Editors,** Polytopes and Discrete Geometry, 2021

761 **Ibrahim Assem, Christof Geiß, and Sonia Trepode, Editors,** Advances in Representation Theory of Algebras, 2021

760 **Olivier Collin, Stefan Friedl, Cameron Gordon, Stephan Tillmann, and Liam Watson, Editors,** Characters in Low-Dimensional Topology, 2020

759 **Omayra Ortega, Emille Davie Lawrence, and Edray Herber Goins, Editors,** The Golden Anniversary Celebration of the National Association of Mathematicians, 2020

758 **Jan Šťovíček and Jan Trlifaj, Editors,** Representation Theory and Beyond, 2020

757 **Kaïs Ammari and Stéphane Gerbi, Editors,** Identification and Control: Some New Challenges, 2020

756 **Joeri Van der Veken, Alfonso Carriazo, Ivko Dimitrić, Yun Myung Oh, Bogdan D. Suceavă, and Luc Vrancken, Editors,** Geometry of Submanifolds, 2020

755 **Marion Scheepers and Ondřej Zindulka, Editors,** Centenary of the Borel Conjecture, 2020

754 **Susanne C. Brenner, Igor Shparlinski, Chi-Wang Shu, and Daniel B. Szyld, Editors,** 75 Years of Mathematics of Computation, 2020

753 **Matthew Krauel, Michael Tuite, and Gaywalee Yamskulna, Editors,** Vertex Operator Algebras, Number Theory and Related Topics, 2020

752 **Samuel Coskey and Grigor Sargsyan, Editors,** Trends in Set Theory, 2020

751 **Ashish K. Srivastava, André Leroy, Ivo Herzog, and Pedro A. Guil Asensio, Editors,** Categorical, Homological and Combinatorial Methods in Algebra, 2020

750 **A. Bourhim, J. Mashreghi, L. Oubbi, and Z. Abdelali, Editors,** Linear and Multilinear Algebra and Function Spaces, 2020

749 **Guillermo Cortiñas and Charles A. Weibel, Editors,** K-theory in Algebra, Analysis and Topology, 2020

748 **Donatella Danielli and Irina Mitrea, Editors,** Advances in Harmonic Analysis and Partial Differential Equations, 2020

747 **Paul Bruillard, Carlos Ortiz Marrero, and Julia Plavnik, Editors,** Topological Phases of Matter and Quantum Computation, 2020

746 **Erica Flapan and Helen Wong, Editors,** Topology and Geometry of Biopolymers, 2020

745 **Federico Binda, Marc Levine, Manh Toan Nguyen, and Oliver Röndigs, Editors,** Motivic Homotopy Theory and Refined Enumerative Geometry, 2020

744 **Pieter Moree, Anke Pohl, L'ubomír Snoha, and Tom Ward, Editors,** Dynamics: Topology and Numbers, 2020

743 **H. Garth Dales, Dmitry Khavinson, and Javad Mashreghi, Editors,** Complex Analysis and Spectral Theory, 2020

742 **Francisco-Jesús Castro-Jiménez, David Bradley Massey, Bernard Teissier, and Meral Tosun, Editors,** A Panorama of Singularities, 2020

741 **Houssam Abdul-Rahman, Robert Sims, and Amanda Young, Editors,** Analytic Trends in Mathematical Physics, 2020

740 **Alina Bucur and David Zureick-Brown, Editors,** Analytic Methods in Arithmetic Geometry, 2019

739 **Yaiza Canzani, Linan Chen, and Dmitry Jakobson, Editors,** Probabilistic Methods in Geometry, Topology and Spectral Theory, 2019

738 **Shrikrishna G. Dani, Surender K. Jain, Jugal K. Verma, and Meenakshi P. Wasadikar, Editors,** Contributions in Algebra and Algebraic Geometry, 2019

For a complete list of titles in this series, visit the AMS Bookstore at **www.ams.org/bookstore/conmseries/**.